PLANT BREEDING IN THE 1990S

PLANT BREEDING IN THE 1990s

Proceedings of the Symposium on Plant Breeding in the 1990s

Held at North Carolina State University
Raleigh, NC
March 1991

Edited by

H. T. Stalker and J. P. Murphy
Department of Crop Science
North Carolina State University
Raleigh, NC 27695-7629

C·A·B International

C·A·B International
Wallingford
Oxon OX10 8DE
UK

Tel: Wallingford (0491) 32111
Telex: 847964 (COMAGG G)
Telecom Gold/Dialcom: 84: CAU001
Fax: (0491) 33508

A catalogue entry for this book is available from the British Library

ISBN 0–85198–717–6

Printed and bound in the UK by Redwood Press Ltd., Melksham

Contents

Modification of plants to tolerate stresses due to diseases and insects

Contributions of biotechnology to plant improvement

Strategies for utilizing unadapted germplasm

Symposium overview

Preface

Plant breeders have made substantial contributions to improvements in quantity and quality of food, feed, and fiber products during the twentieth century. The discipline has been sustained through the close working relationships of scientists with diverse expertise that are employed in both public and private institutions. Only through continued interactions will improved breeding methodologies be conceived and then implemented in the field.

The objective of the Symposium on Plant Breeding in the 1990s was to review the current and emerging methodologies likely to influence the course of plant breeding during the last decade of this century. The program was divided into five sections including (1) the gene base for plant breeding, (2) modification of plants to tolerate environmental stresses, (3) modifications of plants to tolerate stresses due to diseases and insects, (4) contributions of biotechnology to plant improvement, and (5) strategies for utilizing unadapted germplasm.

This volume contains 23 comprehensive and scholarly presentations by experts on these topics. Also, the thought-provoking discussions between conference participants, which were a vital part of the symposium's success, have been included. Abstracts of 75 poster papers are available from North Carolina State University, Department of Crop Science, Raleigh, NC 27695 as Research Report 130.

In addition to the speakers and audience, numerous individuals and organizations contributed to the success of the symposium. The symposium committee comprised of members from the Departments of Crop Science, Genetics, Horticultural Science, and Plant Pathology, North Carolina State University, is especially indebted to the speakers who prepared excellent presentations, to the moderators who presided over discussions, and to the many graduate students who assisted with the program. We sincerely thank Peggy Brantley for her dedicated service and advice during preparations for the symposium and the proceedings. Finally, we are grateful to all members of the symposium audience for their interest and participation in symposium programs.

H. T. Stalker and J. P. Murphy
Department of Crop Science
North Carolina State University

Chapter 1
Plant breeding perspectives for the 1990s

Kenneth J. Frey

Department of Agronomy, Iowa State University, Ames, Iowa 50011

In the past half century, when the human population increased by 3 billion persons, most of the increase in agricultural production has come from increased yields achieved via the application of technology. For example, for the period 1960-1990, annual cereal grain production increased from *ca.* 800 million tons to nearly 2000 million tons, and more than four-fifths of this increase was achieved from greater annual crop yields. Grain production today, if completely utilized without loss, would provide nearly 4000 calories per person per day. Yes, global agricultural production and the science that underpins it have been immensely successful. The technology represented by crop cultivars that have high yield potential and that are well fitted to production environments and possess resistance and tolerance to biotic and abiotic stresses has been a major contributor to this increased agricultural production.

Various studies, mostly conducted on cereals (Duvick, 1986; Fehr, 1984; Frey, 1971; Russell, 1986), have estimated that 50% or more of the increase in crop production has been due to better crop cultivars. Such studies may seem academic and self-aggrandizing for plant breeders but, in the real world, they are important because success is an important ingredient in establishing how future funding and human resources will be distributed. It suffices to say that plant breeding is exciting and on the cutting edge, and the technology that this profession has generated has played an immense role in providing food, feed, and fiber for a relentlessly increasing human population.

With those two paragraphs, I lay the history of plant breeding to rest. Next, I will build a "perspective for plant breeding for the next decade"; i.e., a view of how new techniques, global politics and societal concerns, intellectual property rights, and distribution of funding and human resources may impact plant breeding and define its role in helping to feed hungry people.

BIOTECHNOLOGY

Molecular biology over the past 20 years has greatly expanded knowledge about the genetic architecture of living organisms and the communication system by which a gene or genotype exercises control over phenotype. And soon, we will know how genes (of course, always interacting with environment) control growth, development, and maturation of plants. As a spinoff from research on molecular and cellular biology, several new techniques have emerged that will influence plant breeding. The first of these techniques to be tried by plant breeders was one called **somaclonal variation**. When cell or tissue cultures are grown for extended periods, a burst of genetic variation usually occurs. This genetic variabilty, which is called somaclonal variation (Larkin, 1987), led to proposals for using it as a source of new plant cultivars. That is, a cell or tissue culture established from a good cultivar would be grown, and perhaps mutagenized, and plants regenerated from the culture would be screened to find an even better cultivar. It sounds reminiscent of claims for mutation breeding of 40 years ago. Several unique mutations [e.g., mutants that give resistance to *Helminthosporium oryzae* Breda de Haan of rice (*Oryza sativa* L.) and resistance to the herbicide glyphosate (Ling, 1985; Singer and McDaniel, 1985)], have occurred in plant tissue cultures, but tissue culturing has not led to the creation of many, if any, new crop cultivars. Thus, tissue culturing, if used, probably will be utilized as a source of genetic variation for use in cultivar development, but not as a source of cultivars *per se*. The technique may have value in enhancing desirable gene rearrangements in wide crosses, however.

In vitro **selection** has been proposed as a cheap methodology to screen for variants. That is, the medium in which cells or tissues are grown could be modified so that only cells or calli of a certain variant type would survive and produce regenerated plants. This technique would increase the number of individuals that a plant breeder could screen by several orders of magnitude. It has been used to select herbicide resistant genotypes in cultures of several plants (Singer and McDaniel, 1985). However, only a few plant traits are expressed at the cellular level. Traits of paramount importance to the plant breeder - such as yield, maturity, height, lodging resistance - are not expressed at the cellular or tissue level. Further, a number of variants selected for tolerance to abiotic stresses - such as salinity, acidity, heavy metals - have been ephemeral. *In vitro* selection may have utility in screening for certain mutants, but it is unlikely that it will become a widely used technique in plant breeding programs to select for quantitatively inherited traits.

A third technique from molecular biology is **transformation** or the asexual transfer of genes. Until recently, plant breeders could recombine genes only by crossing plant strains. Over time, the breadth of germplasm available to a breeder of a given species has been extended gradually by special techniques - such as colchicine treatment to double chromosome numbers, use of bridging species crosses, embryo rescue, mentor pollen - but transformation represents a truly phenomenal breakthrough. It provides the possibility for transferring a gene from any organism to any plant species. The skills and techniques involved in discovering, isolating, and cloning a gene and then transforming a plant cell or tissue with it are sophisticated, but they generally are technical in nature. Techniques used for transformation include vectors, electroporation, simple absorption from a gene soup, shooting with microprojectiles that carry the gene, etc. Transformation is a natural method of gene transfer among bacteria, and its potential for improving plants almost defies imagination. Of course, a gene, when placed in a foreign genetic background, may not function at all; it may function in a bizarre way, or its function may not be controllable. To illustrate the possible breadth of gene transfer, consider the following examples. The luminescence gene from the firefly has been transferred to plants and found to function. The *Bt* gene from the bacteria *Bacillus thuringiensis* Berl., which gives resistance to lepidopterous insects, has been transferred from bacteria to cotton (*Gossypium hirsutum* L.); and it confers resistance to several insect pests in its new host. And resistance to glyphosate has been transferred from tobacco (*Nicotiana tabacum* L.) to other plants and found to confer resistance. Without much doubt, transformation will have a major impact on plant breeding in the decade ahead. It seems most useful for transferring traits that are qualitatively inherited but, as more is learned about the inheritance of quantitative traits, it may have impact there as well. At least transformation should make gene transfer among related species more easy and, at best, it may combine genes from all species into a single gene pool.

Plant regeneration from calli or protoplasts is essential if breeders are to capitalize upon somaclonal variation, transformation, and *in vitro* selection. Protocols exist for regenerating plants from protoplasts for most dicotyledonous agricultural species, but for monocotyledonous species, and especially the cereals, regeneration protocols have been difficult to discover. A protocol has been formulated for regenerating plants from protoplasts of rice, so probably protocols for regenerating plants will become available for most, if not all, monocotyledonous species. Plants can be regenerated from pollen grains with subsequent chromosome doubling to the $2x$ level. This technique has been used to develop cultivars of several crop species, such as barley (*Hordeum vulgare* L.) and rice.

Linkage-assisted selection has been an ideal long sought by plant breeders. However, classical genetic linkage and isozyme markers have been little used in crop cultivar development because each has several drawbacks. The nearly perfect system for using markers as an aid for selection would seem to be through restriction fragment length polymorphisms (RFLP). An endonuclease enzyme can be used to cut genomic DNA into pieces or fragments at specific sites recognized by the enzyme. These fragments, which are inherited in Mendelian fashion, can be recognized by using laboratory protocols. If a given DNA fragment can be associated with a plant trait, a plant that carries that RFLP and the trait can be traced and/or selected by identifying the specific DNA fragment or RFLP associated with the trait. The heritability for RFLP-associated traits is virtually 100%. Many endonucleases exist, each of which produces a unique set of fragments, so the number of possible RFLPs is nearly infinite. A RFLP map that covers a plant genome uniformly should be especially useful to (1) detect transformed plants and (2) trace genes with major effects in segregates from crosses. This technique can reduce selection for certain simply-inherited traits to a laboratory exercise. Also, RFLPs may be useful to detect loci that have major effects upon quantitatively inherited traits.

So, one part in the perspective of plant breeding in the 1990s will be the inclusion of biotechniques from research on molecular biology into the plant breeder's set of tools for cultivar development. They promise to make plant breeding more precise, to broaden the gene pool available to each breeding program, and to shortern cultivar development time. Perhaps more important than what these biotechniques can do directly for cultivar development is the knowledge that their use can provide about fundamental plant biology. Transformation, RFLP mapping, and plant regeneration, along with even more sophisticated techniques such as modifying the DNA code within genes, are powerful tools that will permit designing fundamental studies to elucidate how genes control plant growth, physiology, differentiation, development, and phenotype. Knowledge about why and how certain genotypes are tolerant to high and low temperatures, salinity, drought, high soil aluminum, heavy metals, etc. can only increase the precision and expediency of plant breeders' work. However, deciphering the genetic control and physiological mechanisms of plant reactions probably will not be a part of the perspective of plant breeding until sometime beyond the 1990s.

THE PLANT BREEDING ENVIRONMENT

The power that biotechniques provide to the geneticist and plant breeder to literally obliterate natural biological barriers to gene flow, to create new functions within and products from a plant species and, someday, to rewrite the genetic code according to design is rapidly changing the environment within which plant breeding is done. First, the application of biotechniques to cultivar development has brought an immense input of capital and human resources from the private sector of the world economy into plant breeding. This has benefits and debits. Second, because biotechniques permit geneticists and plant breeders to manipulate genes and biology in what might be called "a very unnatural way", politicans and society have become a force that judges "good and bad" of what plant breeders do and create. Changing from an environment in which plant breeding was accepted as neutral to beneficial, to one where it is judged from bad to good (and upon the basis of secondary and tertiary ripple effects of its products) is disquieting, disruptive, and intimidating to plant breeders. Third, the genes, gene combinations, and genotypes that plant breeders and geneticists can create by using the biotechniques are so ingenious and unnatural that they can qualify as truly human-made inventions. These inventions, which are intellectual properties, are subject to ownership, exclusivity, and financial rewards. Literally, this development is causing plant breeders and lawyers to change their relationship from being acquaintances to being partners.

Now, I want to elaborate upon and define the consequences of this new environment for plant breeding.

PRIVATE INDUSTRY INPUT

The presence of private industry in plant cultivar development is not new. Plant selectors (e.g., Luther Burbank) developed and sold cultivars of horticultural crops a century ago. Some present-day plant breeding companies had their origins a half century or more ago. However, the availability of transformation methods, *in vitro* selection procedures, somaclonal variation, and RFLP mapping has brought a virtual explosion of companies coming into plant breeding in the past 15 years. Wholly new companies have been organized to do plant breeding via biotechniques. Established companies, both with and without previous plant breeding experience, have organized biotechnology departments to supplement or do, respectively, plant breeding. National and multinational companies that manufacture products used in agriculture, but which had no previous involvement in plant breeding, have developed or bought subsidiaries to do plant breeding.

The large input of private company effort into cultivar development via use of biotechniques has raised a number of issues that will continue to be debated and can have unpredictable impacts on plant breeding in the 1990s. On a positive note, private industry laboratories are conducting a large amount of basic research in plant biology and especially plant molecular biology. This research is targeted to obtain knowledge that can influence company products, but in general, private industry has been quite responsible in sharing the basic research results from its laboratories. This input into basic plant biology research by private industry, when integrated with the basic research being done in public-sector laboratories, is bringing about a rapid understanding of the genetic code and the regulation of plant differentiation, development, and physiology. Additionally, private industry is providing a sizable amount of funding for research done in public laboratories. Some is provided for generic research on plant biology, but a greater share, even when provided for basic research, is targeted toward certain areas or commodities. Nevertheless, private industry funding of plant biology research, whether the research is done in company or in public-sector laboratories, is having a great impact on the amount of basic research conducted.

The interaction of private industry and public institutions to synergize greater effort and output from research and technology development in plant molecular biology has occurred in several ways. As noted, the funding that private industry provides to public institutions for conducting research and technology development in plant biology has a large impact on the amount and direction of research in the public sector. However, of more influence on the direction of research in the public sector are the contracts that bind individual companies and individual public institutions together in an agreed-upon research and development agenda. Under such contracts, the primary contribution from the company is money, and the primary contributions from the public institution are expertise, unbiased nature and facilities. Most often these contracts provide the company with a preferred position for research results and technology developed from the research and the academic freedom of the public-institution scientists may be somewhat restricted. Two factors - the heavy cost of conducting research on plant molecular biology and the national and international impetus for rapid conversion of research results into technology (in this case, crop cultivars) - make it likely that contractual agreements between private companies and public institutions will increase.

The continued existence of private companies is dependent upon their having unique products for sale. For plant breeding companies, this means having a unique and exclusive cultivar, gene, or plant product. The unique product may result directly from company research, or it may be purchased from another person or company. However, before a private company will invest financial and human resources into products, it must be assured of having exclusive command of the product and probably the intellectual properties that go into its development. Laws have been passed in many countries for "breeders' rights" - that is, for the developer of a cultivar to control and receive compensation for use of that cultivar. In the USA, laws passed by Congress in 1930 and 1970 provided breeders' rights for cultivars of asexual and sexual crop species, respectively. And, in the 1980s, the U.S. Supreme Court declared that living organisms could be protected under utility patent law. The availability of plant variety patenting for asexual plants, plant variety protection for sexual plants, and utility patenting for biological organisms has greatly expanded the protection, ownership rights, and climate for making profits from plant breeding. It is anticipated that all these biotechniques will greatly enhance the development of plant intellectual properties that, when protected and directly and indirectly marketed, can mean profits for private companies.

Three factors - shortage of funding for research, a tendency to greediness by some plant breeders, and the desire to rapidly transfer technology to the consumer - have caused some public institutions with plant breeding programs to mimic private companies in protecting, licensing, and selling the plant intellectual properties that they develop. This is, of course, a very different climate for public plant breeding programs than has existed previously. In the past, these institutions discovered genes and developed cultivars that were shared with all at no cost. More and more, public plant breeding programs are developing business tactics similar to those of private companies. Will this development cause the public institutions to lose their credibility as independent and unbiased brokers of information and germplasm?

There is a concern about whether this new climate of near universal protection for plant intellectual properties by private and public plant breeding programs may restrict germplasm exchange. Already, germplasm flow from public- to private-sector programs involves selling and buying, and germplasm exchange between public plant breeding programs has become more formal. Some argue that exchange of germplasm among plant breeding programs will not be more restrictive but that new mechanisms for exchange will evolve. For sure, these new mechanisms will be more formal and probably most exchanges will occur via sales. There is concern, especially in germplasm deficit countries such as the United States and West Europe, that countries with natural germplasm reservoirs may prohibit the export of their native germplasm stocks. For example, China prohibits the export of certain soybean [*Glycine max* (L.) Merr.] stocks.

Private plant breeding companies almost universally seek an expanded market. The expanded market for an individual company can be obtained by capturing a greater share of an already saturated market, but collectively, private industry can expand only by finding new markets. For multinational companies, this expanded market is being found in second and third world countries. The primary benefits to the receiving countries are (1) better crop cultivars and (2) generally a better infrastructure for the production and distribution of quality seed.

Obviously, the immense and continued investment in plant breeding by private industry is a part of the perspective of plant breeding in the 1990s. It will cause impacts in several ways including greater investment in fundamental research and cultivar development, greater and more formal interaction between private and public plant breeding programs, protection for plant intellectual properties, greater controversy over germplasm exchange, and greater availability of improved cultivars and quality seed in second and third world countries. Whether these impacts are judged to be beneficial and desirable or detrimental and undesirable for humanity will vary by situation and depend upon the experience and viewpoint of the person making a judgment but, nevertheless, they will occur.

The use of certain biotechniques, in particular transformation, to move genes across species, genus, family, and even kingdom barriers has aroused concerns among social scientists, theologians, politicans, and the lay public. The concerns run the gamut from "disturbing the sanctity of life", to "social good vs. evil", to "the possible creation of a Frankenstein". I do not intend to judge the validity of any of these concerns. However, plant breeding, a profession that historically has been considered beneficial for humankind, is being lumped, by some, with other biotechnology "gene jugglers". The profession's activities, products, and workers are being judged by a whole new array of persons and groups, each of whom has his own agenda. This is not necessarily bad for the profession, but it requires study of and response to issues that generally have been of little concern to plant breeders.

As an example, the cause for "farmers' rights" in the plant germplasm arena has several articulate and vociferous spokespersons. Basically, the cause is premised on the assumption that plant germplasm is provided by subsistence farmers in developing countries for free and that genes extracted from this germplasm are used by companies based in the developed world to produce cultivars that, in turn, are sold at a substantial price to farmers in the developing countries where the germplasm originated. The cause espouses farmers' rights as a counterpart to breeders' rights to compensate the farmers who own the "land varieties" that have evolved over centuries of subsistence agriculture. Individual plant breeders may get only peripherally involved in responding to the "farmers' rights" issue, but the plant breeding profession **must** get very involved in the issue because germplasm is the basic material with which plant breeders develop their products. Other examples of issues and concerns that can impact plant breeding could be given.

Without a doubt, the concerns about and judgment of our profession and its products by groups outside of plant breeding and by the lay public will become an increasing part of the perspective of plant breeding in the 1990s. This monitoring of plant breeding activities and products, to a degree, will change the **modus operendi** of plant breeders' work. Plant breeders will need to be more formal in arranging their activities; in some instances, they will need to justify their activities and products to the public; they will need to adhere to guidelines and policies, especially with respect to use of biotechniques; and they will need to present the plant breeders' views in dialogues and forums with antagonists of their profession, activities, and products.

In the discussion on private enterprise in plant breeding, I made a few remarks about protecting plant intellectual properties. Plant variety patenting and plant variety protection have been available for 60 and 20 years, respectively. These cultivar protection laws were constructed specifically for protecting plants and, in general, they are less restrictive than general patents. For example, the Plant Variety Protection Act provides both a "research exemption" and a "farmers' exemption". The research exemption permits varieties protected under the act to be used as parents in crosses for developing additional varieties without permission from or compensation to the originator of the protected variety. The farmers' exemption allows a farmer to sell seed of a protected variety to his neighbors without compensating the originator of the variety. Utility patents, under which many plant intellectual properties - including varieties - are protected, do not provide such exemptions.

Private plant breeding companies, of course, protect all plant intellectual properties that are protectable and salable; and many public institutions are following this pattern. In companies, the intellectual properties created by researchers generally belong to the companies as a condition of employment, and companies have legal departments to carry out protection activities. In public institutions, intellectual properties created by scientists may or may not be the property of the institutions, and few public institutions have adequate and appropriate infrastructure to perform protection activities. Therefore, scientists, including plant breeders in public institutions, must do much of the paperwork associated with protecting an intellectual property.

Historically, plant breeders have used all sources of germplasm in their cultivar development programs without permission from or notification of anyone. This is changing rapidly. Today, few breeders of crops for which there is significant private-sector cultivar development, whether they are employed privately or publicly, would use an elite line in a cross without permission from the line's originator. Groups of breeders working on specific crops are drawing up "breeders' codes" that spell out very specifically what can and cannot be done with intellectual properties. In the next decade, plant breeders will be more formal in their interactions with their colleagues, more paperwork will be required, and research will require more planning, permission, and documentation. This will be another part (some would say unfortunate part) of the perspective of plant breeding in the 1990s.

To now, I have elaborated on items that will be relatively new in the plant breeding perspective for the 1990s. However, let me assure you that the core of plant breeding will remain very much intact during the next decade. The primary sources of genes used in plant breeding will be the primary gene pools for commodity crops. The primary procedure for deriving new genotypes will be via hybridization and segregation. The most expensive part of cultivar development will continue to be field and laboratory testing for quantitatively inherited traits. The most extensive and long-term tasks will be field-testing of candidate varieties for zones of adaptation and consumer acceptability. And the team effort in plant breeding will become more important and complex. Plant breeding, traditionally, has involved cooperation among plant breeders, pathologists, entomologists, physiologists, and food scientists. In the future, biotechnologists certainly will be added to this team effort, and some say that lawyers will be cooperators as well. Another dimension of teamwork will be the increased cooperation among scientists at many sites, sometimes worldwide.

But the profession of plant breeding must change to integrate new biotechniques as tools for the trade; to adopt more formal procedures for research planning, interacting among colleagues, and documentation; to provide greater advocacy for the needs of and the social benefits from the profession; to adapt to protection of plant intellectual properties and the nuances that it entails; and to participate fully in the internationalization of plant breeding.

SOME UNKNOWNS

I have left three subjects that are no less important than those already discussed but, frankly, there is no way to predict the route that each will follow in the next decade. These subjects are (1) who will do the breeding of minor crops, (2) will the profession continue to have an adequate and appropriately trained pool of plant breeders, and (3) will adequate funding be available for plant breeding?

It is axiomatic that private industry will breed varieties for only those crops providing profit. This restricts private industry breeding programs to crops with very large acreages [e.g., maize (*Zea mays* L.), soybeans, wheat (*Triticum aestivum* L.), rice] or to crops with very high income per hectare [e.g., tomatoes (*Lycopersicon esculentum* Mill.), sugar beets (*Beta vulgaris* L.), and truck crops]. There are, however, many so-called minor crops [e.g., oats (*Avena sativa* L.), pecans (*Carya illinoensis* L.), buckwheat (*Fagopyrum esculentum* L.), red clover (*Trifolium pratense* L.), forage grasses, etc.] that will not support breeding programs in private industry. Yet they are important for maintaining agricultural diversity and meeting consumer demands. Probably, public institutions will need to maintain variety development programs for these crops but, according to studies by Brooks and Vest (1985) and Collins and Phillips (1991), many breeding programs for minor crops already have been eliminated or are targeted to be eliminated because of funding problems at public institutions. Experience shows that such breeding programs, once eliminated, are nearly impossible to recover and, if reinstituted, much germplasm and continuity is lost. There is a serious void, a lack of commitment, and no coordination with respect to this problem.

The human resource issue for plant breeding in the 1990s and beyond has several facets: (1) will there be an adequate and continual supply of young people entering the profession to meet demands in public and private sectors for the next 10 to 20 years, (2) will the young people entering the profession be appropriately educated and trained to do the plant breeding of the 1990s and beyond, and (3) where will plant breeders be trained?

Predicting the number of plant breeders that will be needed in the United States for the next decade is an exercise with little accuracy. However, a somewhat meaningful figure could be obtained by integrating data about the age structure of U.S. plant breeders in 1990, past history on how many plant breeders will move on to other activities (e.g., administration), and surveys of employment intentions of public institutions and private companies that do plant breeding. Another concern on the supply side is whether potential scientists can be attracted to the plant breeding profession. Fifteen years ago, I was very encouraged by an influx of erudite young women into plant breeding graduate programs. My experience and sense tells me, however, that the interest of females in plant breeding has waned materially. Minority students never have made up a significant proportion of graduate students in plant breeding programs. A source to fill a void in numbers of plant breeders needed, however, is usually available from newly educated foreign students. Many plant breeding graduate programs traditionally have more than 50% foreign student enrolment, and with the correct inducements, many of these graduates will stay in the United States. Thus, it may well be that the supply of new plant breeders, at least in the United States, will be adequate.

As the profession of plant breeding evolves, so the education of plant breeders must evolve. In fact, the case can be made that the education should lead the evolution. With the rapid development of plant molecular biology and the techniques and knowledge from that field that can be used by plant breeders, there has been a tendency for employers and students to overreact toward assuming that plant breeders of the future would be biotechnologists. Such an assumption, of course, is false. As explained, plant breeding will change gradually and steadily; but the skills and knowledge required of plant breeders in the 1990s will be only marginally different from those for plant breeding of today. The core education and training will remain remarkably stable.

Serious thought needs to be given to where plant breeders will be educated and trained. Several U.S. universities have changed their plant breeding graduate programs to molecular biology and biotechnology graduate programs. In these latter cases, the degree recipients are not equipped knowledgewise or attitudinally to do plant breeding. Thus, a sizable number of graduate student educational slots have been lost to the plant breeding profession. Can private companies provide an experience base to equip students from such graduate programs with the skills and knowledge to make them into bona fide plant breeders? Probably yes, but of far greater concern is the hiring of such persons into plant breeding positions at universities. Young faculty members in academia today usually have free rein to develop research programs. Too often this results in the young scientists continuing whatever they did in graduate school, and some plant breeding graduate educators are lost. This is a very serious matter because attrition of graduate student education and training in plant breeding occurs while universities continue to advertise graduate education in this field.

Future funding for plant breeding as a whole is difficult to predict. Likely, funding of plant breeding in private companies will remain stable or increase. Some companies will fold, some will downsize, and some will be merged; but as a whole, there is a continuous demand by agriculture for improved cultivars of crops, so the size of the private plant breeding industry as a whole should remain constant or grow.

The U.S. Department of Agriculture has retreated from the crop cultivar development arena and, as the cadre of plant breeders that was employed by USDA retire, this retreat will become complete. However, the USDA continues to support research activities that underpin plant breeding. For example, it takes responsibilities for funding most plant germplasm activities from collection to evaluation. Also, USDA funds much research on mapping plant genomes. It is difficult to determine whether the USDA's retreat from cultivar development *per se* has really reduced this agency's funding for plant breeding in its totality - probably not.

The state agricultural experiment stations (SAES) represent the third player in the plant breeding effort. The SAES, in concert with their associated universities, of course, represent the sole source of education in plant breeding. Also, they are the sources of most basic research and knowledge about plant breeding, the laboratories where most germplasm enhancement is done, and the laboratories where most public cultivar development programs reside. SAES funding for basic research in plant breeding comes from state and federal funding, grants and contracts from private companies, and competitive grants from federal agencies and foundations. The only consistent funding in SAES is state and federal funding, and that continues to decline as a percentage of total. Funding for germplasm enhancement and cultivar development in SAES is largely from state and federal sources and commodity groups. I sense that, in total, there is a slow erosion of SAES funding allocated to these latter two activities.

Of course, the subject of breeding minor crops, the human resource base, and the funding for plant breeding are very important issues in the perspective of plant breeding in the 1990s. However, all three are in such a state of flux that it is nearly impossible to forecast their routes in the 1990s. In fact, it is imperative that a thorough study be made to determine if trends of change are apparent for these three issues. This should be followed by setting realistic goals and planning for contributions to and acceptance of responsibility by private industry, USDA, and SAES for the three issues. I would suggest the creation of a task force with appropriate representatives from all three players and a charge to study these issues and prepare a plan for responsibilities to be assigned to the three players so as to assure continued cultivar development for all crops and adequate availability of human resources and funding.

SUMMARY

My perspective for plant breeding in the 1990s is that the primary activities of this profession will change somewhat, but not greatly. Biotechniques, as new tools for the profession, will provide a broader gene pool for breeders to use in cultivar development, and they will lend greater precision to plant breeding. Plant breeders and their knowledge and products will be judged by new groups outside the profession, and this is good for the profession. However, plant breeders will need to articulate their activities more clearly, to seek out and educate their critics and, at times, to become advocates for science and their profession. The activities of plant breeders will become more formalized and better documented as the protection of their intellectual properties becomes more precedential. However, the United States needs to maintain breeding programs for all crops, whether of minor or major importance, to assure an adequate supply of new scientists into the field and to provide adequate funding for basic and applied plant breeding activities. These are now in sufficient disarray that there is an immediate need for an in-depth study of these issues to lay the groundwork for constructing a master plan to assure that the products from plant breeding continue to be available.

ACKNOWLEDGMENTS

Journal Paper No. J-14475 of the Iowa Agriculture and Home Economics Experiment Station, Ames, IA 50011, Project 2447.

REFERENCES

Brooks, H. J.; Vest, G. (1985) Public programs on genetics and breeding of horticultural crops in the US. *HortScience* 20:826-830.

Collins, W. W.; Phillips, R. L. (1991) *Training of future US plant breeders in publicly supported institutions*. Abstracts of the Symposium on Plant Breeding in the 1990s, North Carolina State University, Department of Crop Science Research Report No. 130, p. 4.

Duvick, D.N. (1986) Plant breeding: Past achievements and expectations for the future. *Economic Botany* 40:289-297.

Fehr, W. R. (ed.) (1984) *Genetic contributions to yield gains of five major crop plants*. Crop Science Society of America Special Publication 7, Madison, WI.

Frey, K. J. (1971) Improving crop yields through plant breeding. In: Eastin, J. D.; Munson, R. D. (eds.), *Moving off the yield plateau*. American Society of Agronomy Special Publication 20, Madison, WI, pp. 15-58.

Larkin, P. J. (1987) Somaclonal variation: History, method, and meaning. *Iowa State Journal of Research* 61:393-434.

Ling, D. H. (1985) *In vitro* screening of rice germplasm for resistance to brown spot disease using phytotoxin. *Theoretical and Applied Genetics* 71:133-135.

Singer, S.; McDaniel, C. M. (1985) Selection of glyphosate tolerant tobacco calli and the expression of this tolerance in regenerated plants. *Plant Physiology* 78:411-416.

Russell, W. A. (1986) Contribution of breeding to maize improvement in the United States, 1920s-1980s. *Iowa State Journal of Research* 61:5-34.

PART ONE

THE GENE BASE FOR PLANT BREEDING

Chapter 2
Availability of plant germplasm for use in crop improvement

T. T. Chang

International Rice Germplasm Center, The International Rice Research Institute, P. O. Box 933, Manila, Philippines

INTRODUCTION

Broad-based plant germplasm resources (PGR) are imperative for sound and successful crop improvement programs. Rich and diverse sources also fuel many facets of plant research. For highly successful research and breeding activities, the genetic diversity of experimental materials needs to be sustained to minimize the vulnerability inherent in the growing of uniform and closely related cultivars over wide areas. Genetic diversity becomes more important as cropping intensity and monoculture continue to increase in all major crop-producing regions of the world. With the unabated growth in human population, the world's plant genetic resources need to be secured and more thoroughly exploited to meet the inevitable expansion in future food needs.

A complete array of germplasm in a crop consists of (1) wild relatives, weed races, and landraces in the areas of diversity, (2) unimproved or purified cultivars used earlier in the major production areas or still used in minor areas, and (3) improved germplasm in commercial production and genetic testers from breeding programs and genetic studies. Recent advances in biotechnology have extended the range of usable materials far beyond the related genera of a crop.

Crop germplasm can be made available to the users (agronomists, breeders, entomologists, geneticists, plant pathologists, plant physiologists, crop ecologists, soil scientists, production specialists and growers) on a continuing basis if the interrelated facets of field conservation, multiplication, characterization, rejuvenation, documentation, and distribution are well handled by the germplasm workers. Amply evaluated germplasm along with freely available information will further increase the demand for conserved materials. Duplicate preservation sites will provide added security. The essential components are in Figure 1.

This chapter aims to assess the conservation status of major food crops and to integrate other scientist-oriented components and socio-politico-economic forces with germplasm use, all of which will strengthen the availability of crop germplasm on a sustained basis.

Figure 1 Essential components of a comprehensive rice genetic conservation program. Constraints at an operational level are indicated inside the brackets.

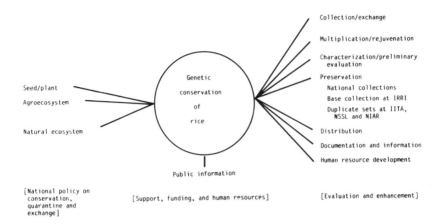

PRESENT CONSERVATION STATUS OF MAJOR FOOD CROPS

Recent surveys indicated that among the major food crops, the relative scope of coverage of varietal collections being conserved may be ranked as follows: cereals > common potato > grain legumes > sorghums and millets > cassava > soybean and peanuts (as oil seeds) > sweet potato > yam (Table 1). Wild relatives of most crops are deficient in coverage, except those of wheat, maize, potato, and tomato. Field collection or re-collection for landraces of many crops in remote areas remains incomplete (Plucknett *et al.*, 1983, 1987; Wilkes, 1983; Lyman, 1984; Chang, 1985d).

Surveys of national and international genebanks yield impressive statistics (Table 2). Estimates of the world's crop collection for wheats total 410,000; grain and oil seed legumes, 260,000; rice cultigens, 250,000; sorghums, 95,000; maize, 100,000; common potato, 42,000; cassava, 14,000; and yams, 10,000. On the other hand, the duplication of accessions within a crop among the various genebanks may vary from 30% (in peanut) to 75% (in wheat). The total holdings of the world's genebanks were estimated at approximately 2,500,000 accessions (Holden, 1984; Plucknett *et al.*, 1987), of which about 1,050,000 accessions may be considered as unique or distinct (Holden, 1984). About 35% or more of the unduplicated world holdings expected to grow to ca. 1,300,000 samples in 1995-96 will be cared for by the nine international agricultural research centers (IARCs) in the Consultative Group on International Agricultural Research (CGIAR) system (van Sloten, 1990). A significant growth in the global figures for various crop

Table 1 Conservation status of major crops.*

Crop	Total accessions in genebanks	Distinct accessions	Wild accessions	% Cultivars uncollected	Major needs[†]
Wheat	410,000	125,000	10,000	10%	E,M
Grain and oil legumes	260,000	132,000	10,000+	30-50%	C,E,M for peanut
Rice	250,000	120,000	5,000	10%	C (wild), E,M
Sorghum	95,000	30,000	9,500	20%	E,M
Maize	100,000	50,000	15,000	5%	M,E
Soybean	100,000	30,000	7,500	30%	C (wild), E
Common potato	42,000	30,000	15,000	10-20%	C,E
Yams	8,200	3,000	60	High	C
Sweet potato	8,000	5,000	550	>50%	C,E

*Source: Data combined from Lyman (1984), Chang (1985c), Plucknett *et al.* (1987), and Williams (1989).

[†]C = collection, E = evaluation, M = maintenance.

Table 2 Estimates of germplasm holdings in major national PGR systems and international centers.

Country/IARC	Categories concerned	Total
U.S.A.	All crops	557,000
China	All crops	400,000
U.S.S.R.	All crops	325,000
IRRI	Rice	86,000
ICRISAT	Sorghum, millet, chickpea, peanut, pigeon pea	86,000
ICARDA	Cereals, legumes, forages	77,000
India	All crops	76,800
CIMMYT	Wheat, maize	75,000
CIAT	Common bean, cassava, forages	66,000
IITA	Cowpea, rice, root crops	40,000
CIP	Potato, sweet potato	12,000

Source: IBPGR (1990), Paroda (1988), Shands *et al.* (1989), Vitkovskij and Kuznetsov (1990), Zhang and Dong (1989).

collections stemmed from more than a decade of intensive field exploration and collection by numerous teams in germplasm-rich areas, beginning in the early 1970s. A forerunner among the cereals in the massive collection activities was the 14-country project on rice germplasm conservation coordinated by the International Rice Research Institute (IRRI), which netted about 43,000 seed samples in Asia (Chang *et al.*, 1989b). Several agencies in West Africa assembled 5300 samples of African rices, wild taxa, and Asian cultivars; these were deposited at the International Institute of Tropical Agriculture (IITA) in Nigeria and the IRRI in the Philippines (Ng, 1991). The International Board for Plant Genetic Resources (IBPGR) was a major funder and coordinator in such worldwide activities. Between 1974 and 1989, IBPGR contributed to the assemblage of 188,400 samples, of which 74,500 came from cereals, and 29,000 samples were forages and food legumes. Root and tuber crops totalled 19,999 samples, and vegetables trailed at 17,000 samples (van Sloten, 1990). Major segments of the collected seed samples were deposited in genebanks of the developed countries and/or the IARCs (Mooney, 1983; Fowler *et al.*, 1988). The fate of other collections varied from one repository to another (Fowler *et al.*, 1988).

One of the shortcomings in such a massive undertaking by diverse groups is the dearth of useful passport data such as site characteristics and ecogenetic features, which later constrains the sorting of duplicates and curtails evaluator's interest in the collected materials (IRRI-IBPGR, 1983; Williams, 1989). Many a rice sample collected by extension workers carried only a vernacular name and the village or district. Thousands of rice seed samples submitted to IRRI were no longer viable at the time of receipt. The genetic quality of seed samples also varied greatly (Chang, 1980). These constraints have led the IARCs to reassess collection strategies and plan future collections more critically (IRRI-IBPGR, 1983; Holden, 1984; Williams, 1984).

Reports of the IBPGR indicated that there were 38 base collections of crops and 50 PGR centers under its "global network" (Williams, 1983; IBPGR, 1986). However, these do not include China, U.S.S.R., and other Communist-block countries. None of the base collections, including that of IRRI, is a complete set for a given crop. Perhaps, the more pertinent questions to ask are: (1) What is the eco-genetic coverage of the collections? (2) How many accessions in the collections are distinct from one another? and (3) What is the proportion of viable accessions with seeds available for use?

Much needs to be done by major genebanks in comparing holdings, culling redundancy, and filling in gaps. The goal is to promote judicious exchange and consolidation among collections so that duplicate storage is provided for every distinct accession (IRRI-IBPGR, 1983; Holden, 1984).

In the case of rice, the 1977 Rice Genetic Conservation Workshop held at IRRI led to the sharing and division of responsibilities among IRRI, IITA, Japan, U.S.A., and other rice-producing countries (IRRI-IBPGR, 1978). The meeting recommended that different centers should further exchange and carefully compare accession lists to minimize the maintenance of obviously duplicate accessions within single collections and to ensure that no distinct accessions or ecostrains are overlooked in the inventorial process. Some progress has been made in these directions. However, a shortage of technical manpower and funding cutbacks at all centers have slowed down the collaborative efforts in recent years (IRRI-IBPGR, 1991).

An earlier IBPGR survey (Hanson *et al.*, 1984) showed that during 1983, there were 39 genebanks with long-term storage facilities. Since then, several Asian countries and two IARCs have acquired modern facilities. An updated breakdown by econo-geographic status and organizational character of facilities having subfreezing storage of varying sizes shows:

25 (43.1%) national genebanks in developed countries (DCs)
 1 (1.7%) regional genebank in a developed country
23 (39.7%) national genebanks in less-developed countries (LDCs)

7 (12.1%) IARCs in less-developed countries
2 (3.4%) regional genebanks in less-developed countries

For a genebank capable of supplying viable and useful germplasm to users, the requisites are (1) comprehensive or representative coverage of the conserved accessions, (2) dependable and cost-efficient preservation facilities, (3) sufficient seedstocks for distribution or exchange, (4) freely available information on the conserved materials, (5) sustained support in funds and personnel, and (6) service capability of the genebank staff. Only a small number of genebanks in the world have all these resources in preservation, multiplication/regeneration, characterization, documentation, and dissemination so as to qualify as effective partners in crop improvement. A genebank's usefulness is enhanced if it is an integral component of a crop research center (Chang *et al.*, 1973). For a large country, a national plant germplasm system is needed to serve many agencies in the country (Burgess, 1971; The Rockefeller Foundation, 1980; Paroda, 1988; Zhang and Dong, 1989). To serve crop researchers worldwide, a global network is essential (Frankel, 1975).

While refrigerated seed storage facilities have been upgraded in several germplasm-rich countries with foreign assistance, the growth in national capabilities remains uneven in terms of trained staff, field facilities for seed production, seed processing prior to storage, and maintenance of the desired storage conditions. In the humid tropics, seed drying is a major bottleneck. For genebanks with a meager funding base, power consumption is a major outlay. The labor-intensive and costly component of seed regeneration is a serious constraint to all genebanks. Consequently, the germplasm workers find difficulties in attaining the goal of providing greater security to the crop collections (Chang *et al.*, 1989a; IRRI-IBPGR, 1991).

GENETIC EROSION INSIDE GENEBANKS

The impressive statistics on the sizes of various crop collections gathered by the Food and Agricultural Organization of the United Nations (FAO) (Gullberg, 1971), IBPGR (Toll *et al.*, 1980 for rice; Ayad *et al.*, 1980a,b for wheat and maize), Lyman (1984), and Plucknett *et al.* (1987) through surveys do not mean that all collections are intact or that one genebank can supply seed or plant material for every accession. The figures produced by such surveys are only "numbers in the books". Mooney (1985) is also skeptical of such numbers.

The security and availability of materials in different genebanks varies greatly. Genetic erosion occurs in all genebanks. Many genebanks in the tropics face rampant genetic erosion because of a lack of refrigerated storerooms, field space, trained personnel, and administrative and financial support. Repeated regeneration in small plots also leads to undesirable effects in terms of genetic identity, population structure, workload and storage space (Chang *et al.*, 1979).

The tragic loss of many Latin American maize races has been described by Goodman (1984). Some of the factors leading to losses and common to other crops are (1) discontinuity in program and/or personnel, (2) human neglect, (3) shifts in program direction or methodology of maintenance, (4) poor storage facilities, (5) disappearance of some records, and (6) lack of periodic monitoring of seed viability.

Similar instances plus the frequent inability of the donor countries to retrieve their materials from the depository or the seedbank concerned have triggered the serious concern and discontent among the donors. The dissatisfaction of the germplasm-donating LDCs first surfaced at the 1981 FAO-UNEP-IBPGR International Conference on Crop Genetic Resources (FAO, 1981). Further agitation by activists and politicians led to the FAO's aborted attempt to establish "an international genebank" under FAO auspices. Member countries in the developing world succeeded in having FAO organize a Commission on Plant Genetic Resources and declare an International Undertaking on Plant Genetic Resources; that, however, was not well received by the DCs. Subsequent meetings of the LDC-dominated Commission has transformed the North (DCs) vs. South (LDCs) confrontation into more conciliatory sessions that now advocate free access to germplasm, recognize "farmers' rights", and set up an "international fund" to promote germplasm conservation. The FAO Commission will continue to meet, serving as an open forum and trustee for the farmers, present and future (FAO, 1989).

In spite of the controversies raised by the above developments, the burden of conserving the world's available plant germplasm still rests with the plant scientists. Unfortunately, most government administrators did not realize the importance of having continuity in genetic conservation operations and management by a small number of trained workers. Such a vital human factor in the preservation phase has been emphasized by several veteran workers (Larson, 1961; Mengesha and Rao, 1982; Goodman, 1984; Chang *et al.*, 1989a).

Even in the International Rice Germplasm Center (IRGC) at IRRI which has been under continuous management by the author, we have lost hundreds of accessions because (1) some accessions were not adapted to adverse environments (high temperature, high precipitation, salinity, zinc deficiency, etc.) in Los Banos, (2) susceptible accessions succumbed on the pest-ridden IRRI farm, or (3) seed viability expired earlier than expected during storage and between germination tests. Such reasons for the loss of registered accessions do not include non-viable seeds collected by extension workers from old (expired) or insect-infested seeds in farmers' bins. We were able to recover only a small number of replacement samples from original donors. Fortunately, the use of tissue culture has helped to revive hundreds of poorly viable accessions (IRRI, 1986). PGR work calls for sustained support, vigilance and loving care from all workers in a program (Chang, 1985b; Chang *et al.*, 1989a).

CONSTRAINTS ON SEED EXCHANGE AND USE

Every germplasm-rich country perceives its indigenous plant germplasm as a national biological treasure, especially in the case of highly valued industrial crops. Fortunately, most countries have now lowered the barriers and are willing to exchange germplasm with other countries and/or the IARCs. The large number of countries that have become members of the FAO Commission on PGR (96) or pledged to adhere to the International Undertaking (89) or both (67) in 1989 attests to the changing trend. However, a number of constraints still persist.

Traditional constraints due to historical and political factors

The exchange of germplasm across national borders sometimes faces political constraints especially between feuding neighbors. But the establishment of the IARCs has markedly facilitated international exchanges since the 1960s.

Our experience has been that when rice researchers freely and repeatedly receive seeds from a source such as IRRI, they will later be more willing to supply their seeds for deposit at IRRI. Among the seed samples of food crops given by China for foreign exchange between 1972 and 1986, IRRI received more than 7000 samples among a total of 16,786 (Zhang and Dong, 1989).

IRRI is unique in enjoying full exchange status with all rice-producing countries, some of which will only give seed to IRRI. This is also a result of nearly three decades of dedicated service to rice researchers and counsel to genebanks in the interest of helping rice growers and the rice consumers.

The relationship between plant breeder's rights adopted by most developed countries and the exchange of germplasm and maintenance of biodiversity

The Plant Breeder's Rights Act has substantially stimulated private plant breeding activities and aroused interest in germplasm. However, the growth of the private sector has markedly reduced some of the public inputs of DCs such as the Land-Grant (State) Colleges of Agriculture in the U.S.A. It is generally recognized that intellectual property protection measures sought by the private sector would restrict free access to useful germplasm and eventually reduce genetic diversity in a crop. Some private breeders are conscious of the need to incorporate genetic diversity in their products (Duvick, 1984) and thus add to the internal genetic diversity in successive new cultivars (Duvick and Brown, 1989).

In the LDCs, the impact of the intellectual property protection systems is mainly on F_1 hybrids and crops of high cash value. The concern of LDCs was aired by Jain (1982).

While the FAO's PGR Commission finds Plant Breeder's Rights provided for under the Union International pour la Protection des Obtensions Vegetales (UPOV) as "Incompatible with the International Undertaking", the national governments of LDCs can counteract the pressure of large multinational seed companies by providing more support to the breeding and research efforts of the public sector. The efforts of the IARCs are also in this direction. In England, the former Plant Breeding Institute was able to pay for its operating expenses by royalties derived from the seed sales of its improved wheat varieties.

Leaders in the seed industry have pointed out that a proper balance may be attained between public and private interests if workers follow common goals and are willing to cooperate and accommodate in a changing world (Duvick and Brown, 1989).

Pressure on PGR programs by sociopolitical activist groups

Among the internationally based sociopolitical activists, the coalition headed by Pat Mooney and Cary Fowler has exerted the largest impact on public opinion, mainly because of their writings (Fowler *et al.*, 1988; Mooney, 1979, 1983), which are forceful allegations that LDCs have national sovereignty over resources and production. Their papers are based on critical analysis, politically charged views, emotional appeals, and an engaging style of writing. But some of their allegations lack scientific insight and accuracy. The main targets are the private seed companies now owned nearly exclusively by multinational corporations that also invest in biotechnology, but the large genebanks and the IBPGR were also accused of helping the seed industry. During the heated debates of the mid-1980s, germplasm flow from some of the LDCs slowed down (MacFadyen, 1985; Sun, 1986).

The debates during the meetings of the FAO Plant Genetic Resources Commission have further added fuel to the controversial issues. Gradually, the South vs. North battle of words simmered down for lack of funds, absence of a substantial action program, and deficiency in scientific leadership. On the other hand, the activists' role also had a positive and significant impact: PGR programs are receiving increased attention and support from the governments concerned and the general public. For instance, the Centro Internacional de Mejoramiento de Maíz y Trigo (CIMMYT) has been mandated by the CGIAR to expand its PGR programs. The USA is once more reassessing its National Plant Germplasm System (NPGS) and has allocated funds to expand the National Seed Storage Laboratory at Ft. Collins. Recent findings of a study group on the NPGS and its recommendation for more effective operations, better coordination, and expanded research have become available (Board on Agriculture-National Research Council, 1991).

Restrictions posed by national plant quarantine regulations

National plant quarantine regulations vary from one country to another. Reports indicate that certain strict plant quarantine measures have sometimes constrained international germplasm exchange, especially in the case of large and frequent shipments when national staff found it difficult to cope with the plants. Physical facilities and staff size are usually the bottlenecks. Insufficient knowledge about the life cycle, host range, and natural history of crop pests and pathogens that are foreign to the importing countries adds to the difficulties (Plucknett and Smith, 1988). Intermediate quarantine, particularly for tropical cash crops, is desirable. Recent experience has also shown that effective communication between the quarantine officers and germplasm workers has paved the way for improving cooperation (IRRI, 1988). Some IARCs have also contributed to the development of quarantine station facilities and staff in their host countries and thus added to the national capabilities for clearing imported germplasm.

HOW CAN GERMPLASM-RICH COUNTRIES SAFEGUARD THEIR PGR AND ENSURE THEIR AVAILABILITY?

The need for every germplasm-rich country to invest more in PGR activities such as field collection, preservation, rejuvenation, evaluation, and use has been repeatedly emphasized. PGR working groups of limited size and inadequate training cannot carry out the mission alone. A national plant germplasm system is needed to enlist the energies of and sustain the inputs from all concerned workers under a nationally based institution but with international perspectives. Less than 30% of the world's nations have formal national PGR programs (Keystone Center, 1990).

Several scientists have pointed out the rather limited use of unimproved crop germplasm from the vast arrays of gene pools being conserved (Peeters and Williams, 1984; Holden, 1984; Frankel and Brown, 1984; Chang, 1985c; Marshall, 1989). The lack of systematic evaluation, meager research on novel sources of germplasm, insufficient ecogenetic studies on germplasm composition, inadequate documentation and information dissemination, and insufficient prebreeding efforts have been mentioned as the major constraints. Multidisciplinary collaboration, interdisciplinary interaction, and information dissemination schemes are needed to fill the gaps before the users become interested in the genetic materials. For plant collections of relatively large size, the core collection approach may facilitate dissemination, evaluation, research and duplicate storage (Brown, 1989a,b; Chang, 1991b; Hodgkin, 1991; Vaughan, 1991).

The crucial role of the germplasm scientist as a member of the team, be it research or breeding, cannot be overlooked. Expertise in choosing and handling novel germplasm will enhance efficacy and reduce the time frame for assessment and use. On the other hand, the conservationists should strive to be a full partner in the team and contribute beyond mere service function. Meanwhile, the contributions of the germplasm workers to crop research and improvement need to be better recognized (IRRI-IBPGR, 1991; Chang, 1991a). Incentives should be augmented so that the workers will find a rewarding career in PGR work (Chang *et al.*, 1989a). The imperative nature of personnel continuity in germplasm management is widely recognized as a pivotal issue (Larson 1961; Mengesha and Rao, 1981; Goodman, 1984). Because genetic conservation is a relatively young and multifaceted technology (Chang, 1985a), more germplasm workers need to be trained on a multi-institutional collaborative scheme (Chang, 1985a; Chang *et al.*, 1989a).

Regardless of variations in national networks, the most basic "grass root" component of every system is the crop (commodity) committees, in which the conservationists and the breeders should collaborate and play the most active roles. The crop committees should enlist the participation of other biological disciplines (crop protectionists, physiologists, etc.) and related disciplines (crop evolutionists, ethnobotanists, crop historians, biostatisticians, economists, etc.). During the formative stage of the crop committees, the conservationists should get out of their "shells" and actively seek the collaboration of others.

A continual dialogue among the different disciplines in a committee and at higher frequencies is essential to sustained cooperation. Periodic meetings and field visits should be held. Technical information should be freely disseminated. Lack of information on evaluated materials and poor communication between conservationists and users have limited the users' (breeders and researchers) interest in the unimproved germplasm (Chang, 1984, 1991a). Plant breeders tend to focus on readily observable traits (Frankel and Brown, 1984). Programs of the IRRI and the IBPGR in standardizing crop research data and disseminating information on collections are representative of efforts of crop-oriented IARCs to bridge these gaps. As a result of multichannel communication, the International Rice Germplasm Center (IRGC) at IRRI has recently added many cellular biologists and molecular geneticists to its clientele and materially aided rapid advances in rice biotechnology by freely furnishing exotic germplasm (IRRI, 1988).

For some of the home-grown cereals, vegetables, fruit trees and nuts, a treasure of knowledge and expertise may be found in the people who continue to grow them from one generation to another. Such growers or their community groups should be included in conservation projects (RAFI, 1986).

In recent years, nearly all of the major rice-growing countries in Asia have established or expanded national PGR programs to save rice and other crops (IRRI-IBPGR, 1991). Several foreign technical assistance agencies [Japan International Cooperation Agency (JICA), U.S. Agency for International Development (USAID), Overseas Development Administration (ODA), Deutsche Gesellschaft Technische Zusammenarbeit (GTZ), and United Nations Development Programme (UNDP)], the Rockefeller Foundation (in China), and the IBPGR have contributed to the physical facilities. Continuing and strong commitment by the national governments is needed to sustain the expanded PGR programs and to fully use the new facilities.

Many LDCs recognize that conserving and using both indigenous and introduced germplasm have led to handsome rewards in upgrading crop production, yield, and quality. The rice crop has perhaps caused the largest impact on such awareness. Outstanding examples of successful use of crop germplasm are well documented in the experience of China (Shen, 1980 on semidwarf rices; Lin and Yuan, 1980 on hybrid rice) and India (Paroda *et al.*, 1988). It is rather ironic that for all the elegant words in defense of the farmers in LDCs, the advocates of "farmers' rights" have overlooked that these farmers are the principal beneficiaries of the Green Revolution which was based on an effective use of germplasm.

The payoff from genebanks on a global basis has been reviewed by Hawkes (1985) and Smith (1987). Larger collections surpass smaller collections in cost effectiveness (Chang, 1989). The cost of power consumption for maintaining IRGC's medium- and long-term storerooms is estimated to be under one U.S. dollar per accession annually.

NEED FOR A MULTI-SECTOR APPROACH

Uncompleted tasks in conserving and using crop germplasm include (1) completing field canvassing and consolidating collections; (2) improving germplasm preservation technology, dissemination, and duplicate storage; (3) expanding systematic evaluation and standardized documentation; and (4) enhancing communication among conservationists, evaluators, breeders, researchers, biotech workers, and statisticians so that the unimproved germplasm and wild species will be more intensively used. Redundancy within collections should be reduced; duplicates of the same crop in different genebanks should be consolidated, and security maintained by duplicate storage sites. Preservation of some recalcitrantly seeded or vegetatively propagated crops will be aided by advances in cell and tissue cultures under slow growth or cryopreservation. Non-destructive monitoring of seed viability needs to be developed. *In-situ* conservation needs to be strengthened to complement *ex-situ* preservation.

Recent controversies about the ownership of and the rights to use unimproved germplasm stem from an incomplete knowledge of the policies and operations of different genebanks. There will be no serious barriers to the supply of conserved germplasm if full communication is achieved among politicians, administrators, scientists of both public and private sectors, social activists, journalists, and the public. The IRRI germplasm bank is widely known for freely supplying seed and relevant information. PGR workers of the IARCs hold the view that they are the custodians of germplasm resources (CGIAR, 1989), and every donor-nation can retrieve its entire collection when needed. During the past decade, the IRGC has returned, upon request, 9608 rice accessions to six donor countries and three states of India. We were encouraged to receive high marks from the influential activists (Fowler and Mooney, 1985; Anon, 1987; Fowler *et al.*, 1988).

The availability of germplasm will depend on increased awareness and support given by public and private sectors to the arduous, and often unglamorous task of genetic conservation. Crop conservationists should exert increased efforts themselves in educating administrators, philanthropic donors, journalists, politicians, social activists, college students, farmers, and the general public about the role of germplasm in human welfare - instead of just talking among themselves. Efforts in this direction are shown in articles by Swaminathan (1986), IRGC (1986), and in Philippine newspaper articles entitled "IRRI's gene bank aids all farmers", "Old plants should not die or fade", "IRRI's humane work", "IRRI: preservation, not confiscation", and "Preserving world's rice varieties has bypassed politics". Newspaper and magazine articles will be more relevant and accurate if the conservationists will furnish the facts and figures of their mission and achievements to the journalists (Kahn, 1985; Anon, 1986; IRGC, 1986; Chang, 1986). We at IRRI have also helped many TV stations to tape germplasm conservation operations. Undoubtedly, other major PGR centers also conduct public information programs.

Full cooperation among DCs and LDCs and between public and private sectors is imperative to safeguard the continuous availability of crop germplasm. Public awareness is sorely needed in all nations so that the national PGR program will receive adequate and sustained support from all sectors of society. The all-out efforts and international cooperation are essential to ensure that crop germplasm can benefit the future generations of mankind by caring for, sharing and pooling resources:

Developing countries (LDCs) serve as:	Developed countries (DCs) serve as:
Sources of diversity	Sources of funding
Sites for seed rejuvenation in home environment	*Ex-situ* preservation and distribution sites
Sites for *in situ* conservation	Training centers

Different geographic regions draw on the crop diversity of other regions to sustain the production of food and industrial crops. Such interdependence has been analyzed by Kloppenburg and Kleinman (1987). Therefore, all countries must cooperate so as to sustain agricultural productivity.

AREAS FOR FUTURE ENDEAVOR

This survey shows that the bulk of the commercially important cultivars of major food crops has been assembled and preserved in various PGR centers, and that the level of security in preservation remains variable. Some of the primitive landraces and many wild relatives for most crops remain to be salvaged and conserved by different methods. The total volume of most crop collections poses difficulties in exploitation for the interested users because of (1) superficial sheer size; (2) redundancy within and between collections; and (3) meager information on their potential usefulness owing to deficiencies in evaluation, related research, information dissemination, and prebreeding. The true diversity in many collections is yet to be assessed. The plant breeder often stands at the end of the line until other disciplines have ascertained the usable genes in the unimproved or raw germplasm. Multidisciplinary and interinstitutional collaboration are urgently needed to fill the gaps and elevate the usefulness of the conserved materials. Unless the potentials of the preserved germplasm are rendered more apparent to the interested users, the collections in genebanks will remain as museum pieces. Germplasm workers themselves should exert efforts toward becoming full partners in germplasm use for the benefit of mankind. The crop-based network approach recently fostered by the IBPGR and collaborating IARCs (IBPGR, 1989; IRRI-IBPGR, 1991) is a step to integrate various disciplines and to add impetus to their activities.

Effective conservation programs will also help to sustain biodiversity and sustainability of the ecosystem. A sound conservation program should have the participation of disciplines beyond the biological fields (Soule, 1985).

For PGR programs overburdened by rapid growth in collection size, increased inputs in financial, physical and human resources are needed by every center. Fund requirements on a global scale have been estimated at U.S. $500 million per annum (Keystone Center, 1990); that is far above the 1982 level of funding at $55 million (Plucknett et al., 1987). Additional inputs from both public and private sectors are necessary (Anon, 1984). The human component in manning PGR programs should be expanded and reinforced to provide improved incentives for this unglamorous service and to permit professional advancement. Decision-makers and the general public should be more fully informed of the crucial role of genetic conservation so that improved support can be assured.

In the light of dwindling resources on all fronts and with the recent emphasis given to biotechnology, competition for both financial and human resources among different disciplines is intensified. Even plant breeders are declining in number (Brooks and Vest, 1985; James, 1990). Past experience indicated that germplasm programs are more liable to suffer cutbacks during times of financial restraints (Reitz, 1976).

The urgency of timely inputs is apparent. On one hand, time for salvaging the remnant unconserved germplasm in many parts of the world is nearly running out; on the other hand, most conserved germplasm has been lightly exploited and used. Evaluation, research, documentation, and prebreeding hold the key to enhanced usage of unimproved germplasm. Genetic conservation and biotechnology can be complementary and synergistic in serving each other, with the participation of related disciplines (Chang and Vaughan, 1991), while conventional plant breeding will continue to play a central role in producing a finished product (Sprague *et al.*, 1980; Chang, 1985c).

It is imperative that all sectors of society and the different disciplines join hands in protecting our most important and irreplaceable biological heritage and in deriving maximum benefit from its use.

REFERENCES

Anon (1984) Conservation and utilization of exotic germplasm to improve varieties. *Report of 1983 Plant Breeding Research Forum.* Pioneer Hi-Bred International, Inc., Des Moines, Iowa.

Anon (1986) IBM outreach: The genetic resources of rice. *Newsweek* 108(8), unpaginated.

Anon (1987) A report on the security of the world's major genebanks. *RAFI Communique*, July issue, 9 p.

Ayad, G.; Toll, J.; Esquinas-Alcazar, J. T. (1980b) *Directory of germplasm collections. 3. Cereals. II. Maize.* IBPGR, Rome.

Ayad, G.; Toll, J.; Williams, J. T. (1980a) *Directory of germplasm collections. 3. Cereals. I. Wheat.* IBPGR, Rome.

Board on Agriculture-National Research Council (1991) *Managing global genetic resources - The U.S. National Plant Germplasm System.* National Academy Press, Washington, D.C.

Brooks, H. J.; Vest, G. (1985) Public programs on genetics and breeding of horticulture crops in the United States. *HortScience* 20:826-830.

Brown, A. H. D. (1989a) The case for core collections. In: Brown, A. H. D.; Frankel, O. H.; Marshall, D. R.; Williams, J. T. (eds.), *The use of plant genetic resources.* Cambridge University Press, Cambridge, pp. 136-156.

Brown, A. H. D. (1989b) Core collections: A practical approach to genetic resources management. *Genome* 31:818-824.

Burgess, S. (ed.) (1971) *The national program for conservation of crop germplasm.* University of Georgia. Athens.

CGIAR (1989) *CGIAR policy on plant genetic resources.* IBPGR, Rome.

Chang, T. T. (1980) The rice genetic resources program of IRRI and its impact on rice improvement. In: *Rice improvement in China and other Asian countries.* IRRI, Los Banos, Philippines, pp. 85-106.

Chang, T. T. (1984) Conservation of rice genetic resources: Luxury or necessity? *Science (Washington, DC)* 224:251-256.

Chang, T. T. (1985a) Principles of genetic conservation. *Iowa State Journal of Research* 59:325-348.

Chang, T. T. (1985b) Preservation of crop germplasm. *Iowa State Journal of Research* 59:365-378.

Chang, T. T. (1985c) Germplasm enhancement and utilization. *Iowa State Journal of Research* 59:399-424.

Chang, T. T. (1985d) Crop history and genetic conservation: Rice - A case study. *Iowa State Journal of Research* 59:425-455.

Chang, T. T. (1986) Rice germplasm resources in meeting world hunger. *Bread and Justice* 59:7-11.

Chang, T. T. (1989) The case for large collections. In: Brown, A. H. D.; Frankel, O. H.; Marshall, D. R.; Williams, J. T. (eds.), *The use of plant genetic resources.* Cambridge University Press, Cambridge, pp. 123-156.

Chang, T. T. (1991a) Rice genetic resources: A practitioner's view. *Forum Applied Research and Public Policy*:(in press).

Chang, T. T. (1991b) Guidelines on developing core collections of rice. In: *Proceedings 1990 rice germplasm workshop.* IRRI-IBPGR, Los Banos, Philippines (in press).

Chang, T. T.; Brown, W. L.; Boonman, J. G.; Sneep, J.; Lambert, H. (1979) Crop genetic resources. In: Sneep, J.; Hendriksen, A. J. T. (eds.), *Plant breeding perspectives.* PUDOC, Wageningen, pp. 83-103.

Chang, T. T.; Dietz, S. M.; Westwood, M. N. (1989a) Management and utilization of plant germplasm collections. In: Knutson, L.; Stoner, A. K. (eds.), *Biotic diversity and germplasm preservation, global imperatives.* Kluwer Academic Publishers, Dordrecht, Boston, London, pp. 127-159.

Chang, T. T.; Dong, Y. S.; Paroda, R. S.; Ying, C. S. (1989b) International collaboration on conservation, sharing, and use of rice germplasm. In: *Progress in irrigated rice research.* IRRI, Los Banos, Philippines, pp. 325-338.

Chang, T. T.; Vaughan, D. A. (1991) Conservation and potentials of rice genetic resources. In: Bajaj, Y. P. S. (ed.), *Biotechnology in agriculture and forestry*, v. 14. Springer-Verlag, Berlin and Heidelberg, pp. 531-552.

Chang, T. T.; Villareal, R. L.; Loresto, G. C.; Perez, A. T. (1973) IRRI's role as a genetic resources center. In: Frankel, O. H.; Hawkes, J. G. (eds.), *Crop genetic resources for today and tomorrow.* Cambridge University Press, Cambridge, pp. 457-465.

Duvick, D. N. (1984) Genetic diversity in major farm crops on the farm and in reserve. *Economic Botany* 38:157-174.

Duvick, D. N.; Brown, W. L. (1989) Plant germplasm and the economics of agriculture. In: Knutson, L.; Stoner, A. K. (eds.), *Biotic diversity and germplasm preservation, global imperatives.* Kluwer Academic Publishers, Dordrecht, Boston, London, pp. 499-513.

FAO (1981) *Report of the FAO/UNEP/IBPGR international conference on crop genetic resources.* FAO, Rome.

FAO (1989) *Report of the Commission on Plant Genetic Resources, third session, Rome, 17-21 April, 1989.* FAO, Rome.

Fowler, C.; Mooney, P. (1985) *The acceptance speech for the Right Livelihood Award.* Rural Advancement Fund International, Pittsboro, NC.

Fowler, C.; Lachkovics, E.; Mooney, P.; Shand, H. (1988) The laws of life. *Development Dialogue* 1988:1-2. Dag Hammarskjold Foundation, Uppsala.

Frankel, O. H. (1975) Genetic resources centres - A cooperative global network. In: Frankel, O. H.; Hawkes, J. G. (eds.), *Crop genetic resources for today and tomorrow.* Cambridge University Press, Cambridge, pp. 473-481.

Frankel, O. H.; Brown, A. H. D. (1984) Plant genetic resources today: A critical appraisal. In: Holden, J. H. W.; Williams, J. T. (eds.), *Crop genetic resources: Conservation and evaluation.* George Allen and Unwin, London, pp. 249-257.

Goodman, M. M. (1984) An evaluation and critique of current germplasm programs. In: *Report of the 1983 plant breeding research forum.* Pioneer Hi-Bred International, Des Moines, IA, pp. 195-249.

Gullberg, U. (1971) World list of germplasm collections: Rice. *Plant Genetic Resources Newsletter* 26:27-35.

Hanson, J.; Williams, J. T.; Freund, R. (1984) *Institutes conserving crop germplasm: The IBPGR network of genebanks.* IBPGR, Rome.

Hawkes, J. G. (1985) *Plant genetic resources - The impact of the international agricultural research centers, CGIAR study paper 3.* World Bank, Washington, D.C. 115 p.

Hodgkin, T. (1991) Improving the utilization of plant genetic resources through core collections. In: *Proceedings 1990 rice germplasm workshop.* IRRI-IBPGR, Los Banos, Philippines (in press).

Holden, J. H. W. (1984) The second ten years. In: Holden, J. H. W.; Williams, J. T. (eds.), *Crop genetic resources: Conservation and evaluation.* George Allen and Unwin, London, pp. 277-285.

IBPGR (1986) *Annual report, 1985.* IBPGR, Rome. 92 p.

IBPGR (1989) The case for crop networks, an IBPGR initiative. *Geneflow,* June issue (unpaginated). IBPGR, Rome.

IBPGR (1990) *Partners in conservation - Plant genetic resources and the CGIAR system.* IBPGR, Rome.

IRGC (1986) *Philippine traditional rice varieties in the IRRI germplasm bank.* IRRI, Los Banos, Philippines.

IRRI (1986) *Annual report for 1985.* IRRI, Los Banos, Philippines, pp. 4-5.

IRRI (1988) *Rice seed health.* IRRI, Los Banos, Philippines.

IRRI-IBPGR (1978) *Proceedings of the workshop on genetic conservation of rice.* IRRI, Los Banos, Philippines.

IRRI-IBPGR (1983) *Rice germplasm conservation workshop.* IRRI, Los Banos, Philippines.

IRRI-IBPGR (1991) *Proceedings of the 1990 rice germplasm workshop.* IRRI, Los Banos, Philippines (in press).

Jain H. K. (1982) Plant breeders' rights and genetic resources. *Indian Journal of Genetics* 42:121-128.

James, N. I. (1990) A survey of public plant breeding programs in the U.S., 1989. *Diversity* 6(2):32-33.

Kahn, F. J., Jr. (1985) The staffs of life. IV. Everybody's business. *The New Yorker,* March 4, pp. 53-75.

Kloppenburg, J., Jr.; Kleinman, D. L. (1987) Seeds and sovereignty. *Diversity* 10:29-33.

Keystone Center (1990) *Final consensus report of the Keystone international dialogue series on plant genetic resources, Madras Plenary Session,* 29 Jan.-2 Feb., 1990, Madras, India. GRCS, Inc., Washington, DC, 38 pp.

Larson, R. E. (1961) Perpetuation and protection of germplasm as vegetative stock. In: Hodgson, R. E. (ed.), *Germplasm resources.* American Association for the Advancement of Sciences, Washington, DC, pp. 327-336.

Lin, S. C.; Yuan, L. P. (1980) Hybrid rice breeding in China. In: *Innovative approaches to rice breeding.* IRRI, Los Banos, Philippines, pp. 35-52

Lyman, J. M. (1984) Progress and planning for germplasm conservation of major food crops. *Plant Genetic Resources Newsletter* 60:3-21.

MacFadyen, J. T. (1985) United Nations. A battle over seeds. *The Atlantic,* November issue, pp. 36-44.

Marshall, D. R. (1989) Limitations to the use of germplasm collections. In: Brown, A. H. D.; Frankel, O. H.; Marshall, D. R.; Williams, J. T. (eds.), *The use of plant genetic resources.* Cambridge University Press, Cambridge, pp. 105-120.

Mengesha, M. H.; Rao, K. E. P. (1982) Current situation and future of sorghum germplasm. In: *Sorghum in the eighties*, vol. 1. ICRISAT, Patancheru, India, pp. 323-333.

Mooney, P. R. (1979) *Seeds of the earth. A private or public resource.* Inter-Parea (Ottawa) and International Coalition for Development Action (London).

Mooney, P. R. (1983) The law of the seed. *Development Dialogue* 1&2:1-172.

Mooney, P. R. (1985) The law of the seed revisited: Seed wars at the Circo Massimo. *Development Dialogue* 1:141-152.

Ng, N. Q. (1991) Rice germplasm exploration and collection in Africa since 1983. In: *Proceedings of the 1990 rice germplasm workshop.* IRRI-IBPGR, Los Banos, Philippines (in press).

Paroda, R. S. (1988) Genetic resources activities in India - Some considerations. In: Paroda, R. S.; Arora, R. K.; Chandel, K. P. S. (eds.), *Plant genetic resources - Indian perspectives.* National Bureau of Plant Genetic Resources, New Delhi, pp. 17-27.

Paroda, R. S.; Arora, R. K.; Chandel, K. P. S. (eds.) (1988) *Plant genetic resources - Indian perspectives.* National Bureau of Plant Genetic Resources, New Delhi, 545 pp.

Peeters, J. P.; Williams, J. T. (1984) Toward better use of genebanks with special reference to information. *Plant Genetic Resources Newsletter* 60:22-32.

Plucknett, D. L.; Smith, N. G. H. (1988) *Plant quarantine and the international transfer of germplasm.* World Bank, Washington, DC.

Plucknett, D. L.; Smith, J. H.; Williams, J. T.; Anishetty, N. M. (1983) Crop germplasm conservation and developing countries. *Science (Washington, DC)* 220:163-169.

Plucknett, D. L.; Smith, N. J. H.; Williams, J. T.; Anishetty, N. M. (1987) *Gene banks and the world's food.* Princeton University Press, Princeton.

RAFI (Rural Advancement Fund International) (1986) *The community seed bank kit.* RAFI, Pittsboro, NC.

Reitz, L. P. (1976) Improving germplasm resources. In: Patterson, F. L. (ed.), *Agronomic research for food.* American Society of Agronomy Special Publication 26. Madison, WI., pp. 85-97.

Rockefeller Foundation (1980) *Crop germplasm conservation and use in China: With special reference to a national plant genetic resources center.* Rockefeller Foundation, New York.

Shands, H. L.; Fitzgerald, P. J.; Eberhart, S. A. (1989) Program for plant germplasm preservation in the United States: The U.S. National Plant Germplasm System. In: Knutson, L.; Stoner, A. K. (eds.), *Biotic diversity and germplasm preservation, global imperatives.* Kluwer Academic Publishers, Dordrecht, Boston and London, pp. 97-115.

Shen, J. H. (1980) Rice breeding in China. In: *Rice improvement in China and other Asian countries.* IRRI, Los Banos, Philippines, pp. 9-30.

Sprague, G. F.; Alexander, D. E.; Dudley, J. W. (1980) Plant breeding and genetic engineering: A perspective. *BioScience* 30:17-21.

Soule, M. E. (1985) What is conservation biology? *BioScience* 35:727-734.

Smith, N. J. H. (1987) Genebanks: A global payoff. *Professional Geographer* 39:1-8.

Sun, M. (1986) The global flight over plant genes. *Science (Washington, DC)* 231:445-447.

Swaminathan, M. S. (1986) Rejoinder: No robbery - Saving and sharing of genetic resources. *Illustrative Weekly of India,* June 29, pp. 50-53.

Toll, J.; Anishetty, N. M.; Ayad, G. (1980) *Directory of germplasm collections. 3. Cereals. III. Rice.* IBPGR, Rome, 20 pp.

van Sloten, D. H. (1990) IBPGR and the challenges of the 1990s: A personal point of view. *Diversity* 6(2):36-39.

Vaughan, D. A. (1991) Core collections for wild relatives of rice. In: *Proceedings of the 1990 rice germplasm workshop.* IRRI-IBPGR, Los Banos, Philippines (in press).

Vitkovskij, V. L.; Kuznetsov, S. V. (1990) The N.I. Vavilov All-Union Research Institute of Plant Industry. *Diversity* 6(1):15-18.

Wilkes, H. G. (1983) Current status of crop plant germplasm. *CRC Critical Review in Plant Science* 1(2):133-181.

Williams, J. T. (1983) *Crop genetic resources for all.* IBPGR, Rome.

Williams, J. T. (1984) A decade of crop genetic resources research. In: Holden, J. H. W.; Williams, J. T. (eds.), *Crop genetic resources: Conservation and evaluation.* George Allen and Unwin, London, pp. 1-17

Williams, J. T. (1989) Plant germplasm preservation: A global perspective. In: Knutson, L.; Stoner, A. K. (eds.), *Biotic diversity and germplasm preservation, global imperatives.* Kluwer Academic Publishers, Dordrecht, Boston, London, pp. 81-96.

Zhang, Y. H.; Dong, Y. S. (1989) Development of research on crop germplasm resources in China. In: Knutson, L.; Stoner, A. K. (eds.), *Biotic diversity and germplasm preservation, global imperatives.* Kluwer Academic Publishers, Dordrecht, Boston, London, pp: 117-125

Chapter 3
Use of genetic variation for breeding populations in cross-pollinated species

Arnel R. Hallauer
Department of Agronomy, Iowa State University, Ames, IA 50011

INTRODUCTION

Plant breeding is the science and the art of effective management of genetic variability to attain desired breeding goals. Breeding goals are identified, and effective breeding methods for developing superior genotypes for target environments are considered. After identifying breeding goals and target environments, breeders determine if adequate genetic variability is available and what breeding method should be used to develop the desired genotypes. In the 1940s and the 1950s, breeding methods for the genetic improvement of maize (*Zea mays* L.) hybrids were examined. Because there was seemingly a plateau of hybrid maize yields, concern arose as to the relative importance of the different types of genetic variability in maize populations and whether the correct breeding methods were being used for the continued genetic improvement of hybrids. The main issue was the relative importance of dominance vs. overdominance effects in the expression of heterosis. Information was therefore needed to determine the genetic variability of maize populations and the relative importance of additive and non-additive (dominance and epistasis) genetic effects.

Because of the differing views regarding the relative importance of the level of dominance in maize, breeding methods were required to emphasize selection for traits having either partial-to-complete dominant effects or overdominant effects. Information was needed about the proportion of phenotypic variability due to genetic causes, or to heritability, and about the relative importance of the additive and non-additive genetic effects contributing to the genetic variability. Mating designs were developed to determine the genetic variability and the level of dominance for different traits in different populations (Comstock and Robinson, 1948; Mather, 1949; Cockerham, 1956). Attention was given to the study of maize populations because of the breeding methods used to develop lines and hybrids.

Estimates of the types of genetic variability and the levels of dominance important in trait expression were reported for different traits in different maize populations. With the exception of some F_2 populations, the additive genetic variance was generally two to four times greater than the variance due to dominance deviations (Hallauer and Miranda, 1988). Average levels of dominance were in the partial-to-complete dominance range for all traits,

37

ranging from 0.2 for number of kernel rows, cob diameter, percentage of oil in the kernel, and number of tillers to 0.9 for grain yield. Estimates of overdominance for grain yield were reported in F_2 populations, but after random intermating of the F_2 populations, partial dominance of alleles was of greater importance (Gardner and Lonnquist, 1959). Estimates of overdominance reported in F_2 populations were evidently those of pseudo-overdominance due to of repulsion-phase linkages.

The genetic studies of traits inherited quantitatively in open-pollinated species established that the genetic variability present in populations was adequate to expect effective selection and that additive genetic effects were the major contributors to genetic variability. Selection studies emphasizing additive genetic effects were conducted to determine response to selection. Because additive genetic effects were of greater importance than non-additive effects, modifications of mass selection (Gardner, 1961) and of ear-to-row selection (Lonnquist, 1964) were suggested for the genetic improvement of maize populations. Mass and ear-to-row selections had been used previously in maize-breeding programs, but the response to selection was inconsistent, especially for grain yield. Earlier mass- and ear-to-row-selection methods were not successful because of inappropriate experimental methods and inadequate isolation, which confounded genetic and environmental effects in selection (Sprague, 1955). In addition to the modified mass and ear-to-row methods, selection methods were evaluated based on the concepts of general (emphasis on additive genetic effects) and specific (emphasis on non-additive genetic effects) combining abilities developed by Sprague and Tatum (1942) including selection emphasizing general combining ability (Jenkins, 1940), selection emphasizing specific combining ability (Hull, 1945), and selection including genetic effects contributing to both general and specific combining abilities (Comstock *et al.*, 1949). These selection methods, developed for the genetic improvement of quantitatively inherited traits in genetically broad-based populations, are collectively designated recurrent selection methods. Modifications of the three basic methods of recurrent selection have been made for the types of testers used and for types of progenies evaluated, but the basic objectives are similar to those of the original methods (Hallauer, 1985). One method of recurrent selection receiving greater attention in the past 20 years was based on the evaluation of inbred progeny (Comstock, 1964). Additive genetic variation among progenies increases with inbreeding. Because quantitative genetic studies indicated that additive genetic variance was the major component of total genetic variance, it seems selection among inbred progenies should be effective in populations of cross-pollinated crop species.

Effectiveness of selection programs conducted in populations of cross-pollinated crop species will be reviewed and discussed in connection with the programs' contributions to plant breeding. Information is available both to make comparisons among different selection methods for response to selection and to determine how the different selection methods contribute to development of cultivars. Different selection methods have received more emphasis in some crop species than in others because of the desired end product (e.g., hybrids vs. synthetics).

MAIZE

Maize breeding methods have evolved around the inbred-hybrid concept proposed by Shull (1909). The two aspects of maize breeding include breeding methods for development of lines and hybrids and for improvement of germplasm sources that can ultimately contribute to the development of lines and hybrids. Development of breeding methods for line and hybrid development preceded recurrent selection methods proposed for germplasm improvement, and open-pollinated cultivars were the primary germplasm sources for development of lines and hybrids before 1950. Pedigree selection either within elite-line F_2 populations or within backcross populations including an elite line as the recurrent parent was commonly used to obtain second-cycle lines (Jenkins, 1978). Genetic progress was made, but the genetic variability was limited to the parental lines included in the crosses. It became evident, however, that continued genetic advance depended upon improved germplasm sources. Recurrent selection programs were initiated to increase the genetic base of breeding programs through systematic genetic improvement of genetically broad-based populations. Pedigree-selection methods would be used to extract lines and hybrids from the populations improved by recurrent selection. Recurrent selection methods were not designed to replace the breeding method used in developing lines and hybrids. Germplasm developed by recurrent selection was to provide additional germplasm sources for pedigree selection methods.

Heterotic groups gradually evolved with the pedigree-selection methods. Shull (1909, 1910) emphasized that not all crosses among lines were highly productive. It was necessary to cross lines and evaluate the crosses to determine which crosses had superior performances. Because heterosis is a function of the differences in allelic frequency between lines and of the level of dominance of alleles influencing the trait, crosses of lines from different germplasm sources were, on the average, superior to crosses of lines from the same germplasm sources. The widely used Reid Yellow Dent-Lancaster Sure Crop heterotic group was determined empirically and has since become the heterotic group used most extensively in the United States. The development of heterotic groups, therefore, had to be considered in the choice of selection methods used in maize breeding.

Choice of selection methods in maize breeding depends upon the objectives of the breeding program. Consideration has to be given to the trait emphasized in selection, the inheritance of the trait under selection, the type of population desired, and the desired degree of enhancement of the heterotic group emphasized in breeding. Hence, the choice of selection method can be either within one population (intrapopulation recurrent selection) or between two populations (interpopulation recurrent selection). Both methods have been used successfully in maize-breeding programs for the traits emphasized in selection. Differences among recurrent selection programs occurred in terms of direct vs. indirect responses observed for the different traits considered in selection.

Intrapopulation selection

Intrapopulation recurrent selection methods have received greater emphasis in maize breeding than in other species because there is a greater range of breeding methods available for different types of traits. Recurrent selection methods have ranged from use of phenotypic mass selection to use of half-sib and full-sib families and inbred progenies as the units of selection. Each method has been used extensively for different traits (Hallauer, 1991).

Mass selection

Mass selection based on the modifications proposed by Gardner (1961) has been used to improve grain yield (Gardner, 1977), to change ear length (Salazar and Hallauer, 1986), kernel size (Odhiambo and Compton, 1987), maturity (Hallauer and Sears, 1972), and to increase number of ears (Lonnquist, 1967). Direct response to selection was obtained in all instances for the trait emphasized in selection. The potential of mass selection for improvement of maize germplasm was reported by Coors and Mardones (1989) for increased number of ears per plant (prolificacy) in the open-pollinated cultivar Golden Glow (Table 1). Increased prolificacy was the primary trait of selection, but parental control, maturity, plant health, root and stalk strength, and grain moisture were also considered in choosing the approximately 300 plants in each cycle of selection. Direct response to mass selection for increased prolificacy was consistent over cycles of selection, and prolificacy increased from 0.94 ears per plant for C_0 to 1.25 ears per plant for C_{12}. Correlated changes for other important traits were favorable for grain yield, grain moisture, and performance index (Table 1). Days to mid-silk, plant height, and root and stalk lodging were stable over cycles of selection. Ear height ($b_1 = 4.4$ cm) increased with selection for increased prolificacy. Although selection emphasized increased prolificacy, the experimental procedures used in mass selection developed a population with improved performance index and acceptable maturity and plant size. Because of the lower heritability for root and stalk streng'ʰ these traits were unchanged with mass selection. Increased number of ears per plant imposes additional stress on roots and stalks because of competition for the products

Table 1 Direct and correlated responses to mass selection for increased prolificacy in Golden Glow maize cultivar and test-crosses after 12 cycles of selection [data are averages of the 2 years although authors indicated errors were heterogeneous between years (adapted from Coors and Mardones, 1989)].

Cycle of selection	Ear no.	Grain Yield (t ha⁻¹)	Grain Moisture (%)	PI*	Days to mid-silk (no.)	Height Plant (cm)	Height Ear (cm)	Lodging Root (%)	Lodging Stalk (%)
C_0	0.94	5.28	34.3	1.56	70	257	110	60	18
C_3	1.04	6.34	34.0	1.90	71	272	128	54	13
C_6	1.12	6.50	33.2	2.00	70	266	126	60	18
C_9	1.20	6.85	32.4	2.14	69	259	117	50	27
C_{12}	1.25	6.80	30.0	2.28	67	252	114	59	30
LSD (0.05)	0.04	0.40	7.0	0.14	1	4	4	10	5
b_1	0.02**	0.29**	--†	0.10**	--†	--†	4.4**	--†	--†
Testcrosses:									
C_0	1.00	7.04	31.9	2.24	68	254	104	48	12
C_6	1.08	7.66	33.8	2.32	70	264	116	49	12
C_{12}	1.15	7.84	32.0	2.50	69	256	114	44	12
LSD (0.05)	0.04	0.30	3.0	0.11	1	4	2	6	3
b_1	0.01**	0.07**	--‡	0.02**	0.1**	--‡	0.8**	--‡	--‡

*Performance index [PI = 100 × (grain yield ha⁻¹ ÷ grain moisture)].

†Estimates were not included because quadratic regression coefficient was significant.

‡Linear regression coefficients were not different from zero.

**Regressions were different from zero at P ≤ 0.01.

of photosynthesis. The study by Coors and Mardones (1989) illustrates the potential use of mass selection in germplasm-improvement programs employing proper controls. Mass selection is an effective selection method for most traits, but often the correlated responses for other traits were undesirable because proper controls had not been used at selection. Salazar and Hallauer (1986) reported that mass selection was effective for divergent ear length. The changes in ear length, however, caused undesirable changes in other agronomic traits: yield decreased with selection for increased and decreased ear length, and plant and ear height and maturity increased with selection for increased ear length. No attention was given to these traits with mass selection for ear length and, consequently, the populations did not contribute germplasm to breeding programs.

Selfed progeny selection vs. half-sib family selection

Selfed progeny selection and half-sib family selection were studied because of the effectiveness of the methods in selecting for additive (Jenkins, 1940) and for non-additive effects (Hull, 1945). Hull (1945) was of the opinion that overdominant effects were important and that a genetically narrow-based tester (inbred line or single cross) should be used to produce half-sib progenies. Jenkins' (1940) method used the population under selection as the tester to produce half-sib progenies. Hull (1945) suggested that if additive genetic effects were of greater importance than non-additive effects, then selection should be based directly on homozygous superiority rather than on half-sib families. Comstock (1964) also concluded that selfed progeny selection was the most promising selection method in the absence of overdominant effects. Selection studies comparing the progress of selfed progeny selection and half-sib family selection did not provide consistent results regarding whether selfed progeny selection or half-sib family selection was superior to the other (Hallauer and Miranda, 1988). Results from two long-term selection studies including comparisons of selfed progeny selection and of half-sib family selection will be discussed to illustrate the types of results obtained and the types of genetic effects affecting response to selection. Observed responses were similar for the studies, but interpretations of the types of genetic effects important in selection differed.

Tanner and Smith (1987) determined response to selfed progeny selection and half-sib family selection after eight cycles of selection in a strain of Krug Yellow Dent (BSK). Increased grain yield was the primary trait of selection. Materials evaluated included populations *per se*, populations selfed, and test-crosses of the populations to BSK C_0 (original population) and the testers used in half-sib family selection. Both direct and indirect responses to selfed progeny and to half-sib family selection were similar after eight cycles of selection (Table 2). The direct response by selfed progeny selection in BSK(S) was 49.1% (6.1% cycle^{-1}), compared with the

Table 2 Observed responses for grain yield to selfed and half-sib progeny selection in BSK and FSA and BSB maize populations.

	BSK*				FS8A and FS8B[†]		
	Populations		Testcrosses			Testcrosses	
Cycle of selection	per se	Selfed	BSKC0	Testers	Cycle of selection	Original	New
		$q\ ha^{-1}$				$q\ ha^{-1}$	
$BSKC_0$	48.7	27.7	48.7	71.1	$(S_2)C_0$	50.6	57.2
$BSK(S)C_4$[‡]	62.1	39.0	61.9	–	$(S_2)C_1$	51.0	58.6
$BSK(S)C_8$	60.1	41.3	65.2	78.6	$(S_2)C_3$	55.4	60.1
Response, %	23.4 (2.9)[§]	49.1 (6.1)	33.9 (4.2)	10.6 (1.3)	Response, %	9.5 (3.1)	5.1 (1.7)
$BSK(HI)C_4$[¶]	52.9	31.9	54.7	–	$(TC)C_1$	52.8	59.4
$BSK(HI)C_8$	64.5	39.1	61.9	80.0	$(TC)C_3$	56.1	63.4
Response, %	32.4 (4.0)	41.2 (5.2)	27.1 (3.4)	12.5 (1.6)	Response, %	10.9 (3.6)	10.8 (3.6)
LSD (0.05)	7.9	7.9	7.9	4.6		2.6	2.7

*Adapted from Tanner and Smith (1987) and the means are for the three testers used in selection (inbred B73 and single crosses WF9 x W22, M14 x B14A).

[†]Adapted from Horner et al. (1989), and the means for test-crosses are an average for FS8A and FS8B test-crosses with original tester, the C_0 populations, and the new tester, an inbred line.

[‡]Selection based on S_1 progenies for cycles 0 to 6 and based on S_2 progenies for cycles 7 and 8.

[§]Percentage gain per cycle of selection.

[¶]Testers included Ia4652 double cross for cycles 1 to 3, two single crosses of Ia4652 for cycles 4 and 5, inbred line Krug 755 for cycle 6, and inbred B73 for cycles 7 and 8.

indirect response of 41.2% (5.2% cycle^{-1}) by half-sib family selection in BSK(HI). The important differences in response to selection by the two methods were in the first four cycles of selection. In BSK(S), 40.8% of the direct response was attained in BSK(S)C_4, compared with 15.2% in BSK(HI)C_4. In BSK(HI), greater response to selection was attained from BSK(HI)C_4 to BSK(HI)C_8. The same trends were observed in the populations themselves and in the population test-crosses. The direct response in BSK(HI) was 12.5%, compared with the indirect response of 10.6% in BSK(S) after eight cycles of selection. Indirect responses of the intermated populations were 23.4% for BSK(S) and 32.4% for BSK(HI). There was no further genetic gain in the populations themselves after four cycles of selfed progeny selection, whereas there was a 21.9% gain with half-sib family selection from cycles four to eight. Tanner and Smith (1987) concluded that the response in BSK(S) in crosses with the testers used for BSK(HI) was primarily for loci with additive genetic effects with partial to complete dominance.

Horner *et al*. (1989) compared S_2 progeny selection with test-cross (TC) selection using inbred testers in two populations, FS8A and FS8B. Four cycles of selection were completed and evaluated in test-crosses to determine response to selection for grain yield (Table 2). Test-crosses of the populations developed after three cycles of test-cross selection were significantly (P < 0.01) better for combining ability than were test-crosses of the populations developed by S_2 progeny selection in three of four comparisons (Horner *et al*., 1989). Averaged over both populations after three cycles of selection, the test-crosses of the populations developed from test-cross selection had greater yields than did S_2 progeny selection, but the differences between methods for the C_3 populations were significant only with the new tester (Table 2). Although significant progress was made by S_2 progeny selection in FS8A and FS8B, Horner *et al*. (1989) concluded that S_2 progeny selection was inferior to test-cross selection for improvement of combining ability. Because of the greater gains in combining ability with test-cross selection and because of the populations themselves, Horner *et al*. (1989) concluded that non-additive gene action, possibly in the overdominant range, was important in FS8A and FS8B.

The two methods of selection, selfed progeny and half-sib family, emphasized selection for different alleles. Tanner and Smith (1987) included crosses of the selected populations, and the BSK(S)C_8 x BSK(HI)C_8 crosses had a yield of 71.1 q ha^{-1}, or a midparent heterosis of 14.1%. Inbreeding depression was reduced in populations developed by selfed progeny selection, and the level of depression seemed consistent in different populations. Lack of continued response in BSK(S) may be due to the fixation of important alleles in as much as Mulamba *et al*. (1983) reported that the genetic variability of BSK(S)C_8 was 55.2% less than the genetic variability of BSK(S)C_0, whereas the genetic variability of BSK(HI)C_8 was only 15.4%

less than that of $BSKC_0$. The reduced rate of response in $BSKC_0$ with selfed progeny selection may be a reflection of reduced genetic variability due to fixation of alleles from selection, genetic drift, or both. Selfed progeny selection is theoretically superior to half-sib family selection, but it has not been consistently superior to half-sib family selection regardless of the measures of comparison used for the two methods. Other studies suggest that test-cross selection is necessary for the long-term improvement of maize germplasm (Hallauer, 1991).

Selfed progeny selection

Selfed progeny selection, usually S_1 or S_2 progenies, has been a popular method of selection for traits other than yield, particularly for increased pest resistance. Mass selection methods were not effective for pest resistance because of poor pest establishment, poor screening techniques, limited genetic variability, plant escapes, and environmental factors causing the heritabilities of the trait to be either low or unpredictable. Williams and Davis (1983), for example, found that two cycles of mass selection for resistance to stalk tunneling by the second-generation southwestern corn borer (*Diatraea grandiosella* Dyar) were ineffective. Lack of response was attributed to the frequency of escapes, and progeny evaluation was recommended to increase level of resistance. Penny *et al.* (1967) found that three cycles of recurrent selection based on S_1 progeny evaluation with artificial infestation were adequate to develop populations with acceptable levels of resistance to the first-generation European corn borer (*Ostrinia nubilalis* Hubner). The contrast in results was for two different maize pests, screening techniques available, and methods of selection (mass vs. S_1). Screening techniques for most plant pests have been developed, and the application of the screening techniques on S_1 progenies effectively reduced the frequency of escapes in selection (i.e., increase of the heritabilities of the traits under selection).

S_1 progeny selection effectively increased level of stalk quality and resistance to European corn borer. Penny *et al.* (1967) selected for resistance to first-generation European corn borer in five synthetic cultivars, and Jinahyon and Russell (1969a) selected for resistance to *Diplodia zeae* (Schw.) Lev., one causal fungus of stalk rot in maize, in Lancaster Sure Crop. Artificial means of infestation and infection were used in both studies. They reported that three cycles of S_1 recurrent selection showed significant improvement for resistance to the respective pest emphasized in selection. Penny *et al.* (1967) did not evaluate populations for yield, but Jinahyon and Russell (1969b) reported that selection for increased resistance to stalk rot did not cause significant changes in yield of populations having greater stalk-rot resistance.

Subsequent studies of resistance to infection by stalk-rot organisms (Devey and Russell, 1983) and of resistance to feeding by European corn borer (Klenke *et al.*, 1986) showed that selection emphasizing improved resistance to plant pests can have negative effects on grain yield. Devey and Russell (1983) found that additional cycles of S_1 recurrent selection in Lancaster Sure Crop effectively improved stalk quality but that there was a 1.2 q ha^{-1} decrease in grain yield with each cycle of selection (Table 3). The changes in yield occurred in the cycles after the initial evaluations by Jinahyon and Russell (1969a,b). The correlated effects for stalk quality (mechanical stalk strength) increased with reduced levels of infection by *D. zeae*, but the improved populations flowered later and had 39.7% less yield than did the unimproved population.

Table 3 Response to S_1 recurrent selection for improved stalk quality in Lancaster Sure Crop (BSL) maize cultivar (adapted from Devey and Russell, 1983).

Cycle of selection	Stalk-rot rating (0.5-6.0)*	Mechanical stalk strength (kg)	Rind puncture (kg)	Days to silk (no.)	Grain yield (q ha^{-1})	Ear length (cm)
BSLC$_0$	3.27	28.7	5.52	74.3	66.5	19.0
BSL(S)C$_1$	3.07	36.1	5.86	76.5	67.2	19.0
BSL(S)C$_2$	2.14	41.0	6.10	76.0	66.3	18.6
BSL(S)C$_3$	2.41	46.6	6.11	77.3	61.8	19.3
BSL(S)C$_4$	2.18	47.9	6.71	76.7	54.4	18.0
BSL(S)C$_5$	1.77	52.3	7.35	77.8	50.2	17.4
BSL(S)C$_6$	1.58	53.2	6.79	78.1	46.9	15.6
BSL(S)C$_7$	1.33	59.6	7.52	79.0	40.1	13.9
b_1[†]	-0.26**	4.7**	0.27**	1.6**	-1.2**	-1.4**

*Rating scale where 0.5 is resistant to infection by *D. zeae* and 6.0 is highly susceptible.

**Significant at $P \leq 0.01$ probability level.

[†]Regression coefficient over cycles of selection.

Klenke *et al.* (1986) evaluated four cycles of S_1 recurrent selection for increased levels of resistance to first- and second-generation European corn borer in the synthetic cultivar BS9. Artificial infestation was used for both generations of the corn borer in selection and in the evaluation of the cycle populations. S_1 recurrent selection effectively reduced levels of infestation for both generations, with a 25.0 and 31.2% reduction in level of infestation to first- and second-generations, respectively (Table 4). Correlated response for grain yield in the selected populations also decreased with increased levels of resistance: 21.1% for increased first-generation resistance and 17.1% for increased second-generation resistance. The effects of S_1 recurrent selection for increased corn-borer resistance also was evident in the control populations: yield decreased 31.1% with no artificial infestation and spraying to control natural infestation. S_1 recurrent selection and artificial infestation increased the heritability of the traits, but the correlated effects on yield were undesirable.

Table 4 Response to S_1 recurrent selection for resistance to first- and second-generation European corn borer in synthetic maize cultivar BS9 (adapted from Klenke *et al.*, 1986).

Cycle of selection	First generation		Second generation		Control* $(q\ ha^{-1})$
	Rating $(1-9)^{\dagger}$	Yield $(q\ ha^{-1})$	Rating $(1-9)$	Yield $(q\ ha^{-1})$	
BS9C$_0$	3.6	60.7	6.4	53.1	69.7
BS9(S)C$_1$	3.6	65.0	5.9	47.5	63.4
BS9(S)C$_2$	2.8	52.4	5.7	44.4	60.3
BS9(S)C$_3$	2.5	56.2	4.4	40.8	55.3
BS9(S)C$_4$	2.7	47.9	4.4	44.0	48.0
LSD (0.05)	0.8	13.8	1.3	13.8	13.8
% Change‡	-25.0	-21.1	-31.2	-17.1	-31.1

*Populations were not infested and sprayed with carbofuran to control natural infestation.

†1 is resistant and 9 is susceptible.

‡Change of C_4 relative to C_0.

Comstock (1964) showed that in the absence of overdominance, selfed progeny selection would be twice as effective as half-sib family selection. A combination of S_1 and S_2 recurrent selections was initiated in the BSTL, BS2, and BS16 populations to determine the effectiveness of selfed progeny recurrent selection for developing elite germplasm sources and the potential use of exotic germplasm in breeding programs (Iglesias and Hallauer, 1990). The protocol used in selection was (1) for pest resistance and agronomic traits at the S_1 generation, (2) to advance selected progenies to S_2 generation, (3) to evaluate S_2 progenies in replicated trials, and (4) to intermate S_1 progenies of selected S_2 progenies to form the next cycle populations. Selection was based on more highly heritable traits at S_1 with yield, stalk and root quality, and grain moisture used in selection among S_2 progenies. After five cycles of selection in BSTL and BS2 and four cycles of selection in BS16, original and selected populations were evaluated to determine response to selfed progeny selection (Table 5). Significant improvement was made in each population *per se* from the C_0 to C_2 and C_3, but no further gains were obtained in the last two cycles of selection in the three populations. Although similar to $BS16C_0$, $BS16(S)C_4$ yielded significantly less than did $BS16C_2$. $BS16C_0$ yielded more than did $BSTLC_0$ and $BS2C_0$, but the response from continued selection in BS16 was not effective. The same trends occurred for the S_1 generations for grain yield, which would be a more direct measure of response to selfed progeny selection than would populations *per se*. Correlated responses for grain moisture and plant height tended to decrease with selection, whereas selection did not effectively increase resistance to lodging. BS16 is a strain of ETO Composite obtained from Colombia, SA that was mass-selected for adaptation to temperate areas. BS2 includes 50% ETO Composite germplasm, and BSTL includes 25% Tuxpeño germplasm. Responses to selfed progeny recurrent selection were similar in the initial cycles of each population, but the effects of continued selection were more evident in BS16.

Selfed progeny selection has an intuitive appeal to maize breeders because of the greater heritability of traits and the practical concerns of developing inbred lines. S_1 progeny recurrent selection improved pest resistance, but correlated effects on grain yield were not desirable (Tables 3 and 4). A possible reason for the correlated changes was that selection for increased pest resistance caused a repartitioning of the products of photosynthesis, particularly if no attention was given to yield in selection (Dodd, 1977). Selfed progeny recurrent selection emphasizing yield and other agronomic traits, however, was not as effective as predicted (Table 5). Evidently, selfed progeny recurrent selection within cross pollinated maize populations should be used on a restricted basis. In all instances, two to three cycles of selfed progeny recurrent selection were effective regardless of whether direct or indirect responses to selection were compared. For pest resistance, three cycles of selfed progeny selection seemed adequate to provide acceptable levels of resistance; further emphasis had to be given to yield and other agronomic traits in subsequent cycles of selection. If levels of pest resistance are monitored in future cycles of selection, obviously susceptible progenies can be deleted. Direct response to selfed progeny selection for yield and other agronomic traits parallel the correlated effects for improved pest resistance. It also seems that selfed progeny selection should be limited to two to three cycles in populations emphasizing selection for yield and other

Table 5 Response to selfed (S_1 and S_2) progeny recurrent selection in BSTL, BS2, and BS16 maize cultivars emphasizing selection for agronomic traits (adapted from Iglesias and Hallauer, 1990).

Population	Grain yield (q ha⁻¹)			Grain moisture (%)			Stalk lodging (5)			Ear height (cm)		
	C_0	C_3	C_5*	C_0	C_3	C_5	C_0	C_3	C_5	C_0	C_3	C_5
BSTL per se†	39.6	52.2	49.8	19.3	18.6	17.8	15	17	13	116	113	110
S_1†	23.4	31.2	30.7	18.3	18.3	18.0	12	15	10	101	104	96
BS2 per se	40.1	51.5	49.1	17.5	18.0	17.8	22	22	21	121	119	104
S_1	24.2	29.5	30.2	17.2	17.7	17.7	17	19	19	106	104	98
BS16 per se	47.1	53.9	44.9	19.8	17.4	16.3	21	13	16	133	103	101
S_1	27.5	26.1	30.0	19.7	17.0	16.9	15	14	15	108	95	86
LSD (0.05)		6.9			0.8			5			6	
Average per se	42.3	52.5	47.9	18.9	18.0	17.3	19	17	17	123	112	105
S_1	25.0	28.9	30.3	18.4	17.7	17.5	15	16	15	105	101	93
LSD (0.05)		5.7			0.5			3			5	

*Cycles of selection for BSTL and BS2 and cycles C_0, C_2, and C_4 for BS16.

†Intermated populations themselves and the bulk S_1 generations of the intermated populations.

agronomic traits and that either half-sib or full-sib family selection should be used in later cycles.

The reason for the seeming ineffectiveness of long-term selfed progeny selection is not clear. The correlated responses to pest resistance could be due to a repartitioning of the products of photosynthesis, but this hypothesis has not been tested directly. Small effective populations could contribute to a reduction in genetic variability, which would restrict continued response to selection (Tanner and Smith, 1987). Too many important loci could be fixed because of genetic drift due to small effective populations. If overdominant effects were present and important, selfed progeny recurrent selection would not be as effective as half-sib family selection (Horner *et al.*, 1989). Although recurrent selection studies comparing selfed progeny and half-sib family selection were not consistent, usually too few cycles of selection were completed to allow definitive conclusions (Hallauer and Miranda, 1988). Interpretations varied regarding the methods of evaluation and the importance of dominance affecting traits. It seems, however, that some form of test-cross (half-sib or full-sib) selection has to be used if long-term selection programs are to contribute to applied breeding programs. Selfed progeny recurrent selection can be used initially (two or three cycles) with a change to test-cross evaluation for future cycles of selection.

Interpopulation selection

Interpopulation-selection schemes were developed to resolve the problem of choice of methods to use because of the concerns about the relative importance of additive and non-additive gene effects. Comstock *et al.* (1949) suggested reciprocal recurrent selection because this method would be as effective as methods emphasizing selection for either additive (Jenkins, 1940) or non-additive genetic effects (Hull, 1945) and would be more effective than these methods if both types of genetic effects were important. Reciprocal recurrent selection methods with their primary use being for grain yield have not been used as extensively as have the intrapopulation methods. Two reciprocal recurrent selection programs evaluated recently will be discussed to illustrate the potential of such methods. Reciprocal recurrent selection based on half-sib families was initiated in 1949 in Iowa Stiff Stalk Synthetic (BSSS) and in Iowa Corn Borer Synthetic No. 1 (BSCB1), and selection based on full-sib families was initiated in 1963 in Iowa Two-ear Synthetic (BS10) and in Pioneer Two-ear Composite (BS11).

Half-sib family selection

Reciprocal recurrent selection for BSSS and BSCB1 was conducted as described by Comstock *et al.* (1949). Half-sib family selection emphasizing grain yield was conducted since its initiation in 1949, but adjustments have been made during the past 40 years for emphasis given to the following: (1) other agronomic traits (e.g., selection indices), (2) effective population sizes (N_e = 10 to 20), (3) plant densities used for evaluation of half-sib progenies (39.0 to 61.8 M ha^{-1}), (4) generation of individuals used to produce half-sib families (S_0 and S_1), (5) methods used to intermate selected progenies of selected families (diallel and bulk entry), and (6) change from use of hand labor to plant and harvest evaluation trials to use of mechanical equipment adapted to plant and harvest experimental plots (Eberhart *et al.*, 1973; Smith, 1983; Keeratinijakal, 1990). Summaries of the number of test sites and replications used for evaluation trials and of the estimates of components from the evaluation trials were given by Hallauer and Miranda (1988).

Keeratinijakal (1990) conducted a comprehensive evaluation of the responses to reciprocal recurrent selection in BSSS, BSCB1, and their crosses after 11 cycles of selection (Table 6). Data were collected for eight traits, and direct and indirect responses were favorable in nearly all instances (data not shown). Grain yield, the primary trait of selection, had a direct response to selection of 2.8 q ha^{-1} (6.6%) per cycle of selection; the indirect response in BSSS and BSCB1 was 0.6 q ha^{-1} cycle^{-1}. Midparent heterosis increased from 25.4% for BSSSC$_0$ x BSCB1C$_0$ to 76.0% for BSSS(R)C$_{11}$ x BSCB1(R)C$_{11}$. Increased grain yield with selection did not occur at the expense of stalk quality or maturity. The percentage of stalk lodging decreased significantly for BSSS, BSCB1, and their cross (Table 6), with the greatest rate of decrease in the population crosses. Days from planting to silking decreased significantly in BSSS and in the cross populations and increased significantly in BSCB1. The differences in days to silk were caused by the 5.9-day difference between BSSSC$_0$ (89.3 days) and BSCB1C$_0$ (83.4 days). Because of the tendency to make pollinations between plants having similar flowering dates, the difference in silking dates between BSSS(R)C$_{11}$ (83.3 days) and BSCB1(R)C$_{11}$ (84.9 days) was only 0.4 days. There was no significant change in grain moisture at harvest of either BSSS or BSCB1 populations with selection, and grain moisture of population crosses increased only 0.06% cycle^{-1} of selection.

Reciprocal recurrent selection in BSSS and in BSCB1 was effective in terms of the objectives of the selection program. Direct response in the population crosses for grain yield was consistent to C$_9$ cycle, with the suggestion of a plateau in the last three cycles of selection (Table 6). Improvement of stalk quality, however, continued in the last three cycles. This continuation may have had a moderating effect on grain improvement (Table 3). Reciprocal recurrent selection effectively improved germplasm sources for line development, which will be reflected in crosses of lines extracted from BSSS(R)C$_n$ and from BSCB1(R)C$_n$. Keeratinijakal (1990) concluded that, based on Smith's (1983) model, increases in grain yield with selection resulted from changes in allelic frequency with additive and dominant effects in BSSS(R) and with primarily dominant effects in BSCB1(R).

Table 6 Direct and indirect response to reciprocal half-sib recurrent selection in BSSS and BSCB1 maize populations and their crosses (adapted from Keeratinijakal, 1990).

Cycle of selection	Grain yield (q ha^{-1})			Stalk lodging (%)			Days to silk (no.)		
	BSSS	Cross	BSCB1	BSSS	Cross	BSCB1	BSSS	Cross	BSCB1
C_0	35.5	42.4 (25.4)†	32.1	19.8	25.0	40.4	89.3	86.0	83.4
C_4	37.6	49.6 (42.8)	31.6	21.4	30.4	32.4	86.4	84.8	86.7
C_7	42.4	60.8 (54.9)	36.1	17.3	19.8	18.3	86.5	84.8	86.1
C_8	43.4	65.9 (72.5)	33.0	13.8	13.9	12.2	86.4	84.9	86.4
C_9	42.5	69.4 (70.9)	38.7	16.0	14.9	10.5	85.3	83.6	84.9
C_{10}	39.8	68.4 (80.5)	35.9	17.6	16.2	11.8	85.3	83.3	84.8
C_{11}	39.2	67.6 (76.0)	37.6	10.9	11.4	9.7	85.3	83.2	84.9
b_1	0.6	2.8**	0.6*	-0.6*	-1.5**	3.1**	-0.4**	-0.2**	0.1**
Gain cycle^{-1}, %	1.7	6.6	1.9	-2.8	-5.1	-7.6	-0.4	-0.3	0.1

*, **Linear regression coefficients were significant at P ≤ 0.05 and 0.01 levels, respectively.

†Values in parentheses are percentage of midparent heterosis.

Indirect responses in BSSS(R) and in BSCB1(R) were similar to the direct response in population crosses if BSSS(R) and BSCB1(R) were adjusted for the effects of genetic drift.

Full-sib family selection

Reciprocal recurrent selection for BS10 and for BS11 was conducted as described by Hallauer and Eberhart (1970). Selection units were full-sib families developed by crossing individual plants of BS10 and of BS11. Grain yield was emphasized in selection, but consideration was given to root and stalk quality and to maturity in the final selections. Eyherabide and Hallauer (1991a) evaluated direct (population cross) and indirect responses (BS10 and BS11) in BS10, BS11, and their crosses after eight cycles of selection.

Direct and indirect responses to selection for grain yield were significant for BS10, BS11, and their crosses, with a direct response of 6.7% cycle^{-1} (Table 7). Midparent heterosis increased from 2.5% (C_0 x C_0) to 28.4% (C_8 x C_8). Selection for increased grain yield was accomplished with significant reductions in percentage of stalk lodging and with similar maturities. There were no changes in days to pollen-shed of population crosses (Table 7); grain moisture did not change in BS10 and was reduced significantly in BS11 and in the population crosses (data not shown). Hence, significantly greater grain yields and significantly less stalk lodging were obtained in BS10, BS11, and their crosses at similar maturities. Eyherabide and Hallauer (1991b) concluded that the increased heterosis for grain yield in the population crosses was caused by accumulation of favorable alleles with additive and dominant effects of BS10 and of BS11 and by the heterozygous condition of loci for which genetic drift caused fixation of alleles in one of the parent populations. Improved yields in BS10 and in BS11 were caused mainly by alleles with additive (BS11) and dominant effects (BS10).

Direct response to reciprocal recurrent selection based on half-sib (BSSS and BSCB1) and full-sib (BS10 and BS11) families was effective (Tables 6 and 7). Rates of direct and of indirect responses were similar for the three traits listed. Grain yield was emphasized in both programs, and the rates of direct response (6.6%, Table 6; 6.7%, Table 7) were greater than those of direct responses for grain yield reported previously for other selection programs (Hallauer and Miranda, 1988). Because of the importance of heterosis in maize breeding, it seems that reciprocal recurrent selection methods should be an important component of maize-germplasm improvement programs. Important alleles with additive and non-additive effects are included in selection, and the importance of the two types of effects may differ between populations. Reciprocal recurrent selection methods are not used as frequently as intrapopulation-selection methods because the former seem more complex. Initially, intrapopulation-selection methods are appropriate

Table 7 Direct and indirect responses to reciprocal full-sib recurrent selection in BS10 and BS11 maize populations and their crosses (adapted from Eyherabide and Hallauer, 1991a).

Cycle of selection	Grain yield (q ha^{-1})			Stalk lodging (%)			Days to pollen shed (no.)		
	BS10	Cross	BS11	BS10	Cross	BS11	BS10	Cross	BS11
C$_0$	41.5	46.5 (2.5)†	49.2	20.7	22.9	22.0	75.6	75.5	77.3
C$_2$	43.8	57.2 (16.7)	51.5	24.0	19.6	15.5	73.1	74.2	74.9
C$_4$	50.9	59.6 (16.5)	48.6	18.8	18.3	15.6	75.8	75.0	76.5
C$_6$	52.4	63.6 (15.8)	54.7	13.4	15.4	11.7	77.0	74.6	74.8
C$_8$	51.3	74.6 (28.4)	55.6	16.1	12.5	12.9	75.4	74.3	75.9
b$_1$	3.2**	3.1**	0.8*	-1.0**	-1.3**	-1.1**	0.2*	--‡	-0.7**
Gain cycle^{-1}, %	7.9	6.7	1.6	-4.8	-5.7	-5.0	0.3	--	-0.5

*, **Indicates that estimate of linear regression coefficient was significant at P $\leq$ 0.05 and 0.01 levels, respectively.

†Values in parentheses are percentages of midparent heterosis.

‡Linear regression coefficient was not significant.

for certain specific traits (e.g., pest resistance, maturity, stalk quality), but reciprocal recurrent selection methods enhance established heterotic groups in the long term. To ensure proper control of standability and maturity, selection indices will need to include other agronomic traits.

Although reciprocal recurrent selection methods are perceived to be more complicated than other selection methods, these methods are no more complex or resource demanding than half-sib recurrent selection in two populations for reciprocal half-sib recurrent selection and full-sib recurrent selection within one population for reciprocal full-sib recurrent selection. The only difference between intra- and interpopulation selection is that progenies are produced between two populations. For practical breeding programs, the two populations included for reciprocal recurrent selection would be populations representative of important heterotic groups. Although selection was conducted in separate pairs of populations, the preliminary results of selection suggest that half-sib (Table 6) and full-sib (Table 7) family selection were equally effective. Some of the advantages and disadvantages of the two methods were discussed by Jones *et al.* (1971) and Hallauer and Miranda (1988).

F_2 populations

The trend to emphasize selection within elite-line crosses has increased rapidly during the past 50 years (Jenkins, 1978), and selection within genetically narrow-based populations is now a major component of maize-breeding programs (Bauman, 1981). Recurrent selection methods within F_2 populations have received limited attention because maize breeders recreate new F_2 populations as elite lines which become available either from within a given program or from other programs. Applied maize breeders do conduct recurrent selection, in the general sense, by creating successive series of F_2 populations based on elite lines derived from previously recycled lines (Duvick, 1977).

Recurrent selection methods have been reported for the following three F_2 populations: half-sib recurrent selection in WF9 x B7 with B14 as tester (Russell *et al.*, 1973), and full-sib recurrent selection in Va17 x Va29 (Genter, 1982) and NC7 x CI21 (Moll, 1991). Moll (1991) compared the effectiveness of recurrent full-sib family selection after 16 cycles of selection in the F_2 population of the single cross NC7 x CI21 and in the open-pollinated cultivar Jarvis Golden Prolific. Both selection studies were designed to provide estimates of genetic variances from the full-sib family test-data for each cycle of selection. Greater grain weight/plant was the primary trait considered in selection. Moll (1991) reported that the gains from selection for grain weight/plant were linear over 16 cycles of selection at a rate of 2.4% cycle[-1] for Jarvis Golden Prolific and 4.5% cycle[-1] for the (NC7 x CI21) F_2 population. The estimate of additive genetic variance of

Jarvis Golden Prolific C_0 population (416 $\pm$ 98) was 48.6% greater than the estimate of the C_1-C_2 cycles of (NC7 x CI21) F_2 population (280 $\pm$ 135). There was no evidence that the additive genetic variance decreased with selection in Jarvis Golden Prolific (466 $\pm$ 93 in C_{14}-C_{16}), whereas the estimate of additive genetic variance after 16 cycles of selection was reduced 52.5% in the F_2 population (Moll, 1991). Accumulated gains for grain weight were greater in the F_2 population of NC7 x CI21 than in Jarvis Golden Prolific even though the initial additive genetic variance was 48.6% less in the F_2 population than in Jarvis Golden Prolific and having decreased 52.5% with selection.

Although the response to full-sib family selection in the (NC7 x CI21) F_2 population was greater than that in Jarvis Golden Prolific, grain weight of C_{15} was nearly one LSD less than that of the original cross (Table 8). General combining ability of the selected cycles of Jarvis Golden Prolific with the $BS13(S)C_3$ tester was also nearly three times greater than that of selected cycles of the (NC7 x CI21) F_2 population. Differences in response in the two populations seemed related to the changes in number of ears and in ear weight. Number of ears and ear weight increased in the (NC7 x CI21) F_2 population - whereas, in Jarvis Golden Prolific, ear weight decreased while number of ears increased (Moll, 1991). The potential of recurrent selection in F_2 populations is evident, but lines were not extracted from the different selection cycles to determine if recurrent selection was effective in developing lines with improved combining ability. As suggested by Duvick (1977), because of the emphasis of pedigree selection in F_2 populations, a comparison of the recurrent selection in (NC7 x CI21) with pedigree selection in (NC7 x CI21) and other single crosses would have been of interest.

Selection within F_2 populations developed from elite lines has been an effective breeding method, but limited information is available regarding amount and types of genetic variation present, sample size required to represent genetic variation present, and effects of genetic recombination from intercrossing within F_2 populations before sampling for line development. Han and Hallauer (1989) reported estimates of genetic variability and levels of dominance in the F_2, as well as after five generations of intermating in a related line (B73 x B84) cross and in an unrelated line (B73 x Mo17) cross. Estimates of overdominance were obtained for grain yield in the F_2 population for both crosses (1.28 for B73 x Mo17 and 1.53 for B73 x B84). After five generations of intermating 250 plants within each F_2 population, levels of dominance were reduced to 0.95 for (B73 x Mo17) and to 0.62 for (B73 x B84). There was no change in the estimates of additive genetic variance with intermating in F_2 of (B73 x Mo17), but the estimate of dominance variance was reduced 26%. The estimate of dominance variance after intermating the F_2 of (B73 x B84) was reduced 42%.

Table 8 Response to full-sib family selection for greater grain weight plant^{-1} in Jarvis Golden Prolific and F$_2$ population of NC7 x CI21 maize population (adapted from Moll, 1991).

Cycle of selection	Jarvis Golden Prolific			Cycle of selection	NC7 x CI21		
	Expt 1* per se	Expt 2† per se	Test-cross‡ g plant^{-1}		Expt 1* per se	Expt 2† per se	Test-cross‡ g plant^{-1}
C$_0$	107.7	114.6	167.7	C$_0$	90.3	85.0	179.1
C$_8$	138.8	148.6	191.3	C$_5$	110.9	108.9	190.7
C$_{10}$	144.7	165.1	199.2	C$_{10}$	119.9	120.1	180.4
C$_{12}$	149.0	167.1	208.2	C$_{15}$	168.9	165.5	197.0
C$_{14}$	153.9	172.5	203.3	NC7 x NC21	--	182.2	--
C$_{16}$	--	178.8	--				
LSD (0.05)	10.3	23.6	10.7		10.5	18.2	9.2
b§	3.4±0.3	4.1±0.5	2.4±8	b§	4.9±0.5¶	5.1±0.5	0.9±0.4

*Data from three locations in 2 years.

†Data from three locations in 1 year.

‡Tester was BS13(S)C$_3$.

§Linear regression on selection cycles.

¶Quadratic term was significant, which indicates response was non-linear.

Covarrubias-Prieto *et al.* (1989) examined the variability among 100 S_1 progenies in the F_2 and after five generations of intermating for the (B73 x Mo17) and (B73 x B84) crosses. For grain yield, there were no significant changes in the estimates of genetic variability among S_1 progenies in either cross. Intermating of F_2 populations, therefore, was not effective for increasing the variability among the 100 S_1 progenies for grain yield. The same trends occurred for 13 other plant and ear traits measured in the F_2 and after intermating. The important difference in the estimates of variability among the 100 S_1 progenies for the related and unrelated line crosses was the magnitude of variance among S_1 progenies. Variability for yield of the unrelated line cross was 106% greater than the estimate for the related line cross (58.5 vs. 28.4).

Estimates of genetic variability within F_2 populations were similar to those for genetically broad-based populations (Hallauer and Miranda, 1988). Responses to recurrent selection methods in F_2 populations were similar to those of other maize populations. It is doubtful, however, that long-term recurrent selection programs within closed F_2 populations will have an impact on maize breeding. Maize breeders continually create new F_2 populations as new elite lines become available. Long-term recurrent selection in WF9 x B7 and NC7 x CI21 was effective, but the genetic variability within the F_2 populations would be limited to the initial lines included in the crosses. F_2 populations included in pedigree breeding are open-ended in the sense that one line is replaced or two new lines are used to form other F_2 populations. Genetic advance of new lines to include in F_2 populations limits interest in long-term recurrent selection within specific F_2 populations.

GRASSES

Genetic variability within forage and turfgrasses has been used effectively to develop cultivars with improved forage yield, forage quality, and pest resistance. Most grasses are cross pollinated, and additive and non-additive genetic variability is substantial enough to permit effective genetic improvement for a range of selection methods. Non-additive genetic effects have not been exploited even though substantial heterosis for yield exists in many grasses (Vogel *et al.*, 1989). Because of the complexities of ploidy levels, the economic importance of different grass species, and the difficulties of developing techniques to produce hybrid seed, the emphasis given to hybrid breeding programs in grasses has not been as intensive as that given to hybrid breeding programs in maize. But the potential of hybrid breeding programs in grasses is evident because of vegetative propagation, self-incompatibility but with cross compatibility, cytoplasmic male sterility, and the potential use of apomictic mechanisms (Hanna and Bashaw, 1987).

Because of the genetic variability available in grass species, direct and indirect responses to improved forage yields have been made. Burton (1989) used modified mass-selection methods very effectively for the improvement of forage yields in Pensacola bahiagrass (*Paspalum notatum* var. *saure* Parodi). Response was obtained in both genetically broad- and narrow-based bahiagrass populations (Table 9). The broad-based population (designated as population A) was a blend of bahiagrass seed collected from 39 Georgia farms. The narrow-based population (designated population B) was derived from 60 selfed progenies of a two-clone F_1 hybrid. Both A and B were closed populations during selection. The mass selection method included eight restrictions to enhance parental control and effectiveness of selection. In population A, average gain per cycle after nine cycles of mass selection was 22.4% (86 g plant^{-1}). The average rate of gain continued through cycle 14 (23.6% cycle^{-1}). The more than fourfold increase in plant yield of population A suggests considerable genetic variability for increased plant yield. Population B had greater yield (1430 g plant^{-1}) after six cycles of selection than did population A after 11 cycles of selection (1324 g plant^{-1}). After three cycles of selection, population B yielded nearly one LSD more than did the F_1 hybrid. Average plant yield response in population B was 10.1% cycle^{-1}, a rate that was only 42.8% of the rate in population A. But average yield in cycles 6 and 7 for population B was nearly equal to average yield of cycle 11 in population A (Table 9).

Different morphological and physiological traits have been studied in grasses to determine their affects on forage yield. Nelson *et al.* (1985) assessed response to bidirectional selection for leaf area expansion rates in tall fescue (*Festuca arundinacea* Schreb.) and the correlated effects of selection on forage yield. Increased forage yields were obtained with recurrent selection for increased leaf-area expansion rate. Leaf-area expansion rate is a component of forage yield, and selection exploited the additive genetic variability of a trait correlated with forage yield. The correlated response for increased forage yield with selection for increased leaf-area expansion rate in tall fescue differs from the selection studies for components of grain yield in maize [Salazar and Hallauer (1986) for ear length and Odhiambo and Compton (1987) for seed size]. Evidently, either the heritability of leaf-area expansion rate and its correlation with forage yield were greater than the components of grain yield in maize or the heritability of forage yield was smaller than that of grain yield of maize. Short and Carlson (1989) practiced bidirectional selection for traits of birdsfoot trefoil (*Lotus corniculatus* L.) that were compatible traits with orchardgrass (*Dactylis glomerata* L.). They reported that it was possible to increase compatability of orchardgrass with birdsfoot trefoil with selection among spaced plants for decreased canopy height, decreased tiller number, and delayed maturity. Vogel *et al.* (1989) emphasized, however, that indirect selection based on physiological traits has not been as effective as direct selection for forage yield.

Table 9 Direct response to modified mass selection methods for improved forage yield in two Pensacola bahiagrass populations A and B (adapted from Burton, 1989) and indirect response for forage yield in tall fescue with selection based on leaf area expansion rate (adapted from Reeder *et al.*, 1984; and Nelson *et al.*, 1985).

	Direct response			Indirect response		
Cycles of selection	Population A g plant^{-1}	Cycles of selection	Population B g plant^{-1}	Cycles of selection	LAER* mm^{2}day^{-1}	Forage yield kg ha^{-1}
C_0	382[†]	C_0	806[†‡]	C_0	123.2	5130
C_6	799	C_1	820	C_1	--	--
C_9	1154	C_2	879	C_2	--	--
C_{11}	1324	C_3	903	C_3	166.8	--
C_{12}	1537	C_5	1138	C_4	189.1	6274
C_{13}	1504	C_6	1430[‡]	C_5	--	--
C_{14}	1646	C_7	1319[‡]			
C_{15}	1655[§]					
LSD (0.05)	108		108			
Response cycle^{-1}, % change	23.6[¶]		10.1[¶]			

*Leaf area expansion rate (Reeder *et al.*, 1984).

[†]Mean green spaced-plant-yield of 100 plants for each cycle.

[‡]One year data with 2-year data for other cycles.

[§]Response cycle^{-1} after 14 cycles of selection.

[¶]Response cycle^{-1} using mean of cycles 6 and 7.

Recurrent selection methods have been used effectively to improve traits in forage crop species (Hallauer, 1981).

SUGARBEET

The first sugarbeet (*Beta vulgaris* L.) hybrids were produced in 1954, and nearly all cultivars presently marketed in the United States and Western Europe are monogerm hybrids. Because root yield of sugarbeets was conditioned primarily by non-additive genetic effects, reciprocal recurrent selection was considered the appropriate method for improvement of germplasm sources for developing parental lines of hybrids. Hecker (1978) reported improved general combining ability of lines isolated from populations undergoing reciprocal recurrent selection, but crosses among those selected were not tested. Two cycles of reciprocal recurrent selection were completed in populations designated A and B. Selection for superior recoverable sucrose concentration was emphasized in AC_0 and BC_0 to form AC_1 and BC_1. Three separate C_2 populations of AC_1 and BC_1 were synthesized based on greater recoverable sucrose, root yield, and sucrose concentration. The six C_2 populations themselves and six crosses of C_2 populations were evaluated to determine response to reciprocal recurrent selection.

Significant ($P < 0.01$) gains were made in population A for recoverable sucrose and root yield. In population B, significant gain ($P < 0.05$) for sucrose concentration was obtained after two cycles of selection, but no gain was made for either recoverable sucrose or root yield (Hecker, 1985). Significant direct response for the traits considered in the synthesis of the C_2 populations was realized for recoverable sucrose (13.6%), root yield (18.7%), and sucrose concentration (5.8%) (Table 10). The greatest recoverable sucrose (17.9%) was obtained from the cross of the two strains selected for root yield, and the crosses that included one strain selected for root yield also had more recoverable sucrose (17.0%). Root yields of crosses of strains selected for recoverable sucrose and for sucrose content, however, were not different from those of C_0 crosses (Table 10). The only crosses having significantly improved root yields were those including at least one strain selected for improved root yield. Root yield was the more important trait, but future selection may have to consider sucrose content to guard against a reduction in sucrose concentration. Increased recoverable sucrose occurred only because of an increase in root yield.

Hecker (1985) concluded that reciprocal recurrent selection may be an effective method for improvement of sucrose production in sugarbeet hybrids. Its greatest merit is selection for both additive and non-additive gene effects. Major problems in use of reciprocal recurrent selection in sugarbeet are the achievement of 100% hybridization in the test-crosses and preservation of maternal genotypes in each cycle of selection. Incorporation of

Table 10 Response to two cycles of reciprocal recurrent selection in crosses of sugarbeet populations A and B for recoverable sucrose, root yield, and sucrose content (adapted from Hecker, 1985).

Crosses Populations	Traits	Recoverable sucrose (RS) Mg ha^{-1}	Root yield (RY)	Sucrose content (S) (%)
AC$_2$ x BC$_2$	RS x RS	6.11*	4.73	16.3**
AC$_2$ x BC$_2$	RY x RY	6.34**	5.34**	15.7
AC$_2$ x BC$_2$	S x S	6.02*	4.37	16.4**
AC$_2$ x BC$_2$	RS x RY	6.29**	5.30**	15.7
AC$_2$ x BC$_2$	RS x S	6.24**	4.94	16.3**
AC$_2$ x BC$_2$	RY x S	6.29**	5.00*	16.1*
AC$_0$ x BC$_0$		5.38	4.50	15.5
LSD (0.05)		0.06	0.47	0.5

*,**Significantly different at P < 0.05 and 0.01, respectively, than AC$_0$ x BC$_0$.

genetic male sterility in each population and cloning as a substitute for selfing offer possible solutions to these two problems.

SUMMARY

Selection methods have been developed to capitalize on the genetic variability within source populations of cross-pollinated crop species. The original source populations for breeding programs were landrace cultivars developed within certain geographic areas. The importance of landrace cultivars in current breeding programs varies among crop species, ranging from little or no use in maize breeding to greater use for some forage grass species. In all instances, however, other germplasm sources were developed, based on selection within the landrace cultivars. The newer germplasm sources include composites, synthetics, and F_2 and backcross populations from crosses of inbred lines. The level of genetic variability within the newer populations was adequate to permit response to selection, and, in some instances, rate of response in F_2 populations was equivalent or greater than rate of response in landraces. Moll (1991) for maize and Burton (1989) for Pensacola bahiagrass conducted parallel selection studies in landrace cultivars and F_2 populations developed from crosses of selected lines: rate of response to selection in the F_2 populations exceeded that in the landrace cultivars, which presumably would have had greater genetic variability.

Inherent genetic variability within populations of cross-pollinated crop species has been recognized, and estimates of the components of genetic variance suggest that additive genetic effects are of greater importance than dominant and epistatic effects. Selection, if conducted appropriately, should be effective. Selection, in nearly all instances, has been effective in populations of cross-pollinated species for the traits emphasized in selection. There are concerns, however, about the correlated responses of other traits if useful germplasm is to be made available for applied breeding programs and the long-term effectiveness of inbred progeny selection. Use of appropriate controls in selection and use of selection indices in selection will reduce the effects of undesirable correlated responses. Inbred line development is emphasized in applied breeding programs, but the direct and correlated responses to inbred progeny selection within maize populations have not been consistent. Theoretically, inbred progeny selection should be twice as effective as half-sib family selection in the absence of any overdominant effects (Comstock, 1964). Although direct comparisons of rate of response for inbred progeny and half-sib family selection are limited, comparative rates of response to selection and genetic interpretations of the differences in response obtained for the two methods have not been consistent. Horner *et al.* (1989), based on differences in general combining ability, suggested that inbred progeny selection was not as effective as half-sib family selection because of possible overdominant effects in the two populations he studied.

Persistent genetic variability in populations of cross-pollinated crop species is maintained by their mode of reproduction, but the sources of genetic variability also may have changed during the past 40 years: traits emphasized in selection, types of environments, and screening techniques for pest resistance, grain and forage quality, grain and plant composition, and stress tolerance (cold, heat, drought, salt, etc.) have changed rapidly as technology has become available. Data collected for grain yield of maize is one example. Before 1970, ear weight from hand-harvested plots was recorded, and samples were taken from harvested ears to adjust for grain moisture and shelling percentage to determine grain yield on an area basis. Data were recorded for standability and ears detached from plants, but all ears were gleaned within plots at harvest. Development of mechanical harvesters for experimental maize trials changed the parameter for yield: only ears on healthy, standing plants were retrieved by the harvesters, and shelled grain weights were recorded. The same genetic factors affecting yield are present whether plots are hand or machine harvested, but different genetic factors would be emphasized by the two methods of harvest. The different emphasis given to genetic factors would occur for other traits as changes in technology become available during selection.

Quantitative genetic studies have revealed that genetic variability for most traits is adequate to expect response to selection. In some instances, screening techniques may not be adequate or sufficiently precise to consistently separate genetic and environmental effects, but this limitation seems more the exception than the rule. Although there is some evidence that genetic variability was reduced in long-term selection (Tanner and Smith, 1987), response after 30 cycles of mass selection for grain yield in "Hays Golden" (C. O. Gardner, 1991, pers. comm.) and 76 generations of divergent selection for protein and oil content (Dudley, 1977) continues. Moll (1991) also found that the genetic variability within the (NC7 x CI21) F_2 population was 51.4% less than that within Jarvis Golden Prolific, but that the F_2 population's response to selection was greater that that of Jarvis Golden Prolific. Adequate genetic variability, therefore, does not seem a restraint on the genetic improvement of cross-pollinated crop species. Rather, it seems we need to define our general objectives and to develop breeding and selection methods that, as Dudley (1977) stated, "most efficiently concentrate the favorable alleles we now have and then proceeding to do it".

ACKNOWLEDGMENTS

Journal Paper No. J-14402 of the Iowa Agric. and Home Economics Exp. Stn. Project 2778.

REFERENCES

Bauman, L. F. (1981) Review of methods used by breeders to develop superior corn inbreds. *Proceedings Annual Corn Sorghum Industry Research Conference* 36:199-208.

Burton, G. W. (1989) Great diversity for increased yield within narrow grass populations. In: *Proceedings of the International Grassland Congress*, Nick, France, pp. 301-302.

Cockerham, C. C. (1956) Analysis of quantitative gene action. *Brookhaven Symposia in Biology* 9:53-68.

Comstock, R. E. (1964) Selection procedures in corn improvement. *Proceedings Annual Corn Sorghum Industry Research Conference* 19:87-94.

Comstock, R. E.; Robinson, H. F. (1948) The components of genetic variance in populations of biparental progenies and their use in estimating average degree of dominance. *Biometrics* 4:256-266.

Comstock, R. E.; Robinson, H. F.; Harvey, P. H. (1949) A breeding procedure designed to make maximum use of both general and specific combining ability. *Agronomy Journal* 41:360-367.

Coors, J. G.; Mardones, M. C. (1989) Twelve cycles of mass selection for prolificacy in maize. I. Direct and correlated responses. *Crop Science* 29:262-266.

Covarrubias-Prieto, J.; Hallauer, A. R.; Lamkey, K. R. (1989) Intermating F_2 populations of maize. *Genetika (Yugoslavia)* 21:111-126.

Devey, M. E.; Russell, W. A. (1983) Evaluation of recurrent selection for stalk quality in a maize cultivar and effects of other agronomic traits. *Iowa State Journal of Science* 58:207-219.

Dodd, J. L. (1977) A photosynthetic stress-translocation balance concept of corn stalk rot. *Proceedings Annual Corn Sorghum Industry Research Conference* 32:122-130.

Dudley, J. W. (1977) 76 generations of selection for oil and protein percentage in maize. In: Pollak, E. *et al.* (eds.), *Proceedings International Quantitative Genetics Symposium*, Ames, IA. 16-21 Aug. 1976. Iowa State University Press, Ames, IA, pp. 459-473.

Duvick, D. N. (1977) Genetic rates of gain in hybrid maize yields during past 40 years. *Maydica* 22:187-196.

Eberhart, S. A.; Debela, S.; Hallauer, A. R. (1973) Recurrent selection in the BSSS and BSCB1 maize varieties and half-sib selection in BSSS. *Crop Science* 13:451-456.

Eyherabide, G. H.; Hallauer, A. R. (1991a) Reciprocal full-sib recurrent selection in maize. I. Direct and indirect responses. *Crop Science* 31:952-959.

Eyherabide, G. H.; Hallauer, A. R. (1991b) Reciprocal full-sib recurrent selection in maize. II. Contributions of additive, dominance, and genetic drift effects. *Crop Science* 31:(in press).

Gardner, C. O. (1961) An evaluation of effects of mass selection and seed irradiation with thermal neutrons on yield of corn. *Crop Science* 1:241-245.

Gardner, C. O. (1977) Quantitative genetic studies and population improvement in maize and sorghum. In: Pollak, E.; Kempthorne, O.; Bailey, J. B., Jr. (eds.), *Proceedings International Conference on Quantitative Genetics*. Iowa State University Press, Ames, pp. 475-489.

Gardner, C. O.; Lonnquist, J. H. (1959) Linkage and the degree of dominance of genes controlling quantitative characters in maize. *Agronomy Journal* 51:524-528.

Genter, C. F. (1982) Recurrent selection for high inbred yields from the F_2 of a maize single cross. *Proceedings Annual Corn Sorghum Industry Research Conference* 37:67-76.

Hallauer, A. R. (1981) Selection and breeding methods. In: Frey, K. J. (ed.) *Plant breeding II*. Iowa State University Press, Ames, pp. 3-55.

Hallauer, A. R. (1985) Compendium of recurrent selection methods and their application. *Critical Reviews in Plant Science* 3:1-34.

Hallauer, A. R. (1991) Recurrent selection in maize. *Plant Breeding Reviews* 9:(in press).

Hallauer, A. R.; Eberhart, S. A. (1970) Reciprocal full-sib selection. *Crop Science* 10:315-316.

Hallauer, A. R.; Miranda Fo, J. B. (1988) *Quantitative genetics in maize breeding*. 2nd ed. Iowa State University Press, Ames.

Hallauer, A. R.; Sears, J. H. (1972) Integrating exotic germplasm into Corn Belt maize breeding programs. *Crop Science* 12:203-206.

Han, G. H.; Hallauer, A. R. (1989) Estimates of genetic variability in F_2 maize populations. *Journal Iowa Academy of Science* 96:14-19.

Hanna, W. W.; Bashaw, E. C. (1987) Apomixis: Its identification and use in plant breeding. *Crop Science* 27:1136-1139.

Hecker, R. J. (1978) Recurrent and reciprocal recurrent selection in sugarbeet. *Crop Science* 18:805-809.

Hecker, R. J. (1985) Reciprocal recurrent selection for the development of improved sugar beet hybrids. *Journal of the American Society Sugerbeet Technology* 23:47-57.

Horner, E. S.; Magloire, E.; Morera, J. A. (1989) Comparison of selection for S_2 progeny vs. testcross performance for population improvement in maize. *Crop Science* 29:868-874.

Hull, F. H. (1945) Recurrent selection and specific combining ability in corn. *Journal of the American Society of Agronomy* 37:134-145.

Iglesias, C. A.; Hallauer, A. R. (1990) Response to S_2 recurrent selection in exotic and semi-exotic populations of maize (*Zea mays* L.). *Journal of the Iowa Academy of Science* 91:4-13.

Jenkins, M. T. (1940) The segregation of genes affecting yield of grain in maize. *Journal of the American Society of Agronomy* 32:55-63.

Jenkins, M. T. (1978) Maize breeding during the development and early years of hybrid maize. In: Walden, D. B. (ed.), *Maize breeding and genetics*. John Wiley and Sons, New York, pp. 13-28.

Jinahyon, S.; Russell, W. A. (1969a) Evaluation of recurrent selection for stalk-rot resistance in an open-pollinated variety of maize. *Iowa State Journal of Science* 43:229-237.

Jinahyon, S.; Russell, W. A. (1969b) Effects of recurrent selection for stalk-rot resistance on other agronomic characters in an open-pollinated variety of maize. *Iowa State Journal of Science* 43:239-251.

Jones, L. P.; Compton, W. A.; Gardner, C. O. (1971) Comparison of full and half-sib reciprocal recurrent selection. *Theoretical and Applied Genetics* 41:36-39.

Keeratinijakal, V. (1990) Evaluation of 11 cycles of reciprocal recurrent selection in BSSS and BSCB1 maize populations. Unpublished PhD thesis, Iowa State University, Ames.

Klenke, J. R.; Russell, W. A.; Guthrie, W. D. (1986) Recurrent selection for resistance to European corn borer in a corn synthetic and correlated effects on agronomic traits. *Crop Science* 26:864-868.

Lonnquist, J. H. (1964) A modification of the ear-to-row procedure for the improvement of maize populations. *Crop Science* 4:227-228.

Lonnquist, J. H. (1967) Mass selection for prolificacy in maize. *Zuchter* 37:185-188.

Mather, K. (1949) *Biometrial genetics*. Methuen, London.

Moll, R. H. (1991) Sixteen cycles of recurrent full-sib family selection for grain weight in two populations of maize. *Crop Science* 31:959-964.

Mulamba, N. N.; Hallauer, A. R.; Smith, O. S. (1983) Recurrent selection for grain yield in a maize population. *Crop Science* 23:536-540.

Nelson, C. J.; Sleper, D. A.; Counts, J. H. (1985) Field performance of tall fescue selected for leaf-area expansion. In: *Proceedings of the International Grassland Congress*, Kyota, Japan, pp. 320-322.

Odhiambo, M. A.; Compton, W. A. (1987) Twenty cycles of divergent mass selection for seed size in corn. *Crop Science* 27:1113-1116.

Penny, L. H.; Scott, G. E.; Guthrie, W. D. (1967) Recurrent selection for European corn borer resistance in maize. *Crop Science* 7:407-409.

Reeder, L. R., Jr.; Sleper, D. A.; Nelson, C. J. (1984) Response to selection for leaf area expansion rate of tall fescue. *Crop Science* 24:97-100.

Russell, W. A.; Eberhart, S. A.; Vega, U. A. (1973) Recurrent selection for specific combining ability in two maize populations. *Crop Science* 13:257-261.

Salazar, A. M.; Hallauer, A. R. (1986) Divergent mass selection for ear length in maize. *Revista Brasileira de Genetica* 9:281-294.

Short, K. E.; Carlson, I. T. (1989) Bidirectional selection for birdsfoot trefoil-compatibility traits in orchardgrass. *Crop Science* 29:1131-1136.

Shull, G. H. (1909) A pure-line method in corn breeding. *American Breeders' Association Report* 4:296-301.

Shull, G. H. (1910) Hybridization methods in corn breeding. *American Breeders' Magazine* 1:98-107.

Smith, O. S. (1983) Evaluation of recurrent selection in BSSS, BSCB1, and BS13 maize populations. *Crop Science* 23:35-40.

Sprague, G. F. (1955) Corn breeding. In: Sprague, G. F. (ed.), *Corn and corn improvement*. Academic Press, Inc., New York, pp. 221-292.

Sprague, G. F.; Tatum, L. A. (1942) General vs. specific combining ability in single crosses of corn. *Journal of the American Society of Agronomy* 34:923-932.

Tanner, A. H.; Smith, O. S. (1987) Comparison of half-sib and S_1 recurrent selection in Krug Yellow Dent maize populations. *Crop Science* 27:509-513.

Vogel, K. P.; Gorz, H. J.; Haskins, F. A. (1989) Breeding grasses for the future. In: *Contributions from breeding forage and turfgrasses*. CSSA Spec. Publ. No. 15. Crop Science Society America, Madison, WI, pp. 105-122.

Williams, P. N.; Davis, F. M. (1983) Recurrent selection for resistance in corn to tunneling by the second-brood southwestern corn borer. *Crop Science* 23:169-170.

Chapter 4
Genetic variation: Its origin and use for breeding self-pollinated species

P. S. Baenziger and C. J. Peterson
Department of Agronomy and ARS-USDA and Agronomy Department,
University of Nebraska, Lincoln, NB 68583-0915

INTRODUCTION: THE BASIS IS IN THE BIOLOGY

Plant breeding methods historically have been differentiated by pollination biology (self or cross). In introducing this chapter and the topics that will be discussed, it is important to understand the significance of these pollination differences. For self-pollinated crops, it is generally difficult to make breeding (used for creating segregating populations), experimental (used for large scale testing), and commercial hybrids. Breeding hybrids are usually single, three-way, and occasionally four-way (double cross) hybrids between inbred lines. Population improvement methods are possible, though rarely used. Gene frequencies for a hybrid with inbred parents vary from 0.0 to 1.0 in 0.25 increments and at most four alleles will be present. These hybrids are used to form segregating populations and it is trivial to allow the populations (and the lines therein) to inbreed rapidly (by selfing). For these populations, the rate of inbreeding is predictable, inbreeding depression is minimal, and natural outcrossing is low (Martin, 1990). Finally, the line *per se* is the basis of evaluation and not the population or hybrids. Hence, self-pollinated crop breeding is often limited by the number of parents in the hybrid, the number of hybrids (segregating populations) that can be made, recombination, selection efficiency, and the number of lines evaluated. The potential number of inbred lines that can be generated is limited by resources, including funding, labor, and land. For some crops (e.g., small grains) individual lines could easily be in the hundreds of thousands each year, while in others (e.g., legumes) the number of lines would be greatly restricted.

In cross-pollinated crops, it is generally easy to make breeding and experimental hybrids, and may be easy to make commercial hybrids. Breeding hybrids may be single, three-way, and double-cross hybrids, as well as population improvement by inbred line, hybrid, or population hybrids. Hence, the gene frequencies will vary from 0.0 to 1.0 with every frequency and number of alleles being possible. While it is easy to make the cross, it is usually difficult or requires labor to inbreed (via selfing or sib matings). The rate of inbreeding is predictable, but due to the various methods of hybridization and their effect on inbreeding, inbreeding is much more variable than in self-pollinated crops. Finally, in cross-pollinated crops, the desired result is often a hybrid or a population and not the line *per se*. Hence cross-pollinated crops are limited by the resources involved in inbreeding and evaluating the desired product. Creating the hybrids and populations is easy.

With the striking differences between cross- and self-pollinated crops, this paper will discuss the following: genetic variation in self-pollinated crops, selection of parental germplasm, hybridization, improving selection efficiency in the progeny, the use of cross-pollinated breeding methods in self-pollinated crops, inbred line performance in diverse environments, and possible future trends in self-pollinated crop breeding. This paper will not document the considerable successes of self-pollinated crop breeding as they recently have been documented elsewhere (Busch and Stuthman, 1990). As the authors' experiences are predominantly with small grains, the majority of examples will be from these crops. The choice of examples is not intended to lessen the considerable contribution that other crops have made toward developing methodology in self-pollinated crops.

ORIGIN AND NATURE OF GENETIC VARIATION

Harlan and DeWet (1971) developed the concept of gene pools, based on the difficulty of hybridization and gene introgression, to describe the genetic resources available to plant breeding programs. The primary gene pool consists of those species that readily hybridize, produce viable hybrids, and have chromosomes that pair and recombine allowing genetic exchange. The secondary gene pool consists of those species which are difficult to use in hybrids due to ploidy differences, chromosome alterations, or genetic barriers. The tertiary gene pool consists mainly of distinct species, subgenera, and genera; characterized by greater difficulties in making hybrids, and fertile progeny are rarely obtained even after repeated hybridizations.

Plant breeders are the largest generators, manipulators, and evaluators of genetic diversity in plants even though they work mainly with the primary gene pool. Within the primary gene pool, the most important component is their own breeding populations and that of their colleagues. The breeding germplasm pool can be extensive, and focuses on generating diversity for improved adaptation, yield, disease and insect resistance, and end-use quality. For example, the annual headrow nursery for the wheat (*Triticum aestivum* L.) breeding program at Nebraska contains almost as many wheat genotypes as does the wheat portion of the USDA National Small Grains Collection. Obviously, the relative diversity within a breeding program will be much less than within a collection. In addition, the nature of breeding is such that most of the genotypic diversity generated within a program will be discarded, hence unavailable to other programs.

For genotypic diversity not already in the breeding program, elite, adapted germplasm from comparable programs within the growing region can be chosen to facilitate their ease of incorporation and utilization. Germplasm from within the breeding program and from comparable programs often constitutes 90% or more of the germplasm used in most small grains crossing blocks. As diversity becomes limiting, breeders will use elite, unadapted germplasm. Many advances in wheat improvement are related to the successes of international germplasm exchange efforts which merged elite gene pools that are geographically and genetically diverse. Using the classical definition of "centers of diversity", breeding programs have become the modern centers of diversity. For example, Peeters (1988) suggested that barley (*Hordeum vulgare* L.) lines from the U.S.A. had the greatest within-country diversity of germplasm originating from the countries (which included Turkey, Iran, Israel, and Ethiopia) represented in an English germplasm collection. As domesticated barley is not indigenous to the U. S., this diversity is a testament to previous germplasm exchanges and utilization.

National and international germplasm collections provide diversity to breeding programs by the systematic collection and distribution of seed stocks. Germplasm collections are extensive (Harlan, 1975; Hawkes, 1981; Frankel and Soule, 1981; Shands, 1990) and most include representatives of the primary, secondary, and tertiary gene pools. For polyploid species such as wheat, these gene pools include the diverse genera and species that are related by evolution (Feldman and Sears, 1981; Kimber, 1984; Sharma and Gill, 1983; Morris, 1983). The relatives of major crops have often co-evolved with the crop pests making them rich sources of useful pest resistance genes (Leppik, 1970). Modern composite crosses that have co-evolved with diseases under natural selection pressure have also been proposed as valuable genetic resources (Allard, 1990).

Where useful germplasm has been identified in domesticated and wild relatives of crops, methods have been developed for germplasm introgression (Stalker, 1980; Lawrence and Frey, 1975; Feldman and Sears, 1981; Sharma and Gill, 1983; Morris, 1983; Kimber, 1984). However, with the value of the secondary and tertiary gene pools for improving wheat (Sharma and Gill, 1983), continued efforts are needed to improve hybridization, recombination, and selective gene introgression. Major limitations exist in reducing the amount of introgressed alien species germplasm so as to lessen the effect of deleterious genes linked to the gene of interest. For example, two very important chromosome translocations involving wheat (A and B) and rye (*Secale cereale* L.) (R) chromosomes (namely 1A/1R, present in varieties grown on an estimated seven million ha in the Great Plains of the U.S.A., and 1B/1R, present in varieties grown on an estimated 25 million ha worldwide) (S. Rajaram, 1990, pers. comm.), which have excellent disease resistance, but are associated with suspect and poor end-use quality, respectively (Dhaliwal and MacRitchie, 1990). The linkage between disease resistance and poor end-use quality genes has limited the utilization of these translocations.

Despite extensive germplasm collections, much of the "available" germplasm diversity has not been successfully used by plant breeders. The relatively little success in using germplasm and genetic diversity in plant breeding can be seen in the narrow genetic base of major crops (National Academy of Science, 1972). In soybeans [*Glycine max* (L.) Merr.], Delannay *et al.* (1983) estimated 10 and seven lines contributed over 80% of the northern and southern U. S. gene pools, respectively. St. Martin (1982) estimated the effective population size for group 00 to group IV soybeans from 1930 to 1980 was between 11 and 15.

If the germplasm is available and methods exist to incorporate it into adapted gene pools, why do populations in most breeding programs and resulting cultivars have little genetic diversity? As described by Goodman (1990), germplasm that is not being used can be grouped into four categories - (1) uncollected, (2) unacquired, (3) unadapted, and (4) unevaluated. For the first two categories it is obvious why the germplasm has not been used. Of the latter two categories, unevaluated germplasm is a critical limitation for use. The true value of germplasm is determined by its contribution to its progeny which may be different from its perceived value. Unfortunately, systematic programs to evaluate germplasm by introgressing it into agronomic types are often lacking. The above-mentioned successes for introgressing chromosome arms of rye were motivated by the potential usefulness of pest resistance genes. Future introgressions will continue to be motivated by attractive genes. There are no systematic programs to determine if other rye chromosomes arms could contribute to wheat improvement. Nor are there dedicated programs to systematically introgress known genes of value into elite germplasm such as backcrossing leaf rust (*Puccinia recondita* Rob. ex Desm.) resistance genes into susceptible cultivars of winter wheat.

Even with the diversity present in germplasm collections, it must be remembered that many traits have little genetic variability [e.g., wheat streak mosaic virus resistance (Lay *et al.*, 1971)] or existing genetic variability has been exhausted (e.g., winterhardiness in wheat). Where genetic diversity is lacking, mutation breeding (Maluszynski, 1990) or genetic transformation technologies (Fraley *et al.*, 1988; Gasser and Fraley, 1989) may provide useful variation. However, mutation breeding has been successfully used only for changing a limited number of traits and requires an efficient mutant identification procedure. Genetic transformation is not routine in many crops and currently cannot transfer complex traits. Hence, hybridization will remain the primary source of introducing variation, particularly for traits that can be identified in the primary and secondary gene pools.

PARENT SELECTION AND LINE PREDICTION

There is no consensus on how best to choose parents that will produce high yielding progenies and parent selection remains one of the critical unanswered questions in plant breeding (Qualset, 1979; Baker, 1984) despite numerous theoretical and empirical studies to identify useful parents. Research on parent selection and predicted line performance can be divided into two major groups. The first group consists of methods to choose parents on the basis of parent performance. The second group includes methods which evaluate the parent on the basis of its progeny.

Methods which are included in the first group are selections based on (1) the midparental value, (2) divergence of coefficient of parentage (Murphy *et al.*, 1986), (3) character complementation [also called the geometric approach (Grafius, 1964, 1965a; Lupton, 1965)], (4) multivariate analysis and parental distances (Bhatt, 1970), and (5) least squares methods, parental complementation, and an ideal genotype (Pederson, 1981). Selection of parents on the midparental value assumes an additive genetic model with the midparent estimating the derived line population mean. It does not include a measurement of genetic variation. Selection of parents on the divergence of coefficients of parentage emphasizes genetic variation between parents, but does not necessarily include an estimate of the parental means. In practice, however, it is likely that lines with high parental values and with low coefficient of parentages will be useful in crosses. The coefficient of parentage represents the minimum genetic relationship as it is based on genes that are identical by descent and does not include genes that are identical in state. Also, estimates of the coefficient of parentage assume the very early parents (for which there may be few pedigree records) are unrelated. Selection on character complementation implicitly involves genetic variation for the characters of interest. Selection using multivariate analysis emphasizes genetic variation between the parents. Selection using least square methods, parental complementation, and an ideal genotype couples genetic diversity with line performance. It also has the advantage of allowing an estimate of correct proportion of each parent in a cross. The types of crosses used in breeding self-pollinated crops will be discussed in greater detail later. The value of these methods for choosing parents is that they use data that can be readily obtained from most performance evaluation trials. As the efficacy of these methods can only be determined by evaluating the progeny from the predicted crosses, the empirical data supporting these methods will be included with the discussion on methods that evaluate parental worth on the basis of their progeny.

Methods which evaluate parents on the basis of their progeny use (1) F_1 data (Suneson and Riddle, 1944; Allard, 1956), (2) F_2 and later generations (Weiss *et al.*, 1947; Atkins and Murphy, 1949; Fowler and Heyne, 1955; Lupton, 1965; Smith and Lambert, 1968; Busch *et al.*, 1974; Cregan and

Busch, 1977; Cox and Frey, 1984), and (3) doubled haploid lines (Reinbergs *et al.*, 1976). All of the these studies include empirical data to prove the efficacy of their predictions. Unfortunately, the data do not provide a consensus. For seed size, a highly heritable trait in lima bean (*Phaseolus lunatus* L. var. *lunatus*), an F_1 diallel of closely related lines was successful for identifying parents (Allard, 1956). Allard argued against using F_2 data because they were not as efficient for identifying dominance relationships and the diallel analysis of F_2s is more sensitive than F_1s to violating the basic assumptions of the diallel analysis [using the methods of Jinks (1956) and Hayman (1954)].

For grain yield, Lupton (1965) used an incomplete diallel in wheat and three methods of cross prediction - (1) dominance relationships and variance and covariance of parental arrays, (2) general and specific combining ability, and (3) the mean and variance of random progeny selections from each cross. Due to environmental and genetic interactions, the dominance relationships were not useful and the array variance and covariance were only useful in identifying crosses with non-allelic interactions. General combining ability analysis was useful because general combining ability and dominance relationships identified in the analysis were less affected by the environment. The variances of random progeny selections were found to be significantly different among populations. Having the mean and the variance of each cross was a useful indicator of parental and cross value. There was general agreement among the three methods for parent and cross prediction despite the difficulties with the first two methods. A number of practical aspects of plant breeding were also discussed by Lupton (1965). He used F_2 and F_3 bulks and not the F_1s due to the labor involved in generating sufficient seed for testing and because previous F_1 tests were unreliable. He produced an incomplete diallel to reduce the number of crosses and progenies that would have to be tested. Thirty-six randomly selected lines per cross were tested in one environment (hence, he could not estimate genotype x environment interactions). Finally, he recognized that, in progenies of some of his crosses between winter and spring wheats, a harsh winter could greatly affect the spring parent and potentially the F_2 bulk performance which in turn could affect the statistical analyses. The authors' experiences in Nebraska agree with Lupton's concern that severe winters will greatly affect F_2 bulk performance of winter wheat lines differing in winterhardiness as well as winter by spring crosses. In general, F_2 bulk performance would be unreliable for predicting parent and cross performance if the bulks underwent significant selection in that generation. Later generations (after the main population changes had occurred) would be better for predicting parent and cross performance.

Smith and Lambert (1968) performed an experiment in barley that was similar in structure to Lupton's (1965) but included topcrosses to genetic male sterile lines, as well as F_2, F_3, and F_4 bulks and randomly selected

F_3-derived F_5 lines. The objective was to test bulk performance in order to identify parent and cross performance. In this study, parents *per se*, midparental values, parental arrays, and early generation bulk performance were all beneficial in identifying parents and crosses with good performance characteristics. Only topcrosses where found to be unreliable, which may have been due to their not being representative of the materials used to evaluate their effectiveness. The topcrosses involved genetic male-sterile lines that were not parents of the bulks or F_5 lines.

Cregan and Busch (1977) studied F_1s of high yielding spring wheat parents, and F_2 through F_5 bulks, plus F_2-derived F_5 lines to identify superior parents and crosses. The F_2 bulk performance and the mean of the F_2 and F_3 bulks grown in the same year (1975) as the F_5 lines were the best predictors of F_5 line performance. F_1 and the other bulk generation yields were also predictive of F_5 line performance. Only the midparental values based on previous year data were not predictive. The study also determined how selection in the predictive generations would have affected retention of high yielding lines in later generations. Using a 25% selection level, the bulk performance was superior to midparent and F_1 performance for predicting which bulks should be retained for selecting high yielding lines. Using a 50% selection level, F_2 bulk and the mean of F_2 and F_3 bulk performance (grown in 1975) was superior to F_1, mean of F_2 and F_3 bulk performance grown in previous years, and the miparental values. However, using the 50% selection level, the superiority of one predictor relative to another was reduced and every predictor of cross performance identified crosses that contained at least 60% of the highest yielding lines. In this study, environmental effects were relatively small.

In other research involving wheat (Fowler and Heyne, 1955), oats (*Avena sativa* L.) (Atkins and Murphy, 1949), and soybeans (Weiss *et al.*, 1947), performance of early generation bulks did not indicate which bulks would be the best source of elite lines.

It is interesting to note that the utility of the above parent and cross prediction methods are verified by randomly derived later generation lines. The line *per se* data, and the cross means and variances based on the derived lines are used to identify which crosses have the most transgressive lines, the highest means, and greatest variances. Randomly derived later generation lines were not suggested as excellent parent and cross predictors because at that stage of the breeding program, the populations have been retained for many generations. The populations are only of interest at this stage if the breeder wishes to resample the population to increase the number of derived lines. Hence, determining the parental or cross value is no longer important. The exception to this conclusion is for crops where doubled haploid lines can be produced rapidly and used to predict the parent and cross value (Reinbergs *et al.*, 1976).

From the above results, it is evident that, while predictors such as mid-parental values can occasionally be used successfully to predict which crosses should be made, once the cross is made, the progeny generations provide more useful information. Of the group of predictors which are useful without progeny tests, the multivariate approach was the most useful. Bhatt (1973) compared crosses made by (1) the conventional method, (2) random method, (3) multivariate methods based on the diversity of the lines, and (4) ecogeographic diversity based on the geographic isolation of parents belonging to the same cluster for genetic divergence. Of these methods, the multivariate method gave the best results for identifying crosses with transgressive segregants followed by the ecogeographic diversity, conventional, and random methods. Although this was only one experiment and the number of parents was limited, it does represent the benefit of having a greater understanding of parent diversity in selecting parents for crosses. As lines having the same parentage and coming from the same ecogeographic region could be placed in different clusters for genetic diversity, the multivariate method should be superior to those parent selection methods based on coefficient of parentage. Coefficients of parentage are not affected by selection among lines within a cross which would be normal in breeding programs. Cox and Murphy (1990) suggested that parents be selected on performance *per se*, then grouped by genetic distance based upon multivariate analysis, and finally chosen by having small coefficient of parentages. Coefficient of parentage, by itself, was not very useful in this study.

One of the limitations of selecting parents on the basis of genetic and ecogeographic diversity is that considerable information must be obtained on each line to determine its genetic diversity relationships. Another method to choose potential parents on ecogeographical diversity would be to select parents from isolated gene pools in geographic regions having similar environments. For crops with extensive worldwide testing, it is possible to identify test sites in diverse regions that are similar (Peterson and Pfeiffer, 1989). If the gene pools are isolated, it would be possible to select parents with little known relationship that perform well under similar climatic conditions, thus coupling diversity and performance.

Once the parents are selected, the breeder will need to determine the contribution of each parent to the resultant cross (Pederson, 1981), the number of crosses that can be handled in the breeding program, whether or not the cross progeny should be intermated, and the population size needed for each cross. The contribution of a parent will depend on its adaptation. If a parent is exotic, unadapted, or a wild species, its contribution in a cross will tend to be small (Lawrence and Frey, 1975; Schoener and Fehr, 1979; Khalaf *et al.*, 1984). If the parents are adapted, most breeders use single, three-way, or double crosses. Double crosses are becoming less popular because they are perceived as potentially having too much variation to be easily used in a breeding program (Jensen, 1988). The breeder also needs to determine how many F_1 seeds are required to represent the gametic array of the parents. Heterogeneous and heterozygous parents require more F_1 seeds.

The number of crosses that can be handled in a breeding program will be decided by the resources needed to make the cross (usually small relative to working with populations), the breeding method, and project resources. Yonezawa and Yamagata (1978) suggested increasing the number of crosses rather than increasing the cross population size when the total number of lines is held constant. They assumed that the breeder was unable to assess the worth of a cross except by its progeny. Baker (1984) provided a quantitative genetic approach to optimize the number of crosses and the number of lines within each cross. The assumptions of the model were as follows: (1) crosses were between random pairs of inbred lines, (2) the total number of lines (the product of the number of crosses by the number of lines per cross) would be 2000, (3) the modified pedigree system was used (Brim, 1966), and (4) epistasis and linkage disequilibrium were absent. The goal was to select the five best lines within the five best crosses and the model predicted 50 to 100 crosses would be needed. Although the number of crosses was determined with assumptions that are not found in practice, it did indicate a minimum number of crosses.

In practice, most breeders make considerably more crosses than 50 and they are made to accomplish specific objectives. With the difficulty in predicting the value of a cross, and if progeny are used in that prediction, it is easier in most programs simply to use one's experience in choosing parents. The number of lines per cross will vary with selection intensity and some crosses will be eliminated from the breeding program. Breeders probably have erred by not using their cross and progeny information more effectively to validate and add perspective to their crossing program.

The last aspect of crossing that should be considered is the effect of linkage on a population. As summarized by Baker (1984), the benefits of intermating to break linkage groups will depend upon whether or not genes of interest are in the coupling or repulsion phase, the tightness of linkage (for chromosome translocations from the secondary and tertiary gene pools there may be no recombination), and time necessary to intermate a population. When the genes are in the coupling phase, the genetic variance increases and the covariance between traits is positive. The opposite is the case for genes in the repulsion phase. It is more likely that each parent will have important genes in both coupling and repulsion phases. Under this circumstance, linkage will not affect the average probability of fixation (Bailey and Comstock, 1976). Breaking favorable linkage blocks could be detrimental in crosses between adapted, elite parents and exotic parents. Bos (1977) found that intermating F_2 plants with or without selection would have negligible to small effects on the subsequent population as determined by the frequency of the optimal genotype. Using genetic male sterility and single cross derived populations, Altman and Busch (1984) found little difference between lines derived from a randomly intermated population before selection and lines derived from the same population that was not intermated.   Intermating after selection (a cycle of recurrent selection) generally has been positive (for review, see Busch and Stuthman, 1990).

SELFING, SELECTION, AND THEIR EFFECT ON PROGENY

Most self-pollinated crop breeding programs use selfing which causes rapid inbreeding for generation advance. Selfing will occur with artificial selection, natural selection, and/or no selection depending upon the breeding method.

The breeding methods used in self-pollinated crops are mass selection, pedigree, bulk, single seed descent, doubled haploidy, backcrossing, recurrent selection, and various combinations of one or more of the methods. These methods have been extensively described in plant breeding text books (Fehr, 1987; Jensen, 1988) and will only be described briefly here. Each method can be highly successful when competently used and the choice should be based upon the breeding objective (assumed to be the development of new high yielding, high quality cultivars with resistance or tolerance to important stresses) and project resources.

Mass selection is most commonly used to rogue off-types from a cultivar, but also may be used to modify plant introductions for improved adaptation. Pedigree breeding is probably the most labor and resource intensive breeding method, which imposes limits on the number of populations that can be manipulated. In contrast, bulk breeding requires little labor and resources; hence, many populations can be manipulated (Florell, 1929). The population undergoes natural and, occasionally, mild artificial selection. A major concern with bulk breeding is that intrapopulation competition among diverse genotypes may lead to the elimination of genotypes that could perform well in pure stands (Early and Qualset, 1971; Khalifa and Qualset, 1974, 1975; Phung and Rathjen, 1976). The single seed descent breeding method (SSD) (Goulden, 1939; Grafius, 1965b; Brim, 1966) requires more labor than bulk breeding, but has the advantage that generations can be rapidly advanced in "off-season" nurseries with little or no selection. The major concern with this method is that, in its purest form, no selection occurs until later generations; hence, labor is lost maintaining undesirable lines.

Doubled haploid breeding (DH) is similar to the SSD in theory (Baenziger *et al.*, 1984; Choo *et al.*, 1985). Doubled haploid breeding requires creating haploids, usually from wide hybridization followed by chromosome elimination (Laurie *et al.*, 1990) or by gamete culture (Kasha *et al.*, 1990). Doubled haploid lines are usually produced by colchicine treatments and are completely homozygous, homogeneous, and inbred. Doubled haploidy is the most complete and rapid method for obtaining homozygous lines and, in theory, they are obtained without selection. For these reasons, it is similar to SSD. One major difference between SSD and DH methods is the opportunity for chromosomal recombination. With single seed descent, effective recombination occurs in every generation of selfing before gene fixation, whereas in doubled haploid breeding, recombination occurs only in the generations of selfing before the haploid is produced. Hence, favorable linkage groups may be maintained during inbreeding. The difference in recombination also has provided a useful tool to differentiate pleiotropic and linked genes (Powell *et al.*, 1985). Doubled haploidy coupled with cytogenetic techniques will allow rapid and simple development of recombinant chromosome lines for genetic studies. Finally, evaluation of homozygous lines is also more precise than heterozygous or heterogeneous line evaluation; hence, SSD and DH line breeding *per se* should be more effective than pedigree breeding (particularly for traits that involve destructive assays).

The last breeding method, backcrossing, is used to transfer one or a few genes into a line that has excellent characteristics except for the trait controlled by the transferred genes (Harlan and Pope, 1922; Briggs and Allard, 1953). The major concern with backcrossing is that while it gives predictable progress there is little opportunity to improve the recurrent parent except for the backcross trait. If a new cultivar is developed during the backcrossing procedure (which includes progeny testing and cultivar seed increase) that is superior to the recurrent parent, the backcrossing efforts may have been wasted. The same concern is also true for transformation which is similar to backcrossing in that one or a few genes are added to a line; the transformed line still requires testing and seed increased.

Two concepts that are important in understanding the effects of selfing and selection are (1) among- and within-line genetic variances for quantitative traits and (2) the level of homozygosity for qualitative traits at multiple loci. Among- and within-line genetic variances are important because they sum to the total genetic variance, and the among-line genetic variance is used to predict genetic gain. In self-pollinated crops which are highly inbred, only additive and epistatic (generally small) genetic variance can be fixed. Dominance genetic variation is present in early generations and can hinder selection. Genetic gain (ΔG) can be expressed as:

$$\Delta G = i \left[\frac{\sigma_g'^2}{\sigma_p y} \right]$$

where i is the intensity of selection, $\sigma_g'^2$ is the among-line genetic variance that can be fixed, σ_p is the square root of the among-line phenotypic variance, and y is the number of years per cycle. For F_2-derived lines in later generations expressing negligible dominance and epistatic variation, the among-line additive genetic variance is $1\sigma_a^2$ and the within-line additive genetic variance is also $1\sigma_a^2$. With each succeeding generation of line derivation, the within-line genetic variance is halved and the among-line genetic variance increases by the reduction in within-line genetic variance. The total additive genetic variance will equal $2\sigma_a^2$. As the among-line genetic variance is directly related to genetic gain, the generation from which lines are derived is critical for predicting gain.

The within-line genetic variance is also important for achieving uniformity standards; and the greater the within-line variation, the more difficult it is to achieve phenotypic uniformity. The difficulties in achieving phenotypic uniformity can be illustrated by examining the proportion of individuals that are expected to be homozygous at all loci which were heterozygous in the F_1:

$$\text{Proportion of homozygous individuals} = \left[\frac{2^n - 1}{2^n} \right]^m$$

where n is the number of generations of selfing and m is the number of heterozygous loci in the F_1. Hence, the fewer the number of selfing generations or the greater the number of F_1 heterozygous loci, the lower the proportion of homozygous lines and the greater the difficulty in selecting phenotypically uniform lines.

Hence, in every breeding method, the generation(s) of selection is critical and early versus late generation selection is still debated. The debate involves the opposing needs of rapidly advancing lines from many populations (which favors both early generation selection and few generations of selection) and of maximizing genetic gain and phenotypic uniformity (which favors later generation selection and more generations of selection). Sneep (1977) compared early generation (F_2) selection with later generation selection using common wheat as a model. He assumed 21 genes (one on each chromosome) controlled a trait of interest. In an F_2 population one plant in 421 is expected to have all the desirable genes in a heterozygous or homozygous state. However, to be sure that the desired plant is in the F_2 population, he recommended growing 1684 F_2 plants. As F_2 plant selection is unreliable, Sneep also suggested evaluating 1684 F_2-derived F_3 families each containing a minimum of 228 plants. If SSD had been used, in an F_∞ population, one plant in 2,097,152 is expected to have all the desirable alleles. On the basis of this calculation, Sneep argued that early generation testing is critical.

As discussed by Baker (1984), testing 1684 F_2-derived F_3 families of 228 plants would mean testing 383,952 plants. It would be virtually impossible to identify the desired phenotype due to environmental influences. Also, the majority of loci would be heterozygous; hence, additional testing and generations of inbreeding would be needed to identify the desired line that is homozygous at each loci. The need for continued selection and the required additional generations of inbreeding detract from the perceived benefits of early generation testing. Finally, if the objective is to accumulate 21 favorable alleles, neither pedigree or SSD would be an appropriate breeding method. Rather, recurrent selection should be recommended.

Each breeding method has its advantages and most breeders use the combination of breeding methods that is determined by their resources and objectives. In early generations, plant selections are made. In later generations, when the within family variation is small, selection is between families. For example, many breeders use culling selection or the elimination of obviously inferior phenotypes in early generations and then rapidly advance their plant selections by SSD or DH to later generations and select among families for traits with low heritability in replicated tests grown in one or more environments.

BREEDING SELF-POLLINATED CROPS AS IF THEY WERE CROSS-POLLINATED

With the success of maize (*Zea mays* L.) breeding, self-pollinated crop breeders have tried to apply cross-pollinated crop breeding methods to improve their various crop species. Their efforts include recurrent selection and the development of hybrids.

Recurrent selection is common in self-pollinated crops because most breeders recycle their elite lines as parents. This is a form of recurrent selection with very long cycle times [6.5 to 10.5 years (McProud, 1979)] and has the advantage that inbred elite lines are usually well characterized without the confounding effects of dominance variation. Recurrent selection with shorter cycle times in self-pollinated crops has been reviewed by Busch and Stuthman (1990) who concluded "recurrent selection can be considered a proven system". It has been successfully used to improve grain yield in oats (Payne *et al.*, 1986) and soybeans (Sumarno and Fehr, 1982; Piper and Fehr, 1987), kernel weight in wheat (Busch and Kofoid, 1982), and heading date in wheat (Avery *et al.*, 1982). Genetic male sterility has been advocated to assist random mating in recurrent selection (Brim and Stuber, 1973; Sorrells and Fritz, 1982) though hand emasculation has also been used.

Recurrent selection in self-pollinated crops has the same problems as was outlined for early generation testing, particularly when selection was based on F_1 plant data. Many of the above researchers have used later generation lines taking advantage of the more comprehensive testing possible with larger quantities of seed. This lengthens each cycle of recurrent selection and still involves selection based upon heterogeneous and heterozygous materials.

With the advent of doubled haploid breeding methods, Griffing (1975) and Scowcroft (1978) illustrated the theoretical advantages of using doubled haploids (pure lines) for recurrent selection. The doubled haploids used as parents in population improvement could also be directly released as new cultivars (Choo *et al.*, 1979). The successful use of doubled haploids in recurrent selection has been supported by experimentation (Foroughi-Wehr and Wenzel, 1990).

With the exception of rice (*Oryza sativa* L.) (Virmani *et al.*, 1982), hybrids in self-pollinated crops have not had the commercial success of hybrid maize or sorghum [*Sorghum bicolor* (L.) Moench]. There are many explanations for this lack of success. The floral structure and pollen biology of many self-pollinated crops do not lend themselves to cross-pollination. Self-pollinated crops do not exhibit high levels of heterosis or inbreeding depression. Cytoplasmic male sterility with adequate restoration genes has been difficult to obtain for some crops. Hybrids must then be produced with complex cytogenetic materials (Ramage, 1965) that slow parent creation, or with gametocides. Finally, many self-pollinated crops have relatively low yields and high seeding rates which make hybrid production costly and reduces the economic return to the producer.

IMPROVED SELECTION EFFICIENCY: THE GOAL OF EVERY BREEDING PROGRAM

With selection so critical to the success of every breeding program, no discussion on the use of variation would be complete without some mention of improving selection efficiency and of effectively using all the information available in breeding decisions. The importance of efficient selection can be readily seen in small grains breeding programs where 1,000,000 or more plants can be grown in 0.5 ha and segregating populations may be planted on many hectares. Obviously, only minimal record keeping and inexpensive and rapid selection assays can be used in populations of this size.

Selection nurseries, as opposed to evaluation nurseries, are used to magnify the differences between genotypes and simplify selection (Brown *et al.*, 1983). The goal of an evaluation nursery is to estimate accurately agronomic performance of experimental lines should they be released as a cultivar. For this reason, early generation populations are grown in selection nurseries. Culling selection for qualitative traits is applied and relies upon natural and artificial causes to magnify phenotypic differences. For example, in the Nebraska wheat breeding program, the F_2 populations are planted at Mead, NE which has the most severe winterkilling of the breeding sites and is artificially infected with *Puccinia graminis* Pers. f. sp. *tritici* Eriks. & Henn., the causal organism of stem rust. As winterhardiness is a recessive trait, homozygous and heterozygous wintertender genotypes are quickly removed from the segregating populations. Similarly, *P. graminis* will greatly reduce the yield and kernel weight of susceptible genotypes. As winterhardiness and stem rust resistance are breeding objectives for Nebraska wheat cultivars, it is reasonable that all of the breeding material be grown at this selection site. If there were different breeding objectives, different selection nursery sites would be appropriate. In addition to culling selection, visual selection with all of its faults (see review by Jensen, 1988) must be applied because no other technique can efficiently evaluate the number of early generation lines (e.g., 45,000 headrows) that some breeding programs create.

Once the number of genotypes has been reduced, more expensive and resource consuming selection assays can be used. These assays can be extremely precise - such as seedling disease screens and biochemical assays using isozymes (Tanksley *et al.*, 1984), restriction fragment length polymorphisms (RFLPs) (Soller and Beckman, 1983; Tanksley, 1988; Lander and Botstein, 1989), group specific primers and polymerase chain reaction for disease infection (Robertson *et al.*, 1991), or relatively imprecise but effective screens for ranking large numbers of genotypes [e.g., sedimentation tests for wheat quality (Moonen *et al.*, 1982)]. The utility of the assay will depend on the breeding objective, its cost, and on the diversity of the material. For example, high molecular weight glutenin proteins have been used to predict baking quality in England (Payne *et al.*, 1987) where the wheats historically have poor quality, but large amounts of diversity exist for quality in breeding populations. However, in Nebraska where the wheats historically have better quality and are less diverse, the high molecular weight glutenin genes are fixed (Graybosch *et al.*, 1990).

Agronomic performance is critical for all crops including those with high premiums for quality. Hence, field assays for performance will continue to be very important. Having evaluation nurseries that accurately represent the possible areas of adaptation is critical. The relationship among testing sites can be determined by statistical techniques (e.g., principal component analysis) using cultivar performance data (Peterson and Pfeiffer, 1989). Also, statistical analyses that provide the greatest precision in differentiating between lines are important. In this area, nearest neighbor analyses are becoming more important as they are able to account for field trends rather than block (incomplete or complete) differences (Stroup and Mulitze, 1991).

Finally most breeding programs create massive data sets with information collected as early as the F_2 generation. Many of these data are used during the year they are obtained, but effectively summarizing the accumulated information over years and locations, and from different generations is a problem for most breeders. As mentioned in the section on choosing parents, the more successful procedures used to identify parental combinations involve or are validated by analyzing progeny. In addition, where the goal is to develop new cultivars, much of the parent-progeny information is not effectively used to determine the underlying genetics for traits of interest. Because selection can be highly successful and efficient, genetic understanding of important traits may suffer. If this comment sounds harsh, consider how many breeders know the levels of resistance that their lines have to diseases and insects without knowing the relatively simple genetics of these traits.

HOMOGENEOUS LINES IN HETEROGENEOUS ENVIRONMENTS

As mentioned earlier, one of the goals of most self-pollinated crop breeding programs is to develop cultivars with phenotypic uniformity. However, these uniform cultivars are grown in diverse environments. The environment includes predictable and unpredictable factors (Allard and Bradshaw, 1964). For some higher value crops, producers will ameliorate environmental deficiencies by the addition of fertilizer, pesticides, irrigation, etc. However, many self-pollinated crops, particularly cereals, are grown in environments unsuitable for other crops and with little environmental buffering. A basic question confronting self-pollinated crop breeders is how to develop uniform lines that are productive in diverse environments. Cultivars with high mean performance are often productive in low, medium, and high production environments with the greatest benefit being seen in the high yielding environments (Adegoke and Frey, 1987; Peterson *et al.*, 1989). However, increasing mean performance has tended to decrease the area adaptation and stability [as determined by the Eberhart and Russell (1966) model] and increase genotype x environment interactions (Peterson *et al.*, 1989).

An alternative method to maintain productivity in diverse environments is to increase the genetic diversity within the cultivar (Allard, 1961). Increasing the within-cultivar genetic diversity can be achieved by a number of different methods. One of the easiest methods is to end selection in early generations to maintain within-line variation. For example, an F_3-derived cultivar will be heterogeneous and have one quarter of the total additive genetic variance within the line (0.5 σ_a^2 within-line, 1.5 σ_a^2 among-line, and 2.0 σ_a^2 for the total additive genetic variance for the population from which the cultivar was derived). Virtually all of the USDA-University of Nebraska winter wheat cultivars are from F_2- (e.g., Brule), F_3- (e.g., Arapahoe and Rawhide), or F_4- (e.g. Centurk) derived lines. Selection in the breeding nursery is for performance and uniformity. Later generations are rogued to remove off-types and obtain acceptable levels of uniformity. However, these lines are heterogeneous as can be readily demonstrated by the high frequency of new releases that have arisen from reselection in these cultivars (Redland from Brule; Siouxland 89 from Siouxland; Baca, Scout 66, and Scoutland from Scout; Rocky and Centurk 78 from Centurk).

A second method is the development of multilines (Jensen, 1952; Borlaug, 1953; Gustafson, 1953). A multiline is a blend of phenotypically similar lines that contribute different useful genes, most often disease or insect resistance genes, to the multiline. Multilines can be composites of isolines (Borlaug, 1953) or of phenotypically similar lines having some parents in common, but not backcrossed to genetic uniformity (Rajaram and Dubin, 1977). Multilines for disease resistance will lessen disease epidemics (Browning and Frey, 1969). A logical extension of multilines is the blending of cultivars which may be considered a multiline but are more commonly called blends or mixtures. As with multilines, the goal of a blend or mixture is to increase within-cultivar diversity. The yield of a mixture is generally lower than its highest yielding component.

FUTURE TRENDS IN BREEDING SELF-POLLINATED CROPS

If the past is the prelude to the future, what will the future hold for breeding self-pollinated crops?

Although not discussed here, most of the benefits of mechanization which allowed breeding programs to reduce labor and increase nursery size have been already achieved. Future improvements will be incremental. Improvements due to improved computerization are continuing with rapid field book generation and complex experimental designs, easier data collection and analysis, and development of long-term databases. New statistical approaches coupled with increased computing power allow more access to these analytical tools by plant breeders. For example, the previous advances with replicated and incomplete block designs have led to statistical techniques that further identify field trends, such as nearest neighbor designs. Nearest neighbor analyses can identify and correct for field trends within experimental blocks (Stroup and Mulitze, 1991). A second example is selection indices (Baker, 1986) that allow multiple traits and multiple year data to be more effectively incorporated into line selection. Critical to the effective selection indices are the determination of the appropriate weighting factors for each trait within the selection index. Optimization models (Johnson *et al.*, 1988a,b, 1990) will use a breeder's ranking of lines to determine objectively the weighting factors for the traits of interest. The weighting factor can be used directly in the selection index, or indirectly to provide insight on how the breeder selects lines by indicating which traits are important (Johnson *et al.*, 1990). Most breeders have an image of what constitutes a good line, but often cannot describe it or provide weighting factors.

Information and useful germplasm remain the driving forces in crop improvement. Hence, one should not underestimate the impact of increased information and germplasm exchanges among breeders due to the improved spirit of international cooperation and ease of communication and travel. However, many of the past germplasm exchanges involved international centers and international nurseries coordinated by national programs. Their efforts are decreasing, as is the commitment to systematic germplasm development and exchange programs.

Transformation, with its ability to access the genetic resources of the biosphere, when coupled with gene identification and regulation, has the potential to greatly increase genetic variation. The complex technologies necessary for the successful use of genetic engineering are formidable. However, great progress has been made in the past and should continue in the future. An area that will need considerable thought for genetic engineering is when is the best time to intercept breeding lines to best couple conventional plant breeding with transformation. Transformation, being similar to backcrossing in its ability to add one or a few genes, will not be useful if genes are added to lines that rapidly become obsolete. Also, genetic engineering will be most useful for transferring genes from the tertiary or more distant gene pools. Once the gene is in the primary gene pool, hybridization and conventional breeding methods (e.g., backcrossing) will be used. It should be remembered that transformation is a technique of accessing previously unavailable germplasm and not a breeding method.

In addition to enhanced computing and statistical techniques, analytical techniques will continue to be improved. The scope of most breeding programs is too large to use biochemical assays in early generations, but for important traits and populations, the techniques may provide useful germplasm and more importantly a better genetic understanding. The improved genetic understanding will be used routinely across populations, whereas the cost of the technique may limit its actual use to very few, well characterized populations.

Another promising area is the increased control over inbreeding, recombination, and linkage. The increased control is coming simultaneously from a number of different research areas. Male sterility and gametocides provide obligate intermating, thus decreasing inbreeding and increasing recombination. Doubled haploidy provides a rapid method for inbreeding with a greatly reduced recombination potential. Hence, self-pollinated crop breeders will be able to choose the level of inbreeding and recombination that is most appropriate for their research.

With the advent of molecular markers such as isozymes and restriction fragment polymorphisms (Soller and Beckman, 1983; Tanksley, 1988; Lander and Botstein, 1989), saturated genetic maps will be available to help breeders identify genes of interest and determine the linkage relationships so that they can rationally decide the level of recombination that is necessary. Remembering the scope of plant breeding programs, molecular markers will be used only in genetic studies, the most elite crosses, and for important traits where direct measurement is very difficult. It should be understood that the value of the marker is dependent upon the choice of populations to analyze and the quality of the empirical data. This is particularly true for linkage studies that identify quantitative traits which should be done in agronomic populations.

As intergeneric and interspecific hybridization barriers are broken, chromosomal chromosomal recombination becomes critical. Repetitive DNA probes allow the translocated segment to be identified and measured. Recombination is greatly reduced in alien chromosome transfers; hence, the physical length of the segment often provides an indication of the number of genes (usually deleterious) that may be linked to the trait of interest on the translocated segment. If the segment is part of a reciprocal translocation, identifying the position of the translocation may indicate what genes have been lost from the recipient plant.

Finally, genetic engineering with its awesome potential to expand the genetic resources of crop plants and tempered with the our limitations of genetic knowledge should not be overlooked for its impact on linkage. Transferred DNA includes many genes (e.g., a marker gene, the regulatory sequence, the gene of interest, etc.). By constructing DNA sequences, extremely tight linkages can be made. If one considers the past efforts to develop systems for hybrid crops that involved complex linkages between male sterile genes and a selectable gene, or between male fertile genes and poorly transmitted chromosomes (Ramage, 1965), most of these procedures failed for two reasons. First, linkage could be constructed only with a limited number of progeny (even in the largest breeding programs), which would not be sufficiently tight to withstand the potential for recombination in production fields that are orders of magnitude larger than a breeding program. Second, linkage with male fertile genes failed because the cytogenetics was complex, slow, and not amenable for routine use. Collecting and tightly linking beneficial genes so that they would segregate as a unit would greatly facilitate plant breeding. The possibilities for tight linkage will provide a rethinking of genetics in plant breeding. These and the above possibilities highlight the need for integrated research teams to use effectively the expanding technologies and resources of plant breeding and genetics.

REFERENCES

Adegoke, A. O.; Frey, K. J. (1987) Grain yield and stability for oat lines with low-, medium-, and high-yielding ability. *Euphytica* 36:121-127.

Allard, R. W. (1956) Estimation of prepotency from lima bean diallel cross data. *Agronomy Journal* 48:537-543.

Allard, R. W. (1961) Relationship between genetic diversity and consistency of performance in different environments. *Crop Science* 1:127-133.

Allard, R. W. (1990) The genetics of host-parasite coevolution: Implications for genetic resources conservation. *Journal of Heredity* 81:1-6.

Allard, R. W.; Bradshaw, A. D. (1964) Implications of genotype-environmental interactions in applied plant breeding. *Crop Science* 4:503-508.

Altman, D. W.; Busch, R. H. (1984) Random intermating before selection in spring wheat. *Crop Science* 24:1085-1089.

Atkins, R. E.; Murphy, H. C. (1949) Evaluation of yield potentialities of oat crosses from bulk hybrid tests. *Agronomy Journal* 41:41-45.

Avery, D. P.; Ohm, H. W.; Patterson, F. L.; Nyquist, W. E. (1982) Three cycles of simple recurrent selection for early heading in winter wheat. *Crop Science* 22:908-912.

Baenziger, P. S.; Kudirka, D. T.; Schaeffer, G. W.; Lazar, M. D. (1984) The significance of doubled haploid variation. In: Gustafson, J. P. (ed.), *Gene manipulation in plant improvement I.* Plenum Press, New York, pp. 385-414.

Bailey, T. B., Jr.; Comstock, R. E. (1976) Linkage and the synthesis of better genotypes in self-fertilizing species. *Crop Science* 16:363-370.

Baker, R. J. (1984) Quantitative genetic principles in plant breeding. In: Gustafson, J. P. (ed.), *Gene manipulation in plant improvement I.* Plenum Press, New York, pp. 147-176.

Baker, R. J. (1986) *Selection indices in plant breeding.* CRC Press, Boca Raton, FL.

Bhatt, G. M. (1970) Multivariate analysis approach to selection of parents for hybridization aiming at yield improvement in self-pollinated crops. *Australian Journal of Agricultural Research* 21:1-7.

Bhatt, G. M. (1973) Comparison of various methods of selecting parents for hybridization in common bread wheat (*Triticum aestivum* L.). *Australian Journal of Agricultural Research* 24:457-464.

Borlaug, N. E. (1953) New approach to the breeding of wheat varieties resistant to *Puccinia graminis tritici. Phytopathology* 43:467.

Bos, I. (1977) More arguments against intermating F_2 plants of a self-fertilizing crop. *Euphytica* 26:33-46.

Briggs, F. N.; Allard, R. W. (1953) The current status of the backcross method of plant breeding. *Agronomy Journal* 45:131-138.

Brim, C. A. (1966) A modified pedigree method of selection in soybeans. *Crop Science* 6:220.

Brim, C. A.; Stuber, C. W. (1973) Application of genetic male sterility to recurrent selection schemes in soybeans. *Crop Science* 13:528-530.

Brown, K. D.; Sorrells, M. E.; Coffman, W. R. (1983) A method of classification and evaluation of testing environments. *Crop Science* 23:889-893.

Browning, J. A.; Frey, K. J. (1969) Multiline cultivars as a means of disease control. *Annual Reviews of Phytopathology* 7:355-382.

Busch, R. H.; Hanke, J. C.; Frohberg, R. C. (1974) Evaluation of crosses among high and low yielding parents of spring wheat (*Triticum aestivum* L.) and bulk prediction of line performance. *Crop Science* 14:47-50.

Busch, R. H.; Kofoid, K. (1982) Recurrent selection for kernel weight in spring wheat. *Crop Science* 22:568-572.

Busch, R. H.; Stuthman, D. D. (1990) Self-pollinated crop breeding: Concepts and successes. In: Gustafson, J. P. (ed.), *Gene manipulation in plant improvement II.* Plenum Press, New York, pp. 21-37.

Choo, T. M.; Christie, B. R.; Reinbergs, E. (1979) Doubled haploids for estimating genetic variances and a scheme for population improvement in self-pollinated crops. *Theoretical and Applied Genetics* 54:267-271.

Choo, T. M.; Reinbergs, E.; Kasha, K. J. (1985) Use of haploids in breeding barley. *Plant Breeding Reviews* 3:219-252.

Cox, D. J.; Frey, K. J. (1984) Combining ability and the selection of parents for interspecific oat matings. *Crop Science* 24:963-967.

Cox, T. S.; Murphy, J. P. (1990) The effect of parental divergence on F_2 heterosis in winter wheat crosses. *Theoretical and Applied Genetics* 79:241-250.

Cregan, P. B.; Busch, R. H. (1977) Early generation bulk hybrid yield testing of adapted hard red spring wheat crosses. *Crop Science* 17:887-891.

Delannay, X. D.; Rodgers, D. M.; Palmer, R. G. (1983) Relative genetic contributions among ancestral lines to North American soybean cultivars. *Crop Science* 23:944-949.

Dhaliwal, A. S.; MacRitchie, F. (1990) Contributions of protein fractions to dough handling properties of wheat-rye translocation cultivars. *Journal of Cereal Science* 12:113-122.

Early, H. L.; Qualset, C. O. (1971) Complementary competition in cultivated barley (*Hordeum vulgare* L.). *Euphytica* 20:400-409.

Eberhart, S. A.; Russell, W. A. (1966) Stability parameters for comparing varieties. *Crop Science* 6:36-40.

Fehr, W. R. (1987) *Principles of cultivar improvement. Vol. 1. Theory and technique.* Macmillan Publishing Company, New York.

Feldman, M.; Sears, E. R. (1981) The wild gene resources of wheat. *Scientific American* 244:102-112.

Florell, V. H. (1929) Bulked population method of handling cereal hybrids. *Journal of the American Society of Agronomy* 21:718-724.

Foroughi-Wehr, B.; Wenzel, G. (1990) Recurrent selection alternating with haploid steps - A rapid breeding procedure for combining agronomic traits in inbreeders. *Theoretical and Applied Genetics* 80:564-568.

Fowler, W. L.; Heyne, E. G. (1955) Evaluation of bulk hybrid tests for predicting performance of pure line selections in hard red winter wheat. *Agronomy Journal* 47:430-434.

Fraley, R. T.; Rogers, S. G.; Horsch, R. B.; Kishore, G. M.; Beachey, R. N.; Turner, N. N.; Fischhoff, D. A.; Dellanay, X.; Klee, H. J.; Shah, D. M. (1988) Genetic engineering for crop improvement. In: Gustafson, J. P.; Appels, R. (eds.), *Chromosome structure and function: Impact of new concepts.* Plenum Press, New York, pp. 283-298.

Frankel, O. H.; Soule, M. E. (1981) *Conservation and evolution.* Cambridge University Press, Cambridge, United Kingdom.

Gasser, C. S.; Fraley, R. T. (1989) Genetically engineered plants for crop improvement. *Science (Washington, D. C.)* 244:1293-1299.

Goodman, M. M. (1990) Genetic and germ plasm stocks worth conserving. *Journal of Heredity* 81:11-16.

Goulden, C. H. (1939) Problems is plant selection. In: Burnett, R. C. (ed.), *Proceedings of the 7th International Genetic Congress (Edinburgh).* Cambridge University Press, Cambridge, United Kingdom, pp. 132-133.

Grafius, J. E. (1964) A geometry for plant breeding. *Crop Science* 4:241-246.

Grafius, J. E. (1965a) A geometry for plant breeding. *Research Bulletin, Michigan Agricultural Experiment Station No. 7.*

Grafius, J. E. (1965b) Short cuts in plant breeding. *Crop Science* 5:377.

Graybosch, R. A.; Peterson, C. J.; Hansen, L. E.; Mattern, P. J. (1990) Relationships between protein solubility characteristics, 1BL/1RS, high molecular weight glutenin composition, and end-use quality in winter wheat germplasm. *Cereal Chemistry* 67:342-349.

Griffing, B. (1975) Efficiency changes due to use of doubled haploids in recurrent selection methods. *Theoretical and Applied Genetics* 46:367-386.

Gustafson, A. (1953) The cooperation of genotypes in barley. *Hereditas* 39:1-18.

Harlan, H. V.; Pope, M. N. (1922) The use and value of backcrosses in small grain breeding. *Journal of Heredity* 13:319-322.

Harlan, J. R. (1975) Our vanishing genetic resources. *Science (Washington, D. C.)* 188:618-621.

Harlan, J. R.; deWet, J. M. J. (1971) Toward a rational classification of cultivated plants. *Taxon* 20:509-517.

Hawkes, J. G. (1981) Germplasm collection, preservation, and use. In: Frey, K. J. (ed.), *Plant breeding II*, Iowa State University Press, Ames, pp. 57-83.

Hayman, B. I. (1954) The theory and analysis of diallel crosses. *Genetics* 39:789-809.

Jensen, N. F. (1952) Intravarietal diversification in oat breeding. *Agronomy Journal* 44:30-34.

Jensen, N. F. (1988) *Plant breeding methodology*. John Wiley & Sons, New York.

Jinks, J. L. (1956) The F_2 and backcross generations from a set of diallel crosses. *Heredity* 10:1-20.

Johnson, B.; Eskridge, K.; Liu, Y.-H. (1990) Use of optimization models for multiple trait selection. *Proceedings 44th Annual Corn and Sorghum Research Conference*, Chicago, IL, pp. 1-13.

Johnson, B. E.; Dauer, J. P.; Gardner, C. O. (1988a) A model for determining weight of traits in simultaneous multitrait selection. *Applied Mathematic Modeling* 12:556-564.

Johnson, B. E.; Gardner, C. O.; Wrede, K. C. (1988b) Application of an optimization model to multi-trait selection programs. *Crop Science* 28:723-728.

Kasha, K. J.; Ziauddin, A.; Cho, U. H. (1990) Haploids in cereal improvement: anther and microspore culture. In: Gustafson, J. P. (ed.), *Gene manipulation in plant improvement II*. Plenum Press, New York. pp. 213-235.

Khalaf, A. G. M.; Brossman, G. D.; Wilcox, J. R. (1984) Use of diverse populations in soybean breeding. *Crop Science* 24:358-360.

Khalifa, M. A.; Qualset, C. O. (1974) Intergenotypic competition between tall and dwarf wheats. I. In mechanical mixture. *Crop Science* 14:795-799.

Khalifa, M. A.; Qualset, C. O. (1975) Intergenotypic competition between tall and dwarf wheats. II. In hybrid bulks. *Crop Science* 15:640-644.

Kimber, G. (1984) Technique selection for the introduction of alien variation in wheat. *Zeitschrift für Pflanzenzuechtung* 92:15-21.

Lander, E. S.; Botstein, D. (1989) Mapping Mendelian factors underlying quantitative traits using RFLP linkage maps. *Genetics* 121:185-199.

Laurie, D. A.; O'Donoughue, L. S.; Bennett, M. D. (1990) Wheat x maize and other wide sexual hybrids: their potential for genetic manipulation and crops improvement. Gustafson, J. P. (ed.), *Gene manipulation in plant improvement II*. Plenum Press, New York, pp. 95-126.

Lawrence, P. K.; Frey, K. J. (1975) Backcross variability for grain yield in oat species crosses (*Avena sativa* L. x *A. sterilis* L.). *Euphytica* 24:77-85.

Lay, C. L.; Wells, D. G.; Gardner, W. S. (1971) Immunity from wheat streak mosaic virus in irradiated Agrotriticum progenies. *Crop Science* 11:431-432.

Leppik, E. E. (1970) Gene centers of plants as sources of disease resistance. *Annual Review of Phytopathology* 8:323-344.

Lupton, F. G. H. (1965) Studies in the breeding of self pollinating cereals. 5. Use of the incomplete diallel in wheat breeding. *Euphytica* 14:331-352.

Maluszynski, M. (1990) Induced mutations - An integrating tool in genetics and plant breeding. In: Gustafson, J. P. (ed.), *Gene manipulation in plant improvement II*. Plenum Press, New York, pp. 127-162.

Martin, T. J. (1990) Outcrossing in twelve hard red winter wheat cultivars. *Crop Science* 30:59-62.

McProud, W. L. (1979) Repetitive cycling and simple recurrent selection in traditional barley breeding programs. *Euphytica* 28:473-480.

Moonen, J. H. E.; Scheepstra, A.; Graveland, A. (1982) Use of the SDS-sedimentation test and SDS-polyacrylamide gel electrophoresis for screening breeder's samples of wheat for bread-making quality. *Euphytica* 31:677-690.

Morris, M. R. (1983) Remodeling crop chromosomes. In: Wood, D. R.; Rawal, K. M.; Wood, M. N. (eds.), *Crop breeding*. American Society of Agronomy, Madison, WI, pp. 109-129.

Murphy, J. P.; Cox, T. S.; Rodgers, D. M. (1986) Cluster analysis of red winter wheat cultivars based upon coefficients of parentage. *Crop Science* 26:672-676.

National Academy of Science. (1972) *Genetic vulnerability of major crops.* National Academy of Science, Washington, DC.

Payne, P. I.; Nightengale, M. A.; Krattiger, A. F.; Holt, L. M. (1987) The relationship between HMW glutenin subunit composition and the breadmaking quality of British-grown wheat varieties. *Journal of the Science of Food and Agriculture* 40:51-65.

Payne, T. S.; Stuthman, D. D.; McGraw, R. L.; Bregitzer, P. P. (1986) Physiological response to three cycles of recurrent selection for grain yield improvement in oats. *Crop Science* 26:734-736.

Pederson, D. G. (1981) A least-squares method for choosing the best relative proportions when intercrossing cultivars. *Euphytica* 30:153-160.

Peeters, J. P. (1988) The emergence of new centres of diversity: Evidence from barley. *Theoretical and Applied Genetics* 76:17-24.

Peterson, C. J.; Johnson, V. A.; Schmidt, J. W.; Mumm, R. F. (1989) Genetic improvement and the variability in wheat yields in the Great Plains. In: Anderson, J. R.; Hazell, P. B. R. (eds.), *Variability in grain yields.* International Food Policy Research Institute, Johns Hopkins University Press, Baltimore, MD, pp. 175-184.

Peterson, C. J.; Pfeiffer, W. H. (1989) International winter wheat evaluation: Relationships among test sites based on cultivar performance. *Crop Science* 29:276-282.

Phung, T. K.; Rathjen, A. J. (1976) Frequency-dependent advantage in wheat. *Theoretical and Applied Genetics* 48:289-297.

Piper, T. E.; Fehr, W. R. (1987) Yield improvement in a soybean population by utilizing alternative strategies of recurrent selection. *Crop Science* 27:172-178.

Powell, W.; Thomas, W. T. B.; Caligari, P. D. S.; Jinks, J. L. (1985) the effects of major genes on quantitatively varying characters in barley. 1. The *GP ert* locus. *Heredity* 54:343-348.

Qualset, C. O. (1979) Mendelian genetics of quantitative characters with reference to adaptation and breeding of wheat. In: Ramanajam, S. (ed.), *Proceedings 5th International Wheat Genetics Symposium, Indian Society of Genetics and Plant Breeding,* Vol. 2, New Delhi, India, pp. 577-590.

Rajaram, S.; Dubin, J. H. (1977) Avoiding genetic vulnerability in semidwarf wheat. *Annals New York Academy of Science* 287:243-254.

Ramage, R. T. (1965) Balanced tertiary trisomics for use in hybrid seed production. *Crop Science* 5:177-178.

Reinbergs, E.; Park, S. J.; Song, L. S. P. (1976) Early identification of superior barley crosses by the doubled haploid technique. *Zeitschrift fur Pflanzenzuechtung* 76:215-224.

Robertson, N. L.; French, R.; Gray, S. M. (1991) Use of group-specific primers and the polymerase chain reaction for the detection and identification of luteoviruses. *Journal of General Virology* 72:1473-1477.

Schoener, C. S.; Fehr, W. R. (1979) Utilization of plant introductions in soybean breeding populations. *Crop Science* 19:185-188.

Scowcroft, W. R. (1978) Aspects of plant cell culture and their role in plant improvement. *Proceedings of a Symposium on Plant Tissue Culture,* May 25-30. Science Press, Peking, China.

Shands, H. L. (1990) Plant genetic resources conservation: The role of the gene bank in delivering useful genetic materials to the research scientist. *Journal of Heredity* 81:7-10.

Sharma, H. C.; Gill, B. S. (1983) Current status of wide hybridization in wheat. *Euphytica* 32:17-31.

Smith, E. L.; Lambert, J. W. (1968) Evaluation of early generation testing in spring barley. *Crop Science* 8:490-493.

Sneep, J. (1977) Selection for yield in early generations of self-fertilizing crops. *Euphytica* 26:27-30.

Soller, M.; Beckman, J. S. (1983) Genetic polymorphism in varietal identification and genetic improvement. *Theoretical and Applied Genetics* 67:25-33.

Sorrells, M. E.; Fritz, S. E. (1982) Application of a dominant male sterile allele to the improvement of self-pollinated crops. *Crop Science* 22:1033-1035.

Stalker, H. T. (1980) Utilization of wild species for crop improvement. *Advances in Agronomy* 33:111-147.

St. Martin, S. K. (1982) Effective population size for the soybean improvement program in maturity groups 00 to IV. *Crop Science* 22:151-152.

Stroup, W. W.; Mulitze, D. K. (1991) Nearest neighbor adjusted best linear unbiased prediction. *American Statistician*:(in press).

Sumarno; Fehr, W. R. (1982) Response to recurrent selection for yield in soybeans. *Crop Science* 22:295-299.

Suneson, C. A.; Riddle, O. C. (1944) Hybrid vigor in barley. *Journal of the American Society of Agronomy* 36:57-61.

Tanksley, S. D. (1988) Molecular mapping of plant chromosomes. In: Gustafson, J. P.; Appels, R. (eds.), *Chromosome structure and function*. Plenum Press, New York, pp. 157-173.

Tanksley, S. D.; Rick, C. M.; Vallejos, C. E. (1984) Tight linkage between a nuclear male-sterile locus and an enzyme marker in tomato. *Theoretical and Applied Genetics* 68:109-113.

Virmani, S. S.; Aquino, R. C.; Khush, G. S. (1982) Heterosis breeding in rice (*Oryza sativa* L.). *Theoretical and Applied Genetics* 63:373-380.

Weiss, M. G.; Weber, C. R.; Kalton, R. R. (1947) Early generation testing in soybeans. *Journal of the American Society of Agronomy* 39:791-811.

Yonezawa, K.; Yamagata, H. (1978) On the number and size of cross combinations in a breeding program of self-fertilizing crops. *Euphytica* 27:113-116.

Discussion

Thomas G. Isleib, Moderator

How can seed companies handle their responsibility of training students for plant breeding? Why not have the universities continue that function?

In my talk this morning, I was very explicit in differentiating between education and training. Our programs at universities have been very responsible in both over the past 50-60 years. There are a few people educated as geneticists who then train as plant breeders after going to work for companies. Bill Brown of Pioneer Hi-Bred International was an example. For reasons I told you this morning, education will continue to be in the university sector. There is no substitute for that even though there are some universities where graduate education and training in plant breeding are now just education. I am rather optimistic that commercial companies can provide this sort of training if they have the proper objectives and proper people to do the teaching. If you need plant breeders and you have erudite people with the proper education, then that approach is fine. My primary concern is that training in the universities is eroding to the point where many people who are being called plant breeders do not have the skills at the time they receive PhD degrees to carry out the activities that plant breeders must perform in the field and greenhouse. The ideal would be to continue both education and training at the universities, but this is eroding.   [K. J. Frey]

How should germplasm conservationists be trained in the future?

There should be some periodic training programs conducted by a consortium of universities, research organizations (like national seed storage laboratories) or even international research centers. We should pool our forces to train germplasm students rather than ask any one institution to carry it on regularly. That's what I think we'll approach in the future.   [T. T. Chang]

Apparently, the international research centers are under pressure from funding agencies to become more biotech-oriented. Is this move political? That is, are the directors of these centers sacrificing efficiency and effectiveness to achieve the public perception of being high tech? Is it scientific condescension to think that the international centers might be better off to go slow on this, especially if the move comes at the expense of traditional plant breeding?

I don't look with suspicion on the biotechnology methods being adopted and used at the IARCs. However, the IARCs should not take the lead in conducting basic research on the molecular biology of plants. They should look for that work to be performed at institutions that are very well equipped in personnel, techniques, and objectives. But the adoption of new methods into the international institute has to occur. They have to take advantage of some techniques as we do in developed countries. There's no dichotomy here whatsoever in the use of biotechnological methods. As soon as biotechniques are proven, can be handled, and are advantageous, then certainly they should be used. They should add to the efficiency and precision with which plant breeding can be done whatever the setting. So basically my division occurs between basic science for developing these techniques or using the techniques that have been developed. I fully recognize that nobody may be doing basic or molecular biology on, for example, cassava. A connection between an IARC and a university that has the capability for doing molecular biology research is the kind of linkage that can do basic biology for developing biotechniques that are commodity-specific.   [K. J. Frey]

I agree with Ken Frey's interpretation that the IARCs should not be deeply involved in biotech research. It's the advanced labs in the developed countries that should do the basic work. We only serve as bridges so that we can pass on the usable technologies to developing countries. Of course, we have an advantage in that we have the rich germplasm to begin the transfer. A cooperative venture is preferable to the IARCs becoming deeply involved in basic research. In fact, our current capabilities are very limited.

[T. T. Chang]

Will the plant breeding community be able to maintain the 4000-calorie per person per day level of production as we approach the coming century? As global population continues to increase to 11 billion by the year 2040, can we project gains of the last 50 years of plant breeding into the future?

I will restrict my response to the case of corn. Since the open-pollinated varieties were replaced by hybrids in the 1930s, different studies have shown about a bushel and a half per year gain that seems to be continuing with no leveling off. For the immediate future I don't think that rate of gain will decrease. With the tremendous expansion in the commercial industry in the last 20 years, there should be superior material coming out of the pipeline now. It should be emerging for at least the next couple of decades, so it should maintain the current rate of gain. A couple of things may offset the genetic gains. Production inputs and cultural practices are changing. In developed countries, governments are advocating reductions in the use of nitrogen fertilizers and herbicides, changes in tillage practices because of soil and wind erosion, etc. In spite of these modifying effects, in corn it looks optimistic to me.   [A. R. Hallauer]

I am not going to be the pessimist on the panel. I feel that we have tremendous opportunities to increase the biological yield of the small grains, particularly in view of the poor job we do of obtaining their current yield potential. We probably have the genetics, when properly managed today, to withstand a large population increase. My biggest concerns are the availability of inputs and the profitability of agriculture. In western Nebraska, 30% of the wheat is not even fertilized. This means that we can make trivial investments to achieve significant advantages. If you look at the population trend and what can be done, one would hope that demand would force the economics to be there.   [P. S. Baenziger]

I share the same optimism. We must remember, however, that increased production is not achieved by genetics alone. It's done with very careful study of the environment and the genetics. We should never look at an increase and say "We did this!" We are instrumental in an agricultural team, and not just the team with whom we cooperate directly. It involves a lot of others including agricultural economists and sociologists. Our tools should include biotechnology as well as engineering and electronics which permit us to work much more precisely in agriculture production. Did you ever think about the potential improvements in efficiency of resource use that could occur if you could have placed in your drill a sensor that would regulate the fertilization rate? We need to increase the precision of agricultural production, not just the precision of plant breeding. We must look a lot farther than just biotechnology as plant breeders to optimize production. On the other hand, I certainly hope that we learn as a human race to control our population growth before we ever get to 11 billion. If we don't, we are in a holding action only. I'm one who believes that pollution will level our population unless we find someway to cut down on population growth.

[K. J. Frey]

What kind of computer program should be used in germplasm management?

This depends on need. The available technology is sufficient for our objectives. My advice is that the biologist should be the boss. You tell the programmer what you need; don't let the programmer decide it for you. I learned this from many gene banks - the architect and the engineers tend to build a white elephant that wouldn't work in our germplasm program; so we have to be the boss. Tell the programmer what should be done, but educate the programmer about the biological needs. Again, this is a matter of working together. On another front, a uniform system of descriptors and descriptor states within a crop or a national PGR system is essential to effective communication.   [T. T. Chang]

The long-term selection experiments for protein and oil in maize are interpreted as the result of concentrating alleles present in the original population. Could we not consider that new variability is continually being produced by mechanisms not yet well understood? Could you review the work of molecular geneticists who are studying the high protein and oil selection program?

I don't know of any molecular geneticists studying the high protein and oil program. It is interesting that we can have persistent genetic variability in such long-term programs. One possible explanation is moving genes or transposable elements. I can relate what we have found in Iowa Stiff Stalk Synthetic which is undergoing intra- and interpopulation selection. The transposable element *Uq* was in the original Stiff Stalk population at a frequency of 19%. Under intrapopulation selection, after 13 cycles, the frequency of that transposable element was 91%. It may be maintaining some genetic variability within the population - whether good or bad. In a companion program, reciprocal recurrent selection using the corn borer synthetic as a tester for the Stiff Stalk, *Uq* was extinct by cycle 5. We concluded on that basis that it was genetic drift or just a random change in frequency of this element within Stiff Stalk. In one case it went from 19 to 91%. In the other it went from 19 to extinction in five cycles of selection. This was recurrent selection based on yield. One would think that oil and protein content would not be quite as complex genetically as yield and, therefore, wonder whether such elements could be operative in these Illinois programs.   [A. R. Hallauer]

Is the improvement of experimental lines relative to "Kharkov" wheat in productive environments - that is, larger beta values, partly a result of improved disease resistance in the more humid eastern areas of the Great Plains?

Surely disease resistance has an impact on the responsiveness of lines to more fertile environments. However, many of our highest yielding types are extremely disease-susceptible. A trait that has changed more than disease resistance is maturity. Kharkov is extremely late and gets caught by drought stress late in the growing season. But Kharkov is no less and no more susceptible than "TAM 107" which is grown widely in the Great Plains. I think it's the straw strength and the maturity that affects Kharkov more than disease susceptibility, because we have many years where the diseases do not have an impact.   [P. S. Baenziger]

How can we change the climate in the field of plant breeding to attract more women and, more importantly, to keep women in the field? Could the alleged reduction of female interest influx and retention in plant breeding possibly be due to the established infrastructure's failure to adapt swiftly enough to meet their unique needs, for example, parenting vs. career responsibilities?

I simply made a statement from experience. If I were to poll 50 or 60 people who have been involved in graduate training in plant breeding, their experiences would mirror what I said this morning. I don't know the reason for the seeming lack of interest among today's women in plant breeding, especially in view of their tremendous interest in the mid-1970s. At one time I had six women as students, very bright people. They stimulated a lot of activity in the project and their productivity was amazing. Several of those women are still in plant breeding today. One or two have left the field, but of course, one expects attrition. Several factors make it difficult to retain women in the profession alluded to in this question. There are demands in terms of the family responsibilities. I have no solution, but I think we must recognize that it is a factor. Perhaps we haven't created the proper infrastructure to meet these demands. I have noticed that there are large numbers of female graduate students and technicians in biotechnology laboratories. Perhaps the reduction of women's involvement in plant sciences conducted in laboratories has not been as great as it has been where you expect those females to put on rubber boots and plod around the fields. That's not an accusation; that's just an observation.   [K. J. Frey]

I think we aren't proactive enough. When I worked at Monsanto, half of the employees were women. I went to a university and Rosslyn Morris, a member of the plant breeding and genetics panel and now retired, was the only woman on our faculty. We now have one woman on our Soils faculty. We simply are not proactive. If biotechnology is attracting women, then our field has failed to convey that most of us have great fun doing what we do. At Monsanto we could certainly attract women at all levels from research leaders to technicians. At the university, we haven't done nearly as well. It is a shared responsibility. We have missed our responsibility somewhere.
[P. S. Baenziger]

While the results for reciprocal recurrent selection population improvements are optimistic, how important is base population development?

It's the core of success in breeding. Our choice of germplasm determines our future options and how much progress we make. I tell my students not to spend a week pushing numbers to determine (1) which is the most efficient method; (2) which one gives greater genetic gain; (3) which one maintains the greatest amount of genetic variability; and (4) what all the predicted gains are - and then run down to the seed room, reach inside, and, because it's chilly, get the first two bags on the shelf and run out to start experiments or research which may take 30 years to complete. Unfortunately, we have had the poorest information on that aspect of reciprocal recurrent selection. There has been a lot of pencil pushing and estimates made, but we've had the minimum of information on how to choose (1) which varieties, cultivars, strains or races have the greatest promise, (2) what's the heterosis between them, (3) what's the inbreeding depression within those populations, or (4) what's the genetic load in those populations. None of that information has been available. Choice of base germplasm is very critical. As Dr. Chang mentioned, we have a lot of germplasm in collections. It is like pieces in a museum - nice to look at but, unless we work with it and try to upgrade it, it is not of much use. We'll never use it in our breeding programs until it's the equivalent of the elite germplasm we have in our breeding programs at the present time.   [A. R. Hallauer]

According to Dr. Baenziger, empirical evidence indicates that intermating within segregating populations is unnecessary to the success and response to selection. How much of that statement is relevant to a specific disomic polyploid, for example? Is this true for all self-pollinated diploids, and finally is this true for all vegetables where so many quality characters must be combined?

Linkage obviously is more important in a diploid as opposed to a hexaploid, especially when there is gene redundancy in the hexaploid whose progenitor species derive from a common ancestor. In the studies that I have seen, when one has made a cross and allowed it to self, there is a level of recombination that occurs at that time. The incremental recombination added by intermating without selection is not of much value unless repulsion linkages predominate. If you intermate, select, and intermate, then you are using recurrent selection, and the recombination is extremely valuable.

Recurrent selection is extremely powerful. If one is going to combine multiple traits, one would be much better off doing some selection and then intermating those favorable genotypes. The key is the addition of the selection cycle rather than randomly intermating after making the cross.

[P. S. Baenziger]

Are the current measures of ecogenetic diversity sufficient to assay for unknown risks, which is one reason for the maintenance of germplasm collections?

As I said earlier, the collection sizes of the major food crops are very impressive and probably more than we can digest. That doesn't mean that we have all the genes we need. Current evaluation methods certainly need refinement for further research. I fear that the global ecosystem is changing - global warming, a rise in the sea level, and higher pollution levels. We are also pushing crop production into marginal areas. This is a necessity. I don't like some of the consequences, but I think we still have to collect some more or even recollect. We are doing two kinds of rescue jobs. One is to rescue what is not collected and to conserve it. The other is to secure what is already conserved but not properly preserved.   [T. T. Chang]

How much loss in diversity occurs in maintaining germplasm collected from diverse environments at a central location?

If the accessions come from very different environments, we could lose quite a bit of the original diversity during repeated regeneration. We try not to do much selection within an accession; but breeders, pathologists, and entomologists like to have pure lines to begin with. A conservationist would rather have germplasm contained in its original population as a bulk. This is one problem that I often face with my users. They even accuse us of mixing seed. We do not mix seed; we simply try to preserve mixtures as they were collected. We have to understand each other and try to make the best use of unimproved germplasm.   [T. T. Chang]

How do we get our message to the people holding the funds?

The plant breeding profession itself must become more proactive. We must become more involved in defending our profession, against those who'd accuse us of inappropriate breeding objectives or of neglecting some clientele groups. We should also be defending our funding. We know better than anybody else what we do. We know the consequences of plant breeding and the value of it. I don't suggest that every plant breeder spend 6 months a year selling himself and the profession, but, collectively, we need to devote time to this. Maybe we should delegate the major responsibility as a profession. Those people who can articulate well are the ones that should be out there selling, devoting more of their time to promoting plant breeding. It is to the advantage of our administrators, as a matter of fact, if they will promote that sort of activity. I'm not suggesting the award of tenure on that basis but that they screen these people for the few who can carry the message well, and then get them out there selling this profession. We need the backing of a majority of the public in order to get our share of funding. We can sway that public if the profession sells itself, sells what it can do, and sells what the fruits of our work mean to that increasing population. Nobody else will do that for us. I don't buy the idea that a majority of the legislators must come from the farm in order to appreciate plant breeding. We need a program of education to make those people who do not come from the farm and who are clearly in the majority in the developed countries understand where their food comes from.   [K. J. Frey]

Chapter 5
Use of genetic variation for breeding forest trees

W. J. Libby

Department of Forestry & Resource Management, University of California, Berkeley, CA 94720

GENETIC VARIATION IN NATURE

Unlike most agronomic crops, most tree species undergoing domestication are still close to their natural evolutionary origins. Furthermore, populations of many of our economically important species still exist in little-changed native stands, and their genetic architecture reflects a cause-and-effect evolutionary relationship to the environments in which they are found and to the natural history to which they were subjected (Critchfield, 1984). Tree breeders and forest geneticists respond to this knowledge and make use of it in several ways.

Breeders of annual crops often question the sanity of those who choose to breed trees, with generation times of decades and harvest cycles that sometimes are measured in centuries. One important use of information on the amounts and patterns of genetic variation in our native species is to provide hope that very large improvements can be made in one or a few generations, as tree populations are deflected from responses to strictly natural forces and are selected and bred to better suit human purposes. This hope has been fulfilled in a heartening percentage of those tree-improvement programs that have bred and deployed one or two generations of improved trees.

Forest geneticists have initiated studies of the genetic architecture of many tree species, and hundreds of such experiments are in progress. Part of this is curiosity driven, and some of our plant-breeding colleagues envy our opportunity to study and understand the genetic organization of our species at that interesting moment when humans began modifying natural population structure. A combination of historical and experimental knowledge has often provided insights into phenomena such as clinal or ecotypic variation in frost or drought tolerance; similarly, data on patterns of among- and within-population genetic variation have facilitated selection for resistance to pests and diseases. Sometimes the distribution of environmental variation across the species' range made little or no causal sense with respect to the patterns of genetic variation observed in genetic architecture experiments; a look at the species' population biology revealed the possibility of rare migrations and/or episodes of bottlenecking as likely explanations. Repeated population extinctions followed by recolonization with founder effects is a common variation on this theme.

An important insight from these early genetic-architecture studies was that very large amounts and proportions of genetic variation are typically maintained within most local populations of trees with respect to a wide range of both qualitative and quantitative traits (Libby and Critchfield, 1986). When isozyme techniques became available, it was soon apparent that most populations of trees had substantially larger amounts of variability in these enzymes, both with respect to numbers of similar loci and to allelic numbers and frequencies at those loci, than did annual plants, which in turn had greater variability than did most animals (Hamrick and Godt, 1990). An attractive hypothesis contrasts the ways in which animals, annual plants and trees deal with environmental variation and the various environmental insults that visit them. Many animals do so by moving, and most annual plants do so by retreating into a seed. But trees, which must stay in one place for many decades, appear to do so by maintaining high levels of genetic diversity within populations, and by evolving multi-locus gene families that each have similar but not identical function. Since most tree species are outcrossing, both of these strategies serve to maintain high levels of functional genetic diversity within individual trees.

NATURAL REGENERATION SYSTEMS

While the title of this paper focuses on tree breeding, it may be useful first to consider forests that are not planted, but which regenerate naturally after harvest. This is particularly important and timely as the recent weight of public opinion is creating laws and regulations that increasingly favor natural-regeneration systems over plantation forestry. Tree breeders are not generally called upon to contribute to planning or operations for natural-regeneration forestry. Forest geneticists, however, are increasingly being asked for advice on such topics as appropriate between- and within-species diversity and about the traits to be used when determining which trees will be harvested and which will be allowed to remain and reproduce.

Clear-cutting

It is common to equate clear-cutting with bad forestry in the popular press of late. Unfortunately, this is a simplistic viewpoint and it fails to take into account the biology of many forest-tree species. Many of our forest species regenerate best, or only, after some disaster such as windthrow, killing flood, landslide, or, most commonly, conflagration. In short, their germinating seeds benefit from or require bare mineral soil and, subsequently, little or no overtopping vegetation. Careful study of many of our apparently unevenly aged forests reveals that, while their component trees are substantially unevenly sized, they are of similar age. Furthermore, and of importance to the topic of this paper, clear-cutting is genetically near-neutral with respect to human-imposed selection. By this I mean that natural regeneration in a clear-cut comes from a combination of (1) resprouting of established genotypes of some species, (2) seeds in or on the ground that germinate following logging, and (3) seeds that blow in or are brought in by birds, squirrels, etc. from the population of trees surrounding the clear-cut that was unaffected by the logging. For many species, particularly those that regenerate best after disasters, a clear-cut most closely mimics nature.

Partial-cutting systems

Three types of partial-cutting harvest-and-regeneration systems are now in use and increasingly in favor with the general public. These are the so-called "**selection system**" (trees are individually selected and removed, generally one per neighborhood, with such harvest occurring every few years for an indefinite period of time), the "**shelterwood system**" (most trees are removed in the initial harvest cut, but enough trees are kept for 3 to 10 more years to shelter the regenerating forest, after which the previous-generation sheltering trees are removed), and the "**seed tree system**" (all but a very few trees per hectare are removed at the initial harvest, and those few are left for 3 to 10 additional years to produce one or more crops of seeds; these previous-generation seed-trees are then removed when a seedling stand is established). Only the selection system produces an uneven-aged forest, whereas the latter two systems produce even-aged stands.

All three partial-cutting systems provide opportunities for human-imposed selection. This genetic leverage is potentially greatest for the seed-tree system and least for the selection system, with effectiveness in the shelterwood system depending on the degree to which the sheltering trees contribute to the regeneration already present at the time of the harvest cut. Selection can be either positive or negative for all traits considered, or it can be positive in varying degrees for a subset of the traits and negative in varying degrees for the remaining traits. As an example, a "**risk rating system**" is basically a financial system that reduces mortality and thus loss due to insects and disease. To the extent that the risk-rating system preferentially removes trees that are infested, it is probably a eugenic selection. But, if payrolls are to be met, or production efficiencies achieved, the cut is often "sweetened" by including large well-formed valuable trees in that harvest. If the stand was originally even-aged, this removal of the faster-growing trees and leaving of the slower-growing and poorly-formed trees on site to reproduce is a dysgenic selection. Even in an uneven-aged stand, this so-called "highgrading", if done consistently, will be dysgenic.

Good examples can be found in various parts of the world of consistently applied eugenic selection under partial-cutting rules. However, it is far more common to find owners and managers responding to short-term goals such as paying taxes, servicing debt, or other components of the bottom line in the private sector, or meeting production quotas in the public sector, and these goals often lead to dysgenic practices (Libby, unpublished data).

A further complication is the **"asymmetry of response"** phenomenon (Falconer, 1989). In general, traits such as high growth-rate and disease resistance that have evolved under natural selection respond relatively little under positive human selection. But human selection in the opposite direction, leaving trees to reproduce that have either grown more slowly or are less valuable because of disease-induced malformation, is likely to result in a substantially greater response in a negative direction. For traits already under strong natural selection, it is a lot easier to genetically degrade a stand than it is to improve it.

This brings us to **genetic conservation**, which is desirable under any circumstances. Under the twin thrusts of aggressive breeding and deployment of relatively few clones or families to plantations, in addition to policy pressures to move more to partial-cutting systems (with dysgenic selection being common), the need for genetic conservation is increasingly compelling. How such genetic conservation will be responsibly done, and how its stable funding can be assured over time periods of perhaps millenia, are topics only beginning to receive attention.

Response to changing environment

One argument for natural regeneration of harvested forests is the high probability that both the harvested population and its naturally regenerated offspring are adapted to that site. Recent concern about the buildup of greenhouse gases (carbon dioxide, methane, etc. trap heat, producing a "greenhouse-like" warming effect), and concern about the cooling effects of increased upper-atmosphere albido from stack gases and dust from the Indian subcontinent, has led to an awareness that climate may change or fluctuate to substantial degrees over the next several decades. As an example of a similar environmental change, when storing water behind Shasta Dam lowered the average temperature of irrigation water in the upper Sacramento River in California, many upper-valley farmers switched from rice (*Oryza sativa* L.) to other crops, and many lower-valley farmers switched from still other crops to rice. Although this was probably annoying to those involved, it was possible to quickly alter the cropping systems. However, for forests that take 60 to 80 years from establishment to harvest, a change such as one in average temperature or other components of the climate cannot be offset by moving the species or adapted populations within this time period.

Other elements of the forest environment are also being changed by human activity. Pollutants such as sulfuric acid from industrial and other sources are raining onto forest sites, changing pH and mineral nutrition in the forests. Various exotic pests and pathogens have been accidentally introduced into or near forests, and new herbaceous and woody competitors are either being purposefully introduced, or are escaping and naturalizing from urban and amenity plantings into the forest. The forest ecosystem has also been changed by hunting out such predators as wolves and cougars.

Under natural regeneration, there will have to be evolutionary adjustments of the previously-adapted populations to their new environments. Forest geneticists are deeply concerned, and are often asked for advice about what is likely to happen to tree populations in such changed environments. How much tree breeders can or should be involved, by inserting selected families into otherwise naturally regenerated forests, is still a matter of conjecture and debate.

ARTIFICIAL REGENERATION SYSTEMS

Natural regeneration of some or all of the resident tree species following a clear-cut is typically uneven, and it often fails entirely. Thus, for both economic and policy reasons, planting has become common or even mandated practice following clear-cutting in many parts of the world.

Plantation environment

Only in rare instances do the propagules planted on a clear-cut come from the trees that previously occupied and evolved on the site. Thus, the assurance that the new population is adapted to the site becomes questionable to unwarranted the more distant the origin of the plants from the plantation.

Interestingly, some foresters were among the first to recognize the importance of seed source. Furthermore, Patrick Matthew (1831, 1860) clearly stated the principles of natural selection and population adaptation and differentiation in response to contrasting environments, the first citation appearing 11 months before Darwin sailed on the *Beagle* (Dr. J. E. Barker, pers. comm.). Published observations on the effect of seed origin on the performance of planted trees go back to 1760 with Duhamel du Monceau's DES SEMIS ET PLANTATIONS DES ARBRES; other foresters and horticulturists subsequently contributed observations and rules pertaining to both the fact and the patterns of hereditary variation among populations within species of trees (Larsen, 1956). Thus, the between-physiographic-region level of genetic variation was one of the first principles of genetics to be used in forestry, not so much to obtain gain in yield and other traits, but to defend against loss from planting of maladapted seedlings in plantations.

Forest geneticists and tree breeders both still devote a great deal of effort to delineation of seed zones (usually some physiographically similar region stratified by elevational bands) and to proposing seed-transfer rules within and between these zones. Lately, as common-garden and other transplant experiments are providing data, we have begun refining these zone delineations and transfer rules by species, and according to intended use (i.e., for direct deployment to plantations or for entry into a breeding program). It is also now recognized that client areas for a seed-orchard's output do not necessarily have the same configuration as origin areas for inclusion of parents in that seed-orchard (Westfall, 1991).

Even if the origins of the plantation trees are very close to the plantation site, either physically or (perhaps more important) ecologically, it is possible that a plantation environment is still substantially different from the environment in a naturally regenerated forest. For example, plantations usually have relatively uniform spacing between trees, whereas natural regeneration tends to be clumpy. Plantations may be of a single species, or, if not, the species mixtures usually differ in some respects from those that occur as a result of natural regeneration. Finally, particularly during the first few years following planting, competition with herbs and brush will often have been substantially reduced in a plantation, compared to the situation with natural regeneration.

For the above reasons, forest geneticists have designed experiments to estimate the nature and magnitude of genotype x environment interactions. Until recently there have been two contending schools of thought, one advocating the deployment of interactive genotypes to their appropriate sites and the other advocating selecting for general-purpose genotypes that would do reasonably well over a variety of sites. As the possibility of climate change has become recognized, the wisdom of deploying broadly-adapted genotypes has become compelling.

Use of wild seeds

Both similarities and differences exist between the gathering and deployment of seeds for use in plantations in forestry and in agriculture. When agricultural domestication began several millenia ago, principles of evolution and genetics were only vaguely understood, but transportation and communication systems were also pretty ineffective then. Thus, without necessarily recognizing the wisdom of doing so, seeds for early agricultural crops were probably gathered from local and reasonably adapted populations. But, in American forestry, in spite of the knowledge of population differences available in Europe, short-term financial considerations frequently outweighed this early biological knowledge.

For example, for many years, most or even all of the ponderosa pine (*Pinus ponderosa* L.) planted in California came from a marginal site with about a dozen low-growing trees whose cones could be harvested from the tops of pickup trucks. Many of these plantations failed, and we are still dealing with those that survived, the planted trees often growing much more poorly than the occasional native wildling that has invaded the plantation.

A perhaps crazier ponderosa pine story comes from New Zealand, where, for a while, in the interests of "buying Empire", they imported their seeds from native ponderosa pine populations in Canada. This was in spite of the fact that parts of the native ponderosa pine range in California are physiographically much more similar to New Zealand than are any in Canada. The awful contrast of these Empire-loyal but maladapted plantations to vigorous research plantings from California was soon apparent. Politics was then replaced with evolutionary biology as the primary decision-maker in the seed-procurement section of the New Zealand Forest Service.

Finally, in a story with importance for North Carolina and other southeastern states, in 1955 the Texas Forest Service opted for cheaper east-coast-source loblolly pine (*Pinus taeda* L.) seeds over the more expensive seeds from local Texas loblolly pine populations. Bruce Zobel, then geneticist for the Texas Forest Service, vigorously objected, arguing the long-term values of adapted seeds versus the short-term cost of those seeds. In 1955, short-term costs still won arguments.

Even in California, which lags far behind the Southeast in its commitment to forests as a renewable resource, things are much better today and wild seeds are carefully tracked from origin to nursery to plantation. For most of our forest species, the wild seeds are picked from above-average trees located near roads, so that these families can be recollected in years with good seed crops. Furthermore, attention is given to the numbers and locations of donor trees for each client area, with goals of maintaining substantial genetic variation in our production plantations, with low levels of inbreeding in case they revert to natural regeneration following harvest (Kitzmiller, 1976).

Classical tree-improvement

During the 1940s, the "Swedish school" was established. This involves selection of trees based on their phenotype in the wild or in a plantation, progeny-testing of these potential parents, and their concurrent clonal establishment in a seed-orchard, to be rogued (selectively thinned) as progeny-test results become available and reliable.

Tree improvement began in earnest in the United States when the Committee on Forest Tree Improvement was established in the southeastern U.S. in 1950, and with the establishment of the Texas Forest Service Industrial Tree Improvement Cooperative in 1951. The Texas program was in part a response to the interest in tree breeding that had been generated by contacts with Ake Gustafsson from Sweden. It was financed through Texas A&M University by contributions from forestry companies. Shortly after the 1955 Texas Forest Service decision on cheaper loblolly pine seed, when the industries in the southeast banded together to form a cooperative similar to the pioneering Texas model, Zobel moved from Texas to North Carolina. Partly because of the faster growth and shorter generation times of southern pines, the North Carolina State University (NCSU) cooperative program soon equalled the Swedish programs and became an important world center for tree-improvement thought and practice.

Two important principles were soon established, based largely on early work at NCSU. The first is that there is enormous genetic variation for most traits of interest within most populations of forest trees. Thus, having earlier established a defensive strategy based on our understanding of between-region genetic variation, we now could focus on aggressively pursuing gain in important traits by selecting within populations.

The second principle is that only one tree (parent) per stand (neighborhood) should be entered in a production seed-orchard, no matter how many other good trees might be found nearby. This was particularly apparent and important in stands of loblolly pine, whose natural history often involved the colonization of burned sites or abandoned fields from relatively few founder trees, leading to a high degree of relatedness within such stands. However, it is a principle with applicability to naturally reproduced populations of other tree species, and it has been adopted worldwide in most classical tree-improvement programs. A high proportion of the pollinations that produce wild seeds in naturally regenerated populations involve near neighbors, which are often relatives. Therefore, the trees in most plantations established with seedlings from such wild seeds have various levels of inbreeding; in most tree species, inbreeding depression results in substantial losses of growth vigor.

It was soon apparent that plantations from seed-orchards that were established with the only-one-parent-per-stand rule were performing substantially better than were plantations from wild-seed collections. Some of that gain was due to the phenotypic selections that brought the parents to the seed-orchards, and to progeny-test results that rogued out the poorer parents. But it was also due in large measure to a virtual elimination of inbreeding among these offspring that led to such outstanding results in the first generation.

Our population-architecture experiments were able to estimate the relative proportions of genetic variation that existed among regions, among stands within regions, among families within stands, and within families. But there is little theory that allows one to predict how these components might behave when breeding lines are constructed by combining between-stand genetic variation with within-stand genetic variation. This is not so surprising, because most agronomic populations were hundreds or even thousands of generations beyond these origins when quantitative genetic theory made these questions interesting.

If differentiation among stands is due entirely to drift, then the allocation of additive genetic variation among stands is a function of "F" (Wright, 1969), which measures changes in allocation of heterozygosity from that expected in a panmictic population. In this case, additive genetic variation in a breeding line drawing parents from different stands will be larger than that in the stands, but not as large as the sum of the within-stand plus among-stand components by an amount that is $1/(1 + F)$. This implies that some "original" allelic frequencies will have been randomly changed in the subpopulations, but that they remain constant over the entire region and would be reconstituted in a synthetic recombined population. This seems unlikely.

If new (or originally very rare) alleles have become common in some stands (particularly if due to selection), then the reconstituted population is richer in alleles, or at least in additive genetic variation, than was the original population.

If all of the genetic variation is not additive, then the absolute and relative amounts of additive and non-additive genetic variance will change as allelic frequencies change (Falconer, 1989). Since genetic variation among stands is almost always due to differences in allelic frequencies (polyploidy provides a theoretical exception), then combining them will of necessity change allelic frequencies.

As members of different stands intermate in the seed-orchards and recombine over several generations in the breeding lines, there will be a short-term period of gametic-phase disequilibrium and a longer-term period of linkage disequilibrium as population-specific chromosomes are recombined. To the degree that these disequilibria are in coupling ($+ + + +$ and $----$) or repulsion ($+ - + -$ and $- + - +$) phases, the total genetic variances will be initially larger (coupling) or smaller (repulsion) than if unlinked, and they will approach the unlinked values as the disequilibria decay.

The above five paragraphs present some things that forest geneticists and tree breeders have to think about that probably do not bother most breeders of domesticated plants.

Mating and test designs

As classical tree-breeding commenced, tree breeders naturally drew on the well-established theories and practices of plant breeding and even animal breeding for mating and field-test designs. There are some differences in biology between agronomic crops and forest trees that led tree breeders to modify some of these and largely abandon others. Among the important biological differences are the generation time, the typical number of offspring per parent, the size of the organism, and the variability of the sites on which they are grown. Much of our progress over the past four decades is reviewed in Namkoong and Kang (1990).

Mating designs

An early workhorse for tree improvement was the hierarchical design popularized by the work of Ralph Comstock and Cotton Robinson at NCSU in

Raleigh, NC. It fits the biology of corn (*Zea mays* L.) nicely, and it does a good job of estimating genetic components of variance and a reasonable job of progeny-testing. In chickens and corn, each male (the rare sex) is typically mated to several females (the common sex). The mechanics of making pollinations in the crowns of large trees often led tree breeders to mate several males onto each female, thus making the males the common sex and the females the rare sex. The hierarchical design is not so good for the breeding line, as the genes of members of the the rare sex are much more frequent in subsequent generations than are those of the members of the common sex, leading to a reduction of effective population number. An advantage is that one can develop a greater selection differential if the better genotypes are assigned to the rare sex. This latter strategy is possible in most tree species, as few tree species are dioecious.

Because of the great reproductive potential of the abundant male organs and the many separate female organs on a typical monoecious tree, large diallels are biologically possible as well as theoretically attractive. While a few forest-genetic programs have created complete diallels in the 8x8 to 12x12 range, most tree breeders have used partial and/or disconnected diallels that include many more parents per unit of effort. These are used for both variance-component estimation and family-index selection.

Large rectangular factorials of 4x*N* or 5x*N* dimension, where *N* is an unspecified number in general but always an integer in a given program and can be large, have been popular for progeny-testing. Sets of small disconnected square factorials are effectively used for variance-component estimation and for family-index selection in breeding lines. With 2x2 and 4x4 factorials, it is possible to use each parent as a male in half the crosses and as a female in the other half, thus keeping organelle DNA variability in the breeding line. Unlike angiosperms, most conifers have patriclinal inheritance of chloroplast DNA, and coast redwood (*Sequoia sempervirens* D. Don) inherits mitochondrial DNA patriclinally as well (Harrison and Doyle, 1990).

Mike Lerner's (1958) THE GENETIC BASIS OF SELECTION was published at about the time that mating designs for tree improvement were being considered and developed. The question was asked, "Can chickens really teach us much about trees?". In at least some respects, the answer is "Yes!". One important "chicken design" that we borrowed was single-pair matings, which keeps the maximum number of parents in the breeding line per unit of effort (Libby, 1972).

Polycross designs are also used, usually as a mixture of 10 to 20 unrelated pollens crossed onto a set of unrelated females. This provides an efficient and fair progeny-test of those females, and it can also be effective for estimation of additive genetic variances. However, since it loses pedigree of the offspring, it is not useful for most other purposes. For example, tree breeders are wary of the so-called "55 effect". By chance, in several tree-improvement programs, males numbered "55" when used in pollen mixture sire a very large proportion of the offspring selected from the polycross family.

Tree breeders often yield to the temptation of using open-pollinated families for various purposes, mostly because they are quick and initially cheap to produce. They sometimes inaccurately call open-pollinated sibs "half-sibs", I suppose because it impresses their managers. This can lead to overstated accuracies in progeny-tests and upward biases for variance-component estimates. Open-pollinated families from seed-orchards or plantations are various combinations of half-sibs, full-sibs and possibly selfs; worse, if the open-pollinated families are from a naturally regenerated stand, they have various levels of inbreeding in addition to selfing, which further complicates interpretation of variance components and biases comparisons among these families.

Sublining is a recent idea that has been adopted for tree-breeding. In this strategy, the initial breeding line or population (made up of only one parent per natural stand) for a particular client-area or seed-orchard is (sometimes arbitrarily) divided into sublines, each of which is then bred without genetic exchange from any of the others. Following the principle earlier established for natural stands, if no more than one parent per subline is drawn for the current seed-orchard, production seedlings from such orchards will be free of inbreeding depression for an unlimited number of generations (van Buijtenen and Lowe, 1979).

Field-test designs

For many agronomic crops, the focus is on plot yield rather than on the yield of individual plants. Following this lead, many early field tests with trees consisted of short row-plots of four to 25 trees, or small square plots of 16 to 36 trees.

In forestry, the cost of raising an individual test-tree is high, and therefore field-test designs should be efficient. Individual-tree performance is of interest, both to estimate variance components and as the unit of selection for breeding. Furthermore, unit-area yield is not accurately estimated by using small plots. For example, if one finds that two border rows must be sacrificed to buffer against adjacent-plot competition effects, then only the one center tree provides yield data in a 5x5 square plot; a row-plot provides no data that is free of such competition effects.

For precompetition selection or evaluation of juvenile traits, single-tree plots or non-contiguous plots are now commonly used in tree breeding. When the test is on an irregular or patchy site, the efficiency of such designs can be usefully increased by an analysis employing a data adjustment based on neighbor performance (Loo-Dinkens, 1991). For growth-and-yield evaluation, or for within-family selection, very large plots are now recommended (Foster, 1991). With two border rows, the smallest plot with half the trees being useful for data analysis is a 192-tree, 12x16 rectangular plot, and some organizations are using quarter-hectare (or even larger) plots (Libby, 1987b).

Clonal and family forestry

Agronomists often deploy pure lines, or cross pure lines to produce and deploy uniform hybrid cultivars. Although these are produced by sexual propagation, and are like huge self families or full-sib families, they are essentially genetically identical within lines or cultivars. At the current stage of forest-tree breeding, even self or full-sib families typically are segregating for much genetic variability, and open-pollinated or polycross families are segregating to an even greater extent. The method used to achieve genetic uniformity in forestry is cloning.

Family forestry

"Family forestry" is a step in the direction of achieving the uniformity and predictability of agronomic lines or cultivars. The concept and practice of family forestry has largely been developed within the past few years, with some earlier pioneering by the Weyerhaeuser Company in the southeastern U.S. The families that are deployed have usually been previously tested, and they may be open-pollinated (in a seed-orchard, not in the wild), or control-pollinated polycross or full-sib families. These are collected and raised separately, and the latter two types are sometimes produced separately. They may be deployed to particular compartments or sites as single families, or as prescribed family mixtures. The New Zealand tree-improvement group in Rotorua, and their ProSeed offspring, have the most experience with this new kind of forestry (Carson, 1986).

Family forestry may deploy seedlings, or it may deploy propagules that have been vegetatively multiplied from expensive seedlings or from plant material in short supply. The propagation is most often done at a juvenile maturation state, but it can sometimes be advantageous to propagate at an early adolescent maturation state. Family forestry is used because of the excessive expense associated with cloning, or because cloning has difficulties or unresolved uncertainties (most of which are related to maturation effects) making cloning unsuitable for production deployment.

Clonal forestry

This involves much more than the use of vegetative propagules in forestry. It has as qualifying criteria the deployment of relatively few well-tested and well-understood clones in ways that allow substantial increases in management efficiency and effectiveness. It also allows the deployment of clones that would rarely survive (such as short fat trees) in a genetically heterogeneous mixture, or that would rarely occur (such as trees that produce few or no sexual organs) in zygotic forestry.

Clonal forestry allows the capture of the best traits derived from both the additive and non-additive components of genetic variability. We are coming to appreciate that it can capture not only favorable segregants in transmission genetics, but some developmental genetic variability as well. For example, under conditions of relatively fertile sites in New Zealand and Australia, radiata pines (*Pinus radiata* D. Don) propagated at an early-adolescent maturation state grow with substantially better form than do the same genotypes as seedlings or when clonally propagated at a juvenile maturation state (Gleed, 1992). While these trees have exactly the same DNA library, it seems likely that parts of it are being read differently at these two maturation states, and that it is possible and practicable to select for the more favorable constellation of gene expression within a clone or a set of clones.

Historically, clones were produced as rooted cuttings (Ohba, 1992; Zsuffa *et al.*, 1992), and sometimes as grafts for urban and Christmas-tree purposes (Kleinschmit *et al.*, 1992). More recently, some clones are being produced by tissue-culture techniques (Gleed, 1992; Talbert *et al.*, 1992) and by somatic embryogenesis (Becwar, 1992). Production-level deployment of tissue-culture clones of radiata pine species in New Zealand and of eucalyptus species [particularly river red-gum (*Eucalyptus camaldulensis* Dehn. Longbeak)] in California is being done as of 1987 and 1991, respectively, by Tasman Forestry Limited and Simpson Timber Company.

Family forestry seems to be an advance over classical tree-improvement; it is promising for short-term genetic gain and for achieving some uniformity. It overcomes some of the difficulties and uncertainties of cloning, and can capture some maturation-related advantages by vegetative multiplication at the appropriate maturation state. Yet, the advantages of clonal forestry, if achievable, seem much greater, and there are presently many more programs practicing clonal forestry (Ahuja and Libby, 1992) than are practicing family forestry. One of the advantages of clonal forestry is the ability to better focus on elements of harvest index, such as growing shorter fatter trees. Allocation of biomass within the tree crop, as opposed to our earlier focus on increasing the adaptedness of our plantations, is only recently gaining attention (Libby, 1987a).

PLANTATION DIVERSITY

While the genetic diversity maintained in any forest stand or plantation is a matter of concern, it comes into sharpest focus with respect to questions about the numbers and relatedness of clones deployed in a local area. Early theoretical modeling, backed by related experiments in agronomy, indicated a perhaps surprising result, namely, that mixtures or mosaics of relatively few clones, each different from the others in many ways, seem safer than deploying a continuum of genotypes with very small differences along the resulting continuum of biotypes (Libby, 1982). At our current level of knowledge, the mosaic deployment seems, on balance, to offer more advantages than does a deployment in intimate mixture (Ahuja and Libby, 1992).

BIOTECHNOLOGY

The promise and promises of biotechnology are attractive for forestry. One of the major apparent advantages is to be able to modify and select new tree lines in a very short time, as compared to the time needed using conventional methods.

Gene insertions

Many agronomic crops have well-known pure lines with many excellent qualities that can be further improved by one or a few specific qualitative changes. In contrast to introgressive substitution of the desired genes, these changes can perhaps be produced by cleanly inserting one or a few specific genes using biotechnological techniques, without disrupting the rest of a highly selected genome.

Forest trees differ from such cases in several important respects. At present, tree breeding has produced no such highly selected well-understood pure lines, and few clones are available that are valuable enough and sufficiently understood to be candidates for gene-by-gene fine-tuning. Typical breeding lines are still in their first to third generations, and most have abundant genetic variability that is in early stages of being sorted. The precise addition of one or two genes to this already rich genome would generally add very little to the near-term value of the line.

There are a few exceptions to the above cautious view. One is the possibility of inserting a genetic element that turns off the development of sexual organs. This would have several beneficial effects. First, a sexual part of such clones could be kept separate and fully fecund, so that these clones could then contribute to ongoing breeding lines. Second, the substantial proportions of photosynthate that are allocated to male and female organs would be partly and perhaps (in some clones) mostly redirected to harvestable wood in the lower part of the bole (Libby, 1987a). Finally, small *in situ* genetic conservation reserves that are or that become surrounded by large plantations of domesticated trees of the same species are in danger of genetic swamping. But if the domesticated trees produce little or no pollen and few or no seeds, then small genetic conservation reserves will be contaminated to a lesser and perhaps acceptable degree.

Hybrids

The role of hybrid trees remains uncertain in forestry. Currently, interspecific hybrids seem to be useful in poplars (*Populus* spp.) and eucalypts, but hybrids have had only limited usefulness in the many other combinations of races and species of trees that have been tried. Biotechnology offers the possibility of combining completely unrelated taxa, and some of these may prove to be remarkably useful; but my prognosis is that the great majority of these hybrids will be little more than curiosities.

Genes in culture

Trees may contribute genes that are useful for producing industrial or medicinal chemicals in culture, or perhaps when inserted in other organisms. A current example is the possibilty of finding the tissue or genes that produce taxol, a chemical with cancer-fighting properties. Then, instead of cutting and grinding up yew (*Taxus* spp.) trees, a practice that may encourage poaching some yew species to the point of endangerment or even extinction, this valuable chemical may be produced in yew tissue-culture, or perhaps in a surrogate organism such as a yeast.

Biotechnology in the service of breeding

In addition to direct applications of biotechnology, the techniques developed in this and other branches of technology can be applied to make selections and to characterize or identify trees better or more accurately. Only a few examples from a much longer list are given below.

Selection

Much controversy currently swirls around the possibility of marker-assisted selection, and a conference is scheduled on this very topic at Gatlinburg, Tennessee in June 1991. One problem with forest trees is that the linkage relationships that make such selection possible among hybrid derivatives of pure-line crops are liable to be frustratingly inconsistent among forest stands and even among families in a stand. But this may not be the first time that bright solutions are offered to solve apparently serious problems.

On the other hand, a technological advance spawned by cold-fusion research is now available for biological research. Sensitive microcalorimetry seems to hold much promise not only for evaluating the response of well-known clones to various treatments or environmental variables, but also for screening for clones that are above average in growth potential, or are able to grow in unusual circumstances (Anekonda *et al.*, 1990).

Identification

The ability to evaluate mating systems (Adams and Birkes, 1989), check for contaminations (Wheeler *et al.*, 1991), and fingerprint clones is already proven and well-accepted (Cheliak, 1992).

REFERENCES

Adams, W. T.; Birkes, D. S. (1989) Mating patterns in seed orchards. In: *Proceedings 20th Southern Forest Tree Improvement Conference*, Charleston SC, pp. 75-86.

Ahuja, M. R.; Libby, W. J. (eds.) (1992) *Clonal forestry: Genetics, biotechnology and application.* Springer Verlag, Berlin, Germany (in press).

Anekonda, S. T.; Criddle, R. S.; Libby, W. J. (1990) The use of microcalorimetry for early culling. In: *Proceedings joint meeting WFGA & IUFRO Work. Part. S2.02-05, 06, 12 and 14,* 20-24 Aug. 1990, Olympia WA. Weyerhaeuser Co., Tacoma, WA, 11 pp.

Becwar, M. (1992) Somatic embryogenesis and clonal forestry. In: Ahuja, M. R.; Libby, W. J.(eds.), *Clonal forestry: Genetics, biotechnology and application.* Springer Verlag, Berlin, Germany (in press).

Carson, M. J. (1986) Advantages of clonal forestry for *Pinus radiata* - Real or imagined? *New Zealand Journal of Forestry Science* 16:403-415.

Cheliak, W. (1992) Clone identification. In: Ahuja, M. R.; Libby, W. J. (eds.), *Clonal forestry: Genetics, biotechnology and application.* Springer Verlag, Berlin, Germany (in press).

Critchfield, W. B. (1984) Impact of the Pleistocene on the genetic structure of North American conifers. In: Lanner, R. M. (ed.), *Proceedings 8th North American Forest Biologists Workshop*, Utah State Univ., Logan, UT, pp. 70-118.

Falconer, D. S. (1989) *Introduction to quantitative genetics.* 3rd ed. Longman Scientific & Technical/John Wiley & Sons, NY, 438 pp.

Foster, G. S. (1991) Estimating yield: Beyond breeding values. In: Fins, L.; Friedman, S.; Brotschol, J. (eds.), *Handbook of quantitative forest genetics.* Kluwer, Dordrecht, The Netherlands (in press).

Gleed, J. (1992) Development of plantlings and stecklings of radiata pine. In: Ahuja, M. R.; Libby, W. J.(eds.), *Clonal forestry: Genetics, biotechnology and application.* Springer Verlag, Berlin, Germany (in press).

Hamrick, J. L.; Godt, M. J. W. (1990) Allozyme diversity in plant species. In: Brown, H. D.; Clegg, M. T.; Kahler, A. L.; Weir, B. S. (eds.), *Plant population genetics, breeding and genetic resources.* Sinauer Assoc., Sunderland, MA, pp. 43-63.

Harrison, R. G.; Doyle, J. J. (1990) Redwoods break the rules. *Nature (London)* 344:295-296.

Kitzmiller, J. H. (1976) *Tree improvement master plan for the California region.* USDA Forest Service, San Francisco, CA, 123 pp.

Kleinschmit, J.; Khurana, D. K.; Gerhold, H. D.; Libby, W. J. (1992) Past, present and anticipated applications of clonal forestry. In: Ahuja, M. R.; Libby, W. J. (eds.), *Clonal forestry: Genetics, biotechnology and application.* Springer Verlag, Berlin, Germany (in press).

Larsen, C. S. (1956) *Genetics in silviculture.* Oliver and Boyd, London, England, 224 pp.

Lerner, I. M. (1958) *The genetic basis of selection.* John Wiley & Sons, New York, 298 pp.

Libby, W. J. (1972) Efficiency of half-sib and bi-parental populations for selection for next generation of improvement. *Industrial Forestry Association Tree Improvement Newsletter* 16:3-8.

Libby, W. J. (1982) What is a safe number of clones per plantation? In: Heybroek, H. M.; Stephen, B. R.; von Weissenberg, K. (eds.), *Resistance to diseases and pests in forest trees*. Pudoc, Wageningen, The Netherlands, pp. 342-360

Libby, W. J. (1987a) *Do we really want taller trees? Adaptation and allocation as tree-improvement strategies*. The H.R. MacMillan Lecture in Forestry, Univ. of British Columbia, Vancouver, Canada, 15 pp.

Libby, W. J. (1987b) Testing and deployment of genetically engineered trees. In: Bonga, J. M.; Durzan, D. J. (eds.), *Cell and tissue culture in forestry*. Martinus Nijhoff, Dordrecht, The Netherlands, pp. 167-197.

Libby, W. J.; Critchfield, W. B. (1986) Patterns of genetic architecture. *Annales Forestales* 13:77-92.

Loo-Dinkens, J. (1992) Field test design. In: Fins, L.; Friedman, S.; Brotschol, J. (eds.), *Handbook of quantitative forest genetics*. Kluwer, Dordrecht, The Netherlands (in press).

Matthew, P. (1831) *On naval timber and arboriculture - with critical notes on authors who have recently treated the subject of planting*. Adam Black, Edinburgh, Scotland. 391 pp.

Matthew, P. (1860) Nature's law of selection. *Gardener's Chronicle and Agriculture Gazette*, Apr 7:312-313.

Namkoong, G.; Kang, H. (1990) Quantitative genetics of forest trees. *Plant Breeding Reviews* 8:139-188.

Ohba, K. (1992) Clonal forestry with sugi (*Cryptomeria japonica*). In: Ahuja, M. R.; Libby, W. J. (eds.), *Clonal forestry: Genetics, biotechnology and application*. Springer Verlag, Berlin, Germany (in press).

Talbert, C.; Ritchie, G. A.; Gupta, P. K. (1992) Conifer vegetative propagation: An overview from a commercial perspective. In: Ahuja, M. R.; Libby, W. J. (eds.), *Clonal forestry: Genetics, biotechnology and application*. Springer Verlag, Berlin, Germany (in press).

van Buijtenen, J. P.; Lowe, W. J. (1979) The use of breeding groups in advanced generation breeding. In: *Proceedings 15th Southern Forest Tree Improvement Conference*, pp. 59-65.

Westfall, R. D. (1991) Developing seed transfer zones. In: Fins, L.; Friedman, S.; Brotschol, J. (eds.), *Handbook of quantitative forest genetics*. Kluwer, Dordrecht, The Netherlands (in press).

Wheeler, N. C.; Adams, W. T.; Hamrick, J. L. (1991) Pollen distribution in wind-pollinated seed orchards. In: Askew, G.; Blush, T.; Bramlett, D. L.; Bridgewater, F. E.; Jett, J. B. (eds.), *Pollen management handbook #2*. USDA Forestry Service, Agriculture Handbook, Washington DC.

Wright, S. (1969) *Evolution and the genetics of populations. Vol. 2. The theory of gene frequencies*. Univ. Chicago Press, Chicago, IL, 511 pp.

Zsuffa, L.; Sennerby-Forsse, L.; Weisgerber, H.; Hall, R. B. (1992) Strategies for clonal forestry with poplars, aspen and willows. In: Ahuja, M. R.; Libby, W. J. (eds.), *Clonal forestry: Genetics, biotechnology and application*. Springer Verlag, Berlin, Germany (in press).

Chapter 6
Predictive methods for germplasm identification

R. W. Allard

Department of Genetics, University of California, Davis 95616

It is axiomatic that, to survive, any living system must be adapted to the environments it is likely to encounter. Adaptedness in nature is a very complex trait that hinges on a multitude of morphological and physiological characteristics affected by alleles of many loci. Adaptedness is still more complex under cultivation. In addition to ability to carry out essential biological processes efficiently, successful cultivars must also satisfy various production requirements of farmers and meet numerous quality preferences of end users; these attributes of production and quality often run counter to essential biological functions, so complicating the task of the breeder. Despite complexities and complications it is clear that a great deal of genetic progress has been made in improving adaptedness, productivity, and quality during the thousands of years our major crops have been grown in agricultural environments. Landraces of our major crops perform agricultural tasks much better than their wild progenitors and modern cultivars are far more successful in this respect than the landraces they replaced. Figure 1 provides numerical support for the latter assertion. The average yield of corn (*Zea mays* L.) in the United States was about 25 bu acre^{-1} from the 1860s through the 1930s but by the late 1980s yields were about five times higher. During the period of open-pollinated cultivars the rate of gain in yield was about 1/50 bu acre^{-1}yr^{-1}. However, during the period of double-cross hybrids the rate of gain had increased 50-fold to 1 bu acre^{-1}yr^{-1} and in the period of single-cross hybrids rate of gain increased to 1.8 bu acre^{-1}yr^{-1}. Various studies (Duvick, 1984) indicate that somewhat more than half the nearly 100 bu acre^{-1} increase in yield since 1930 can be attributed to genetic improvement and the remainder to improved cultural practices. Bradley *et al.* (1988) and Troyer (1990) credit much of the rapid gain in recent years to improved breeding procedures that include testing in more locations and more years than in earlier periods: the new hybrids have "better defensive traits" that lessen their vulnerability to stresses and hence contribute to wider adaptedness.

Figure 2 gives an idea of the effects of natural selection in itself on the adaptedness of genetically variable populations of cultivated barley (*Hordeum vulgare* L. ssp. *vulgare*, abbreviated *H.v.*) and cereal legumes (*Phaseolus vulgaris* L. and *P. lunatus* L.). The five populations shown in this figure were all synthesized from crosses between a few cultivars (four in the case

Figure 1 U.S. corn yields and kinds of corn - 1866 to present; b values (regressions) indicate gain in bushels acre^{-1}year^{-1} (A. F. Troyer, 1990, pers. comm.).

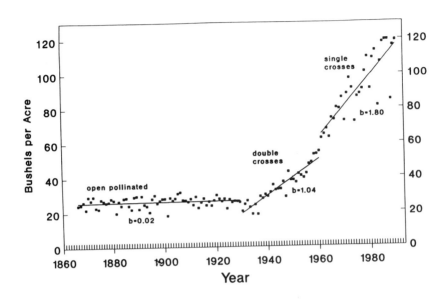

of population 43) to many cultivars (6200 in the case of CCXXI). In each case a bulk of sufficient numbers of seeds to produce > 30,000 mature plants was sown in a large plot, grown under standard agricultural conditions, harvested in mass without conscious selection, and each successive generation was propagated from a random sample of seeds from the previous harvest. Seed yields of various generations of the populations and the three to five most popular commercial cultivars of the time, included as controls, were determined in 13 different growing seasons (1960-1963, 1965-1969, 1976-1982). In the earliest generations seed yields of all of the populations were significantly lower than those of the control varieties. However, yields increased rapidly during the next 10 to 20 generations and they have continued to increase, but at a slower rate (about 1/2 to 1% per year) up to the present. These results indicate that natural selection alone was responsible for useful evolutionary changes in all of these genetically diverse experimental populations.

I now turn to my main purpose which is to examine, in retrospect, the genetic changes that have occurred in our major crop species in response to hundreds of generations of natural and man-directed selection for high and steady performance and high quality in agricultural environments. I will examine genetic change at two levels. First, I will characterize changes in

Figure 2 Grain yields of experimental barley (CCII, CCV, CCXXI) and cereal legume (22, 43) populations determined in replicated yield trials. Yields of the experimental populations are expressed in each of three periods as percentages of the yields of three to five commercial varieties grown as checks (from Allard, 1988).

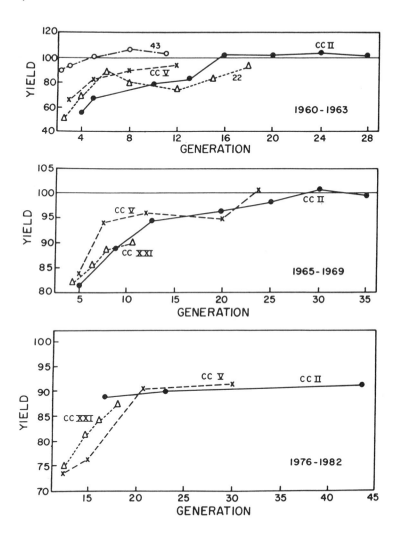

the frequencies of discretely identifiable alleles of loci affecting morphological, disease resistance, allozyme, and restriction fragment variants that occurred as wild species were converted into landraces and then into modern cultivars. In most cases the selection practiced could not have been directed at specific alleles because it is only within the last few decades that the technology needed to identify the alleles has been available. Second, I will discuss the processes by which the surviving alleles were assembled into the multilocus combinations that are associated with the superior adaptedness and performance of elite cultivars. Genetic change at both the allelic and genotypic levels will be illustrated using data from barley, an inbreeder, and corn, an outbreeder. Finally, I will discuss the greater intralocus genetic diversity of *Avena barbata* Pott ex Link, a tetraploid wild oat, relative to its diploid ancestors, *A. hirtula* Lag. and *A. wiestii* Steud., and the advantages in adaptedness associated with the increased genetic diversity. I have a great deal of data to present and, because the data speak for themselves, I will emphasize presentation of data rather than interpretations.

ALLELIC AND GENOTYPIC DIVERSITY IN BARLEY

Allelic diversity in barley

Table 1 gives numbers of alleles encountered in two samples of wild barley (*H. vulgare* ssp. *spontaneum* C. Koch, abbreviated *H.s.*) and in cultivated barley. The main conclusion from both samples is the same: numbers of alleles of the 25 or so loci monitored decreased dramatically in the progression from primitive materials to the most advanced modern cultivars. Table 2 illustrates this decrease in allelic diversity in terms of *Rrn1* and *Rrn2*, two thoroughly studied loci that govern ribosomal DNA (rDNA) variability in barley (Saghai Maroof *et al.*, 1984, 1990; Allard *et al.*, 1990). Allele *112*, the most frequent among the 12 alleles of *Rrn1* in *H.s.*, is nearly fixed in *H.v.* This is the most common pattern observed in barley - alleles present in high frequency in *H.s.* are also usually the most frequent in *H.v.* and the infrequent alleles of *H.s.* are usually not present in *H.v.* The pattern for locus *Rrn2* is much less commonly observed. Allele *107*, which is the most frequent among the eight alleles of *Rrn2* in *H.v.* (f ≈ 0.78), is also the most frequent allele in cultivated barleys from Mediterranean climates (f usually > 0.50) but it is only moderately frequent in *H.v.* (f ≈ 0.30) on a world-wide basis. However, allele *104*, the most frequent allele of *Rrn2* (f ≈ 0.65) in *H.v.* worldwide, is infrequent in *H.s.* (f ≈ 0.02). Allele *104* provides one of the most clear-cut examples of an infrequent allele in *H.s.* that has become dominant in *H.v.* Table 2 also shows that all other moderately frequent or infrequent alleles of *Rrn2* were absent or infrequent in *H.v.*

The population dynamics of genetic change

Barley is a highly favorable organism for the study of within-population dynamics of genetic change (Allard *et al.*, 1972), and I will now use barley data to identify the forces responsible for the observed changes in allelic and genotypic frequencies that occurred during the domestication of this species.

Table 1 Numbers of alleles of commonly studied loci encountered in two samples of wild barley (*H.s.*) and cultivated barley (*H.v.*).

	No. of loci	No. of alleles	Mean no. of alleles/ locus	Relative no. of alleles/ locus
		Sample A*		
H.s. (Israel)	19	77	4.05	100
H.v. (12 Iranian landraces)	25	56	2.24	55
H.v. composite cross 21 (F_{17})	20	34	1.70	42
H.v. composite cross 34 (F_4)	20	36	1.80	44
		Sample B		
H.s. (Middle East)[†]	20	103	5.15	100
H.v. (Middle East)[‡]	20	55	2.75	53
H.v. (CCII, F_7-F_9)	26	41	1.58	31
H.v. (CCII, F_{53})	26	37	1.42	28
H.v. (CCV, F_6)	25	39	1.56	31
H.v. (CCV, F_{35})	25	36	1.44	28
H.v. (CCXXI, F_4)	25	42	1.68	33
H.v. (CCXXI, F_{22})	25	38	1.52	30
H.v. (9 California cultivars)[§]	25	36	1.44	28

*Sample A: Brown and Munday (1982); Sample B: Kahler and Allard (1981), Kahler *et al.* (1981), Saghai Maroof *et al.* (1984, 1990), Allard (unpublished data). Sample A included only allozyme loci. Sample B included loci affecting allozyme and restriction fragment variants in *H.s.* and allozyme, restriction fragment and morphological variants in the composite crosses.

[†]Israel, Lebanon, Syria, Jordan, Turkey, Iran, Afghanistan.

[‡]18 landraces from Lebanon, Syria, Jordan, Turkey and Iran.

[§]Mean number of alleles cultivar^{-1} was 27.2 (many cultivars were polymorphic, especially for *Est* loci *1*, *3*, and *4*).

A feature that became apparent during studies of Mendelian inheritance is that patterns of allelic frequency change in segregating families are good predictors of the probable usefulness of alleles. As an example, in F_2 families in which the predominant allele *112* and any one of the infrequent alleles of *Rrn1* were segregating, expected 1:2:1 ratios were nearly always distorted; allele *112* was consistently in excess and the infrequent alleles were

Table 2 Alleles of rDNA loci *Rrn1* and *Rrn2* in wild barley (*H.s.*) from Israel and Iran and from a worldwide sample of cultivated barley (*H.v.*) (from Saghai Maroof *et al.*, 1990).

Locus and allele	Wild barley No.	Freq.	Cultivated barley No.	Freq.
		Rrn1		
108a	29	0.06	0	0.00
108	100	0.19	0	0.00
109	68	0.13	0	0.00
110	28	0.05	0	0.00
111	10	0.02	3	0.02
112	273	0.53	187	0.98
113	4	<0.01	0	0.00
114	3	<0.01	0	0.00
115	0	0.00	0	0.00
116	2	<0.01	0	0.00
117	1	<0.01	0	0.00
118	0	0.00	0	0.00
Total	518	1.00	190	1.00
		Rrn2		
100	6	0.01	2	0.01
101	0	0.00	3	0.02
102	4	<0.01	2	0.01
103	4	<0.01	0	0.00
104	13	0.02	116	0.65
105	25	0.05	0	0.00
106	69	0.13	2	0.01
107	427	0.78	53	0.30
Total	548	1.00	178	1.00

consistently in deficiency (Allard *et al.*, 1990). Thus, even under the conditions of little or no plant-to-plant competition in which the inheritance studies were conducted, infrequent alleles behaved as semilethals or subvitals relative to the predominant wild-type alleles. Furthermore, when populations formed by bulking seeds produced by F_2 families were advanced into F_3, F_4, and later generations under conditions of competition, the wild-type alleles rapidly became predominant and the infrequent alleles were rapidly eliminated. This was also the case (Table 3) in large genetically variable populations, e.g., Composite Cross II (CCII), a population that was synthesized from 28 cultivars selected to represent the major barley growing areas of the world (Harlan and Martini, 1929). In CCII, allele *112* of *Rrn1* had become virtually monomorphic and all other alleles present in the parents of this population had virtually disappeared by generation F_8. All originally infrequent alleles of *Rrn2* had also virtually disappeared by F_8. However, allele *104*, the most frequent allele of *Rrn1* worldwide (and in the parents of CCII) decreased in frequency from f = 0.66 to f = 0.30 by generation 53, whereas allele *107*, which is the most frequent allele of Mediterranean climates (including that of California), increased in frequency from f = 0.26 to f = 0.70.

Table 3 Frequencies of alleles of rDNA loci *Rrn1* and *Rrn2* in wild barley (*H.s.*) and in the 28 parents and various generations of composite cross II (CCII). Alleles present in frequency <0.01 in CCII are not listed (from Saghai Maroof *et al.*, 1984, 1990).

Allele	Wild barley	Parents	F_8	F_{13}	F_{23}	F_{45}	F_{53}
				Generation of CC II			
			Rrn1				
110	0.05	0.07		0.01			
111	0.02	0.07	0.02				
112	0.53	0.86	0.98	0.99	1.00	1.00	1.00
			Rrn2				
101	0.00	0.02					
102	<0.01	0.02					
104	0.02	0.66	0.49	0.48	0.41	0.46	0.30
106	0.13	0.04	0.01		0.02		
107	0.78	0.26	0.50	0.52	0.57	0.54	0.70

Additional examples of the above three main patterns of allelic frequency change in CCII are given in Table 4. The patterns can be summarized as follows. The most common pattern observed, that of continued high frequency in all generations of the predominant allele of originally monomorphic or nearly monomorphic loci, suggests that predominant alleles contribute to high adaptedness in virtually all environments in which barley is cultivated, including the Davis, California environment. The brittle/non-brittle

Table 4 Patterns of allelic frequency change in barley composite cross II (CCII) for three categories of loci (from Allard, 1988).

Locus	Allele	Frequency in parents	Frequency in later generations of CCII
		Nearly monomorphic loci[*]	
Pgd2	Null	0.96	>0.99 (F_6)
Lemma color	White	0.96	>0.99 (F_6)
Brittle rachis	Non-brittle	1.00	1.00
		Weakly polymorphic loci	
Est2	2.7	0.89	0.95 (F_{53})
Row number	6 row	0.75	>0.99 (F_{13})
Awn	rough	0.89	>0.99 (F_{19})
		Highly polymorphic loci	
Est1	1.0	0.29	0.95 (F_{53})
Est3	5.4	0.36	0.92 (F_{53})
Est4	6.6	0.32	0.91 (F_{53})
Aleurone	Blue	0.36	0.78 (F_{53})
Rachilla	Short	0.61	0.93 (F_{53})
Dentate lemma	Dentate	0.71	0.96 (F_{53})

*Most loci are monomorphic or nearly monomorphic in cultivated barley. Wild barley is monomorphic for brittle rachis and cultivated barley is monomorphic for non-brittle rachis.

rachis loci provides an example of the occasional loci for which alleles that are nearly monomorphic in the wild are not present in cultivated barley. A closely linked pair of complementary loci, Bt1 and Bt2, produce brittle rachis when they are present in double-dominant condition. Wild barley is monomorphic for Bt1 and Bt2 whereas cultivated barley is monomorphic for bt1 and bt2. Bt1-Bt2 confer high adaptedness as a seed dispersal mechanism in wild barley and whereas b2-b2 confer high adaptedness in cultivation as a harvest aid. The second pattern, that of weakly polymorphic loci, differs from that of the nearly monomorphic loci in two respects: (1) the predominant allele is usually present in slightly lower frequency and (2) the predominant allele approaches fixation at a slower rate. Figure 3 illustrates the pattern of change for such loci in CCII over a period of more than 50 generations. The third and least frequent pattern is that of highly polymorphic loci (two or more alleles present in intermediate frequencies). For two

Figure 3 Allelic frequencies of *Est 2* in various generations of CCII (from Allard, 1988).

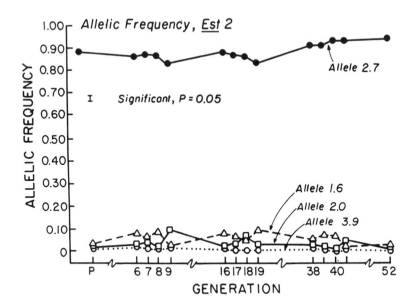

(rachilla hair length, dentate lemma) of the six loci of this type (given in Table 4) the allele that was most frequent in the parents of CCII approached fixation by generation F_{53}. In the four remaining cases an allele that was originally second most frequent ultimately became most frequent in the California environment. In these six cases allelic frequency changes were not monotonic; rather, as illustrated in Figure 4, changes in allelic frequencies often shifted dramatically in amount and in direction from year to year. Detailed analyses showed that these shifts were significantly correlated with environment (particularly variations in amount of rainfall).

Figure 5 gives four patterns of allelic frequency change in CCII for 29 loci that govern resistance vs. susceptibility to Californian races of scald disease caused by *Rhynchosporium secalis* (Oud.) Davis (Webster *et al.*, 1986). Only one among the 28 parents of CCII (4%) was resistant to pathotype 72; also the frequency of resistant plants remained at low levels in all generations. This population behavior is not surprising for two reasons. First, even though pathotype 72 is capable of overcoming nearly all of the 29 resistance alleles in CCII, it infects only sporadically and it usually detracts little from host yield even when it produces symptoms on the host. Second, in studies of the effects of the host-resistance vs. the host-susceptibility allele based on head-to-head comparisons of highly isogenic lines repeated in many different seasons, it was found that the resistance allele regularly had

Figure 4 Allelic frequencies of *Est 1* in various generations of CCII (from Allard, 1988).

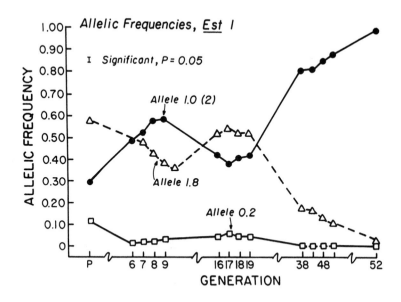

significant deleterious pleiotropic effects on the viability and performance of the barley host. The resistance allele was always at a selective disadvantage in the host when pathotype 72 was present and it was nearly always at a disadvantage even when pathotype 72 was present. The host-pathogen inter-relationships of more than half the 29 pathotypes of *R. secalis* with their corresponding resistance alleles of barley followed this pattern (Allard, 1990). Pathotype 16, in contrast, often killed seedlings and severely damaged older plants; consequently it is very important that barley plants carry the allele for host resistance to pathotype 16. Fortunately, this host resistance allele was found in studies of numerous pairs of resistant vs. susceptible isogenic lines tested in several years to have little if any detrimental effect on performance in the absence of pathotype 16. It is therefore not surprising that the alleles that provide resistance to this pathotype increased to high frequencies in CCII. Resistance to pathotype 16 has not been found in *H.s.*; thus the presence in 11 of the 28 parents (39%) of CCII of alleles that confer resistance to this pathotype suggests that such alleles have arisen by mutation in cultivated barleys on multiple occasions and that these resistance alleles increased in frequency because they were favored by selection under conditions of cultivation (McDonald *et al.*, 1989; Allard, 1990). The coevolutionary interactions of pathotypes 40 and 61 with their corresponding resistance-susceptibility alleles resemble those of pathotype 16 more closely

Figure 5 Observed and expected frequencies of plants resistant to four pathotypes of *R. secalis* in various generations of CCII: A, pathotype 16; B, pathotype 40; C, pathotype 61; D, pathotype 72. Solid lines show observed frequencies and dashed lines expected frequencies assuming neutrality (no selective advantage) of resistance vs. susceptibility alleles (from Allard, 1990).

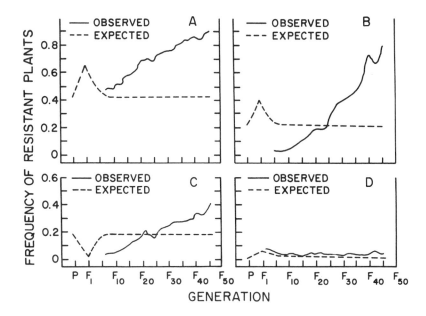

than those of pathotype 72 (Figure 5). Alleles that provide resistance to pathotypes 16, 40 and 61 protect barley against nearly all severely damaging pathotypes present in the Davis environment; these three resistance alleles all became moderately to highly frequent in CCII which may be the reason that scald has been seen only rarely on more than occasional plants in this population.

Pleiotropy vs. linkage

The closeness and the consistency of the associations between the alleles discussed above and adaptedness suggested the alleles monitored might in themselves have played significant roles in adaptive processes (see Dobzhansky, 1927). This led to the notion that a straightforward Mendelian approach based on determining frequency changes over generations for numerous simply inherited and easily scored variants, combined with precise assays of the effects of these loci on various quantitative characters, might provide genetic information useful in characterizing the genetic basis of adaptive change. The program began in 1948 with six easily-classified characters of the spike of barley (two-row vs. six-row, rough vs. smooth awn, long vs. short haired rachilla, blue vs. white aleurone color, black vs. white lemma color, dentate vs non-dentate lemma veins). A second category of easily classified characters - resistance vs. susceptibility to five patho-types of scald disease - was added in the 1950s. A third category - seven enzyme loci detectable by starch gel electrophoresis - was added in the 1960s and a fourth category - ribosomal DNA alleles detectable by Southern blot analysis - was added in the 1980s. The effects of individual alleles of each locus on various quantitative traits, including yield, were determined by extensive testing of numerous pairs of highly isogenic lines descended from single barley plants isolated from various generations of experimental popu-lations (review in Allard, 1988). Comparisons between alternative homo-zygotes revealed that all the loci monitored had statistically significant addi-tive effects on several to many quantitative traits; thus, each locus moni-tored, in addition to being a locus for its discrete descriptive effect, was also a locus for several quantitative traits. Consistent associations were found in various generations between frequent alleles and superior adaptedness and superior reproductive capacity (e.g., high yield and larger numbers of ker-nels per plant). It was concluded that quantitative character differences associated with different alleles of the loci monitored are pleiotropic effects of the alleles themselves; if linked loci governing quantitative characters are involved they are so tightly linked to the loci monitored that for practical purposes each locus monitored behaved as a single complex locus during the evolution of cultivated barley. These results raise an important question: Do the loci monitored act independently of one another in their effects on reproductive capacity, survival ability and other components of fitness that affect adaptedness and high performance, or do they interact epistatically in complex ways? Stated another way, must superior alleles be assembled into synergistic multilocus combinations to achieve superior adaptedness? I now turn to results indicating that the answer to this question is yes.

Analysis of multilocus combinations

Although the simultaneous analysis of associations among several loci is complicated (Allard, 1988; Zhang *et al.*, 1990; García *et al.*, 1991), the main features of interactions involving more than two loci simultaneously can be illustrated with simple plots of changes in the frequency of multilocus gametic types. Figure 6 gives changes in 4-locus gametic frequencies that occurred for four allozyme loci in CCII grown at Davis and at Bozeman, Montana (Allard, 1988). Although 13 of the 16 possible 4-locus gametic types were present in the early generations, only two gametic types, 2112 and 1221, increased significantly in frequency at Davis in any sequence of generations: gametic type 2112 declined in frequency from F_7 to F_{20} (a

Figure 6 Multilocus gametic frequencies for *Est 1*, *Est 2*, *Est 3* and *Est 4* in various generations of CCII grown in Davis, California and Bozeman, Montana. With four loci and two alleles per locus, there are 16 gametic types coded 1111, 1112, . . . 2222 (1 indicates the originally most frequent allele and 2 indicates the originally second most frequent allele at each locus) (from Allard, 1988).

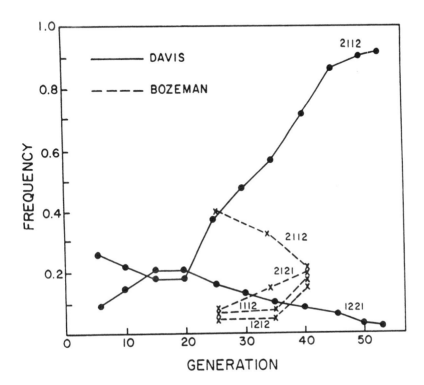

series of generally dry years) but increased steadily in frequency thereafter (generally wet years) to f = 0.95 in F_{53} whereas gametic type 1221, after an early increase in frequency, declined steadily thereafter to f = 0.02 in F_{53}.

When the Davis population was in generation F_{26}, a large sample of seeds was sent to Bozeman, Montana where Dr. E. S. Hockett grew the population for 19 generations from F_{27} to F_{46}. In the continental climate of Montana, gametic type 2112 (the favored genotype in Mediterranean climates) declined rapidly in frequency and gametic types 1221, 1112 and 1212 increased in frequency. It is not clear which, if any, among these three frequently found types (Figure 6) might ultimately have become predominant in CCII in Montana in the way that 2112 has become predominant in California. In this connection, all commercial varieties of the Central Valley of California carry gametic type 2112 whereas 2121 is the most common gametic type of North Plains and Prairie Province varieties we have examined.

A particularly informative study of the four esterase loci involved a Davis population that was descended from a single F_1 hybrid plant obtained by crossing two of the less well-adapted parents of CCII. Expressed in coded form, the two founding gametic types of this population were 1122 and 2211. By generation F_4, single-crossover gametes were detected between loci 1, 2 and 3 (these three loci are very tightly linked). By generation F_9, double-crossover gametes had appeared and, by generation F_{21}, gametic type 2112 (a double-crossover type) had become the most frequent in the population (f = 0.33). It is clear that, despite the very tight linkage, segregation and recombination rapidly broke down the original four-locus structure in this population and that selection reassembled the component parts into the adaptively superior 2112 gametic type.

Another feature of evolutionary change relates to the number of loci that are involved in interlocus associations in various generations of the experimental barley populations (Allard, 1988). In early generations the 20 or more polymorphic loci monitored in CCII were clustered in five or six groups of three or four loci each, among which the four esterase loci made up one of the groups of four. However, during the middle generations, a number of disassociations occurred and new associations formed and, by the late generations, there were two clusters, one of six and one of eight loci. The picture of genetic change that emerges is one in which increases in adaptedness were correlated with the development of clusters of alleles of loci affecting many different morphological and physiological characteristics and gradual coalescence of clusters into a few large synergistically interacting complexes. Clearly the processes involved in organizing the population genotype into superior multilocus genotypic combinations are bewilderingly complex, and it is therefore not surprising that analyses based on quantitative characters have been largely unsuccessful in helping us understand the genetic basis of the evolution of multilocus systems.

ALLELIC AND GENOTYPIC DIVERSITY IN CORN

Changes in allelic frequencies

Table 5 gives the changes that occurred over time in numbers of alleles at 23 allozyme loci in corn. In total 226 alleles have been described at these 23 loci in the wild teosintes and corn, an average of 9.83 alleles locus^{-1}; however, only 77, 49, and 33 alleles were found in three sets of increasingly advanced inbred lines. Thus, allelic diversity was reduced to 34%, 22% and then to 15% of its original level in the progression from the wild teosintes and landraces, to 342 inbreds, to the 30 most popular U.S. inbreds, and then to the six inbreds that are included in the parentage of virtually all U.S.

Table 5 Number of alleles encountered at 23 allozyme loci in corn and teosinte and in three sets of inbred lines of corn (from Goodman and Stuber, 1980, 1983; Doebley *et al.*, 1984, 1985; Smith *et al.*, 1985a,b).

Locus	Total teosinte and maize	342 inbreds	30 most popular inbreds	6 most widely used inbreds*
Acp1	12	4	4	3
Adh1	6	3	1	1
Cat3	7	4	2	1
Ep	8	6	1	1
Est1	9	5	2	2
Glu1	27	9	6	2
Got1	9	2	2	1
Got2	8	2	2	1
Got3	6	1	1	1
Idh1	5	3	2	1
Idh2	7	3	2	2
Mdh1	14	4	2	1
Mdh2	19	5	4	1
Mdh3	8	2	2	2
Mdh4	11	2	1	1
Mdh5	12	4	2	1
Me	4	1	1	1
Mmm	2	2	1	1
Pgd1	9	3	2	2
Pgd2	4	2	2	2
Pgm1	14	2	1	1
Pgm2	16	4	3	2
Phi1	9	4	2	2
Totals	226	48	77	33
Mean	9.83	3.30	2.09	1.43
Per cent of total	100	34	22	15

*Lancaster types, C103, Oh43, Mo17; Reid types, B73, B37, A632.

single-cross hybrids. It is widely accepted that Mexico was the cradle of domestication of corn and that North American corn was derived from the Mexican landraces. Doebley *et al.* (1985) observed a total of 163 alleles, a mean of 7.09 alleles locus^{-1}, in a sample of 94 collections representing 34 Mexican races of corn. Table 6 shows that 18 (11%) of these alleles were present in overall frequencies between 0.78–0.99 in the Mexican collections; 17 of these 18 alleles were observed in all 94 of the Mexican collections and

Table 6 Frequency of alleles in Mexican races of corn and in the 30 most popular and six most widely used sets of inbreds (from Goodman and Stuber, 1980; Doebley et al., 1985).

Allelic frequency in Mexican races	Mean no. collections in which observed	No. alleles in frequency class	No. alleles in: Set of 30 lines	Set of 6 lines	No. alleles monomorphic in: Set of 30 lines	Set of 6 lines
0.99-0.78	93.9 (93-94)*	18	18	18	7	14
0.68-0.21	85.1 (71-94)	10	10	6	0	0
0.19-0.05	35.7 (11-67)	13	10	5	0	0
0.05-0.01	12.2 (2-16)	25	9	4	0	0
<0.01	1.7 (1-11)	97	2	1	0	0
Totals		163	49	34	7	14

*Range (in parentheses).

one was observed in all except one of the 94 collections. These same alleles also occur in closely similar high frequencies in 31 Bolivian races of corn (Goodman and Stuber, 1983) and in the teosintes (Doebley *et al.*, 1984). All 18 of these frequent and ubiquitous alleles were also observed in the 30 most popular public inbred lines and in all of the six inbred lines that occur in the parentage of virtually all U.S. commercial corn hybrids.

Alleles that were present in overall frequencies <0.01 stand in sharp contrast. Most of the 97 alleles in this category were present in only one or two of the 94 Mexican collections (mean of 1.7 alleles collection^{-1}) and only two of these alleles survived into the 30 most popular inbreds and only one into the set of six most widely used inbreds. The 48 remaining alleles were present in the Mexican (and also Bolivian) races in intermediate frequencies. All of the 10 alleles present in frequencies 0.21-0.68 and 10 of the 13 alleles present in frequencies 0.05-0.19 persisted into the 30 most popular inbreds; however, only about half of these alleles survived into the set of six most widely used lines and none became fixed in these advanced lines. None of the 25 alleles present in the Mexican races in overall frequency 0.01-0.05 was present in more than a few of the 94 collections and none became frequent in the sets of 30 and six inbred lines.

Table 7 gives the allozyme genotypes of the six inbred lines that are included in the parentage of virtually all commercial U.S. corn hybrids (only the nine loci that are polymorphic in the widely used inbreds are listed in Table 7). Among these nine polymorphic loci, eight are fixed in the three Lancaster-type lines and eight are polymorphic in the three Reid-type lines. Thus, the three Lancaster-type lines are monomorphic for 22 of the 23 loci and polymorphic for only one locus whereas the three Reid-type lines are monomorphic for 15 but polymorphic at eight loci. For 20 of the 23 loci the most frequent allele in the Mexican races was also the most frequent allele in the set of six widely-used inbred lines and for the three remaining loci the second most frequent allele of the Mexican races became predominant. Only three of the infrequent alleles of the Mexican races were found, and none was frequent, in the set of six most widely used inbred lines.

In summary, allelic frequencies, as well as the frequency of appearance of alleles in landraces are highly reliable predictors of survival ability, and hence probable usefulness in corn. As in barley, all alleles that were present in high frequency in landraces were also present in high frequency in the most advanced materials. The corn data also indicate that alleles present in intermediate frequencies, particularly the higher intermediate frequencies, frequently appear in heteroallelic combinations in elite hybrids. In contrast, alleles that are infrequent (f < 0.05) are rarely found in advanced germplasm and in no case did such alleles reach high frequency in the most elite inbreds. The reduction in allelic variability that has occurred in corn has usually been represented as erosion of valuable germplasm. However, the data for the precisely identifiable alleles given in Tables 5-7 indicate that the

Table 7 Allozyme genotypes of nine polymorphic loci* in the six inbred lines that are included in the parentage of virtually all commercial U.S. corn hybrids (from Goodman and Stuber, 1980).

Inbred line	Idh2	Mdh3	Pgd1	Pgd2	Locus Pgm2	Phi1	Est8	Glu1	Acp1
Lancaster-type									
C103	6	6	3.8	5	4	4	4	6	3
Oh43	6	6	3.8	5	4	4	4	6	2
Mo17	6	6	3.8	5	4	4	4	6	4
Reid-type									
B73	4	3.5	3.8	5	4	4	5	7	2
B37	6	6	2	5	8	4	5	7	2
A632	6	6	3.8	2.8	4	5	4	7	4

*The 14 remaining among the 23 loci are monomorphic (fixed) for the most common allele in all six of the most widely used inbred lines.

recurring cycles of selection to which corn has been subjected led to increases in the frequency of the prevalent alleles and to elimination of infrequent alleles. Thus, the reduction in allelic variability that has occurred in corn is perhaps more appropriately ascribed to purifying selection than to genetic erosion.

Genotypic organization in corn

The corn data indicate further that substantial reorganization occurred at the genotypic level during the evolution of advanced inbreds and hybrids and that this reorganization, in fact, ran largely counter to expectations based on the widely held proposition that heterosis is due to advantage of heterozygotes over homozygotes. Under the proposition of heterozygote advantage it is expected that the parental inbreds of successful hybrids will diverge in allelic composition so that a maximum number of loci will be heterozygous in the hybrids. However, among the nine polymorphic loci of Table 7 this was the case only for locus *Glu1*; the three Lancaster (L)-type and three Reid (R)-type inbreds are monomorphic for alleles *6* and *7*, respectively, so that all among the nine possible F_1 hybrids among the L- x R-type inbreds are expected to be *Glu1-6/Glu1-7* heterozygotes. However, fewer of the other F_1 hybrid combinations produce heterozygotes. Overall, among the 207 possible allelic combinations among the L- x R-type hybrids of Table 7, only 39 (19%) are heterozygous, whereas 168 (81%) are homozygous combinations. Thus, the data of Table 7 indicate that genetic uniformity (homoallelism) is advantageous for the great majority of loci, i.e., that two copies of the prevalent allele at these loci leads to wide adaptedness and high performance and that heteroallelic interactions between alleles of the same locus only occasionally lead to performance superior to that of the best homoallelic combination.

It can also be seen from Table 7 that none of the nine L- x R-type F_1 hybrids carried a full quota of the 33 alleles present in the set of six most widely-used inbred lines; three each among the nine possible F_1 hybrids have 18, 17, or 15 alleles (84, 81, or 78%, respectively, of the full quota of 33 alleles). Thus the data of Table 7 suggest that it is more important that the few most frequent alleles be present rather than just large numbers of alleles, i.e., that the quality is more important than the quantity of allelic variability.

ALLELIC AND GENOTYPIC DIVERSITY IN *AVENA*

I now turn to comparisons of tetraploid *Avena barbata* (*A.b.*) with its diploid ancestors for the insights the differences among these taxa provide concerning the relationship between kinds and amounts of genetic diversity and adaptedness. *A.b.* is a highly self-pollinated annual grass derived by polyploidization from the highly self-pollinating diploid *A. hirtula-A. wiestii* complex (Rajahathy and Thomas, 1974). *Avena hirtula (A.h.)* and *A. wiestii (A.w.)* are, respectively, the Mediterranean and desert ecotypes of a single biological species. F_1 hybrids between *A.h.* and *A.w.* regularly form seven bivalents in meiosis and they are fully fertile. *Avena barbata (A.b.)* regularly forms 14 bivalents in meiosis; homoeologous chromosome pairing is suppressed by a simple genetic system that restricts pairing to homologous chromosomes of the same genome. Recent studies of segregation of allozyme variants (Hutchinson *et al.*, 1983) show that *A.b.* behaves as a fully diploidized tetraploid, i.e., that pairing is fully preferential within each of the two sets of seven pairs of chromosomes and that no pairing occurs between chromosomes of the homoeologous sets.

A.h. and *A.w.* are indigenous throughout the Mediterranean Basin where they are found in sparse and more or less disjunct stands. *A.b.* is much more widely distributed: it is very common throughout the Mediterranean Basin across the Middle East to Nepal. It thrives under a wide range of ecological conditions and often occurs in massive stands numbering many millions of individuals in undisturbed sites, along roadsides, and in cultivated fields. It has also been a highly successful colonizer in Mediterranean-type climates throughout the world. *A.b.* is unquestionably much more widely adapted and much more successful in covering wide ranges more or less continuously then its diploid ancestor(s). It has often been postulated that the successes of polyploids are due to heterosis associated with ability to breed true for a highly heterozygous state in which allelic differences within loci are fixed through chromosome doubling, i.e. to "fixed heterozygosity". Recent studies (García *et al.*, 1991) show that fixed intralocus allelic variability has indeed contributed importantly to the superior adaptedness of *A.b.* and that this was a major factor in the evolution of this enormously successful tetraploid from the gene pool of the less successful diploid complex; however, these studies also showed that heterozygosity in itself contributes very little if anything to superior adaptedness in *A.h.*, *A.w.* or *A.b.*

Table 8 gives the number of alleles, single-locus genotypes and multilocus genotypes observed in collections from Israel, collections made in a narrow range of habitats where all three taxa often occur in mixed stands, and in collections made in a much wider range of habitats in Spain (*A.w.* has not been observed in Spain). The main thing to note in this table is that numbers of alleles and numbers of genotypes were much larger in the tetraploids than in the diploids.

Table 9 gives genotypic frequencies in the collections from Israel and Spain for nine representative allozyme loci. Three loci, *Pgm1*, *Got2* and *Mdh2*, which have long been known to be monomorphic or very nearly monomorphic worldwide for allele *100* (coded *1*), were monomorphic in all samples of the diploids (coded *11*) and tetraploid (also coded *11*). This result suggests that the *1* allele of each of these three loci confers superior

Table 8 Number of alleles and genotypes for 14 allozyme loci in tetraploid *Avena barbata* (*A.b.*) and its diploid ancestors, *A. hirtula* (*A.h.*) and *A. wiestii* (*A.w.*) (from García *et al.*, 1989, 1991).

	Plants	Alleles	Single-locus genotypes	14-locus genotypes
Diploids				
Israeli *A. wiestii*	800	20	30	55
Israeli *A. hirtula*	334	13	13	7
Spanish *A. hirtula*	754	40	40	107
Tetraploid				
Israeli *A. barbata*	384	29	46	77
Spanish *A. barbata*	4453	51	80	427

adaptedness in all genetic backgrounds, diploid or tetraploid, and in all environments worldwide in which *A.b* occurs. The few other alleles observed at these loci have always been rare and at a disadvantage, even when segregating in families grown under conditions of minimal competition - this result indicates that infrequent alleles, as in barley and other species, usually have adverse effects on adaptedness and that their fate is rapid elimination from populations in which they arise sporadically by mutation.

A new feature appears in locus *Got1* - namely weak polymorphism in the collections of *A.w.* from Israel and *A.h.* from Spain; alleles *2* and *3*, although always less frequent than allele *1*, appear to improve overall population adaptedness in the occasional sites where they are polymorphic in the diploids. One possible explanation is that these sites are environmentally heterogeneous on a microgeographical scale, including some patches that favor genotype *11*, and some patches that favor either duplex genotype *22* and/or *33*. Another possible explanation is that, in some environments, the reproductive rates of the duplex genotypes *11*, *22*, and/or *33* are higher when these genotypes grow in intimate association with one another than when they grow in pure stands (Allard and Adams, 1969). Another feature shows up in locus *Pgi1* in Spanish *A.b.*, namely the appearance of three heteroallelic quadriplexes, *100 100 103 103* (coded *13*), *100 100 104 104* (coded *14*), and *100 100 105 105* (coded *15*). However, all three of these heteroallelic quadriplexes are infrequent and it seems likely that their contributions to adaptedness are small. Thus loci *Got1* and *Pgi1* both resemble the three completely homoallelic loci in that each features a single

Table 9 Genotypic frequencies in samples of *A.w.*, *A.h.* and *A.b.* from Israel and Spain. Genotypes present in frequencies <0.02 in all samples are not listed (from García *et al.*, 1991).*

Locus	Code	*A.w.* Israel	*A.h.* Israel	*A.h.* Spain	*A.b.* Israel	*A.b.* Spain
Pgm1,Got2,Mdh2	*11*	1.00	1.00	1.00	1.00	1.00
Got1	*11*	0.75	1.00	0.95	1.00	1.00
	22	0.09		0.05		
	33	0.16				
Pgi1	*11*	1.00	1.00	0.48	1.00	0.88
	22			0.17		
	44			0.31		
	55			0.04		
	13[†]					0.02[†]
	14[†]					0.04[†]
	15[†]					0.06[†]
Mdh1	*22*	1.00	1.00	1.00		
	12[†]				1.00[†]	1.00[†]
Pgd2	*11*	1.00	1.00	1.00	0.48	0.12
	12[†]					0.83[†]
	23[†]				0.51[†]	0.05[†]
Mdh3	*11*	0.80	0.73			
	22	0.20	0.07	0.86	0.02	0.10
	66			0.14		
	12[†]				0.95[†]	0.84[†]
	34[†]				0.03[†]	0.06[†]
Lap1	*11*	0.90	1.00	0.39	0.19	0.02
	22			0.42	0.02	0.12
	33			0.02		0.10
	44	0.10		0.17		
	12[†]				0.23[†]	0.25[†]
	13[†]				0.30[†]	0.05[†]
	23[†]				0.21[†]	0.30[†]
	24[†]					0.13[†]
	34[†]					0.02[†]

*Pgm1, *Got2* and *Mdh1* were monomorphic for a single genotype (allele *100*, *100* in the diploids and (allele *100 100*, *100 100*) in the tetraploid, coded *11* in both cases. Heteroallelic quadriplexes (e.g., alleles *100 100*, *101 101*) are coded *12*.

[†]Denotes heteroallelic quadriplex.

predominant allele that is monomorphic in the great majority of diploid as well as tetraploid populations.

Still another feature, a very important feature, shows up in locus *Mdh1* - a single heteroallelic quadriplex *1122* (coded *12*) of this locus is monomorphic worldwide in the tetraploid; allele *2* of this locus is monomorphic in the diploids but allele *1* has not been found in the diploids. Clearly allele *2* is superior to allele *1* in the diploids, but in the tetraploid these two alleles complement each other such that two copies of each allele *1* and allele *2* confer adaptedness superior to that of four copies of either allele *2* (or allele *1*). The three remaining loci were included to represent a number of loci that are polymorphic for two or three alleles in the diploids and also feature extensive and complex patterns of polymorphism for several homoallelic and heteroallelic quadriplexes in the tetraploid. Note that five of the loci of Table 9 (*Pgm1*, *Got2*, *Mdh2*, *Got1*, *Pgi1*) were monomorphic or nearly monomorphic for a single homoallelic quadriplex in a wide range of environments. Thus, wide adaptedness was associated exclusively, or almost exclusively, with genetic uniformity for these five representative loci. However, heteroallelic quadriplexes were more frequent than homoallelic quadriplexes for the many loci represented in Table 9 by loci *Mdh1*, *Pgd2*, *Mdh3* and *Lap1*. A single heteroallelic quadriplex (coded *12*) was fixed for locus *Mdh1*; thus, wide adaptedness for this locus is associated with genetic uniformity (monomorphism for the *1122* quadriplex) but also with genetic diversity in the form of the heteroallelism of this quadriplex. The situation regarding genetic uniformity vs. genetic diversity is more complex for loci *Pgd2*, *Mdh3*, *Lap1* - most individual populations are polymorphic for several different homoallelic and several different heteroallelic quadriplexes of these loci. Thus, favorable interactions among alleles at the same and different loci appear to be very common in nature; however, if such interactions are to be exploited, some mechanism must be available to allow the favorably interacting alleles to become associated and some mechanism must be available to prevent their disassociation by segregation and recombination. Mating systems featuring predominant selfing have been effective in many species, including barley and other diploids, in building up and preserving favorable interlocus (epistatic) combinations of alleles (Allard, 1975, 1988; Saghai Maroof *et al.*, 1990; Pérez de la Vega *et al.*, 1991). Polyploidy followed by diploidization is a mechanism by which favorable intralocus combinations of alleles can be assembled and protected from disassociation, as shown by the data for *A.b.* given in Table 9. There is also some evidence at the allelic level (R. W. Allard, unpublished data) that this mechanism has been effective in tetraploid wheat. Vegetative reproduction is still another possible mechanism; such reproduction (e.g. by means of tubers in potatoes) is expected to hold favorable allelic combinations together between cycles of sexual reproduction.

SUMMARY AND CONCLUSIONS

What does all of this have to do with the title that was assigned to me -
"Predictive methods for germplasm identification"? The main finding relevant to identifying useful germplasm is that the frequencies of alleles of
presumably random samples of loci affecting discretely recognizable morphological, disease resistance, allozyme and restriction fragment variants are
highly correlated with adaptedness and productivity. The alleles of several
species that have been studied in detail fit into one of three disjunct classes.
Alleles of Class I include those alleles that are predominant (present in
frequency ≈ 0.80 or higher) in a high proportion of accessions from a broad
range of environments. These ubiquitous and frequent alleles are also nearly
always monomorphic or nearly so in the most advanced cultivars from all
ecogeographical regions in which the species grows - this suggests that each
of these predominant alleles makes a significant contribution to wide adaptedness and high productivity in nearly all genetic backgrounds and in nearly
all agricultural environments. Nevertheless, these all-purpose wild-type
alleles probably do not rate highest priority in the sampling of accessions to
identify potentially useful germplasm. This is because such alleles would
almost inevitably have been included in all samples of germplasm involved
in domestication and it seems likely that nearly all alleles of this type will
already be present in the breeding stocks of virtually all growing areas.
Obviously, however, gene bank managers and breeders should remain alert
to such alleles because each discovery of an allele of this type not already
present in local breeding stocks offers promise of opportunity for significant
improvement in adaptedness and productivity.

Class II alleles are those that are absent or infrequent in nearly all accessions. The population behavior of such alleles during the conversion of wild
materials into landraces and advanced cultivars, and also in various kinds of
experimental populations, has consistently been that of semilethals or subvitals. Mutation introduces such alleles into populations at rates of approximately 1×10^{-6} locus^{-1} generation^{-1} and the fate of any single semilethal or
subvital is nearly always rapid elimination from the population (Dobzhansky,
1955). Occasionally, such alleles (e.g., endosperm mutants such as the
sugary alleles of corn) affect quality or other characteristics in useful ways.
In general, however, these infrequent alleles offer little or nothing of value
in crop improvement programs.

Class III alleles are those that are present in intermediate to high frequencies (say $f = 0.20$-0.80) in the accessions and advanced cultivars from many
but not all ecogeographical regions. The population biology of such alleles
indicates that their contributions to adaptedness and performance often differ
in different ecogeographical regions, as well as from year to year within
given ecographical regions, and from one genetic background to another.
Data from barley indicate that alleles of this third class frequently enter into

favorable interlocus (epistatic) interactions in diploids whereas data from *A.b.* indicate that such alleles, in addition, also frequently enter into favorable intralocus heteroallelic interactions that increase allelic diversity and also improve adaptedness. Thus, alleles of Class III appear to be especially frequent contributors to increased allelic diversity. Consequently, among the three classes of alleles, those of class III should probably be assigned highest priority in the sampling of germplasm.

Experienced plant breeders often contend that the information provided by traditional methods of characterizing germplasm is inadequate for predicting the potential value of accessions in crop improvement programs. The observation that the frequencies, and the within-population dynamics of alleles of loci governing discretely recognizable variants, are closely correlated with adaptedness, performance and other important quantitative characters suggests an efficient and cost-effective method of evaluating accessions for such difficult-to-measure characters. If an accession carries one or more alleles of the first and/or third classes described above, and these alleles are not present in the breeding stocks of a given ecogeographical region, the chances are good that this accession will be of value in crop improvement programs of that region. Fortunately, frequencies of alleles of loci governing discretely recognizable morphological, allozyme, and restriction fragment variants in germplasm collections, and in genetically enhanced populations, can usually be determined quickly, precisely and relatively inexpensively. This is also the case for resistance to seriously damaging races of pathogens, particularly if specific pathotypes are available for gene-for-gene tests of host resistance vs. pathogen virulence. The rapidly growing arrays of restriction fragment variants, including random amplified polymorphic DNA variants (Williams *et al.*, 1990), appear to me to offer outstanding opportunities for identifying promising alleles for introgression into breeding stocks.

The task of identifying and utilizing superior germplasm at the genotypic level is unfortunately much less straightforward and much more difficult than at the allelic level. This is because having superior alleles in breeding stocks is not enough in itself; superior alleles must be assembled into superior combinations at the intralocus level to exploit favorable heteroallelic interactions and also at the interlocus level to exploit favorable epistatic interactions among alleles of different loci (Allard, 1988; Pérez de la Vega *et al.*, 1991; García *et al.*, 1991). These are substantial complications because, as experienced breeders are acutely aware, numbers of genotypes increase exponentially with numbers of loci and numbers of alleles per locus; the consequences are that large numbers of cycles of segregation and recombination are required and that population sizes must be large to guarantee production of the most useful genotypes. Furthermore, and no less troublesome, laborious and expensive testing for yield and other quantitative characters is required over many years at many locations to determine the value of novel genotypes (Bradley *et al.*, 1988).

Each of the nine tables and six figures above were included to illustrate one or more of the above points. All of the 36 or so alleles that were ultimately incorporated into modern Californian barley cultivars were present in the Middle Eastern landraces of barley but progress in assembling the favorable alleles into superior combinations was slow until hybridization among selected genotypes became prevalent in the 1920s and 1930s. These crosses greatly increased opportunities for segregation, recombination, and the production of new genotypes, thus enhancing selective progress. Composite Cross II, which was made up of the 378 possible intercrosses among the 28 cultivars selected to represent the major barley growing areas of the world (the parents of CCII were probably the first "core" collection), played an important role in this process (Allard, 1988, 1990). Much of the progress made in this period in barley resulted from exploitation of favorable epistatic combinations of alleles of different loci, held together by restriction of recombination due to the mating system of predominant selfing. However, no intralocus heteroallelic combinations have been observed in barley which is not surprising because no adequate mechanism exists in diploids to prevent segregation from breaking up favorable associations once they have formed. The situation in corn appears to be more or less parallel to that of barley. All ultimately successful alleles of the 23 loci monitored (Tables 5-7) were in place in the open-pollinated varieties but the breeding method of the time - mass selection - was inefficient and the rate of progress in assembling the 33 most successful alleles of the 23 loci into superior combinations was slow. When more effective breeding methods based on double-cross hybrids were adopted progress increased dramatically (Figure 1) and when still more efficient single-cross procedures, combined with superior methods of evaluation of hybrids, were adopted in the 1960s, rates of progress nearly doubled again. However, few heteroallelic intralocus combinations were observed (Table 7) indicating that most of the progress was due to increases in the frequency of superior homoallelic combinations and/or to exploitation of favorable epistatic interactions among alleles at different loci. It should be noted that crossing homozygous inbreds to produce single-cross cultivars provides protection of favored interlocus, as well as favored intralocus combinations of alleles, from disassociation due to segregation and recombination, but only for a single generation.

Breeding in barley and corn, as well as in other major crops, has increasingly focused on crosses among elite materials and rates of progress indicate not only that this strategy has been successful but also that there has been little, if any, slowing of progress due to reduction of exploitable genetic variability (Figures 1, 2). Among the 25 or more loci that have been monitored over time in experimental populations of barley at least 15 loci ($\approx 60\%$) remain conspicuously polymorphic in the most advanced cultivars of California. Assuming only two alleles per locus the number of possible homozygous genotypic combinations for this particular set of polymorphic loci exceeds 30,000. If these 25 loci are representative, the number of polymorphic loci in the gene pool of advanced materials must be large and the total number of novel possible genotypes must be enormous. It consequently seems unlikely that readily exploitable genetic variability will soon be exhausted in this or other similar sets of barley breeding stocks. This also appears to be the case for the Reid and Lancaster gene pools of corn. Prospects for rapid and long-continued advances in adaptedness, performance and quality seem still better as we introgress more and more favorable Class I and Class III alleles into breeding stocks and assemble them into increasingly superior intralocus and interlocus interactive combinations in our major crop species.

REFERENCES

Allard, R. W. (1975) The mating system and microevolution. *Genetics* 79:115-126.

Allard, R. W. (1988) Genetic changes associated with the evolution of adaptedness in cultivated plants and their wild progenitors. *Journal of Heredity* 79:225-238.

Allard, R. W. (1990) The genetics of host-pathogen coevolution: Implications for genetic resource conservation. *Journal of Heredity* 81:1-6.

Allard, R. W.; Adams, J. P. (1969) The role of intergenotypic interactions in plant breeding. *Proceedings XII International Congress of Genetics* 3:349-370.

Allard, R. W.; Kahler, A. L.; Weir, B. S. (1972) The effect of selection on esterase allozymes in a barley population. *Genetics* 72:489-503.

Allard, R. W.; Saghai Maroof, A.; Zhang, Qifa; Jorgensen, R. A. (1990) Genetic and molecular organization of ribosomal DNA (rDNA) variants in wild and cultivated barley. *Genetics* 126:743-751.

Bradley, J. P.; Knittle, K. H.; Troyer, A. F. (1988) Statistical methods in seed corn product selection. *Journal Production Agriculture* 1:34-38.

Brown, A. H. D.; Munday, J. (1982) Population genetic structure and optimal sampling of land races of barley from Iran. *Genetica* 58:85-96.

Dobzhansky, Th. (1927) Studies on the manifold effect of certain genes in *Drosophila melanogaster*. *Zeitschrift fur Induktive Abstammungs-und Vererbungslehre* 43:330-388.

Dobzhansky, Th. (1955) A review of some fundamental concepts and problems of population genetics. *Cold Spring Harbor Symposium of Quantitative Biology* XX:1-15.

Doebley, J. F.; Goodman, M. M.; Stuber, C. W. (1984) Isoenzymatic variation in *Zea* (Gramineae). *Systematic Botany* 9:203-218.

Doebley, J. F.; Goodman, M. M.; Stuber, C. W. (1985) Isozyme variation in the races of corn from Mexico. *American Journal of Botany* 72:629-639.

Duvick, D. N. (1984) Genetic contribution to yield gains in U.S. hybrid maize, 1930-1980. In: Fehr, W. R. (ed.), *Genetic contribution to yield gains of five major crop plants. Special publication 7.* Crop Science Society of America, Madison, pp. 15-47.

García, P.; Morris, M. I.; Sáenz-de-Miera, L. E.; Allard, R. W.; Pérez de la Vega, M.; Ladizinsky, G. (1991) Genetic diversity and adaptedness in tetraploid *Avena barbata* and its diploid ancestors, *A. hirtula* and *A. wiestii*. *Proceedings of the National Academy of Sciences USA* 88:1207-1211.

García, P.; Vences, F. J.; Pérez de la Vega, M.; Allard, R. W. (1989) Allelic and genotypic composition of ancestral Spanish and colonial Californian gene pools of *Avena barbata*: Evolutionary implications. *Genetics* 122:687-694.

Goodman, M. M.; Stuber, C. W. (1980) Genetic identification of lines and crosses using isoenzyme electrophoresis. *Annual Corn Sorghum Research Conference Proceedings* 35:10-31.

Goodman, M. M.; Stuber, C. W. (1983) Races of maize. VI. Isozyme variation among races of maize in Bolivia. *Maydica* 28:169-187.

Harlan, H. V.; Martini, M. L. (1929) A composite hybrid mixture. *Journal of the American Society of Agronomy* 21:407-490.

Hutchinson, E. S.; Price, S. C.; Kahler, A. L.; Morris, M. I.; Allard, R. W. (1983) An experimental verification of segregation theory in a diploidized tetraploid: esterase loci in *Avena barbata*. *Journal of Heredity* 74:381-383.

Kahler, A. L.; Allard, R. W. (1981) Worldwide patterns of genetic variation among four esterase loci in barley (*Hordeum vulgare* L.). *Theoretical and Applied Genetics* 59:101-111.

Kahler, A. L.; Heath-Pagliuso, S.; Allard, R. W. (1981) Genetics of isozyme variants in barley. II. 6-phosphogluconate dehydrogenase, glutamate oxalate transaminase, and phosphatase. *Crop Science* 21:536-540.

McDonald, B. A.; McDermott, J. M.; Allard, R. W.; Webster, R. K. (1989) Coevolution of host and pathogen populations in the *Hordeum vulgare-Rhynchosporium secalis* pathosystem. *Proceedings of the National Academy of Sciences USA* 86:3924-3927.

Pérez de la Vega, M.; García, P.; Allard, R. W. (1991) Multilocus genetic structure of ancestral Spanish and colonial Californian populations of *Avena barbata*. *Proceedings of the National Academy of Sciences USA* 88:1202-1206.

Rajhathy, T.; Thomas, H. (1974) Cytogenetics of oats (*Avena* L.). *Genetic Society of Canada Miscellaneous Publication No. 2*, Ottawa.

Saghai Maroof, M. A.; Allard, R. W.; Zhang, Qifa (1990) Genetic diversity and ecological differentiation among ribosomal DNA alleles in wild and cultivated barley. *Proceedings of the National Academy of Sciences USA* 87:8426-8490.

Saghai Maroof, M. A.; Soliman, K. M.; Jorgenson, R. A.; Allard, R. W. (1984) Ribosomal DNA (rDNA) spacer-length (sl) variation in barley: Mendelian inheritance, chromosomal location and population dynamics. *Proceedings of the National Academy of Sciences USA* 81:8014-8018.

Smith, J. S. C.; Goodman, M. M.; Stuber, C. W. (1985a) Genetic variability within U.S. maize germplasm. I. Historically important lines. *Crop Science* 25:550-555.

Smith, J. S. C.; Goodman, M. M.; Stuber, C. W. (1985b) Genetic variability within U.S. maize germplasm. II. Widely-used inbred lines 1970-1979. *Crop Science* 25:681-685.

Troyer, A. F. (1990) A retrospective view of corn genetic resources. *Journal of Heredity* 81:17-24.

Webster, R. K.; Saghai Maroof, M. A.; Allard, R. W. (1986) Evolutionary response to barley composite cross II of *Rhynchosporium secalis* analyzed by pathogenic complexity and gene-by-race relationships. *Phytopathology* 76:661-668.

Williams, J. G. K.; Kubelik, A. R.; Livak, K. J.; Rafalski, J. A.; Tingey, S. C. (1990) DNA polymorphisms amplified by arbitrary primers are useful as genetic markers. *Nucleic Acids Research* 18:6531-6535.

Zhang, Qifa; Saghai Maroof, M. A.; Allard, R. W. (1990) Worldwide pattern of multilocus structure in barley determined by discrete log-linear multivariate analysis. *Theoretical and Applied Genetics* 80:121-128.

Chapter 7
Identification of useful germplasm for practical plant breeding programs

A. B. Maunder

Dekalb Plant Genetics, Route 2, Lubbock, Texas 79415

Plant breeding progress, regardless of crop, depends on the selection and utilization of the most appropriate germplasm and the particular breeding method or methods which can be applied to this germplasm. Significant gains in yield, as well as yield stability through the incorporation of defensive traits, reflects successes of sizeable magnitudes. Nevertheless, the rate of yield gain in corn (*Zea mays* L.) and sorghum [*Sorghum bicolor* (L.) Moench] continues to decrease; concern is often expressed about the presence of yield plateaus, and our efficiency in new variety or inbred development must be scrutinized in relation to both germplasm and breeding methodologies utilized.

Corn and sorghum have both become hybridized crops since 1930, with 1990 yields increasing some 471 and 455%, respectively, over this 60-year period. Wheat (*Triticum aestivum* L.), though essentially all varietal, has increased some 191% during this same period, and soybeans [*Glycine max* (L.) Merr.], also varietal, by 159%. Genetic gain in corn reported by Duvick (1977) and Russell (1974) for the four decades since the introduction of hybrids ranged from 57 to 63%. As suggested by Hallauer (1981), the significant yield increases for corn and sorghum in the 1960s and 1970s was likely related heavily to changes in management practices - for example, farmers' use of single cross hybrids and higher population densities in corn, and almost total acceptance of hybrids in sorghum. Are we on a yield plateau? Frey (1971) suggested that yield plateaus may exist momentarily for a certain species in a given region, but he sees no evidence that a worldwide plateau exists for any species which is receiving significant research attention. However, plateaus should be expected when breeding programs face new pests or attempt breakthroughs such as related to yield or nutritional quality.

The efficiency of developing superior parent lines or varieties is of primary importance in a breeding program. Lindstrom (1939) reported that only 677 of the 27,641 lines (2.4%) from 24 public maize programs were useful; but all were defective for one or more traits, including poor yields. Hallauer and Miranda (1981) suggested that there have likely been one million inbred lines advanced to test-crossing from 1939 to 1981. Only 0.01% has been used to any significant extent in commercial hybrids. This figure of one in 10,000 for corn also seems reasonable when applied to a commercial program involved in the improvement of hybrid sorghum. In comparison to Lindstrom's estimate of 2.4%, the 0.01% may seem low, but the latter estimate more realistically reflects the elite level of a more intensive breeding effort (consequently a less likely probability) at which breeders are attempting to isolate significantly improved parental inbreds. It seems appropriate then to ask whether the current choice of germplasm or breeding approach is the most effective or efficient? This ratio obviously gives support to the critics who suggest that breeding relies too heavily on random chance rather than science.

EXAMPLES OF USEFUL GERMPLASM

Before discussing sources of potentially useful germplasm, examples of breeding progress affecting both offensive and defensive traits seems appropriate.

Yield Improvement

Rice (*Oryza sativa* L.) and wheat made significant yield gains during the past four decades as illustrated by the green revolution for Third World countries. The rice cultivar IR-8 with shorter height, stiffer straw, and photoperiod insensitivity produced yields double those of commonly grown cultivars in tropical environments. Borlaug *et al.* (1969) believed that the semidwarf wheats accounted for much of the quadrupling of yields in Mexico from 1955 to 1969. A similar testimonial can be made for the wheat cultivar Gaines, which showed a 74% increase over the commonly grown cultivar Burt on dryland conditions and 27% more grain under irrigation (Frey, 1971). Both Gaines and the Mexican semidwarf wheats trace much of their superiority to a 1946 Japanese semidwarf introduction, "Norin 10", and progeny from its cross to Brevor, which had up until that time been considered a short cultivar. By 1966, a soft red winter wheat, "Blueboy", and a hard red winter wheat, "Sturdy", were the first of their class to be released with this semi-dwarf character and significantly improved yields (Dalrymple, 1980). Sturdy obtained its dwarfing genes from "Sen Seun 27", a 1947 Korean introduction, which offered another source of dwarfing genes (Atkins *et al.*, 1967). Although Norin 10 has had a worldwide influence on wheat improvement, it was lacking in disease resistance, was handicapped by a short coleoptile, and often expressed male sterility which allowed for outcrossing. Credit for successfully using this germplasm must go to breeders such as Borlaug of CIMMYT and Vogel of Washington State University.

Following the successful release of Sturdy, K. B. Porter (Texas A&M University), using only semi-dwarf, unreleased germplasm from Kansas outside his own breeding material, effectively concentrated on yield with outstanding releases of "TAM101", "TAM105", and "TAM107", the latter with stem rust resistance and thereby going farther north. The TAM101 and TAM105 cultivars likely accounted for 20% of the hard red winter acreage in 1984. Porter (1990, pers. comm.) suggested using only elite germplasm when yield was the most significant trait. He also felt that the evaluation environment, as well as interaction of new germplasm with current material, was critical. The success of Porter and workers in other Great Plains states who developed new improved semidwarf varieties (Table 1) was probably responsible for the slow acceptance of hybrid wheat.

Table 1 Semi-dwarf area as a proportion of total wheat area in the United States (Dalrymple, 1986).

Year	Semi-dwarf area (ha)	Proportion of total wheat area (%)
1964	651,000	2.9
1969	1,540,300	7.0
1974	6,376,600	22.1
1979	9,052,000	31.3
1984	18,815,000	58.7

Probably no example of successful use of corn germplasm has received more reference than the Iowa Stiff Stalk Synthetic (BSSS). This population was established by George Sprague in 1933-34, with its components primarily restricted to lines expressing resistance to stalk breakage. Hallauer and Miranda (1981) indicated BSSS had only average yield, above average stalk quality, vigorous plant type, dark green leaf coloration, good ear size, no distinctive features for pest resistance, and full season maturity for the Central U.S. Corn Belt. That it produces lines with above-average combining ability with other elite lines (especially the Lancaster type) is obvious by progeny such as inbred lines B14, B37, and B73. Although BSSS appears uniform in phenotype, Hallauer and Miranda (1981) indicated that hybrids produced among elite lines selected from the synthetic (e.g., B37 x B73) often express high yields. This indicates there is substantial genetic variability for yield within BSSS.

The diversity existing between the flint and dent corns has recently been exploited successfully. Hybrids using Argentine flints benefit from the diverse germplasm of the dents just as we saw in the U.S. between the New England flints and the Southeastern dents. The significant heterosis attributed to the temperate dent germplasm may result from more efficient physiological processes. Also, the shallow flint kernel types benefit from recombination with the deep dent kernel types. As regards to temperate x tropical germplasm, much progress in performance relates to improved plant architecture, such as lower ear placement, shorter plant height, and decreased photoperiod sensitivity (as has been reported throughout the subtropics). In the more tropical areas, breeders must rely on a higher percentage of tropical germplasm because of adaptation requirements, especially disease resistance. Maize hybrids in tropical and subtropical areas have been improved by 25 to 40% during the past 10-15 years with the introgression of a wider germplasm base (N. G. Robinson, 1990, pers. comm.); Table 2 illustrates gains possible by system of hybridization.

Table 2 Percentage yield advantage by type of hybrid over the variety Suwan 1 in Thailand (N. G. Robison, 1990, pers. comm.).

Type of hybrid	Yield (% of Suwan 1)	Reps of data
Topcross	121.8	171
4x	128.1	93
3x	136.4	90
2x	146.8	27

Sorghum yields doubled shortly after the release of hybrid cultivars in 1957 (Figure 1). At least half the early hybrids were dependent on common varieties as opposed to parental lines developed for hybrid use. Webster (1976) suggested that, by 1960, the genetic potential of sorghum was nearly exhausted because prehybrid cultivars traced essentially to only 29 plant introductions. Fortunately, by the mid-1960s, a new generation of hybrids with 25% higher yields were developed, in large part because of the introgression of plant introductions from Sudan and Nigeria. These unique food types imparted agronomic traits such as drought tolerance, disease resistance, stiff stalk, improved combining ability, and increased time in GS-3 (bloom to maximum dry weight) resulting in heavier test weights. Much of the yield fluctuation since the late 1960s related to greenbug [*Schizaphis graminum* (Rodini)] damage and the resulting emphasis on insect breeding, available moisture, and reduced cultural inputs such as irrigation.

Figure 1 U.S.A. sorghum yields from 1930 to 1990 in bushels acre[-1] (D. R. Krieg, 1990, pers. comm.).

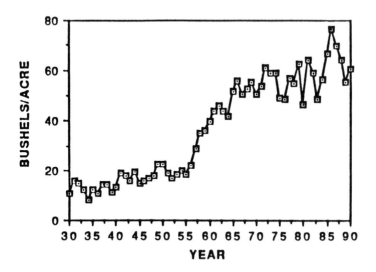

Resistance to insects

Resistance to insects has become a popular topic as environmental concerns increase, as well as a need to reduce input costs. For example, much public effort has been spent on the development of Hessian fly-resistant wheats at Purdue University and likewise considerable public and private research has worked towards reducing losses to the European corn borer (*Ostrinia nubilalis*) in maize. One year's savings ($17 million) through planting varieties resistant to the Hessian fly more than offsets the entire cost of developing these varieties for the past 50 years (National Academy of Science, 1972). Also, during the past 22 years, sorghum has benefitted measurably from breeding efforts for resistance to both the greenbug and the sorghum midge [*Contarinia sorghicola* (Coquillet)], a more serious pest worldwide. Currently, considerable breeding efforts are aimed at reducing losses by the sorghum chinch bug (*Blissus leucopterus*). A weedy sorghum, *S. virgatum* (Hack.) Stapf., provided the source of resistance to greenbugs in the line KS-30, which was first released to breeders in 1969 by H. L. Hackerott (pers. comm.). Both SA-7536-1 and IS809 were released also as germplasm resistant to greenbug biotype C. A plant introduction from Spain, P.I. 264453, and a Russian line, "Capbam", were also available to breeders but, because they were photoperiod-sensitive and tall, the lines received only limited attention. Later, however, when in 1980 the new E biotype emerged across the Great Plains, these latter two sources proved resistant. For those who had worked with them, only 3 years (as compared to 8 years for C) were needed to once again have resistant hybrids. Experience with the greenbug reconfirms that working diverse sources of resistance improves the chance of keeping up with changing biotypes.

Sorghum midge resistance, which is conditioned by a recessive gene and more difficult to screen, also traces to plant introductions primarily from Ethiopia and Brazil. In contrast to greenbug-resistant hybrids, which may actually have benefitted yield in the absence of the insect, sorghums with midge resistance have generally exhibited such deleterious effects as small seed, limited drought tolerance, late maturity, and reduced levels of combining ability. By working with diverse sources of resistance, the breeder can hopefully raise the economic injury level from the current 5-10 insects per panicle.

Tolerance to drought

Most stress-tolerant cultivars trace their origin to geographic areas of similar environmental conditions. "Turkey", the first hard red winter wheat of significance in the United States, and grown from 1919 to 1939 on more acres than any wheat (Dalrymple, 1988), excelled for its superior yield, in large part due to both drought tolerance and winterhardiness. Scout, the fifth most widely grown variety, possessed some Turkey germplasm and also exhibited yield stability, in large part due to drought tolerance.

Sorghum, a crop respected for drought tolerance, has improved that image with the development of hybrid cultivars which produce more yield per unit of water. As pointed out by Allard (1969), selection for specific attributes of drought resistance has not been very successful. Rather, breeders have focused attention for general adaptation to aridity. Of the locally adapted varieties compared to world collection material, more progress is expected through future use of introductions. Rosenow and Clark (1987) developed usable inbreds possessing "stay-green" (or non-senescence) from converted plant introductions. Sorghums from northeast and west Africa and the Middle East have given more diffusive resistance, greater root development, increased dormancy, and heat tolerance to modern cultivars. DK-46, a widely accepted drought-tolerant hybrid, additionally has the unique attribute of osmoregulation along with stay-green. Its diverse pedigree utilizes temperate germplasm for stalk quality along with a minimum of tillering and introduced germplasm for stay-green and diffusive resistance. For drought tolerance, where germplasm selection is critical, the ability to identify and select through appropriate screening takes on equal significance.

Improved nutritional aspects

Sorghum improvement perhaps best illustrates the use of germplasm to improve feed value. At the time of hybrid introduction in 1956, this crop was considered to have only 88% of the nutritional value of corn. By 1984, however, the National Research Council (1984) showed that the nutritional value of flaked sorghum had increased to 97% of corn for the net energy of maintenance, 96% for net energy of gain, and 97% for percentage total digestible nutrients. Maunder (1971) suggested that the new hetero-yellow or pure yellow hybrids, which were likely to be grown on more than 50% of the U.S. acreage, would maintain the 10% improved feed value illustrated in numerous evaluations to that date. This yellow germplasm from Nigeria and Sudan began to be widely used in commercial hybrids in 1964. The exact reason for its improved nutritional value is yet to be determined; but increased digestibility, lower tannin levels, an improved array of amino acids, better palatability, and perhaps a more corneous endosperm are likely contributors to this obvious improvement. Most recently, Brandt *et al.* (1989) at Kansas State University found no differences in dry matter intake, average daily gain, or feed conversion between steers fed steam-flaked Milo (sorghum) or steam-flaked corn.

Disease Resistance

Downy mildew [*Sclerospora sorghi* (Kulk) Weston and Uppal] of grain sorghum was observed by Nider *et al.* (1969) in the systemic form in four provinces in Argentina during the 1968-69 growing season. By screening a wide range of cultivars at three locations, it was possible to show resistance in the three principal types of sorghum, i.e., grain, silage, and grazing. By 1970, mildew was regarded as a significant problem in the U.S. Fortunately, the Zerazera sorghums of Ethiopia carried resistance which, through conversion, provided inbreds such as TAM428 and Tx430 (Tx2536 x IS12661C). Both of these inbreds have gained widespread use in the United States and Mexico.

Downy mildew essentially wiped out the corn acreage in Thailand. The "Suwan-1" synthetic, which included Philippine germplasm for mildew resistance as well as ETO (from Estación Tulio Ospina) and Tuxpeño breeding materials, has become the primary synthetic cultivar in Thailand. Adapted hybrids now rely on topcrosses with Suwan-1 or utilize inbreds derived from this significant source material.

Adaptation

Introgressing temperate and tropical germplasm frequently illustrates the need for adaptation. Subtropical maize areas can benefit from single crosses of temperate x tropical germplasm, such as materials from southern Brazil (30°S) or northern Mexico (26°N). The 15-20°N-S latitude tropical maize zone for these countries, however, needs one or more backcrosses to the tropical parent. Perhaps the greatest hindrance to wider use of tropical maize germplasm in temperate zones is the photoperiod response (N. G. Robinson, 1990, pers. comm.). Sorghum, also, appears to adapt based on the percentage of tropical germplasm in the hybrid; hybrids with more than 50% tropical germplasm perform best under low or tropical latitudes, whereas hybrids with 50% or less tropical germplasm perform well in southern Texas. In Argentina, a similar relationship is evident, with the Chaco province of the north requiring a greater level of tropical germplasm in the hybrids. In wheat, the early use of Turkey and Scout, as already pointed out, related to their tolerance to winter kill and drought stress, both traits essential for adaptation to the Great Plains.

Adaptation genes have, frequently and of necessity, been bred into crops. A good example relates to the reduction or elimination of silk delay in corn, which has been considered a prime reason for yield increases during the 1960s and 1970s (B. Tsotsis, 1991, pers. comm.). Avoidance of silk delay was a selection criterion when doubling planting density and producing hybrids at much higher fertility levels. As regards to soil fertility, IR-8 rice outyielded typical cultivars by 20% at low levels of nitrogen, but by three-fold at a rate of 132 kg ha^{-1} (Frey, 1971).

Standability/height

Often newer cultivars such as IR-8 perform to their potential in high productivity environments. An important innovation has been the use of dwarfing genes, such as derived from Norin-10, in the semidwarf wheats, with Gaines and Sturdy being two examples. As a general rule, the cereals have benefitted from the shorter plant height for standability in higher yielding environments. Most sorghums in the world collection carry dominant genes for tallness, as did the early introductions into the United States. By 1961, however, commercial hybrids were being grown with one allelic pair of height genes heterozygous and the other three pairs recessive. Generally, today's hybrids have only one allelic pair homozygous dominant for height.

Cytoplasmic-Genetic Male Sterility

Crops that can be readily sterilized through an interaction of non-restoring nuclear genes and sterile cytoplasm such as corn, sunflower (*Helianthus annuus* L.), sorghum, wheat, cotton (*Gossypium hirsutum* L.), and rice have at times required the use of wild or weedy relatives for the source of cytoplasm. Such was true for rice, sunflower, cotton, and wheat, the latter exploiting the cytoplasm of the tetraploid species *T. timopheevi* L. Sorghum offers an opportunity to isolate numerous sources of sterile cytoplasm from wild or exotic types, which hopefully can diversify a system which currently is without alternatives for hybrid production.

Summary

These examples give strong support to the use of exotic germplasm, to populations with favorable genetic variability and, when possible, to the use of single crosses for significant increases in productivity, including necessary defensive traits. The source and most effective selection of germplasm, however, will determine the efficiency and potential accomplishments from the appropriate breeding methodology applied.

SOURCES OF USEFUL GERMPLASM

Choice of germplasm to be used in a practical plant breeding program may be the most critical decision facing the breeder, although the choice of breeding method and system of evaluation can be equally important. Goodman (1985) further pointed out that choice of materials is perhaps most critical to the success of the program, but one that can be readily and cheaply changed. When one reviews the literature or reflects on plant breeding courses, however, far more time is devoted to breeding methods than to the selection of useful germplasm. Hallauer and Miranda (1981) stated that the germplasm initially selected could very well serve as the basis of the breeding program for the lifetime of the breeder. Obviously, maximum potential improvement to be attained via breeding rests on this choice of germplasm, while the actual breeding approach will determine how much of the potential can be realized. Choice of materials will depend on the objectives of the breeder, demands of the seed industry, and consumer preference. Often cultivars, lines, or populations of value to one program will not be appropriate to another merely because of environmental considerations or established materials already in use.

Breeding progress, as measured by genetic gain in yield, is not a smooth progression, but shows severe fluctuations which suggests that the breeder needs to put more emphasis on appropriate germplasm selection and adequate methods of evaluation to improve stability of performance. For genetic gain, we must recognize the importance of selecting the low frequency of favorable recombinants, a frequency dependent on the presence of appropriate genetic variability within the original population. Therefore, the number of families being worked will generally be more productive than intensive selection within a few families, unless population size is significantly increased. Most population sizes in pedigree breeding or random mating populations are inadequate for either transgressive segregation or the rearrangement of a significant number of favorable alleles. While it is recognized with applied maize improvement that second cycle breeding of elite lines is universally practiced in U.S. agriculture, Hallauer and Miranda (1981) suggested that the apparent yield plateau of maize in the 1950s (about 2520 kg ha^{-1}) stimulated the development of population improvement programs. Shortly thereafter, yields once again made a significant increase with germplasm better adapted to higher plant populations, increased levels of fertilizer and, of course, a significant factor, by 1970 some 75% of the corn belt acreage was planted to single cross hybrids (National Academy of Science, 1972).

Before discussing sources of germplasm more specifically, several apparent conclusions relative to this topic seem appropriate. Second cycle (pedigree breeding) and backcross breeding characterize an applied program and would be considered short-term approaches requiring more elite source material, as opposed to intermediate or long-term activities. Also, breeding in the classical definition remains a numbers game which implies maximization of efficiency of all three components, i.e., germplasm selection, breeding methodology, and evaluation. Success in a program, whether it is hybrid or variety oriented, suggests maximum genetic diversity of parental material to either maximize heterosis or provide more variability for additive improvement. It is much easier to start a program with diversity than to try and develop it.

The classical approach to maize breeding in temperate regions has been to produce hybrids from crosses between inbreds originating from the Reid group with inbreds from the Lancaster group, whereas under tropical situations Tuxpeño by ETO or Cateto flints are productive. In hybrid sorghum, the initial, and widely used, heterotic pattern was Kafir x Milo. However, the heavy use of plant introductions suggests that no obvious pattern exists, with most breeding programs choosing parental inbreds based more on fertility restoration characteristics than defined heterotic groups. This brings up an interesting approach to diversity in that the restorer sorghums commonly trace to the center of origin (also the source of sterile cytoplasm), whereas the non-restorers are more frequently found away from the center of origin. Besides Milo, the Feteritas and Hegari types of restorers are to be found nearer the center of origin, whereas the Kafirs trace to southern Africa. A recent collection of some 50 cultivars from this latter area, when screened for restoration, indicated only three, or 6%, of this material carried restorer genes. Certainly, when choosing breeding material, it is essential to know, or soon discover, the heterotic group to which it best fits and to avoid crossing at random and subsequently mixing combining ability patterns.

In summary, an applied breeder must understand the predominant heterotic action of the source material in his program. Testcrossing is the only method to evaluate this trait, as phenotypic selection may be misleading. My experience with sorghum suggests that additive gene action is perhaps more important with later maturing germplasm, while non-additive gene action tends to be more important in the earlier maturity hybrids. Heritability/combining ability studies can be helpful when selecting the best germplasm, but their worth may be limited to the actual parents of the diallel and not apply to the remaining source material of a particular program. The more a breeder knows concerning the parental origin of lines, varieties, populations, and introductions, the more likely he can make effective and efficient use of this new germplasm. Where possible in a hybrid program, avoid bringing in germplasm having an origin which has combined the principal diversity groups. This often happens by using varieties as inbreds once a crop is hybridized. Obviously, the same rule applies to avoid mixing the sterile-restorer system, if at all possible, where hybridization is dependent on male sterility.

A proper strategy for meeting short-term goals in an applied breeding program is to emphasize elite germplasm with sufficient genetic variability for yield and quantitative traits such as standability and heat and drought tolerance supported by a backcross program utilizing the best known sources of insect and disease resistance. Even with this "elite" approach, the breeder will benefit by using germplasm with a diverse background. Where the necessary genes for improvement are not present in the program's breeding material, the threshold level for gain will be equally limited. If genetic gain cannot be maintained because of lack of variability, the breeder must recognize when to incorporate exotic material. This may be the single most important analysis a breeder makes! The percentage of useful exotic germplasm is dependent on the particular species and the breeding materials in the program. Also, bringing in a few defensive traits from exotic germplasm does not really expand genetic variability.

Exotic germplasm

The broad definition of exotic cultivated germplasm includes all sources of unadapted germplasm - domestic, temperate, and tropical. Most crops at some stage in their improvement seem to have lost enough genetic variation as to require an outside infusion from an exotic germplasm source, whether it be from plant introductions from the world collection or materials adapted to a different environment. To broaden the genetic base, new lines may originate from (1) indigenous material, (2) indigenous crossed with exotic sources, or (3) wholly exotic sources (National Academy of Science, 1972).

Cox (1991) indicated that, until the 1930s, all wheat produced in the U.S. was harvested from some 28 foundation introductions or by direct selection from them. New plant pests, as well as lodging, provided the incentive for wheat breeders to look outside this country for parental germplasm. Webster (1976) reached the conclusion that the genetic potential of the sorghum germplasm being used for crop improvement was almost exhausted by 1960. In the same paper, however, he reminded us that many inbred parent lines were actually hybrid derivatives of the heterotic pattern Kafir x Milo which is not necessarily a favorable situation for utilizing diversity. With only 29 original sorghum introductions, of which 20 were sorgo or sweet forage types (Martin, 1936), the need for outside variability was obvious. Maunder (1971) recognized Dr. O. J. Webster for initially introgressing Kaura germplasm of Nigeria into adapted U.S. lines. This yellow endosperm germplasm, along with Korgi from Sudan, has been responsible for an entirely different germplasm composition of second generation sorghum hybrids. Currently, all commercial sorghum hybrids in the DEKALB Plant Genetics product line contain a percentage of exotic germplasm introduced since 1950. Cox *et al.* (1988) indicated that 75% of the publicly released sorghum inbreds registered in CROP SCIENCE from 1960 to 1986 had at least some exotic parentage.

P. Cregan (1991, pers. comm.) indicated that eight to nine soybean introductions form the genetic background for midwestern maturities of this crop. Essentially all U.S. hybrid corn, with a few notable exceptions, traces its ancestry to the combination of New England flints with southern dents. Goodman (1985) reported that only about 4% of the U.S. maize acreage is being planted with hybrids containing any non-U.S. germplasm, and he further suggested that foreign exotic germplasm accounts for less than 1% of the U.S. maize germplasm base. In response to a survey on potential genetic vulnerability, Duvick (1981) reported 83% of respondents obtain both pest and stress resistance in maize from adapted, elite inbred lines. A trend exists, however, to utilize more exotic germplasm in a long-term approach to increase genetic diversity. For the short term, the maize breeder has moved increasingly to second-cycle breeding with elite germplasm. For greater efficiency and in a competitive environment, he has also gone to earlier generation testing. Now the question the breeder faces relates to when to exploit the exotic germplasm pool.

A few generalizations on the use of exotic germplasm seem in order before commenting specifically about use of the world collection for source material. The percentage of exotic germplasm acceptable in adapted lines will vary with genotype, as well as with geographic and/or environmental situation. As a general rule, less (rather than more) exotic background seems appropriate because the breeder can accept only a minimum number of undesirable characteristics. Krull and Borlaug (1970) concluded that there are many desirable genes in the world collection of wheat, but most of the variability is detrimental and most wild types are of low yield potential. They reminded us that the challenge is to incorporate desirable characteristics into a cultivar without negative progress from the undesirable material. Including exotic germplasm offers the best approach to reducing genetic vulnerability, while at the same time broadening the genetic variability available for selection. Finally, for the breeder wishing to embark into this vast and untapped resource, the best source of information, such as descriptors, and often even seed of the more elite plant introductions, is likely to be at the state experiment station level, where some semblance of a crop curator more likely functions.

Sorghum world collection conversion program

Sorghum has benefitted in recent years from a world collection which has been estimated to be in excess of 30,000 accessions. Perhaps more than two-thirds of this collection is photoperiod-sensitive for short-day length, and a high percentage of accessions carry dominant genes for height. Thus, the much needed variability for crop improvement was not readily being utilized prior to the inception of the sorghum conversion program in 1963 by Stephens *et al.* (1967). This well-defined process currently involves some 1433 plant introductions with 423 being finished and released conversions, each having four or five backcrosses. Initially, materials to be converted included a sampling of J. D. Snowden's 65 groups of the sorghum species plus several of the Meridian, Mississippi high sugar lines (D. T. Rosenow, 1990, pers. comm.). Current determination of some 30 to 50 new entries per year relates to (1) suggestions from the Crop Advisory Committee, (2) Zerazera lines from Ethiopia known for their foliar resistance and, perhaps most importantly, (3) cultivars having useful country of origin descriptors.

The process of conversion for sorghum involves crossing a daylength-insensitive, four-dwarf (all height genes recessive), temperate genetic stock by the tall, daylength-sensitive plant introduction during the fall generation under short days in Puerto Rico. The F_1 hybrid is selfed during the winter in Puerto Rico, again under short-day conditions. The segregating F_2 population is grown in Texas under temperate conditions (34°N) with an adequate population (1500 or more plants) to recover short, early (insensitive) plants. The process then becomes one of backcrossing with the above three-generation cycle using the recurrent plant introduction as the male. The plant introduction is used as a female prior to completion of backcrossing to assure the presence of the alien cytoplasm in the final conversion. Five F_3 families of the fifth backcross generation are grown next to the original plant introduction in Puerto Rico and the family most uniform and similar to the original is selected for release.

Perhaps as significant as the conversion program *per se* has been the wide screening of the converted lines by breeders, pathologists, entomologists, nutritionists, and soils specialists in both public and private breeding projects to provide descriptors relating to insect and disease resistance, drought tolerance, nutritional characteristics, and acid and salinity soil response. Combining ability data are limited except where these conversions have been introgressed into adapted germplasm - a good example being R. Tx430, developed from converted PI 276837 (IS12661C) as a non-recurrent parent. F. R. Miller (1989, pers. comm.) indicated that R. Tx430 is perhaps the single most widely used male in hybrid sorghum seed production. Tx623, also derived from IS12661C, is the female parent of "Hageen-Dura 1", the first sorghum hybrid developed for Sudan (Africa). TAM428, a close relative to IS12610C, has been widely used as a male parent in Mexico. In our own program, 47-3589c, a converted plant introduction from Sudan and of limited value agronomically (high tannin, marginal yield) when crossed to an elite and adapted line, provided much improved combining ability and resistance to anthracnose [*Colletotrichum graminicola* (Cesati) Wilson], downy mildew, and rust (*Puccinia purpurea* Cooke). Hybrids using this line adapt well from 35°N to 35°S in the western hemisphere.

Although sorghum has been the principal recipient of benefits from conversion, D. T. Rosenow (1991, pers. comm.) suggests G. W. Burton and W. W. Hanna are considering a similar breeding approach for millet [*Pennisetum glaucum* (L.) R. Br.]. J. C. McCarty and J. N. Jenkins (1990, pers. comm.) started a program in 1976 to introduce day-neutral genes into some 2000 accessions of primitive cotton. Of this material, some 311 accessions screened to date have shown beneficial insect and disease resistances, many with multiple pest resistance. In maize, Crossa *et al.* (1990) suggested a conversion program that includes the introgression of heterotic patterns found among tropical race collections into new commercial varieties or populations to exploit the heterosis among racial collections. N. G. Robinson (1990, pers. comm.) speculated that the greatest hindrance to wider use of tropical maize germplasm in temperate zones is the photoperiod response, and he suggested a systematic conversion program would be productive. However, he did remind those not readily familiar with tropical maize that a conversion program will be more difficult than for sorghum because of the lack of public tropical inbreds since public germplasm consists of bulks or populations.

Introgression of exotic germplasm

As an alternative to a conversion program, or when working with species not affected by a height or photoperiod genetic barrier, introgressing exotic germplasm can be an intermediate-term breeding approach for incorporating needed traits or to increase genetic variability in the breeding program. According to Kronstad (1986), PI 178383 from Turkey was susceptible to leaf rust (*Puccinia recondita* Rob. ex Desm.) and had weak straw, but contained resistance to four races of stripe rust (*Puccinia striiformis* West), 35 races of common bunt [*Tilletia caries* (D.C.) Tal. and *T. foetida* (Wallr.) Lito], and 10 races of dwarf bunt (*T. controversa* Kuhn). Besides tolerance to flag smut (*Urocystis tritici* Koern.) and snow mold (*Fusarium* and *Typhula* spp.), it is considered the standard for evaluating seedling emergence. Here again is an example of an apparently undesirable line by first appearance being a parent to many of the cultivars now grown in the Pacific

Northwest. Nider (1990, pers. comm.) showed that wheat yields in Argentina increased by 12.67 kg ha^{-1}yr^{-1} from 1910 to 1975; following the introgression of Mexican germplasm the rate of increase rose to 27.58 kg ha^{-1}. The first extensively grown pure line wheat cultivar of hybrid origin was the hard red spring, Marquis, released in 1912 with an origin of the Indian landrace, "Hard Red Calcutta" x the European landrace, "Red Fife". By 1929, Marquis was grown on 90% of the northern Great Plains spring wheat area (D. Marshall, unpublished data). An on-going introgression effort (Kronstad, 1986) involves the winter x spring wheat improvement program between CIMMYT and Oregon State University. Selection and breeding took place affecting each class of wheat in its environment, and the resulting increased genetic variability enhanced the improvement of both spring and winter cultivars.

A primary concern to the breeder using exotic material relates to the percentage of introduced germplasm present in the adapted or finished line. In maize, inbreds derived from temperate x tropical crosses are directly useful in the subtropics without backcrossing, but much depends on the selection environment, and these inbreds frequently require a minimum of one backcross when used in the tropics. The sorghum PI 264453, which is resistant to both the original biotypes C and E of greenbug, has the typical height and photoperiod limitation. However, rather than converting, the line was introgressed into an elite line with the BC_2 to BC_4 progenies being highly acceptable and equal in performance to the elite line. This suggests that 88-97% adapted and 3-12% exotic germplasm are acceptable levels for utilization by a parental inbred. Frey *et al.* (1983) suggested that with cultivated cereals the base populations should contain 3-12.5% wild germplasm to allow the greatest short-term progress. However, they believed that for longer term objectives 6-25% wild germplasm would be optimum for introgression.

One must remember that the interaction of exotic with adapted germplasm is not a constant relationship, and obviously population size after each backcross will be a significant factor for selecting desirable genotypes. We attempted to better understand the optimum percentage exotic germplasm by introgressing a "wild sorghum" (actually a grassy sudangrass) into the male lines of eight commercial hybrids and comparing these pseudo-isogenic lines at BC_0 to BC_3 at four locations. Five of the hybrids had their strongest yield effect in the BC_3, or with 6% exotic background in one parent; two had high yields at BC_2; and one fluctuated between the BC_2 and BC_3 depending on the test environment. Five of the eight hybrids had, regardless of backcrossing, a negative heterotic response. The data suggested the premise that introgression levels will depend on genotypes of both the elite and exotic lines. Not to be overlooked when introgressing exotics is the need to immediately classify the material for its fertility restoring-non-restoring ability, in order to work with the appropriate elite material.

In maize, Albrecht and Dudley (1987), using an Illinois synthetic composite and a South African composite, found the highest genetic variance estimates for yield were obtained with 25% exotic germplasm. This BC_1F_2 generation foundation population appeared to be the most favorable for gain from selection for grain yield. A statistical study by Bridges and Gardner (1987) was conducted to determine whether the F_2 vs. the first backcross to the adapted parent was the better foundation population in which to begin selection in adapted by exotic maize. Calculated results suggested that (1) the F_2 is better for both long- and short-term selection goals when the adapted x exotic populations perform the same, (2) the BC_1 is better for short-term gain assuming the adapted parent is superior, (3) the BC_1 is superior for long-term gain when the adapted population is superior due to more loci with favorable alleles present, and (4) the F_2 is better for long-term selection when the adapted line is superior due to presence of favorable alleles at loci with large effects.

Populations

A long-term breeding strategy (such as illustrated by the Iowa Stiff Stalk Synthetic) should contain greater genetic variation than that present in typical applied breeding programs with narrow elite germplasm bases. An applied program in less developed agricultural areas may very well want to emphasize population breeding with appropriate germplasm. If recurrent selection is the method used for improvement, Hallauer and Miranda (1981) suggested the populations used should have adequate and useful genetic variability, high mean yield, and that heterosis be expressed in the cross of two populations undergoing recurrent selection. They further stressed the need for stalk quality, if populations are to be useful as breeding material and if inbred lines are to be of value in hybrids.

Population improvement by utilizing genetic male sterility has been practiced rather extensively with sorghum during the past 20 years. Although such populations are of real value for specific objectives - such as insect, disease, quality, or soil adaptation - using them for improved levels of heterosis has not been very successful. These populations are used in the context of being intermediate- and long-term approaches. In addition, these populations frequently have failed to be recombined for sufficient generations, have low mean yields, and are subject to dominance of both height and maturity genes. These populations best resemble a reference population for specific traits in a very heterogeneous background.

Conventional breeding

Pedigree or second cycle breeding and backcrossing

Until now, major reliance in maize improvement has been placed on short-term procedures and objectives with the primary emphasis on parental diversity and orderly development and distribution of superior hybrids (National Academy of Science, 1972). Jenkins (1978) emphasized the geometric trend towards pedigree and backcross breeding in maize. He reported that, in 1936, 2% of public inbreds were derived from second cycle crosses, some 20% in 1948, 26% in 1952, 40% in 1956, and 50% in 1960. Since 1960, most public releases have been second-cycle lines developed by pedigree and backcrossing methods of selection, with the same likely true for proprietary lines (Hallauer and Miranda, 1981). This suggests the genetic base is narrowing and would obviously benefit from recurrent selection and the introgression of exotic germplasm as mentioned earlier. Even with second-cycle breeding using elite lines, real gains can be made if these elite lines are as unique as possible and if a maximum number of families is used in selection nurseries. Although our sorghum program combines modified reciprocal recurrent selection with pedigree or second-cycle breeding, we plan to work some 750 different segregating families in 1991.

Backcross breeding requires known source material as the non-recurrent parent to be followed by a routine procedure which can be varied depending on the type of inheritance of the trait to be improved. Even here the choice of germplasm can be significant. For example, in sorghum the transfer into sterile cytoplasm by using a related sterile, and one without an opposite and recessive seed color, can greatly expedite the process, allowing early generation test-crossing. When pedigree or second-cycle crosses do not appear to be producing usable segregating populations, then a backcross of a bridging nature to the more divergent type can allow progress. This limited back-crossing may be a reversal of the above-mentioned introgression but can accomplish the desired objective. Again, no one rule applies and one cannot emphasize enough the importance of knowing the breeding material.

Recurrent selection

As pointed out by Hallauer and Miranda (1981), recurrent selection should be integrated with applied breeding programs. To do so effectively, it is necessary to recombine superior progenies selected for their yield or from test-crosses. Populations must be created with a known and usable heterotic pattern. For instance, sorghum requires both a fertility restorer population of known variability and a non-restorer population; both require not only desired traits but also complementary factors where dominance is essential. The program should use unrelated elite lines as testers; testers should be selected from the complementary heterotic group related to the opposite population. When a program serves more than one environmental niche, more than one set of populations is appropriate. Also, a new elite source of germplasm can and should be added to the populations where appropriate. Generally, the frequency of superior progenies is expected to increase in later cycles. An effective recurrent program will provide selections which can be used directly in hybrids or as improved cultivars. Every breeder realizes the need to improve his odds in the development of a superior inbred or overcoming limitations to increasing genetic gain which emphasizes variability of both germplasm and also the best breeding approach.

Intercrossing opportunities from divergent germplasm improvement programs

Because heterosis is vital to a hybrid program, the apparent breeding plan should emphasize maintaining separation between heterotic groups until the final test-cross or hybrid is produced and evaluated. However, other opportunities exist for gains and may be more productive as in the following discussion.

International x domestic

As both public and private programs become increasingly global, greater opportunities exist for effective germplasm exchange. Lines from geographical diverse regions that have similar latitude and elevation are most likely to be compatible breeding materials. As an example, sorghum of the Argentine pampa tends to perform in a relatively similar fashion to sorghum of central and south Texas; prevalent pests are common to both areas. This relationship has been beneficial to more efficient plant improvement in both countries. The best commercial cultivar of soft winter wheat - Rendezvous, carrying foot rot (*Pseudocercosporella herpotrichoides*) resistance - originated at the Plant Breeding Institute in England but adapts well to the Pacific Northwest. Differences still exist with unacceptable tannin levels in Argentina sorghum, and red grain in the introduced wheat rather than the more desirable white (A. Encinas, 1990, pers. comm.). More often than not, however, germplasm brought into a program from an international activity should be regarded as exotic and likely requires some degree of introgression.

Public x private

Combining germplasm in hybridized crops of the private sector with public released lines has frequently allowed increased heterosis or additional defensive traits. Elite inbreds such as B73 in corn or Tx430 in sorghum often exhibit sufficient general combining ability to be widely used with proprietary lines, and they subsequently provide more diversity in the commercial crop. The public x private combination frequently is successful because of the long-term program of the public sector, such as with population improvement or exotic introgression; both approaches bringing in more genetic variation than the elite x elite pedigree situation. When the public lines are derived from very narrow objectives, a situation which is not all that uncommon, breeding success is not likely unless the private line contains a very high level of favorable alleles.

Exchange between programs

Breeders within the same organization, whether public or private, would be remiss not to share elite inbreds or, for that matter, even early generation segregating material. Unfortunately, plant variety protection could greatly reduce this germplasm exchange at the public level, even though the practice has often been of great benefit to the producer. In our program, I can recall at least three significant hybrids utilizing males from the Nebraska station crossed onto females from our Texas location. Fortunately, breeders, like germplasm and breeding methods, exhibit considerable variability. Thus, their approach to breeding and the materials involved can be expected to generate productive opportunities between different programs.

MORE EFFECTIVE SELECTION OF GERMPLASM

A multitude of factors, both environmental and genetic, have contributed to higher yields in U.S. crops. Some inputs, however, may have reached levels where their effect will be stabilized or could actually be reduced (fertilizer and irrigation). Also, hybridization at the single cross level may be the final or fixed plateau for gaining heterosis through a production technique. As environmental concerns are increasing, we need to utilize genetic approaches for pest control, a constant theme of alternative agriculture. These challenges all suggest more demand for genetic gain to insure continued progress in productivity.

More effective selection and efficient use of germplasm must be a challenge of the '90s. As already mentioned, Hallauer and Miranda (1981) speculated that only one in 10,000 S_2 or S_3 lines evaluated in the previous 40 years was eventually used to any extent in commercial hybrids. This estimate of only 0.01% reflects the reduced number of inbreds for single cross hybrids and that fewer elite lines are being used more extensively. A question posed to breeders must be: "Can the choice of germplasm and subsequent breeding and testing methods significantly increase this frequency?"

Improved communication concerning best source material

Probably the most frequent complaint for not using exotic germplasm in a breeding program is the unavailability of helpful descriptors. Most breeders would benefit from combining ability data but would gladly settle for details of less costly traits exhibited by a plant introduction, such as unique characteristics in its natural habitat, whether the traits are for drought, disease, nutritional value, or yield. Certainly, the suggested addition of crop curators (National Academy of Science, 1991) could go a long way towards not only providing descriptors, but making them generally available. Public agencies are to be commended for their efforts with sorghum, but even here we have only scratched the surface of desired information. Journal articles such as "Heterotic patterns among Mexican races of maize" by Crossa *et al.* (1990) represent improved and helpful communications on germplasm. C. M. Rick, commenting about germplasm and its use (Hawkes, 1981), stressed the value of careful notes taken at the time of collecting. Wild species found in stress habitats may possess the genes sought in programs involved with breeding for resistance to drought, waterlogging, soil salinity, and various pests. The identification of accessions or geographical regions with high frequencies of desired alleles was suggested by Frey *et al.* (1983) to make utilization of wild cereal germplasm more efficient. Kronstad (1986) also encouraged a more systematic approach to evaluate agronomic attributes of the accessions currently in many of the collections, as well as a more effective means of disseminating available data on a worldwide system. Finally, and more emphatically, breeders should keep in mind Kronstad's statement: "Those breeding wheat will acknowledge that if it were not for the sharing of information and germplasm, the current yield levels would never have been attainable."

Sophistication of determining descriptors for quantitative as well as qualitative data

The increased activity and accomplishments of biotechnology this past decade can serve a most worthwhile purpose as a tool for more effective germplasm selection. A Teweles multi-client study indicated some $2 billion has been spent on plant biotechnology since 1980 (SEED INDUSTRY, Nov. 1990). Because breeders work on the chromosome segment level, heterosis can and has been utilized without the knowledge of its more exact basis. RFLPs can help identify areas of the plant chromosomes that relate to such quantitative traits as combining ability and drought tolerance (Abelson, 1990). Isozymes can provide the breeder a tool to track genes - for example, in the backcrossing process resulting in more effective selection.

Increased investment in applied plant breeding

The genetic base of our food, feed, and fiber crops should be broadened through the development and implementation of new genetic sources equal or superior in performance to lines and hybrids now in commercial use. To avoid vulnerability, and open new opportunities for plant improvement, programs will require increased investment in applied plant breeding. Exotic germplasm as well as other longer term breeding efforts can also improve yield stability.

SUMMARY

Germplasm utilized with appropriate breeding techniques and combined with cultural improvement has provided yield increases up to sixfold over the past 60 years. Adequate genetic variability within elite lines, populations, and from exotic sources combined with both short- and long-term approaches should continue to allow for genetic gain and avoid lengthy yield plateaus. Finally, increased and improved exchange of information about source material as well as more sophisticated approaches to its identification will result in more effective selection of germplasm.

REFERENCES

Abelson, P. H. (1990) Hybrid corn. *Science (Washington, DC)* 249:837.

Albrecht, B.; Dudley, J. W. (1987) Evaluation of four maize populations containing different proportions of exotic germplasm. *Crop Science* 27:480-486.

Allard, R. W. (1969) Some observations on breeding for drought resistance in plants. In: Stickley, T. S. (ed.), *Man, food, and agriculture in the Middle East.* American University, Beirut, Lebanon, pp. 459-471.

Atkins, I. M.; Porter, K. B.; Merkle, O. G. (1967) Registration of Sturdy wheat. *Crop Science* 7:406.

Borlaug, N. E.; Narvaez, I.; Aresvik, O.; Anderson, R. G. (1969) A green revolution yields a golden harvest. *Columbia Journal World Business Report.*, Sept.-Oct., pp. 9-19.

Brandt, R. T., Jr.; Kuhl, G. L.; Kastner, C. L. (1989) Utilization of steam-flaked milo or corn by finishing steers. In: *Proceedings 16th biennial grain sorghum research utility conference*, Lubbock, TX, 19-23 Feb. National Grain Sorghum Producers Association, Abernathy, TX, p. 64.

Bridges, W. C., Jr.; Gardner, C. O. (1987) Foundation populations for adapted by exotic crosses. *Crop Science* 27:501-506.

Cox, T. S. (1991) The contribution of introduced germplasm to the development of U.S. wheat cultivars. In: Shands, H. L.; Wiesner, L. E. (eds.), *Use of plant introductions in cultivar development - Part I.* Crop Science Society of America, Madison, WI.

Cox, T. S.; Murphy, J. P.; Goodman, M. M. (1988) The contribution of exotic germplasm to American agriculture. In: *Seeds and sovereignty: Debate over the use and control of plant genetic resources.* Duke University Press, Durham, NC, pp. 114-144.

Crossa, J.; Taba, S.; Wellhausen, E. J. (1990) Heterotic patterns among Mexican races of maize. *Crop Science* 30:1182-1190.

Dalrymple, D. G. (1980) Development and spread of semi-dwarf varieties of wheat and rice in the United States - An international perspective. *USDA-ERS Report Number 455*. U.S. Government Printing Office, Washington, DC.

Dalrymple, D. G. (1986) *Development and spread of high-yielding wheat varieties in developing countries*. U.S. Agency for International Development, Washington, DC.

Dalrymple, D. G. (1988) Changes in wheat varieties and yields in the United States, 1919-1984. *Agricultural History* 62:20-36.

Duvick, D. N. (1977) Genetic rates of gain in hybrid maize yields during the past 40 years. *Maydica* 22:187-196.

Duvick, D. N. (1981) Genetic diversity in corn improvement. In: Loden, H. D.; Wilkinson, D. (eds.), *Proceedings 36th annual corn sorghum research conference*, Chicago, IL, 9-11 Dec. American Seed Trade Association, Washington, DC.p. 48-60.

Frey, K. J. (1971) Improving crop yields through plant breeding. In: Eastin, J. D.; Munson, R. D. (eds.), *Moving off the yield plateau*. American Society of Agronomy Special Publication 20, Madison, WI, pp. 15-58.

Frey, K. J.; Cox, T. S.; Rodgers, D. M.; Bramel-Cox, P. (1983) Increasing cereal yields with genes from wild and weedy species. In: *Proceedings XV international congress of genetics*. Oxford & IBH Publishing Company, New Delhi, Bombay, Calcutta, pp. 51-68.

Goodman, M. M. (1985) Exotic maize germplasm: Status, prospects, and remedies. *Iowa State Journal of Research* 59:497-527.

Hallauer, A. R. (1981) Selection and breeding methods. In: Frey, K. J. (ed.), *Plant breeding II*. Iowa State University Press, Ames, pp. 3-55.

Hallauer, A. R.; Miranda, J. B. (1981) *Quantitative genetics in maize breeding. 1st edition*. Iowa State University Press, Ames.

Hawkes, J. G. (1981) Germplasm collection, preservation and use. In: Frey, K. J. (ed.), *Plant breeding II*. Iowa State University Press, Ames, pp. 57-84.

Jenkins, M. T. (1978) Maize breeding during the development and early years of hybrid maize. In: Walden, D. B. (ed.), *Maize breeding and genetics*. Wiley, New York, pp. 13-28.

Kronstad, W. E. (1986) Germplasm: The key to past and future wheat improvement. In: Smith, E. L. (ed.), *Genetic improvement of yield in wheat*. Special Publication 13. Crop Science Society of America, Madison, WI, pp. 41-54.

Krull, C. F.; Borlaug, N. E. (1970) The utilization of collections in plant breeding and production. In: Frankel, O. H.; Bennett, E. (eds.), *Genetic resources in plants - Their exploration and conservation*. Blackwell Scientific Publishers, Oxford, U.K., pp. 427-439.

Lindstrom, E. W. (1939) Analysis of modern maize breeding principles and methods. In: *Proceedings seventh international genetics congress*. Cambridge University Press, Edinburgh, pp. 191-196.

Martin, J. H. (1936) Sorghum improvement. In: *Yearbook of agriculture*. U.S. Government Printing Office, Washington, DC, pp. 523-560.

Maunder, A. B. (1971) Agronomic and quality advantages for yellow endosperm sorghums. In: Sutherland, J. I.; Falasca, R. J. (eds.), *Proceedings 26th annual corn sorghum research conference*, Chicago, IL, 14-16 Dec. American Seed Trade Association, Washington, DC, pp. 42-53.

National Academy of Science (1972) *Genetic vulnerability of major crops*. Print and Publ. Office, Washington, DC.

National Academy of Science (1991) *Managing global genetic resources - The U.S. plant germplasm system*. National Academy Press, Washington, DC.

National Research Council (1984) *Nutrient requirements of beef cattle*. Washington, DC.

Nider, F.; Maunder, A. B.; Krull, C. F. (1969) Occurrence of downy mildew in Argentina. *Sorghum Newsletter* 12:3.

Rosenow, D. T.; Clark, L. E. (1987) Utilization of exotic germplasm in breeding for yield stability. In: *Proceedings 15th biennial grain sorghum research utility conference*, Lubbock, TX, 15-17 Feb. National Grain Sorghum Producers Association, Abernathy, TX, pp. 49-50.

Russell, W. A. (1974) Comparative performance for maize hybrids representing different eras of maize breeding. In: Wilkinson, D. (ed.), *Proceedings of the 29th annual corn sorghum research conference*, Chicago, IL, 10-12 Dec. American Seed Trade Association, Washington, DC, pp. 81-101.

Stephens, J. C.; Miller, F. R.; Rosenow, D. T. (1967) Conversion of alien sorghums to early combine genotypes. *Crop Science* 7:396.

Webster, O. J. (1976) Sorghum vulnerability and germplasm resources. *Crop Science* 16:553-556.

Discussion

Todd C. Wehner, Moderator

What evidence is there that wood yield is improved if the genes for sex are turned off?

Since we have not turned off the genes for sex yet, there is no evidence for yield increases. [W. J. Libby]

Are there any people brave enough to breed redwood trees for lumber production?

We have a substantial program going on in our redwood tree-improvement cooperative. It is an important species in California. Many people don't realize that coast redwood is the fastest growing temperate conifer in the world. It has enormous productivity, about 10 times that of the average American conifer. [W. J. Libby]

In spite of their simple usage, is there any evidence that the isozymes are contributing to (or associated with) some vital physiological functions?

There is a great deal of evidence that isozyme loci have physiological effects that contribute to adaptive changes. Although such physiological effects have usually been attributed to one or more loci postulated to be linked to the isozyme loci, it is now clear that alleles of loci governing enzyme, morphological, physiological, color, disease resistance, and restriction fragment variants themselves usually have pleiotropic effects on more than one and often many quantitatively distributed characters such as seed number, time to heading, height, yield, and survival ability. It is not easy to distinguish between pleiotropy and linkage. However, data from comparisons of alternative alleles in numerous pairs of highly isogenic lines, tested in numerous seasons and locations, have established that the physiological effects are due to the "marker" loci themselves. However, if physiological effects are in fact due to linked loci, the linkages are so tight that for practical purposes each "marker" locus behaves as a single complex locus over very large numbers of generations. [R. W. Allard]

Do you see a point at which a particular species will reach saturation of beneficial genes using molecular biological techniques, for example, so that there is no more room for other factors to operate, resulting in a decline in yield or other quantitative traits?

We have enough variability that is unused at this time that it will be a long time before saturation. I would like to point out, though, that, for the other systems to be effective for biotechnology, it is essential that the breeder be able to provide biotechnologists with the most elite sources of germplasm, or it is likely they are going to be faced with some of the same problems the breeder had 30 or 40 years ago and the resulting limited progress. Developing the highest level of elite materials for the biotechnology programs is essential, but I do not think that in the near term we should worry about running out of useful genes with the vast amount of untapped germplasm in the world collection. The question remains, "which is more efficient, conventional breeding or biotechnology?"

Regarding one of our biggest problems, especially as we talk about sustainable agriculture or alternative agriculture, I would like to mention parasitic *Striga* spp. in Asia and Africa, which is deleterious to corn, sorghum, and other crops. This may be a situation where some of the new techniques (e.g., genetic transformation) will be able to allow us to reduce some of the hazards where herbicides are unavailable or unwanted. So I think there are some cases where there are, in fact, exceptions to available or existing variability. [A. B. Maunder]

Is it possible to select for and manage the maximum tree product yield while still maintaining natural biodiversity in forest ecosystem?

We really do not know whether or how breeders will put things into a natural ecosystem and - if you can imagine things like climate change, acid rain, pests, pathogens, and competitors effecting a substantially modified ecosystem - then it is conceivable that we will attempt some insertions. We have a lot of thinking and planning to do on how we might insert human-bred or human-directed things into what we might call an "ongoing, near-native ecosystem". It is not something we should do quickly, but I certainly do not want to rule it out. I think "maximum" yield is not ecologically or economically wise even in wholly artificial ecosystems, and it is probably impossible with "natural" biodiversity maintained or even approximated. [W. J. Libby]

The sorghum conversion program seems to have been a rather long-term and labor-intensive procedure. How many converted lines are found in the pedigrees of released cultivars?

The answer may be none, in that the use of the world collection frequently involves partially converted lines and cultivars, such as TAM 428 and TAM 430 to name two. Texas 623, also, only used partially converted lines in its parentage. Those are three examples of very extensively used cultivars that occupy a lot of land; so, if the question can be taken in the broad sense, I would say that several have made a major impact worldwide.

[A. B. Maunder]

Besides gene frequencies and desirable alleles, what other predictive methods are available for germplasm identification, for example, in an applied breeding program at an international research center?

The most widely used method of identifying superior germplasm is to determine which cultivars have the best performance record under farm conditions in your environmental situation and use these elite cultivars as parents in your breeding program. Regarding untapped germplasm, most attempts to identify potentially useful alleles or groups of alleles have been based on tests, in the local environment, of accessions obtained from gene banks. Ordinarily a few plants of each of a large number of accessions are grown and each accession is "characterized" for general adaptedness, yielding ability, response to various stresses such as high or low temperature, drought, and reaction to local races of diseases and insects. These are usually quantitatively distributed traits, each with low heritabilities, especially if the evaluation tests are repeated in different years. Unfortunately, the "characterization" data obtained from such tests have proved to be ineffective in separating locally adaptive dimensions from dysgenic dimensions of the continuously distributed phenotypic variation, i.e., they have been ineffective in identifying superior germplasm. All progress in plant breeding (as in any evolutionary process) stems from differences among individuals. Although these differences are often slight and difficult to measure directly, selection recognizes those attributes that contribute to survivability and increases such attributes in frequency over generations. Stated another way, what really matters in evolutionary terms is the frequency at which an allele (or group of alleles) is represented in future generations. If an allele has become frequent in many different populations from any major ecogeographical region, and particularly if it is frequent in several ecogeographical regions, the prospects are good that this allele (or group of alleles) will be useful in your region. Fortunately, frequencies of alleles governing discretely recognizable morphological, allozyme, restriction fragment and disease resistance alleles can be determined quickly and relatively inexpensively in germplasm collections, in genetically enhanced populations, and also in adapted local populations (enumeration data have favorable statistical properties). If an allele is frequent somewhere, and it isn't in your stocks, give it a try by introgressing it into your breeding materials. If an allele is consistently rare, it isn't likely to be useful in your situation. In the vernacular, if it works, go after it; if it doesn't, don't bother.

[R. W. Allard]

Contrary to the maize data shown, data from rice improvement from the International Rice Research Institute indicate that a yield plateau has been reached and yield may actually be declining. What is causing it and is it going to continue?

The increases in the yield of the recent rice cultivars have not been as dramatic as they were during the green revolution, so there is a feeling that a yield plateau has been reached. But that is not true when you compare the yield from various international nurseries. They show that there has been at least a 15 to 20% yield increase in several cultivars. That is in spite of the emphasis on selecting a very short growth duration of 100 to 105 days (as compared to 130 to 140 days for 'IR8'). Yield of new cultivars is slightly less but, when you compare their productivity, it is almost 50% higher than IR8. There has been a lot of emphasis on selecting a shorter growth duration because of the possibility of growing two or three crops in the tropics and the yield is slightly less. There are cultivars with the same growth duration as IR8 which have 15 to 20% higher yield.   [Anon from audience]

As it is currently practiced, is there very much risk of reduced genetic diversity due to widespread use of super trees and clones over large areas?

It is still too early in tree improvement and dissemination for reduced diversity to be a widespread problem. One example of a problem occurred in Yugoslavia about 10 or 12 years ago when they planted hybrid-poplar clone "I214" over their three major river drainages and based an entire industry on that one clone, which then came down with three epidemic diseases. But, mostly, I think we still have an enormous amount of genetic diversity in most of our forests, and most breeders are aware of the fact that we want to maintain that level of diversity.   [W. J. Libby]

You stated that, if alleles are found at a very low frequency, they are in fact probably not favorable alleles; however, curators of gene banks attempt to save everything possible. Do you think it is worth it?

In general, no. However, I suggest that some rare alleles may be worth saving, e.g., alleles that have favorable effects on product quality, alleles that code for medically useful products, disease resistance alleles that appear to have arisen recently under conditions of cultivation.   [R. W. Allard]

If one of only 23 alleles studied was consistently heterozygous, then is there a very small percentage of loci contributing to heterosis or was the sample biased?

More precisely the data I cited showed that, among 23 loci studied by the North Carolina corn group, 14 were homoallelic in the most widely grown elite U.S. single crosses, 8 were homoallelic in most of these elite single crosses, and only 1 locus was heteroallelic in all of these elite single crosses. I have no reason to suspect that the North Carolina sample of loci was biased; indeed, data from about a dozen additional loci in my own project give parallel results. I am thus led to the conclusion that only a small percentage of corn loci contribute to heterosis. Heteroallelism also appears to make little contribution to heterosis and wide adaptedness in other diploids from which appropriate data are available. However, more than 50% of loci are heteroallelic in tetraploid *Avena barbata*. This is apparently because tetraploidy, followed by diploidization, combined with a mating system featuring predominant self-fertilization, provides an effective mechanism by which favorably interacting heteroallelic combinations could be assembled in this polyploid and also protected from disassociation due to segregation and recombination.   [R. W. Allard]

Is the exotic germplasm breeding effort currently on the increase or decrease in the U.S. private industry for sorghum?

The industry is pretty much divided in their breeding approaches and obviously this relates closely to available funding. As for the five or six large programs, I am confident that there is a fair amount of effort in this regard but perhaps not as much as in the past. Obviously, the converting or introgressing of exotic sorghum germplasm is a costly procedure. So, as to whether it is on the increase or decrease, I could only relate to our own program; we are running on about the same level. We try to run about 60% on the pedigree-type approach (or elite x elite), 20% on introgression (not on conversion), and 20% on recurrent selection.   [A. B. Maunder]

Given that clear-cutting is the harvest method of choice for some tree species, how can you, as a plant breeder, deal with the environmental hazards which go along with that procedure?

This is a big topic, perhaps one for another whole symposium. I think that you probably recognize that, as a tree breeder, I certainly like clearcuts (or at least conflagrations) so that I can go and plant something where the forest used to be. On a properly done clearcut, the problems are mostly aesthetic, in that it offends people's eyes to see the apparent devastation. In fact, environmental problems such as downstream siltation and erosion are in some ways greater in the various selection or partial-cut systems (assuming you want the same amount of wood for the various things for which we use wood) simply because you have to enter those forests more often and sometimes with more damaging equipment if you do not clear-cut. Now, to answer your question, we can and do breed trees that more quickly develop a "near natural" ecosystem following a clearcut, fire, or other forest-clearing devastation.

[W. J. Libby]

Considering the widespread exchange of germplasm in the developed areas of the world, how does this relate to the diversity of the gene base and the need to deploy genes for disease resistance?

This question has two main parts. <u>First</u>, what effect has the exchange of germplasm had on the diversity of the gene base? Materials taken to different areas ordinarily quickly diverge from each other in allelic content without much decrease in diversity within areas - thus exchange of germplasm increases total genetic diversity of the gene base worldwide. Population numbers are very large in cultivation, often many orders of magnitude larger than mutation rates per generation, so that large numbers of novel mutations appear per locus per generation. Differential survival of different novel mutants in different areas has led to continuing increases in total diversity worldwide.

<u>Second</u>, what effect does exchange of germplasm have on deployment of resistance? Recent studies of host-pathogen systems show that the coevolution of the genetic systems of the host and pathogen is complementary - resistance alleles which protect against locally evolving races of the pathogen increase in frequency, thus enhancing the value of local host populations and decreasing the value of alien populations as sources of alleles for resistance to local pathotypes. More generally, as genetic divergence increases between the gene pools of different areas, the more likely breeders are to rely on local elite x elite parents and the less likely to use local elite x exotic parents in their crosses.

[R. W. Allard]

I see some real advantages to evaluate germplasm from distinct environments whether they be Europe vs. the United States, or Nebraska vs. Texas. Those two states are as different as two countries, and I think that this type of situation has led to some extremely favorable heterotic responses. The reason is that programs have different objectives and greatly different environments, and breeders are making selections in them. We make a mistake in doing too much of our selection work in a very narrow range of environments. We test over a wide range, but we do not select over a wide enough range. We then get the benefit of selection over a wide range by using materials from different environments and often complementary resistance factors.

[A. B. Maunder]

Do you see an increase in breeding efforts on tolerance to abiotic stress as a way to continue yield gains (for example, high yield under low soil fertility, or improved nutrient efficiency)? Do we

need to be dealing with abiotic tolerance for yield improvement in the future?

Yes. I am continually noticing that lack of adaptation very often is a sign of a poor genotype x soil interaction. The breeder very definitely needs a close effort with the soil scientist.   [A. B. Maunder]

Are breeders directing their programs at fragile environments such as the Cerrado in Brazil?

I think that the current decade is one of alternative or sustainable agriculture and the environment, whereas the 1980s were certainly one of biotechnology. We would be remiss in not thinking in terms of the need for increased variability in plant breeding programs to encounter an environment such as the Cerrado. We should take care of such questions on alternative agriculture, but the plant breeder has more pressure on him now than ever before to take care of pest resistance and that sort of thing.   [A. B. Maunder]

How do you regard the destruction of tropical forests in South America? Will the genetic diversity of trees and other plants regenerate when the cattle farmers move on?

I regard the destruction of tropical forests not just in South America, but perhaps with greater alarm in other places such as Madagascar, Indonesia or the Philippines as an absolute crisis situation. While there are problems in the Amazon and other South American forests, I think we are still in its early stages there and, while one can say we are losing some species, the places that are really scary are the latter ones I have mentioned. I think it is one of those things that people will look back on and say that was a disaster visited upon earth by 20th century civilization. Let me put that a different way. Almost any science - including tree breeding, plant breeding, or molecular biology - that we do today, we could probably do better tomorrow because we are going to know more. The one screaming exception to that is the conservation of natural populations and, for that matter, natural ecosystems. If we do not do it today, there is not going to be a tomorrow for them, so I view the problem as being very serious.   [W. J. Libby]

Dr. Allard's data suggest that alleles that are rare in nature are rarely useful in cultivar development programs. Yet we are all aware of the importance of rare alleles in germplasm collections for the plant breeder. How do you reconcile these apparently antagonistic facts with respect to conservation of plant genetic resources?

In attempting to answer this question, let me first confess that I am an instinctive conservationist who has no quarrel with the proposition that we should collect and save as much genetic variability as realistically possible. A very real difficulty is that most gene banks are now burdened with more accessions than they can maintain in viable condition, let alone evaluate and distribute to users. There is urgent need for effective ways to reduce the number of accessions in gene banks while improving the quality of the germplasm sent to users. What really matters in evolutionary terms is the frequency of an allele in future generations - if an allele promotes adaptedness in a given environment, it survives and increases in frequency; if it doesn't promote adaptedness, it goes to low frequency or is eliminated. I have no difficulty reconciling these apparently antagonistic "facts" with respect to conservation of plant genetic resources. Cases are few and far between in which rare alleles have been useful in plant breeding and most plant population geneticists and experienced plant breeders recognize this.

[R. W. Allard]

Is the effect obtained by heteroallelism the same as obtained by parallel spindle formation?

Parallel spindle formation is a cytologically observable physical phenomenon. In meiosis it affects the distribution of chromosomes, and hence alleles, to the poles in ways that can lead to fixed heteroallelism. Thus, it may have the same effect as diploidization. However, allelic data will be required to determine whether parallel spindle formation does in fact lead to heteroallelism. I am unaware of such data if they exist.   [R. W. Allard]

PART TWO

MODIFICATION OF PLANTS TO TOLERATE ENVIRONMENTAL STRESSES

Chapter 8
Mechanisms for obtaining water use efficiency and drought resistance

J. S. Boyer

College of Marine Studies, University of Delaware, Lewes, DE 19958

A prevalent problem in crop production is shortage of water (Boyer, 1982). Plants require large amounts for their growth, and shortages limit not only the size of the plant but also the development of various plant parts. The effects can be so large that in some instances production is reduced to near zero.

Early reproductive development is especially vulnerable to water deficits (Salter and Goode, 1967; Claassen and Shaw, 1970). The losses often are irreversible, and much of the lost productivity during the drought of 1988 in the U.S. Midwest was caused by these early reproductive effects. However, the effects are not confined to individual years. Statistics show that more insurance indemnities are paid for crop loss from drought than from any other kind of loss (Clampet, 1982). Most of these payments are for crops with valuable reproductive tissues.

The traditional solution to water shortages has been irrigation. However, supplies of high quality water are diminishing and municipalities in many areas are competing for the same water, so this option is becoming less available even if farmers can justify the large capital costs of the equipment and the expense of pumping the water (Boyer, 1982). The trend is unlikely to reverse, and an increasing interest is being directed toward plants capable of yielding well in water-deficient conditions. Using genetic manipulation or altered cultural practices to achieve this goal could be less degrading of soils and water supplies and more cost effective than irrigation.

A number of methods exist for improving crop productivity when water shortages occur. If we include irrigation as an option but recognize the limits on its use, the methods can be classified in three categories - (1) increasing the efficiency of water delivery, (2) increasing the efficiency of water use by the plants, and (3) increasing the drought resistance of the plants.

Most agricultural research has been devoted to improving water delivery. Transporting water with minimal evaporation, preventing runoff, storing water in catchments, and timing irrigation to the needs of the plant have been successful in improving productivity per unit of water pumped. There are estimates that this approach can reduce the water pumped by at least a half while maintaining high levels of production (Bordovsky et al., 1974).

However, by comparison, only a small amount of scientific effort has been devoted to how plants cope with water shortages and whether water use efficiency and drought resistance can be improved. It is known that species such as pineapple [*Ananas comosus* (L.) Merr.] use less water than others for similar dry matter production (Joshi *et al.*, 1965; Neales *et al.*, 1968; Hanks, 1983), and that some climates extract less soil water than other climates for the same crop output (Briggs and Shantz, 1914; Tanner and Sinclair, 1983; Hanks, 1983). Thus, genetic variation in water use exists between species, and crop culture can be relocated to reduce water inputs. However, changing the species to be grown or relocating to a new climate may not be possible for the individual farmer. What then are the prospects for improving water use efficiency within species, or protecting against yield loss in a particular climate when irrigation is not possible? This chapter is devoted to the progress made with these alternative approaches. For more detail, the reader is encouraged to read reviews by Bradford and Hsiao (1982), Hanson and Hitz (1982), Taylor *et al.* (1983) and O'Toole and Bland (1987).

WATER USE EFFICIENCY IN PLANTS

Plants lose water as they fix CO_2 from the air. The loss is inevitable because the CO_2 must dissolve in water before it becomes available to the cells. The wet cell surface must be exposed to the atmosphere inside the leaf, and evaporation will occur. As a result, the photosynthesizing cells dehydrate to varying degrees depending on how rapidly the water evaporates and how readily the lost water can be replaced.

In marine plants exposed to air, water absorption awaits the next tide and the plants have little control over evaporation or rehydration. In land plants, water generally is absorbed from the soil, the shoot is covered with a waxy layer, and the stomata regulate water loss as CO_2 is fixed. Therefore, land plants have a much greater degree of control of evaporation and water acquisition than their marine counterparts, and water is conserved to a greater degree. Depending on the leaf anatomy and physiology, the dry matter produced per unit of water used can vary. This variation in water use efficiency often forms the basis for genetic control of plant productivity with limited water supplies.

Nevertheless, it is important to recognize that CO_2 diffuses down a concentration gradient to the leaf interior, and the air in the intercellular spaces inside the leaf has a concentration lower than in the air outside. Likewise, water diffuses outward along a decreasing gradient in humidity, and the humidity of the air outside is generally lower than inside the leaf. The lower the external humidity, the faster the evaporation will be when all other factors are constant. No amount of genetic improvement will change that fact. What can be expected from genetic improvement of water use efficiency is a greater dry matter production or economic yield per unit of water used in a particular evaporative environment.

Because this two-way diffusion is the basis of water use efficiency in plants, water use efficiencies have been estimated from gas exchange by measuring the ratio of CO_2 fixed to water transpired (e.g., Brown and Simmons, 1979; Robichaux and Pearcy, 1984). The carbon dioxide molecule contributes relatively heavy atoms of carbon and oxygen to the plant and thus contributes the bulk of the dry mass that is accumulated. Water supplies hydrogen atoms and these are so light that they contribute only a small amount to the plant dry mass. Thus, the water use efficiency defined as the dry mass produced from a unit of water is approximated by the ratio of CO_2 gained to water lost.

While this approach gives valuable insight into the physiologic and metabolic controls that might operate during photosynthesis and transpiration, it is clear that water use efficiency involves much more. The mass of the plant is determined by long-term net CO_2 fixation which also is affected by the respiratory losses at night and by non-photosynthetic organs in the plant. It is affected by temperatures that alter each of these factors as well as the long-term partitioning of dry mass between photosynthetic and non-photosynthetic organs. There is direct evaporation from the soil in addition to transpiration from the plants. Figure 1 shows a comparison of water use efficiency measured by gas exchange (instantaneous water use efficiency) and the water use efficiency for the whole growing season measured from total dry mass and total water used in tomato (*Lycopersicon esculentum* Mill.) and its wild relative *L. pennellii* (Cor.) D'Arcy. The relationship is poor because of the additional factors affecting dry mass accumulation (Martin and Thorstenson, 1988). Therefore, agriculturally relevant water use efficiencies should be based on long-term experiments involving whole plants.

Figure 2 shows that a typical field measurement of water use efficiency gives a linear relationship between season-long yield and water use. The same is true for total dry matter production (Hanks, 1983). This striking feature suggests that yield is proportional to the dry matter production and in turn to transpiration because of the link between photosynthesis and transpiration through stomatal behavior. Under field conditions, this link is strong because the radiation input is completely absorbed after the canopy closes and, in a given climate, the input is partitioned in a constant proportion between energy for transpiration and energy for photosynthesis. The proportion differs between species, climates, and from year to year (Hanks, 1983). The linear relationship is always present, however, and the linearity indicates that the water use efficiency does not change as the availability of water varies.

It has been argued that water use efficiencies should not be expressed as absolute dry mass gained per unit of water mass used but should be

Figure 1 Relation between the CO_2 fixed:H_2O transpired and the season-long dry mass accumulation:water mass used in tomato. Panels A, B, and C are for plants grown at 100% (open symbols), 50% (shaded symbols), and 25% (closed symbols) of soil field capacity, respectively. Triangles are for *Lycopersicon esculentum*, squares for *L. pennellii*, and circles for the F_1 hybrids. None of the relationships are significant at the $P < 0.05$ level. Note that the water use efficiency measured for the whole season is highest for *L. pennellii*, lowest for *L. esculentum*, and generally intermediate for the hybrids (redrawn from Martin and Thorstenson, 1988).

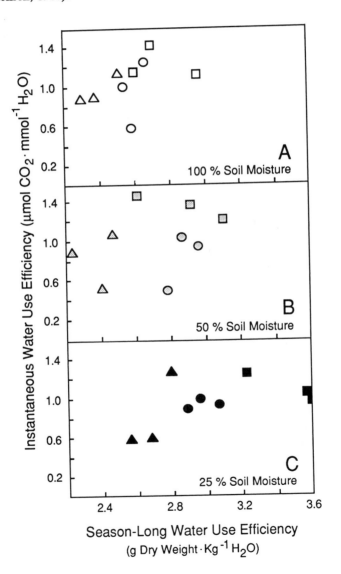

Figure 2 Relation between grain yield and evapotranspiration for four barley cultivars in Logan, Utah. The slope of the solid line gives the water use efficiency (2.1 g dry weight kg^{-1} water used). The yield was suppressed by an early freeze at the highest water use, and these data are slightly below the line. The x-intercept probably represents evaporation from the soil under the crop. For total dry mass instead of yield, the relationship also is a straight line but of steeper slope (redrawn from Hanks, 1983).

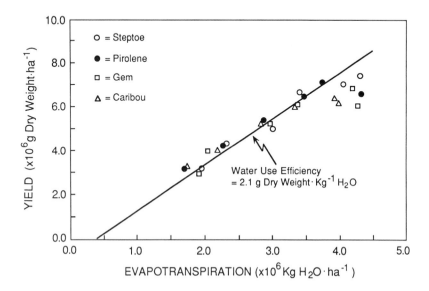

normalized for evaporative demand (de Wit, 1958; Tanner and Sinclair, 1983) and the potential productivity of the crop (Hanks, 1983). Thus, the efficiency becomes the fractional dry mass gained for the fractional water used, where the fractional dry mass is expressed relative to the maximum dry mass produced with optimum water and the fractional water use is expressed relative to the potential transpiration at the site. This normalizes for differences between genotypes and environments, and many differences in water use efficiency disappear. The approach has the advantage that for a water use of, say, half the potential transpiration at the site, a dry mass gain of half the maximum would be predicted. This simplification eases the job of predicting the impact of water shortages.

However, from an agricultural standpoint, it is usually more important to know the absolute dry mass of the crop for an absolute unit of water used because farm income is based on the absolute dry mass and expense is based on the absolute amounts of water used. A farmer contemplating whether to irrigate semi-arid land needs to have high absolute production of dry mass to justify pumping larger amounts of water than a farmer in a humid region.

Even better, he should know the absolute production of marketable yield, which may be only a part of the total dry mass. Therefore, absolute water use efficiencies are often more useful than normalized ones.

Considering for the moment the differences in water use efficiency between species, plants exhibiting C_3 photosynthesis (most crops) tend to have lower water use efficiencies than those with C_4 photosynthesis such as maize (*Zea mays* L.), and sorghum [*Sorghum bicolor* (L.) Moench.] in a particular environment (Briggs and Shantz, 1914; Brown and Simmons, 1979; Tanner and Sinclair, 1983; Robichaux and Pearcy, 1984; Kawamitsu *et al.*, 1987). The C_3 plants have no CO_2 concentrating metabolism whereas C_4 plants possess an internal CO_2 pump that pre-fixes CO_2 into C_4 compounds such as oxaloacetate and transports these compounds to the CO_2-fixing site where the CO_2 is released. Thus, the CO_2 is present at higher concentration around the CO_2-fixing site. As a result, stomata need not be as wide open as in C_3 plants, which conserves water, but the CO_2 concentrations are high around the fixation site (Brown and Simmons, 1979; Robichaux and Pearcy, 1984; Kawamitsu *et al.*, 1987). By contrast, for the same rate of CO_2 fixation, C_3 plants must open their stomata widely and water use becomes large.

A more extreme example is in plants with Crassulacean Acid Metabolism. Pineapple is the prime example and it can pre-fix CO_2 into C_4 compounds during the night and close its stomata during the day. This markedly reduces water loss. The CO_2 is released during the day inside the leaf and, because of the closed stomata, remains there in high concentration. The CO_2 is then fixed in the usual way by C_3 enzymes. The estimates of water use efficiency for pineapple are about 20 g dry mass kg^{-1} water and for C_3 plants are 1 to 2 g dry mass kg^{-1} water (Briggs and Shantz, 1914; Joshi *et al.*, 1965; Neales *et al.*, 1968) depending on the evaporative environment.

What about differences within species? G. D. Farquhar and his colleagues at the Australian National University, Canberra, demonstrated differences in water use efficiency between genotypes of wheat (*Triticum aestivum* L.), peanut (*Arachis hypogaea* L.), barley (*Hordeum vulgare* L.) and other crops using the $\delta^{13}C$ technique (Hubick *et al.*, 1986; Condon *et al.*, 1987, 1990; Brugnoli *et al.*, 1988; Bowman *et al.*, 1989; Hubick and Farquhar, 1990). The technique measures the ^{12}C and ^{13}C isotope content of plant tissue in comparison with that in air. The isotopes occur naturally in air and, because the ^{12}C isotope is lighter, it diffuses faster. Also, ribulose-1,5-bisphosphate carboxylase fixes the lighter isotope faster. The two effects cause the leaf dry mass to become impoverished in ^{13}C. The unused ^{13}C builds up in the intercellular spaces of the leaf and diffuses out according to the extent of stomatal opening. This in turn is correlated with the water use efficiency.

The $\delta^{13}C$ technique makes it possible to survey a large number of plants with a simple analysis of the leaf tissue. Differences integrate the conditions over which the plant was grown. Analyzing the entire shoot indicates the water use efficiency for the time required to grow the shoot whereas analyzing only leaf starch indicates the water use efficiency during the photosynthetic period necessary to accumulate the starch. Thus, one may integrate over long or short times with this method.

The success of the method suggests that differences in water use efficiency not only exist within individual species but might be incorporated into breeding programs, although this is still in its infancy. When water is optimally available, the $\delta^{13}C$ measurements show the largest differences and thus the most promise. A significant amount of variability is sometimes present in the data, but it is becoming clearer that selecting for extremes in $\delta^{13}C$ will select for extremes in water use efficiency. Thus, mass selection programs seem warranted especially to identify elite lines.

Martin and Thorstenson (1988) used this technique to show that differences in water use efficiency were present between the domestic tomato species and *L. pennellii* and their hybrids. Figure 3 shows that the differences in water use efficiency were detectable in $\delta^{13}C$ data between the parents and the hybrids. The domesticated parent had the lowest efficiency and the wild parent the highest efficiency with the hybrids showing intermediate behavior. Because the species could be crossed, it was possible to correlate the differences in water use efficiency with restriction fragment maps of the tomato DNA (Martin *et al.*, 1989). Three loci were found to be predictors of the variation in water use efficiency in field-grown tomato. This landmark effort indicates that water use efficiencies were determined by relatively few genetic loci and implies not only that agriculturally relevant differences exist but that they can be genetically manipulated in a simple fashion.

It is surprising that a complex trait like water use efficiency should be controlled by only a few genetic loci. Thus far it has not been determined whether each locus corresponds to more than one gene and it remains possible that the trait is in fact complex. Despite this situation, further studies of the genetic basis for differences in water use efficiency seem warranted, and it is likely that the differences will be heritable.

DROUGHT RESISTANCE

Plants exhibiting improved growth with limited water are considered to resist drought more effectively regardless of how the improvement occurs. Increases in water use efficiency are one form of increased drought resistance. Other forms may not affect the water use efficiency but are nevertheless important. For example, increased rooting depth may increase the amount of water available and the dry mass in the same ratio, thus

Figure 3 Relation between $\delta^{13}C$ composition of the leaf tissue and the water use efficiency for the whole growing season. Symbols are as in Figure 1. Note that the correlations between $\delta^{13}C$ composition and water use efficiency are better than for the ratio of CO_2 fixed:H_2O transpired in Figure 1. The correlations in panels A and B are significant at the $P < 0.01$ and $P < 0.05$ levels, respectively (redrawn from Martin and Thorstenson, 1988).

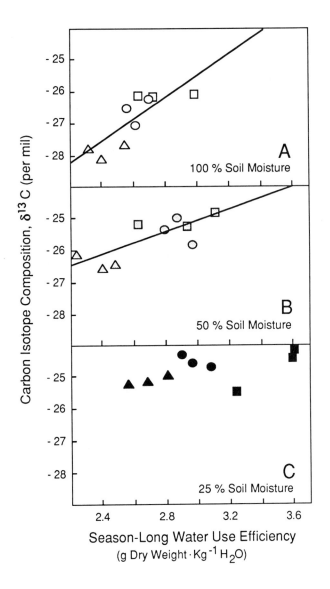

increasing absolute productivity but leaving the water use efficiency unaltered. Others having a similar effect are increased cuticle thickness, decreased organ senescence, improved production of a harvestable organ or substance, and osmotic adjustment (Morgan, 1984) which may be defined as any change in cell solute concentration beyond that caused by dehydration-rehydration effects.

Among these factors, some may be considered to be mechanisms of dehydration avoidance, others to be forms of dehydration tolerance. Improved rooting depth and increased cuticle thickness delay dehydration whereas increased production of a harvested plant part is a form of tolerance. In general, mechanisms that delay dehydration are usually structural and may require partitioning of dry mass differently to produce more roots, thicker cuticle and so on. On the other hand, mechanisms involving tolerance of dry conditions often do not require significant repartitioning of dry mass. An example is osmotic adjustment which occurs because dry mass normally used to synthesize new cells is not used for that purpose and instead accumulates in the cells (Meyer and Boyer, 1972, 1981) or is deposited in fewer or smaller cells (Sharp *et al.*, 1990). Only a brief interruption in biosynthesis is necessary to accomplish this (Meyer and Boyer, 1981), and the increased concentration of solutes that results is present only under dry conditions. In other words, there is little cost to the plant when water is scarce and no cost when water is plentiful.

More examples can be given, but the main principle is that crop improvement under conditions of limited water should consider more than water use efficiency alone. Consideration should be given to the likely water deprivation conditions and plant modification should be sought that might alleviate those conditions.

It is often considered that crop improvement programs for drought can be based on high productivity when water is plentiful. This reasoning is based on the idea that, for a given climate, water use efficiency will be highest when dry matter production is highest. Because productivity is often linearly proportional to water use (Figure 2), high productivity of dry matter will carry over to drought conditions. However, it is clear that many opportunities will be missed if superior selections are based only on high productivity when water is plentiful. Characters such as osmotic adjustment are called into play only during a water deficit. Others may be present normally, but persist better during a water deficit. Without plant selection under water-deficient conditions, these traits will be missed.

How should a crop improvement program be designed to include improved performance under water-limited conditions? Water is so ubiquitously involved in growth and metabolism that identifying specific targets at first seems impossible. Moreover, the multitude of possible targets implies that effects might involve enormous numbers of genes, and improvements might be only incremental or, worse still, may cause inevitable problems at other genetic loci.

However, there are some examples of successful approaches from which we can learn. Each has resulted in a clear increase in crop performance under water limited conditions. Jensen and Cavalieri (1983) described a maize program involving field testing at various locations varying in water availability (Figure 4). Genotypes were identified having most combinations

Figure 4 Stability regressions of four maize hybrids grown at various locations throughout the U.S. over 3 years. The dashed line indicates the mean yield for all hybrids at each location. The solid line shows the yield of an individual hybrid for comparison. A) Hybrid 3323, B) hybrid 3377, C) hybrid 3358, D) hybrid 3388. Except for B), the slopes of each hybrid regression differed significantly from the slope of the dashed line (P < 0.01). The r^2 values were between 0.67 and 0.82 for the regressions of the four hybrids. Regressions were formed for 399 genotypes and, in most cases, for over 500 sites (redrawn from Jensen and Cavalieri, 1983).

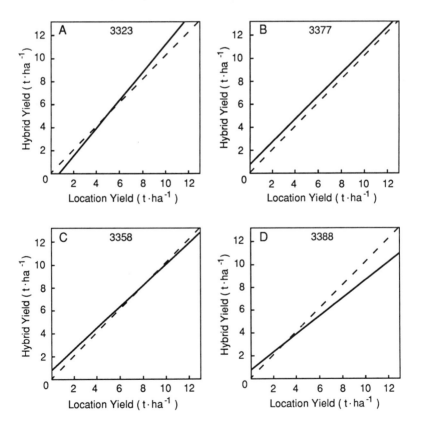

of yield performance under optimum and water-deficient conditions: high yield in both conditions (Hybrids 3377 and 3358 in Figure 4), high yield in optimum conditions but low yield under deficit (Hybrid 3323 in Figure 4), and low yield in optimum conditions but high yield under deficit (Hybrid 3388 in Figure 4). The first classification is the preferred one and the last classification seems worthy of some consideration. Wright and Jordan (1970) showed rapid improvement in the establishment of boer lovegrass (*Eragrostis curvula* Nees) selected for seedling growth in dehydrated soil. The character that appeared most improved was the thickness of the cuticle covering the shoot tissues of the seedlings (Hull *et al.*, 1978). With these selections, the establishment of grasses on the western range became more reliable. Burton *et al.* (1954; 1957) demonstrated the value of deep rooted Bermuda grass [*Cynodon dactylon* (L.) Pers.] in contrast to more shallow rooted types, and with this change he increased the productivity of pasture in humid regions subjected to sporadic drought. Hall and Grantz (1981) selected early flowering cowpeas [*Vigna unguiculata* (L.) Walp.] that escaped late season drought. This is a common way of avoiding water limitations that develop toward the end of a growing season. Morgan (1983) selected wheat for superior osmotic adjustment (Morgan, 1984) and observed improved yields under dryland conditions that were at no cost to yield in optimum conditions.

These successes are quite diverse in their approaches but include common concepts that allow us to formulate two broad principles for improving drought resistance. First, in every case, water limitation was present in a realistic form during the selection procedure. To improve seedling establishment, seeds were germinated in dehydrated soil (Wright and Jordan, 1970) or, for improved reproduction, plants were selected for earliness during late season drought (Hall and Grantz, 1981). Second, in most cases, an intimate knowledge of the physiology and biology of the crop was required. The value of deep rooting (Burton *et al.*, 1954, 1957) or osmotic adjustment (Morgan, 1983) was recognized, and selection for these single traits increased the rate at which superior genotypes could be found.

Approaching the problem this way ensured that drought-adaptive factors were called into play and had an opportunity to express themselves. Thus, the presence of water limitation avoided the possibility that important factors might be missed. It also avoided the plight of crops yielding well in favorable environments but "crashing" in water-limited environments.

Most of the studies reduced the problem to a few specific traits that could be evaluated easily. This reduced the number of selections that had to be made and decreased the range of conditions necessary to evaluate elite germplasm. In this way, not only was progress accelerated but financial resources did not need to be large.

In only one case was there an exception to this concept (Jensen and Cavalieri, 1983), and these investigators used grain yield from about 500 field replications to evaluate their maize genotypes in a major and costly effort. Their demonstration that improvements in drought performance need not sacrifice productivity in favorable conditions is the most compelling test of this principle. The number of replications, genotypes, and field sites was so large that the principle is now beyond doubt (Figure 4). Moreover, because they ranked for whole season yield performance, problems associated with particular parts of the life cycle would have been identified. This principle of productivity gain without sacrifice in favorable conditions was confirmed by the selections of Morgan (1983) for superior osmotic adjustment. His tests showed that the new selections yielded as well as cultivars in commercial production and at the same time lost less yield when water became limited.

In summary, drought resistance studies can be expected to bring about significantly improved plant performance in practicing agriculture where water limitations are common. Success usually relies on two rules - (1) make selections under realistic drought conditions and (2) reduce the problem to specific traits that have relevance for the particular drought problem. A corollary is that the selection need not involve only total dry matter production per unit of water used. Selecting for other aspects of performance such as improved growth of only the harvested part of the crop can lead to sizeable increases in productivity even when total dry matter production is unaltered.

REVERSING DROUGHT-INDUCED LOSSES IN REPRODUCTIVE DEVELOPMENT

Selecting for improved performance of the harvested component is especially important for reproductive crops. The total U.S. land devoted to agriculture is 423×10^6 ha of which 149×10^6 ha are harvested. Reproductive crops (grain, fruit, nut, vegetable) account for 115×10^6 ha of the harvested portion or 78%. Therefore, reproductive crops comprise a large fraction of the harvest in U.S. agriculture.

Although reproduction is more inhibited than other stages of development when water is limited, it has only recently received much scientific attention. D. Aspinall and his coworkers from the Waite Agricultural Institute in Adelaide were some of the first to recognize the need to investigate the problem (Damptey *et al.*, 1978) and other investigators have begun to enter the field. We can describe their efforts so far with a brief summary in individual crops.

In maize, losses in reproductive activity were reported to be caused by megagametophyte sterility (Moss and Downey, 1971), asynchronous floral development (Herrero and Johnson, 1981), and non-receptive silks (Lonnquist and Jugenheimer, 1943) depending on when dehydration occurred. However, when gamete and floral development were normal and plants were hand-pollinated, reproductive failure still occurred and could be induced by dehydration of only a few days (Westgate and Boyer, 1986). The loss was caused by irreversibly arrested embryo development (Westgate and Boyer, 1986). The effect was correlated with low photosynthetic reserves in the maternal plant (Westgate and Boyer, 1985). Because photosynthesis was inhibited during the treatment, the lack of reserves could have caused embryo starvation. However, endogenous growth regulators could have blocked transport or inhibited embryo growth, and effects of low water potentials or water contents directly on embryo growth and metabolism also could have been involved.

Westgate and Thomson Grant (1989) observed that the sugar content of the embryos was not significantly different in hydrated and dehydrated maize plants but concluded that the flux of photosynthate might differ. Myers *et al.* (1990) showed an inhibition of endosperm cell division by high abscisic acid (ABA) levels 5 to 10 days after fertilization. ABA levels rise when maize plants are subjected to dehydration (Beardsell and Cohen, 1965).

In other crops such as wheat and barley (Morgan, 1980; Saini and Aspinall, 1981, 1982; Saini *et al.*, 1984), drought during microsporogenesis caused pollen sterility. Hydrated plants whose stems were fed ABA (Saini *et al.*, 1984) or whose shoots were sprayed with ABA (Morgan, 1980) showed a similar pollen abortion thus implicating high ABA levels during dehydrating conditions. However, increasing CO_2 pressures around wheat plants overcame some of the reproductive losses (Gifford, 1979), which also may implicate photosynthesis. On the other hand, in rice (*Oryza sativa* L.), dehydration of the soil caused especially severe dehydration of reproductive tissues, and death and bleaching of florets followed (O'Toole *et al.*, 1984). The cuticle is only poorly developed on the floral tissues of rice and may have been insufficient to prevent excessive dehydration (O'Toole *et al.*, 1984).

Therefore, in various crops there is increasing evidence for metabolic and growth regulator effects and some direct dehydration effects that might account for the susceptibility of early reproduction to water limitation. So far, the evidence is correlative and there is none to show a causal effect of any of the factors. Photosynthesis, CO_2 and ABA are implicated but each might act in concert or separately, depending on the crop.

What is needed is direct evidence for causal relationships and their interactions, which is not a simple matter in view of the complexity of reproduction. Our laboratory has recently attempted to develop some of this understanding by taking a simple approach to the problem. The finding that a few days of water deprivation can prevent all grain development (Westgate and Boyer, 1986) establishes a window of susceptibility in maize. Using this window, we developed a system (Boyle *et al.*, 1991a) to feed stems a nutrient solution to allow conditions in the plant to be varied at will. A small well was made in the stem and was filled with nutrient solution. The well was sealed with a serum cap and a reservoir was attached to a needle extending through the cap so that the well could be continuously supplied by the reservoir. We fed a modified medium normally used for *in vitro* culture of maize kernels (Gengenbach, 1977; Cobb and Hannah, 1983). The medium was made more concentrated in sugars, and no agar was used. This allowed photosynthate as well as other metabolites to be supplied to the plants in a form that could support embryo growth. By feeding the stems, the nutrients entered the main storage tissue for photosynthetic reserves and could potentially act as reserves. In maize, each plant normally accumulates 5 g of dry mass day^{-1} during reproduction, and the dehydration largely prevented this activity. Therefore, the fed solution needed to provide at least 5 g of dry mass without hydrating the plants. By concentrating the medium, it was possible to feed 5 g in only 30 ml of solution. This was much less water than the plants were extracting from the soil each day (200 to 300 ml plant^{-1} day^{-1} during the severest dehydration period). We were able to supply this for several days by initiating a new feeding site each day. Thus, reserves could be maintained at levels usually supported by photosynthesis without rehydrating the plants.

Figure 5 shows that withholding water from the soil dehydrated the plants (decreased the water potential, Ψ_w) and inhibited photosynthesis of the leaves (Boyle *et al.*, 1991b). Both returned to near-control levels when water was resupplied several days later. Infusing the culture medium into the stems did not alter this behavior (low Ψ_w + M), and infusing the same amount of water without nutrients was similarly without effect (low Ψ_w + W). This result allowed us to test whether the altered reserve status affected reproductive development without disturbing the water status of the plants.

It should be noted that these treatments began at early silking and allowed normal anthesis. The silks could be hand-pollinated at the minimum leaf water potential so that we could separate effects on floral development from those on gametic transmission and later events, and test whether dehydration altered the later events in reproduction. We designed the experiments this way because failures in early events might be correctable but reproduction could still be blocked by later events. Our finding that undeveloped zygotes were always present (Westgate and Boyer, 1986) indicates that the early events proceeded normally and later events had been blocked.

Figure 6 confirms this result (Boyle *et al.*, 1991b). Whereas the controls yielded well (Figure 6A), withholding water as in Figure 5 virtually eliminated grain production (Figure 6B). Reproduction was maintained at near-control levels when the plants were infused with complete medium each day as low water potential developed (Figure 6C). Infusing the same amount of water alone showed no activity (Figure 6D). This indicates that it was possible to maintain reproductive development in dehydrated plants without rehydrating the plants. The experiment showed that water was available to the embryos in the dehydrated plants and that reproduction had been blocked

Figure 5 Leaf water potentials (A) and photosynthesis (B) at various times after withholding water from soil-grown maize during early reproduction. The controls (C) were kept at high water potential by watering daily. The plants at low water potentials (low Ψ_w) had water withheld from the soil on day 1 (-W). Some plants at low Ψ_w were infused (+I) with medium (low Ψ_w + M) beginning on day 1 and ending on day 5. Other plants at low Ψ_w were infused with the same amount of water but no other constituents of the medium (low Ψ_w + W). Plants were rewatered on day 7 (W). Plants were hand-pollinated on day 5 when water potentials were low or, in some experiments, on day 7 after rewatering. Photosynthesis measured as CO_2 fixation was monitored in the same leaves used to sample Ψ_w.

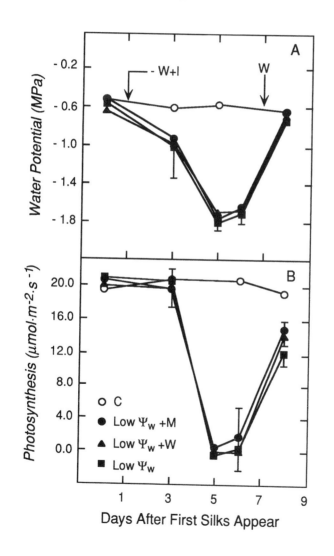

Figure 6 Grain yield at maturity for soil-grown maize plants subjected to low water potentials as in Figure 5. Treatments were adequately watered controls (A), low water potential plants from which water was withheld for 5 days (B), low water potential plants with stem infusion of medium (C), and low water potential plants with stem infusion of water (D) in the same amount as in C. All plants were rewatered on the sixth day. Plants were hand-pollinated. Grain weight of the plants infused with medium at low water potentials was about 80% of the weight of the controls.

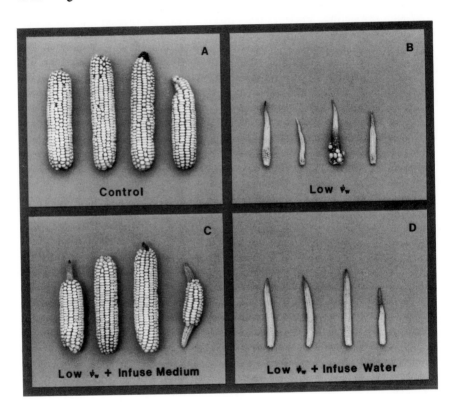

by some other substance(s) that the maternal plant failed to supply. Feeding the complete medium supplied the substance(s).

Subsequent experiments showed that auxin and cytokinin could be deleted from the medium without inhibiting the recovery (Boyle *et al.*, 1991b) and sucrose had activity when added alone (M. G. Boyle, 1990, pers. comm.). However, the quantitative contribution of sucrose remains unknown, and the other components of the medium have not been tested for activity. The effects of ABA also have not been investigated, although the results of Saini and Aspinall (1981, 1982), Saini *et al.* (1984) and Myers *et al.* (1990) indicate that some effects might be expected.

This type of experiment offers the promise of eventually identifying components that are required for reproductive development in plants and which may be lacking when drought occurs. Supplying these nutrients by stem infusion is, of course, impractical under agricultural conditions but it might be possible to search for genotypes or cultural conditions that would allow large amounts of the missing constituents to be present at the right time. This might protect against drought effects on reproduction, at least those caused by short periods of dehydration of the shoot.

It also demonstrates that reproductive development can vary from zero to nearly normal during drought without significant effects on the total dry matter production, since the drought lasted only a few days and photosynthesis returned to near normal levels soon after water was resupplied. This implies that successful protection of reproductive development may be possible under otherwise debilitating drought conditions.

CONCLUSIONS

The new techniques and findings regarding differences in water use efficiency in crops and the relationship of this efficiency to the plant genome promise that genetic methods may be employed for minimizing water use by crops. In addition, there is an increasing number of examples of improvements in drought resistance that are based on enhanced productivity of plant parts likely to be useful for maintaining viability and reproduction. Because so much agricultural production relies on plant reproduction, this approach has had large effects. Central to these efforts has been the use of realistic drought conditions, and selection for a small number of traits that might be valuable for crop performance with limited water. Importantly, it is now clear that successful improvement of drought performance can come at no sacrifice to performance under favorable conditions.

Recent advances in understanding reproductive failure at low water potentials may provide new ways of protecting this phase of development against loss. So far, the experiments show that later phases of the process do not fail because of direct effects of water availability but rather because an essential substance is not supplied that would normally be supplied by the maternal plant or because there is an inhibitory substance supplied in large quantity by the maternal plant. Thus, the problem increasingly appears to be a chemical one rather than a physical one, and it is hoped that the substances involved may soon be identified. At this stage, the findings mostly apply to short periods of dehydration, but the principles may prove to be more broadly applicable when more information becomes available.

REFERENCES

Beardsell, M. F.; Cohen, D. (1975) Relationships between leaf water status, abscisic acid levels, and stomatal resistance in maize and sorghum. *Plant Physiology* 56:207-212.

Bordovsky, D. G.; Jordan, W. R.; Hiler, E. A.; Howell, T. A. (1974) Choice of irrigation timing indicator for narrow row cotton. *Agronomy Journal* 66:88-91.

Bowman, W. D.; Hubick, K. T.; von Caemmerer, S.; Farquhar, G. (1989) Short-term changes in leaf carbon isotope discrimination in salt- and water-stressed C_4 grasses. *Plant Physiology* 90:162-166.

Boyer, J. S. (1982) Plant productivity and environment. *Science (Washington, DC)* 218:443-448.

Boyle, M. G.; Boyer, J. S.; Morgan, P. W. (1991a) Stem infusion of maize plants. *Crop Science* 31:1241-1245.

Boyle, M. G.; Boyer, J. S.; Morgan, P. W. (1991b) Stem infusion of liquid tissue culture medium prevents reproductive failure in maize plants at low water potentials. *Crop Science* 31:1246-1252.

Bradford, K. C.; Hsiao, T. C. (1982) Physiological responses to moderate water stress. In: Lange, O. L.; Nobel, P. S.; Osmond, C. B.; Ziegler, H. (eds.), *Encyclopedia of plant physiology (New Series), Vol. 12B, Physiological plant ecology, II.* Springer-Verlag, Berlin, pp. 263-324.

Briggs, L. J.; Shantz, H. L. (1914) Relative water requirement of plants. *Journal of Agricultural Research (Washington, DC)* 3:1-63.

Brown, R. H.; Simmons, R. E. (1979) Photosynthesis of grass species differing in CO_2 fixation pathways. I. Water-use efficiency. *Crop Science* 19:375-379.

Brugnoli, E.; Hubick, K. T.; von Caemmerer, S.; Wong, S. C.; Farquhar, G. D. (1988) Correlation between the carbon isotope discrimination in leaf starch and sugars of C_3 plants and the ratio of intercellular and atmospheric partial pressures of carbon dioxide. *Plant Physiology* 88:1418-1424.

Burton, G. W.; DeVane, E. H.; Carter, R. L. (1954) Root penetration, distribution and activity in southern grasses measured by yields, drought symptoms and P^{32} uptake. *Agronomy Journal* 46:229-233.

Burton, G. W.; Prine, G. M.; Jackson, J. E. (1957) Studies of drouth tolerance and water use of several southern grasses. *Agronomy Journal* 49:498-503.

Claassen, M. M.; Shaw, R. H. (1970) Water deficit effects on corn. II. Grain components. *Agronomy Journal* 62:652-655.

Clampet, G. L. (1982) *Agricultural statistics.* U.S. Government Printing Office, Washington, DC.

Cobb, B. G.; Hannah, L. C. (1983) Development of wild type, *shrunken* 1 and *shrunken* 2 maize kernels grown *in vitro. Theoretical and Applied Genetics* 65:47-51.

Condon, A. G.; Farquhar, G. D.; Richards, R. A. (1990) Genotypic variation in carbon isotope discrimination and transpiration efficiency in wheat. Leaf gas exchange and whole plant studies. *Australian Journal of Plant Physiology* 17:9-22.

Condon, A. G.; Richards, R. A.; Farquhar, G. D. (1987) Carbon isotope discrimination is positively correlated with grain yield and dry matter production in field-grown wheat. *Crop Science* 27:996-1001.

Damptey, H. B.; Coombe, B. G.; Aspinall, D. (1978) Apical dominance, water deficit, and axillary inflorescence growth in *Zea mays*: The role of abscisic acid. *Annals of Botany (London)* 42:1447-1458.

de Wit, C. T. (1958) Transpiration and crop yields. In: Institute of Biological and Chemical Research on Field Crops and Herbage, Wageningen, The Netherlands. *Verslagen Landbouwkundige Onderzoekingen* 64(6):1-88.

Gengenbach, B. G. (1977) Development of maize caryopses resulting from *in vitro* pollination. *Planta* 134:91-93.

Gifford, R. M. (1979) Growth and yield of CO_2-enriched wheat under water-limited conditions. *Australian Journal of Plant Physiology* 6:367-378.

Hall, A. E.; Grantz, D. A. (1981) Drought resistance of cowpea improved by selecting for early appearance of mature pods. *Crop Science* 21:461-464.

Hanks, R. J. (1983) Yield and water use relationships: An overview. In: Taylor, H. M.; Jordan, W. R.; Sinclair, T. R. (eds.), *Limitations to efficient water use in crop production*. American Society of Agronomy, Madison, WI, pp. 393-411.

Hanson, A. D.; Hitz, W. D. (1982) Metabolic responses of mesophytes to plant water deficits. *Annual Reviews of Plant Physiology* 33:163-203.

Herrero, M. P.; Johnson, R. R. (1981) Drought stress and its effect on maize reproductive systems. *Crop Science* 21:105-110.

Hubick, K.; Farquhar, G. (1990) Carbon isotope discrimination and the ratio of carbon gained to water lost in barley cultivars. *Plant Cell and Environment* 12:795-804.

Hubick, K. T.; Farquhar, G. D.; Shorter, R. (1986) Correlation between water-use efficiency and carbon isotope discrimination in diverse peanut (*Arachis*) germplasm. *Australian Journal of Plant Physiology* 13:803-816.

Hull, H. M.; Wright, L. N.; Bleckmann, C. A. (1978) Epicuticular wax ultrastructure among lines of *Eragrostis lehmanniana* Nees developed for seedling drouth tolerance. *Crop Science* 18:699-704.

Jensen, S. D.; Cavalieri, A. J. (1983) Drought tolerance in U.S. maize. In: Stone, J. F.; Willis, W. O. (eds.), *Plant production and management under drought conditions. Developments in agricultural and managed-forest ecology #12*. Elsevier Science Publishers, New York, pp. 223-236.

Joshi, M. C.; Boyer, J. S.; Kramer, P. J. (1965) Growth, carbon dioxide exchange, transpiration, and transpiration ratio of pineapple. *Botanical Gazette* 126:174-179.

Kawamitsu, Y.; Agata, W.; Miura, S. (1987) Effects of vapour pressure difference on CO_2 assimilation rate, leaf conductance and water use efficiency in grass species. *Journal of the Faculty of Agriculture of Kyushu University* 31:1-10.

Lonnquist, J. H.; Jugenheimer, R. W. (1943) Factors affecting the success of pollination in corn. *Journal of the American Society of Agronomy* 35:923-933.

Martin, B.; Nienhuis, J. King, G.; Schaefer, A. (1989) Restriction fragment length polymorphisms associated with water use efficiency in tomato. *Science (Washington, DC)* 243:1725-1728.

Martin, B.; Thorstenson, Y. R. (1988) Stable carbon isotope composition ($\delta^{13}C$), water use efficiency, and biomass productivity of *Lycopersicon esculentum*, *Lycopersicon pennellii*, and the F_1 hybrid. *Plant Physiology* 88:213-217.

Meyer, R. F.; Boyer, J. S. (1972) Sensitivity of cell division and cell elongation to low water potentials in soybean hypocotyls. *Planta* 108:77-87.

Meyer, R. F.; Boyer, J. S. (1981) Osmoregulation, solute distribution, and growth in soybean seedlings having low water potentials. *Planta* 151:482-489.

Morgan, J. M. (1980) Possible role of abscisic acid in reducing seed set in water-stressed wheat plants. *Nature (London)* 285:655-657.

Morgan, J. M. (1983) Osmoregulation as a selection criterion for drought tolerance in wheat. *Australian Journal of Agricultural Research* 34:607-614.

Morgan, J. M. (1984) Osmoregulation and water stress in higher plants. *Annual Review of Plant Physiology* 35:299-319.

Moss, G. I.; Downey, L. A. (1971) Influence of drought stress on female gametophyte development in corn (*Zea mays* L.) and subsequent grain yield. *Crop Science* 11:368-372.

Myers, P. N.; Setter, T. L.; Madison, J. T.; Thompson, J. F. (1990) Abscisic acid inhibition of endosperm cell division in cultured maize kernels. *Plant Physiology* 94:1330-1336.

Neales, T. F.; Patterson, A. A.; Hartney, V. J. (1968) Physiological adaptation to drought in the carbon assimilation and water loss of xerophytes. *Nature (London)* 219:469-472.

O'Toole, J. C.; Bland, W. L. (1987) Genotypic variation in crop plant root systems. *Advances in Agronomy* 41:91-145.

O'Toole, J. C.; Hsiao, T. C.; Namuco, O. S. (1984) Panicle water relations during water stress. *Plant Science Letters* 33:137-143.

Robichaux, R. H.; Pearcy, R. W. (1984) Evolution of C_3 and C_4 plants along an environmental moisture gradient: patterns of photosynthetic differentiation in Hawaiian *Scaevola* and *Euphorbia* species. *American Journal of Botany* 71:121-129.

Saini, H. S.; Aspinall, D. (1981) Effect of water deficit on sporogenesis in wheat (*Triticum aestivum* L.). *Annals of Botany* 48:623-633.

Saini, H. S.; Aspinall, D. (1982) Sterility in wheat (*Triticum aestivum* L.) induced by water stress or high temperature: Possible mediation by abscisic acid. *Australian Journal of Plant Physiology* 9:529-537.

Saini, H. S.; Sedgley, M.; Aspinall, D. (1984) Developmental anatomy in wheat of male sterility induced by heat stress, water deficit or abscisic acid. *Australian Journal of Plant Physiology* 11:243-254.

Salter, P. J.; Goode, J. E. (1967) *Crop responses to water at different stages of growth.* Commonwealth Agricultural Bureau, Farnham Royal, Bucks, England.

Sharp, R. E.; Hsiao, T. C.; Silk, W. K. (1990) Growth of maize primary root at low water potentials. II. Role of growth and deposition of hexose and potassium in osmotic adjustment. *Plant Physiology* 93:1337-1346.

Tanner, C. B.; Sinclair, T. R. (1983) Efficient water use in crop production: Research or re-search. In: Taylor, H. M.; Jordan, W. R.; Sinclair, T. R. (eds.), *Limitations to efficient water use in crop production.* American Society of Agronomy, Madison, WI, pp. 1-27.

Taylor, H. M.; Jordan, W. R.; Sinclair, T. R. (1983) *Limitations to efficient water use in crop production.* American Society of Agronomy, Madison, WI.

Westgate, M. E.; Boyer, J. S. (1985) Carbohydrate reserves and reproductive development at low water potentials in maize. *Crop Science* 25:762-769.

Westgate, M. E.; Boyer, J. S. (1986) Reproduction at low silk and pollen water potentials in maize. *Crop Science* 26:951-956.

Westgate, M. E.; Thomson Grant, D. L. (1989) Water deficits and reproduction in maize: Response of the reproductive tissue to water deficits at anthesis and mid-grain fill. *Plant Physiology* 91:862-867.

Wright, L. N.; Jordan, G. L. (1970) Artificial selection for seedling drought tolerance in boer lovegrass (*Eragrostis curvula* Nees). *Crop Science* 10:99-102.

Chapter 9
Genetic basis of plant tolerance of soil toxicity

J. Dvořák[1], E. Epstein[2], A. Galvez[1], P. Gulick[3], and J. A. Omielan[2]

[1]Department of Agronomy and Range Science, University of California, Davis, CA 95616; [2]Department of Land, Air and Resources, University of California, Davis, CA 95616; and [3]Department of Biology, Concordia University, Montreal, Canada

Understanding of the genetic and physiological mechanisms by which plants cope with adverse soil conditions is critical for the development of efficient strategies for breeding stress tolerant cultivars. Soil conditions stressful to plants range from deficiencies of nutrient elements to excessively high (toxic) concentrations of elements or salts (Epstein, 1990). Deficiencies may often be remedied by the application of appropriate fertilizers, but removal or fixation in forms unavailable to plants of toxic constituents is often not feasible. It is for such situations that the development of toxicity stress tolerant plants is important.

Although adaptation of plants to various forms of soil toxicity has been described, not all forms of toxicity are equally important for crop production. Some are caused by localized excesses of trace elements that may be of little or no significance for crops as a whole. Other conditions are more widespread. Crop tolerance of those conditions is an important breeding objective. Aluminum toxicity and soil salinity are examples of the latter category. Nevertheless, an understanding of the physiological and genetical control of tolerance of all types of soil toxicity is needed to form a complete picture of how the tolerance of adverse soil conditions evolves. Such an understanding, in turn, may provide clues for the genetic improvement of plants grown under such conditions.

Stress caused by toxic soil conditions may be divided into two broad categories according to the complexity of factors causing them - (1) stress caused primarily by a single factor and (2) stress caused by a combination of factors. It is reasonable to expect that the complexity of the physiological and genetic basis of the tolerance would reflect the complexity of the stress.

The tolerance of an excess of heavy metal ions in soil is an example of the first category, although combinations of factors may occur to create a complex situation. Heavy metal toxicities occur both naturally and as a result of human activities. For example, many soils in the humid tropics are acidic and as a result have high concentrations of aluminum, manganese, or both in their soil solutions. As for human activities and their effects, lead from leaded gasoline tends to affect plants along highways, and acid rain resulting from industrial processes may solubilize heavy metals in soils. Plants have been shown to adapt to toxic concentrations of zinc, copper, lead, and other metal ions (for review, see Antonovics *et al.*, 1971). Even when the stress was caused by the excess of a single ion most studies indicated that the inheritance of tolerance vs. susceptibility was complex (Wilkins, 1960; Bröker, 1963; Urquhart, 1971; Cartside and McNeilly, 1974; Walley *et al.*, 1974; Brown and Devine, 1980). At least in some instances, however, the inheritance of the adaptation to toxic concentrations of heavy metal ions may depend on as few as one or two major genes, and the activity may be modified by other genes (MacNair, 1977, 1979, 1983). Future studies may show this to apply also to some of the instances mentioned earlier. Partial tolerance was observed in plant populations growing under non-toxic conditions and was concluded to be a component of natural variation in the gene pools of many species (Walley *et al.*, 1974). Consequently, natural or artificial selection can be effective in selecting tolerant genotypes from overall non-tolerant populations. Perhaps the best evidence that tolerance of this type of environmental stress can be significantly enhanced by a single gene affecting a simple physiological mechanism was provided by the introduction of the human metallothionein gene into *Brassica* and *Nicotiana* and the demonstration that it enhanced cadmium tolerance in the transgenic plants (Misra and Gedamu, 1989).

Earlier in this century the idea prevailed that heavy metal tolerance is not highly specific; a plant tolerant of, say, zinc was considered likely to be tolerant of copper and other heavy metals as well (Ernst, 1982). More recently, however, examples have come to light of genotypes that are specifically tolerant of a given heavy metal ion but not of another (Walley *et al.*, 1974; Cox and Hutchinson, 1980; Graham, 1984). If ion-specific tolerance were the prevalent genetic basis of heavy metal tolerance the genetic system conferring tolerance of stress caused by several ions would become necessarily more complex.

Another instance of soil toxicity caused by the excess of a single element is boron toxicity. In the U.S.A. this stress occurs locally on irrigated land in certain areas of the San Joaquin Valley in California often along with excess of selenium. A recent study demonstrated a tolerance in cereals which is physiologically manifested by partial exclusion of this element (Nable, 1988). The genetic basis of this tolerance is not known.

Toxicity due to aluminum solubilized in acid soils is another example of soil toxicity caused by a single major factor. Very large areas in the humid tropics are affected. Both intraspecific and interspecific variations in aluminum tolerance have been found. The genetic basis of intraspecific variation has been investigated in corn (*Zea mays* L.) (Rhue *et al.*, 1978), barley

(*Hordeum vulgare* L.) (Stolen and Andersen, 1978) and wheat (*Triticum aestivum* L.) (Foy *et al.* 1974; Polle *et al.*, 1978; Aniol and Gustafson, 1984), and several other species. While several of these studies indicated a simple genetic basis for the variation, more detailed studies revealed a complex picture. For instance, rye (*Secale cereale* L.) is more aluminum-tolerant than is wheat. Wheat x rye hybrids express this tolerance, but to a much diminished extent (Aniol and Gustafson, 1984). This limited tolerance is conferred on wheat by at least three rye chromosomes (Aniol and Gustafson, 1984). A number of wheat chromosomes, particularly those of the wheat D genome, were shown to affect the expression of the tolerance by rye chromosomes (Gustafson and Ross, 1990).

Stress caused by soil salinity, typically associated with irrigated agriculture in arid or semi-arid environments, is an example of a toxic soil condition inherently affecting a plant in a number of ways (Epstein, 1983; Läuchli and Epstein, 1990); stress may be caused simultaneously by toxic concentrations of several ions, reduced water potential, high pH, and unavailability of specific nutrients. For that reason it would be expected that the genetic control of salt tolerance is complex. Studies comparing cultivars differing in salt tolerance indicated a complex inheritance but suggested that breeding progress can be achieved (Akbar *et al.*, 1972; Akbar and Yabuno, 1977; Moeljopawiro and Ikehashi, 1981; Azhar and McNeilly, 1988). Yeo and Flowers (1986) concluded that in rice (*Oryza sativa* L.) salt tolerance depends on a number of components and that improvements can be best achieved by selecting for each component individually and pyramiding them in a breeding program. The parallelism between the complexity of salt stress and the complexity of genetic response is indirectly indicated by experimental evidence showing that both within and among species the tolerance of the excess of individual ions may be controlled by different gene systems (Wu, 1981; Ashraf *et al.*, 1989; Omielan *et al.*, 1991).

Most of what is known about the genetic basis of intraspecific variation in salt stress tolerance comes from studies of contrasting phenotypes using the methods of traditional quantitative genetics. This approach is limited in that it does not allow application of genetics as a tool for physiological studies aimed at the elucidation of salt tolerance mechanisms. Nevertheless, several attempts have been made to bridge this gap by employing mutants, isogenic populations, synthetic amphiploids constructed from salt sensitive and tolerant species, and alien single-chromosome addition and substitution lines (Abel, 1969; Dvorák *et al.*, 1985; Dvorák and Ross, 1986; Grumet and Hanson, 1986; Warne and Hickok, 1987; Gorham *et al.*, 1986, 1987; Shah *et al.*, 1987; Dvorák *et al.*, 1988; Forster *et al.*, 1990; Omielan *et al.*, 1991).

The physiological mechanisms by which plants tolerate salt stress can be divided into two broad categories - (1) partial exclusion of ions which are in excess in the environment (the exclusion may be accompanied by synthesis of organic-compatible solutes to depress the internal solute potential) and (2) inclusion of ions as osmotica into the vacuoles and accumulation of compatible solutes in the cytoplasm (for reviews see Flowers *et al.*, 1977; Greenway and Munns, 1980; Epstein, 1983; Stavarek and Rains, 1983; Gorham *et al.*, 1985; Tal, 1985; Läuchli and Epstein, 1990).

Cereals, which are the most important food source of mankind, and their salt-tolerant wild relatives fall into the first category, that of "ion excluders". Early studies focused on the role of compatible solutes, such as betaines, accumulated under salt stress in cereals (Wyn Jones and Storey, 1981). Grumet and Hanson (1986) constructed barley isopopulations differing in the constitutive levels of glycinebetaine and were able to show that the isopopulation with constitutively high glycinebetaine levels maintained low solute potentials under salt stress. However, the most conspicuous feature of the physiological response of cereals and their salt-tolerant relatives to salt stress is the partial exclusion from the shoots, and, hence, from the photosynthetic apparatus, of ions that are in the excess (Storey *et al.*, 1985; Gorham *et al.*, 1986; Shah *et al.*, 1987; Schachtman, *et al.*, 1989).

Two genetic systems have been used to investigate the association between the reduced Na^+ accumulation in shoots and salt tolerance. One system exploited the contrast in Na^+ accumulation between the tetraploid macaroni wheat, *T. turgidum* L. (genome formula AABB) and the hexaploid bread wheat, *T. aestivum* (genome formula AABBDD). The latter species shows lower Na^+ accumulation and higher ratio of accumulated K^+/Na^+ under salt stress than the former species (Shah *et al.*, 1987). As expected from the genome formulas of the two species, this character is controlled by the D genome (Shah *et al.*, 1987). Comparison of Na^+ accumulation in shoots of disomic substitution lines in which each of the D-genome chromosomes of *T. aestivum* was individually incorporated into *T. turgidum* indicated that the trait is controlled by genes on a single chromosome, 4D (Gorham *et al.*, 1987, 1990). However, no evidence has been provided that this chromosome substitution enhances salt tolerance in *T. turgidum*. To pursue this question, chromosome 4D was homologously recombined with chromosome 4B and K^+/Na^+ ratios in shoots under salt stress were determined in the population of 29 recombinant lines (J. Dvorák, J. Gorham, and G. Wyn Jones, unpublished data). The recombinant lines either had the high K^+/Na^+ ratio characteristic of the *T. turgidum* disomic substitution line 4D(4B) and euploid *T. aestivum* or the low K^+/Na^+ ratio characteristic of *T. turgidum*; no intermediate phenotypes were observed. This phenotypic effect mapped entirely to a specific, distal region of the long arm of chromosome 4D. This indicated that the superior Na^+ exclusion in *T. aestivum* is controlled by a single locus.

Lophopyrum elongatum (Host) Löve ($2n=2x=14$), genome E, and its close relative *L. ponticum* (Podp.) Löve ($2n=10x=70$) were used to develop the second system that has been employed in the study of the genetics of ion uptake and accumulation under salt stress in cereals and their relatives. Both species are highly salt tolerant (Elzam and Epstein, 1969; McGuire and Dvořák, 1981) and amphiploids or partial amphiploids from hybridization with wheat have been produced for both of them. They show salt tolerance enhanced relative to wheat but reduced relative to the *Lophopyrum* parents (Dvořák *et al.*, 1985; Dvořák and Ross, 1986).

To determine the genetic control of the salt tolerance of *L. elongatum* as expressed in wheat, each of the seven chromosomes of *L. elongatum* was individually added to the chromosome complement of wheat cultivar Chinese Spring and substituted for each of the Chinese Spring homoeologues (Dvořák, 1980; Dvořák and Chen, 1984; Tuleen and Hart, 1988). The salt tolerance of the seven addition lines and their diallel crosses, 13 ditelosomic addition lines and 20 of the 21 possible disomic substitution lines was evaluated in solution culture (Dvořák *et al.*, 1988) and that of disomic substitution lines also in field experiments (Omielan *et al.*, 1991). The solution culture experiments showed that the enhanced salt tolerance of the amphiploid Chinese Spring x *L. elongatum* is due to genes acting largely additively. The greatest effect was associated with *Lophopyrum* chromosome 3E and smaller effects with chromosomes 2E, 4E, and 7E (Dvořák *et al.*, 1988). Several complementary interactions were detected, one involving chromosome 6E. Field experiments with the set of disomic substitution lines at two levels of salinity and control conditions in 1988, 1989, and 1990 substantiated the results obtained in solution cultures but indicated that minor enhancement of salt tolerance was also associated with chromosomes 1E and 5E (Figure 1). Salt tolerance in *L. elongatum* was thus shown to be controlled either singly or in an interaction by all seven chromosomes, clearly indicating a complex, multigenic inheritance. Yet salt tolerance in wheat could be significantly improved by incorporation of individual *L. elongatum* chromosomes. The additive effect of chromosome 3E accounted for about 50% of the difference between Chinese Spring and the salt-tolerant amphiploid (Figure 1). Limited data are available suggesting that this effect is controlled by genes on the short arm of chromosome 3E (Dvořák *et al.*, 1988).

The amphiploid differs from Chinese Spring by lower accumulation of Na^+ and enhanced accumulation of K^+, and consequently by higher K^+/Na^+ ratios, under salt stress (Storey *et al.*, 1985; Schachtman *et al.*, 1989; Table 1). Although a similar tendency was described in other salt-tolerant amphiploids relative to their salt-sensitive wheat parents and in *T. aestivum* relative to the even more sensitive *T. turgidum* (Gorham *et al.*, 1986; Shah *et al.*, 1987; Schachtman *et al.*, 1989), no evidence has been provided to substantiate the relationship between salt tolerance and K^+/Na^+

Figure 1 Overall performance (average of grain yield, biomass, seed weight, plant height, and tiller number as % of the unstressed controls) of the amphiploid, 'Chinese Spring' x *L. elongatum* (AgSC) and disomic substitution lines in which the specified Chinese Spring chromosomes (lower row below the horizontal axis) were replaced with their *L. elongatum* homoeologues (upper row below the horizontal axis) relative to Chinese Spring (zero line) at intermediate and high field salinities in 1988, 1989, and 1990 (*specifies missing data in 1990) (from Omielan *et al.*, 1991).

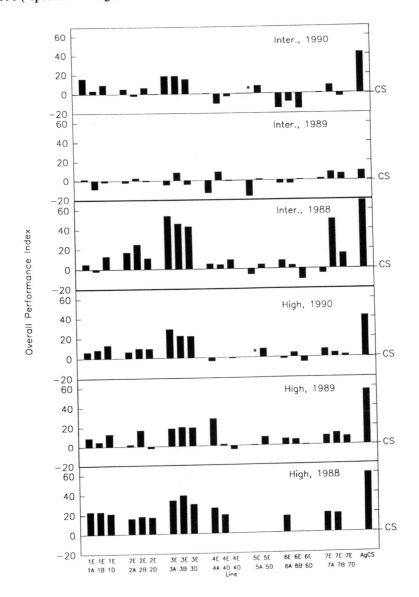

ratio. Therefore, to scrutinize this relationship, the accumulation of Na^+, K^+, Mg^{2+}, Ca^{2+} and Cl^- in the flag leaves of Chinese Spring, the amphiploid, and the disomic substitution lines salt-stressed in the field was determined and compared with their field salt tolerance (Omielan *et al.*, 1991). Data for 1989 and 1990 have been reported (Omielan *et al.*, 1991); the 1990 data are shown in Tables 1 and 2.

Under control conditions the amphiploid accumulated the same amounts of Na^+ and K^+ as did Chinese Spring. When salt-stressed, however, it accumulated far less Na^+ than Chinese Spring and more K^+ than Chinese Spring and, thus, had higher K^+/Na^+ ratios. Exclusion of Na^+ appeared to be enhanced by all seven *L. elongatum* chromosomes and the accumulation of K^+ by six of the seven chromosomes (Table 1). The greatest effect on Na^+ exclusion and K^+ accumulation was associated with chromosome 3E. Interestingly, *Lophopyrum* chromosome 4E, while enhancing Na^+ exclusion and K^+ accumulation in DS4E(4A^a) did not have this effect in DS4E(4D). Both 4E(4D) sibling disomic substitution lines accumulated more Na^+ and less K^+ than Chinese Spring (Table 1). A likely reason for this lack of compensation of 4E for 4D is that these two homoeologues chromosomes bring about Na^+ exclusion in different ways. This is indicated by the following evidence. While chromosome 4E excluded about 2 to 3 μmol g^{-1} dry matter of Na^+ per each μmol g^{-1} dry matter of accumulated K^+, chromosome 4D excluded Na^+ and K^+ approximately stoichiometrically (Table 1). The same was true for the disomic substitution lines with chromosomes 1E, 3E, and 7E and for the amphiploid (Table 1). In the disomic substitution lines 3E(3A), 3E(3B), and 3E(3D) and the amphiploid this relationship was even more distorted toward exclusion of Na^+ (Table 1).

In addition to these effects the amphiploid diminished Ca^{2+} and Mg^{2+} uptake under salt stress. This exclusion was controlled by chromosomes 1E and 3E but the activity of these two chromosomes appeared to be opposed by the activity of chromosomes 2E and 6E which enhanced accumulation of both divalent cations (Table 2).

Finally, the amphiploid partially excluded Cl^- under salt stress. The disomic substitution lines possessing chromosome 1E, 3E, 4E and 6E had diminished Cl^- concentrations in the flag leaf whereas 2E and 7E caused an increase in Cl^- accumulation. Thus, the partial Cl^- exclusion in the amphiploid is controlled by a system of genes with opposing effects, not unlike the genetic control of aluminum tolerance in wheat (Gustafson and Ross, 1990).

To indicate the validity of these relationships, and assess their potential as selection tools, correlation coefficients were calculated between grain yield, either absolute or as the percentage of the unstressed controls, and the concentrations of ions in the flag leaves at the time of anthesis (Table 3). Strong negative correlations occurred between grain yield or biomass and

Table 1 Mean concentrations of Na$^+$ and K$^+$, sum of Na$^+$ plus K$^+$, and K$^+$/Na$^+$ ratio in dry weight of flag leaves of Chinese Spring (CS), the amphiploid CS x L. *elongatum* (AgCS), and disomic substitution (DS) lines in control plots, and at the intermediate and high levels of salinity in 1990 (from Omielan et al., 1991).

							$mol\ g^{-1}$					
		Na^+			K^+			$Na^+ + K^+$			K^+/Na^+	
Line	Cont.	Int.	High	Cont.	Int.	High	Cont.	Int.	High	Cont.	Int.	High
Chinese Spring	9	274	522	335	372	379	344	646	901	49.3	1.4	0.8
AgCS	8	55*	96*	345	413	591*	353	468*	687*	46.0	7.7*	6.7*
1E(1A)	11	168*	533	342	294*	380	353	462*	913	45.7	1.8	0.7
1E(1B)	6	155*	354*	333	365	487*	339	520*	841	57.9	3.0	1.5
1E(1D)	5	179*	409	358	368	402	363	547*	811	70.0	2.1	1.1
2E(2A)	6	277	532	419*	416	436	425*	693	968	73.6	1.5	0.8
2E(2B)	8	241	381*	478*	632*	570*	486*	873*	951	59.9	2.7	1.5
2E(2D)	5	165*	399*	404	406	441	409	571	840	78.6*	2.5	1.1
3E(3A)	5	56*	107*	337	396	435	342	452*	542*	66.5	7.3*	4.1*
3E(3B)	7	111*	195*	312	357	505*	319	468*	700*	44.6	3.6*	2.7*
3E(3D)	8	77*	166*	386	435*	566*	394	512*	732*	48.8	6.0*	3.5*
4E(4A[a])	9	146*	303*	295	408	502*	304	554*	805	35.6	2.9	1.6
4E(4D)[†]	25*	372*	645*	309	222*	265*	334	594	910	12.7*	0.6	0.4
4E(4D)	18*	353*	702*	308	240*	228*	326	593	930	20.5*	0.7	0.3
5E(5D)	7	139*	420	468*	411	452	475*	550*	872	65.4	3.7*	1.1
6E(6A)	5	176*	492	368	374	406	373	550*	898	70.1	2.3	0.9
6E(6B)	7	213	372*	413*	394	366	420*	607	738*	71.2	2.0	1.0
6E(6D)	5	165*	454	428	391	405	433	556	859	93.8*	2.4	0.9
7E(7A)	6	122*	373*	403	421	478*	409	543*	851	72.5	3.6*	1.3
7E(7B)	7	120*	471	335	389	441	342	509*	912	54.3	3.7*	1.0
7E(7D)	12	222	466	423*	407	427	435*	629	893	37.6	1.9	0.9

*Values significantly different from CS at the 5% probability level.

[†]Since DS4E(4B) is sterile another DS4E(4D) was used.

Table 2 Mean concentrations of Ca^{2+} and Mg^{2+} and Cl^- in dry weight of flag leaves of Chinese Spring (CS), the amphiploid CS x _L. elongatum_ (AgCS), and disomic substitution lines in the control plots and at the intermediate and high levels of salinity in 1990 (from Omielan _et al._, 1991).

Line	Ca^{2+}			Mg^{2+} ($\mu mol\ g^{-1}$)			Cl^-		
	Cont.	Int.	High	Cont.	Int.	High	Cont.	Int.	High
Chinese Spring	172	128	94	109	119	111	99	115	98
AgCS	161	111	60*	85*	108	93	82	98	71*
1E(1A)	193	110	74	101	87	82*	77	82*	100
1E(1B)	195	117	71*	102	105	90*	91	91*	94
1E(1D)	193	132	96	106	108	111	84	80*	70*
2E(2A)	204	152*	112	110	101	124	130	145*	114
2E(2B)	174	59*	46*	102	65*	64*	180*	163*	125*
2E(2D)	202	149	104	103	127	121	141*	124	98
3E(3A)	160	109	83	82*	91	98	87	68*	41*
3E(3B)	186	117	68*	101	111	94	77	75*	62*
3E(3D)	186	116	76	104	111	101	88	83*	60*
4E(4A)[a]	129*	115	87	81*	104	105	81	86*	68*
4E(4D)	182	119	98	128	148	117	76	70*	60*
4E(4D)	211*	134	94	138*	110	110	78	75*	65*
5E(5D)	138	111	77	110	110	99	116	97	94
6E(6A)	184	164*	115	107	130	125	98	103	84
6E(6B)	183	137	109	119	126	125	86	82*	54*
6E(6D)	208	155	91	106	121	94	127	78*	122
7E(7A)	179	129	83	112	137	114	117	86*	85
7E(7B)	165	143	104	107	134	121	136*	101	93
7E(7D)	182	141	90	117	135	116	143*	123	101

*Values significantly different from Chinese Spring at the 5% probability level.

leaf Na^+ concentrations and positive ones between grain yield or biomass and leaf K^+ concentration and leaf K^+/Na^+ ratio. Weak negative correlations occurred between grain yield or biomass with leaf Cl^-, Ca^{2+} and Mg^{2+} concentrations. Leaf Na^+ concentrations correlated positively with leaf concentrations of Ca^{2+}, Mg^{2+}, and Cl^-. Leaf K^+ concentrations correlated negatively with those of Ca^{2+} and weakly so with those of Mg^{2+}. All these correlations were stronger at the high salinity level than at the intermediate level. This indicated that the correlations reflected the responses of the plants to salt stress, not merely the relations among ions in the plant.

The strong negative correlations between grain yield or biomass under salt stress and either leaf Na^+ concentration or the magnitude of the leaf K^+/Na^+ ratio in this isogenic system of substitution lines can be considered as evidence of a functional relationship between these variables. Although these correlations do not prove causality - it is possible that the Na^+ exclusion and K^+ accumulation are a consequence of salt tolerance rather than a primary cause - they can be exploited in breeding for salt tolerance.

There are two implicit points in these correlations that are important for breeding for salt tolerance - (1) the correlations were equally strong when salt tolerance was measured in terms of absolute grain yield or biomass as in terms of the percentage of the unstressed controls and (2) the correlations were stronger at the high-stress level than the intermediate-stress level (Table 3). These points imply that (1) if sufficiently high salt stress is applied, salt tolerance in wheat and its close relatives can be indirectly assessed by measuring the level of accumulated Na^+ or the K^+/Na^+ ratio in the flag leaf at anthesis and (2) it is not necessary to use unstressed controls. The latter point would be important for selections on a single-plant basis. A corollary is that selection for high productivity under high salt stress is automatically a selection for salt tolerant genotypes. Finally, Omielan *et al.* (1991) demonstrated that there is no *a priori* penalty for salt tolerance in the absence of salt stress and that, therefore, breeding for salt tolerance is unlikely to conflict with other breeding objectives. This conclusion is based on the fact that DS3E(3A) outyielded Chinese Spring not only under salt stress but also under control conditions. A number of other disomic substitution lines and the amphiploid performed equally as well as did Chinese Spring under control conditions.

Although the salt tolerance of wheat has been considerably increased by the incorporation of the *L. elongatum* genome, and to a lesser extent of individual *L. elongatum* chromosomes, the expression of *Lophopyrum* salt tolerance in the wheat genetic background is lower than in *Lophopyrum* itself, which tolerates exceedingly high levels of salinity (Elzam and Epstein, 1969; McGuire and Dvorák, 1981). The expression of salt tolerance in wheat, *L. elongatum*, and the amphiploid requires a period of acclimation (J. Dvorák,

Table 3 Correlation coefficients between specified variables at the intermediate (lower half) and high (upper half) levels of salinity in 1990 (from Omielan et al., 1991).

	Grain yield	% Grain yield†	Biomass	% Biomass†	Na+	K+	K+/Na+	K++Na+	Ca²⁺	Mg²⁺	Cl⁻
Grain yield	—	0.98**	0.72**	0.67**	-0.76**	0.46*	0.81**	-0.70**	0.34	-0.18	-0.42*
% Grain yield†	0.87**	—	0.72**	0.69**	-0.74**	0.46*	0.81**	-0.68**	-0.40	-0.24	-0.38
Biomass	0.62**	0.59**	—	0.96**	-0.59**	0.40	0.75**	-0.50**	-0.31	-0.15	-0.22
% Biomass†	0.42**	0.49*	0.91**	—	-0.58**	0.40	0.70**	-0.50**	-0.30	-0.15	-0.22
Na+	-0.52**	-0.49*	-0.46*	-0.42*	—	-0.72*	-0.82**	0.82**	0.42*	0.30	0.38
K+	0.13	0.02	0.14	0.13	-0.37	—	0.59**	0.20	-0.50**	-0.39	0.15
K+/Na+	0.66**	0.57**	0.57*	0.46*	-0.84**	0.39	—	-0.68**	-0.44**	-0.30	-0.34
K++Na+	-0.38	-0.44*	-0.31	-0.28	0.63**	0.50**	-0.45*	—	0.18	0.11	0.67**
Ca²⁺	-0.23	-0.36	-0.19	-0.17	0.16	-0.41*	-0.33	-0.19	—	0.88**	-0.18
Mg²⁺	-0.12	-0.22	-0.11	-0.18	0.21	-0.30	-0.21	-0.06	0.58**	—	-0.29
Cl⁻	-0.14	-0.23	-0.04	-0.01	0.18	0.61**	-0.14	0.68**	-0.07	-0.21	—

*,**Significant at the 5% and 1% levels of probability, respectively.
†These variables are expressed as a percentage of the unstressed control.

unpublished) during which time the expression of a number of genes is either elevated, reduced, or induced *de novo* in the roots (Gulick and Dvorák, 1987). Eleven genes that show elevated expression in salt-stressed roots of *L. elongatum* were isolated as cDNA clones from mRNA extracted 6 h after the initiation of the acclimation of *L. elongatum* to 250 mM NaCl (Gulick and Dvorák, 1990). The expression of these genes during the acclimation period was examined in *L. elongatum*, Chinese Spring, and their amphiploid. The 11 genes were expressed as a single or possibly two coordinately regulated groups with a maximum expression 6-12 h after the beginning of the acclimation period in *L. elongatum* (e.g., see Figure 2). Of these 11 genes the expression of four, measured by the mRNA levels, was lower in the amphiploid than in *L. elongatum* and further lower in wheat (e.g., see

Figure 2 Quantitation of the time-course of the expression, measured as levels of mRNA, of the ES18 gene family in roots during plant acclimation to 250 mM NaCl. The expression at each time-point was determined by densitometry of autoradiograms of hybridization intensity of the ES18 clone with total root-RNAs slot-blot-immobilized in the same membrane and simultaneously hybridized with the probe. Note that the maximum expression of the ES18 gene family is more than one order of magnitude higher in *L. elongatum* than it is in Chinese Spring and that the maximum expression in the amphiploid is disproportionately lower. Note also the different expression patterns among the three genotypes.

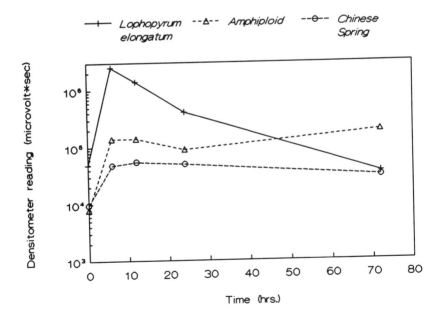

Figure 2). The remaining seven genes were expressed equally in the three genotypes (Table 4). Thus, the expression of every gene that was expressed less in Chinese Spring than in *L. elongatum* was reduced in the amphiploid. The reductions could come from different gene regulation in wheat and the amphiploid or simply from dilution of the *Lophopyrum* gene transcripts by the wheat genetic background, or both. The diminished levels of mRNA of these genes in the amphiploid relative to *L. elongatum* are likely to reduce the physiological traits which these genes control, unless more mRNA is produced in *L. elongatum* than actually needed. Although more mRNA may be transcribed than is actually needed for the full function of a single gene, it is very unlikely that this would be happening for many genes. Thus, if a trait requires expression of a number of genes, it is almost certain that at least some of them are not overtranscribed in the diploid and reduced levels of mRNAs of those genes would then become a limiting factor for the expression of the trait as a whole in the amphiploid. Therefore, if the behavior of this set of genes is indicative of the expression of other genes during the acclimation to salt stress, the mere fact that the acclimation and growth under salt stress are based on a number of genes predicts that salt stress

Table 4 Maximum relative expression, measured as levels of mRNAs relative to their levels in Chinese Spring, of specified salt-inducible genes during the acclimation of Chinese Spring, *L. elongatum*, and their amphiploid to 250 mM of NaCl.

| Gene | *Maximum level of mRNA relative to Chinese Spring* | | |
	Chinese Spring	*Amphiploid*	*L. elongatum*
ES2	1.0	8.2	21.4
ES3	1.0	1.4	3.7
ES4	1.0	0.9	1.0
ES14	1.0	1.0	1.4
ES15	1.0	1.0	1.3
ES18	1.0	4.4	47.1
ES28	1.0	1.3	1.1
ES32	1.0	1.0	0.9
ES35	1.0	2.9	4.4
ES47	1.0	1.4	1.3
ES48	1.0	1.7	1.4

tolerance will be reduced in the amphiploid relative to *L. elongatum*. For the same reason, aluminum tolerance in wheat, derived from rye, is also expected to be reduced as observed (Gustafson and Ross, 1990) because it also requires the expression of a number of genes.

CONCLUSIONS

(1) Comparison of the genetic mechanisms for the tolerance of toxic soil conditions showed that the complexity of the genetic basis of the adaptation to a specific stress parallels the complexity of the effects of the stress on a plant; the genetic control of the tolerance of salinity or aluminum appears more complex than the tolerance of the excess of heavy metals other than aluminum, particularly if only a single element is in excess.

(2) Salt stress tolerance in salt tolerant *L. elongatum*, when dissected with the aid of wheat lines possessing single *L. elongatum* chromosomes, appears to be controlled by genes on most of the chromosomes and, hence, must show a complex inheritance.

(3) Physiologically, *L. elongatum*, like wheat, is an ion excluder. In a number of instances exclusion of different ions is controlled by different chromosomes.

(4) In spite of the complex inheritance, genes with major effects on salt tolerance or its physiological components can be found; the K^+/Na^+ discriminating locus on wheat chromosome 4D and the major effect of *L. elongatum* chromosome 3E are examples.

(5) Strong correlations may be found between plant stress tolerance and the expression of specific physiological traits; exclusion of Na^+ from shoots or high K^+/Na^+ ratios in shoots of wheat and related grasses are examples.

(6) These correlated physiological traits may be exploited as selection tools in the breeding for salt stress tolerance.

(7) Salt stress tolerance of wheat may be readily enhanced by introgression of genes from salt-tolerant wheatgrass genera, such as *Lophopyrum*. *Lophopyrum elongatum* chromosome 3E is the most promising source.

(8) Molecular evidence indicates that the expression of oligogenically or polygenically controlled traits in alien genomes must be expected to be diminished in their hybrids and amphiploids with wheat. Since most types of tolerances of soil toxicities are oligogenically or polygenically controlled their expression should be expected to be diminished when incorporated into wheat.

REFERENCES

Abel, G. H. (1969) Inheritance of the capacity of chloride inclusion and chloride exclusion by soybean. *Crop Science* 9:697-698.

Akbar, M.; Yabuno, T. (1977) Breeding for saline-resistant varieties of rice. IV. Inheritance of delayed-type panicle sterility induced by salinity. *Japanese Journal of Breeding* 27:237-240.

Akbar, M.; Yabuno, T.; Nakao, S. (1972) Breeding for saline resistant varieties of rice. III. Response of F_1 hybrids to salinity in reciprocal crosses between Jhona 349 and Magnolia. *Japanese Journal of Breeding* 25:215-220.

Aniol, A.; Gustafson, J. P. (1984) Chromosome location of genes controlling aluminum tolerance in wheat, rye, and triticale. *Canadian Journal of Genetics and Cytology* 26:701-705.

Antonovics, J.; Bradshaw, A. D.; Turner, R. G. (1971) Heavy metal tolerance in plants. *Advanced Ecological Research* 7:1-85.

Ashraf, M.; McNeilly, J.; Bradshaw, A. D. (1989) The potential for evolution of tolerance to sodium chloride, calcium chloride, magnesium chloride and seawater in four grass species. *New Phytologist* 112:245-254.

Azhar, F. M.; McNeilly, T. (1988) The genetic basis of variation for salt tolerance in *Sorghum bicolor* (L.) Moench seedlings. *Plant Breeding* 101:114-121.

Bröker, W. (1963) Genetisch-physiologishe Untersuchungen über die Zinkverträglichkeit von *Silene inflata* Sm. *Flora Jena* 153:122-156.

Brown, J. C.; Devine, T. E. (1980) Inheritance of tolerance or resistance to manganese toxicity in soybeans. *Agronomy Journal* 72:898-904.

Cartside, D. W.; McNeilly, T. (1974) Genetic studies in heavy metal tolerant plants. I. Genetics of zinc tolerance in *Anthoxanthum odoratum*. *Heredity* 32:287-299.

Cox, R. M.; Hutchinson, T. C. (1980) Multiple metal tolerances in the grass *Deschampsia cespitosa* (L.) Beauv. from the Sudbury smelting area. *New Phytologist* 84:631-647.

Dvorák, J. (1980) Homoeology between *Agropyron elongatum* chromosomes and *Triticum aestivum* chromosomes. *Canadian Journal of Genetics and Cytology* 22:237-259.

Dvorák, J.; Chen, K. C. (1984) Phylogenetic relationships between chromosomes of wheat and chromosome 2E of *Elytrigia elongata*. *Canadian Journal of Genetics and Cytology* 26:128-132.

Dvorák, J.; Edge, M.; Ross, K. (1988) On the evolution of the adaptation of *Lophopyrum elongatum* to growth in saline environments. *Proceedings National Academy of Science USA* 85:3805-3809.

Dvorák, J.; Ross, R. (1986) Expression of tolerance of Na^+, K^+, Mg^{2+}, SO_4^{2-} ions and sea water in the amphiploid of *Triticum aestivum* x *Elytrigia elongata*. *Crop Science* 26:658-660.

Dvorák, J.; Ross, K.; Mendlinger, S. (1985) Transfer of salt tolerance from *Elytrigia pontica* (Podp.) Holub to wheat by the addition of an incomplete *Elytrigia* genome. *Crop Science* 25:306-309.

Elzam, O. E.; Epstein, E. E. (1969) Salt relations of two grass species differing in salt tolerance. I. Growth and salt content at different salt concentrations. *Agrochimica* 13:187-195.

Epstein, E. (1983) Crops tolerant of salinity and other mineral stresses. In: Nugent, J.; O'Connor, M. (eds.), *Better crops for food*. Pitman, London, pp. 61-72.

Epstein, E. (1990) Roots: New ways to study their function in plant nutrition. In: Hashimoto, Y.; Kramer, P. J.; Strain, B. R. (eds.), *Measurement techniques in plant science*. Academic Press, New York, pp. 291-318.

Ernst, W. H. O. (1982) Schwermetallpflanzen. In: Kinzel, H. (ed.), *Pflanzenökologie und mineralstoffwechsel.* Verlag Eugen Ulmer, Stuttgart, pp. 472-506.

Flowers, T. J.; Troke, P. F.; Yeo, A. R. (1977) The mechanism of salt tolerance in halophytes. *Annual Review of Plant Physiology* 28:89-121.

Forster, B. P.; Phillips, M. S.; Miller, T. E.; Baird, E.; Powell, W. (1990) Chromosome location of genes controlling tolerance to salt (NaCl) and vigor in *Hordeum vulgare* and *H. chilense. Heredity* 65:99-107.

Foy, C. D.; Lafever, H. N.; Schwartz, J. W.; Fleming, A. L. (1974) Aluminum tolerance of wheat cultivars related to region of origin. *Agronomy Journal* 66:751-758.

Gorham, J.; Forster, B. P.; Budrewicz, E.; Wyn Jones, R. G.; Miller, T. E.; Law, C. N. (1986) Salt tolerance in the Triticeae: Solute accumulation and distribution in an amphiploid derived from *Triticum aestivum* cv. Chinese Spring and *Thinopyrum bessarabicum. Journal of Experimental Botany* 37:1435-1449.

Gorham, J.; Hardy, C.; Wyn Jones, R. G.; Joppa, L. R.; Law, C. N. (1987) Chromosomal location of a K/Na discrimination character in the D genome of wheat. *Theoretical and Applied Genetics* 74:584-588.

Gorham, J.; Wyn Jones, R. G.; Bristol, A. (1990) Partial characterization of the trait for enhanced K^+-Na^+ discrimination in the D genome of wheat. *Planta* 180:590-597.

Gorham, J.; Wyn Jones, R. G.; McDonnell, E. (1985) Some mechanisms of salt tolerance in crop plants. *Plant and Soil* 89:15-40.

Graham, R. D. (1984) Breeding for nutritional characteristics in cereals. *Advances in Plant Nutrition* 1:57-102.

Greenway, H.; Munns, R. (1980) Mechanisms of salt tolerance in nonhalophytes. *Annual Review in Physiology* 31:149-190.

Grumet, R.; Hanson, A. D. (1986) Genetic evidence for osmoregulatory function of glycinebetaine accumulation in barley. *Australian Journal of Plant Physiology* 13:353-364.

Gulick, P. J.; Dvorák, J. (1987) Gene induction and repression by salt treatment in roots of salinity-sensitive Chinese Spring wheat and the salinity-tolerant Chinese Spring x *Elytrigia elongata* amphiploid. *Proceedings National Academy of Science USA* 84:99-103.

Gulick, P. J.; Dvorák, J. (1990) Selective enrichment of cDNAs from salt-stress-induced genes in the wheatgrass, *Lophopyrum elongatum*, by the formamide-phenol emulsion reassociation technique. *Gene* 95:173-177.

Gustafson, J. P.; Ross, K. (1990) Control of alien gene expression of aluminum tolerance in wheat. *Genome* 33:9-12.

Läuchli, A.; Epstein, E. (1990) Plant responses to saline and sodic conditions. In: Tanji, K. K. (ed.), *Agricultural salinity assessment and management.* American Society of Civil Engineers, New York, pp. 113-137.

MacNair, M. R. (1977) Major genes for copper tolerance in *Mimulus guttatus. Nature* 268:428-430.

MacNair, M. R. (1979) The genetics of copper tolerance in the yellow monkey flower, *Mimulus guttatus.* I. Crosses to nontolerants. *Genetics* 91:553-563.

MacNair, M. R. (1983) The genetic control of copper tolerance in the yellow monkey flower, *Mimulus guttatus. Heredity* 50:283-293.

McGuire, P. E.; Dvorák, J. (1981) High salt-tolerance potential in wheatgrasses. *Crop Science* 21:702-705.

Misra, S.; Gedamu, L. (1989) Heavy metal tolerant transgenic *Brassica napus* L. and *Nicotiana tabacum* L. plants. *Theoretical and Applied Genetics* 78:161-168.

Moeljopawiro, S.; Ikehashi, H. (1981) Inheritance of salt tolerance in rice. *Euphytica* 30:291-300.

Nable, R. O. (1988) Resistance to boron toxicity among several barley and wheat cultivars: A preliminary examination of the resistance mechanism. *Plant and Soil* 112:45-52.

Omielan, J. A.; Epstein, E.; Dvorák, J. (1991) Salt tolerance and ionic relations of wheat as affected by individual chromosomes of salt-tolerant *Lophopyrum elongatum*. *Genome* (in press).

Polle, E.; Konzak, C. F.; Kittrick, J. A. (1978) Visual detection of aluminum tolerance levels in wheat by hematoxylin staining of seedling roots. *Crop Science* 18:823-827.

Rhue, R. D.; Grogan, C. O.; Stockmeyer, E. W.; Everett, H. L. (1978) Genetic control of aluminum tolerance in corn. *Crop Science* 18:1063-1067.

Schachtman, D. P.; Bloom, A. J.; Dvorák, J. (1989) Salt-tolerant *Triticum* x *Lophopyrum* derivatives limit the accumulation of sodium and chloride ions under saline-stress. *Plant Cell and Environment* 12:47-55.

Shah, S. H.; Gorham, J.; Forster, B. P.; Wyn Jones, R. G. (1987) Salt tolerance in the Triticeae: The contribution of the D genome to cation selectivity in hexaploid wheat. *Journal of Experimental Botany* 38:254-269.

Stavarek, S. J.; Rains, D. W. (1983) Mechanisms for salinity tolerance in plants. *Iowa State Journal of Research* 57:457-476.

Stolen, O.; Andersen, S. (1978) Inheritance of tolerance to low soil pH in barley. *Hereditas* 88:101-105.

Storey, R.; Graham, R. D.; Shepherd, K. W. (1985) Modification of the salinity response of wheat by the genome of *Elytrigia elongatum*. *Plant and Soil* 83:327-330.

Tal, M. (1985) Genetics of salt tolerance in higher plants. *Plant and Soil* 89:199-226.

Tuleen, N. A.; Hart, G. E. (1988) Isolation and characterization of wheat - *Elytrigia elongata* chromosome 3E and 5E addition and substitution. *Genome* 30:519-524.

Urquhart, C. (1971) Genetics of lead tolerance in *Festuca ovina*. *Heredity* 26:19-33.

Walley, K. A.; Khan, M. S. I.; Bradshaw, A. D. (1974) The potential for evolution of heavy metal tolerance in plants. I. Copper and zinc tolerance in *Agrostis tenuis*. *Heredity* 32:309-319.

Warne, T. R.; Hickok, L. G. (1987) Single gene mutants tolerant to NaCl in the fern *Ceratopteris*: Characterization and gene analysis. *Plant Science* 52:49-55.

Wilkins, D. A. (1960) The measurement and genetic analysis of lead tolerance in *Festuca ovina*. *Report of the Scottish Plant Breeding Station* 85-96.

Wu, L. (1981) The potential for evolution of salinity tolerance in *Agrostis solonifora* L. and *Agrostis tenuis* Sibth. *New Phytologist* 89:471-486.

Wyn Jones, R. G.; Storey, R. (1981) Betaine. In: Paley, L. G.; Aspinall, D. (eds.), *Physiology and biochemistry of drought resistance in plants*. Academic Press, Sydney, pp. 171-204.

Yeo, A. R.; Flowers, T. J. (1986) Salinity resistance in rice (*Oryza sativa* L.) and a pyramiding approach to breeding varieties for saline soils. *Australian Journal of Plant Physiology* 13:161-173.

Chapter 10
Mechanisms for obtaining freezing stress resistance in herbaceous plants

Jiwan P. Palta

Department of Horticulture, University of Wisconsin, Madison, WI 53706

INTRODUCTION

Temperature stress is a major factor affecting plant growth and development and ecological distribution of plant species. Injury to plants exposed to freezing temperatures results in significant losses in plant productivity (Boyer, 1982). Just in the last two winters, the losses from frost damage to citrus and vegetable crops in Florida, Texas and California were over a billion dollars. Problems associated with frost damage to cultivated plants are largely man-made. In recent times we have pushed the cultivation of many crops to marginal geographic areas in terms of the occurrence of frost. Therefore, improvement of freezing stress resistance has become more important to sustain successful cultivation of crops in these areas.

The topic of low temperature stress is vast and encompasses chilling stress, untimely frosts during the growing season (primarily in the spring and fall), and freezing stress during winter. Many plants have evolved to survive these stresses. Several recent books and reviews have presented discussions on the survival mechanisms for low temperature stresses (Levitt, 1980; Li and Sakai, 1978, 1982; Li, 1987; Sakai and Larcher, 1987; Lyons et al., 1979). In this chapter the discussion is limited to the mechanisms of survival to freezing stress in herbaceous plants. Within that topic a major portion of materials presented is focused on the mechanism of injury and of cold acclimation in these plants when extracellular ice is present in the plant tissue. A brief discussion is presented on the inheritance of cold acclimation and freezing tolerance and on screening techniques. To illustrate some of the points, I give examples from our own work on potato (*Solanum* spp.) species and onion (*Allium cepa* L.) bulb tissue. These materials have several advantages. Onion bulb tissue can survive freezing temperatures (with ice in tissue) and the inner epidermal cell layer which can be easily removed is an ideal material for cellular studies. Within tuber-bearing potatoes we have some diploid species that vary considerably in freezing tolerance and in ability to acclimate (Li and Palta, 1978; Li et al., 1979), thus making this material ideally suited for physiological and genetic studies.

219

TERMINOLOGY: TYPES OF SURVIVAL MECHANISMS IN PLANTS

Freezing stress resistance

In a broad sense, the term freezing stress is used to define temperatures below 0°C that restrict a plant from realizing its genetic potential for growth, development and productivity. Freezing stress resistance is the ability of the plant to maintain its functions and survive freezing temperatures.

Two survival mechanisms have been distinguished for plants exposed to freezing stress. These are **freezing tolerance and freezing avoidance**. By freezing tolerance, we mean tolerance of extracellular ice and by freezing avoidance we mean avoidance of extracellular or intracellular ice. There is a general misconception among growers that freezing itself (i.e., formation of ice in the tissue) is damaging. In reality, plants have varying degrees of tolerance to ice in the extracellular space. Repeated freezing and thawing of grass during fall and spring is a good example of freezing tolerance as a mechanism of survival. As will be clear later in this chapter, freezing tolerance is a heritable trait and there exists variability in freezing tolerance in germplasm.

Some plant parts are able to survive freezing temperatures by avoiding the formation of ice. This mechanism has been found to be important in overwintering seeds, buds and xylem ray parenchyma cells (see reviews by Burke *et al.*, 1976; Li and Sakai, 1978, 1982; Sakai and Larcher, 1987). These tissues avoid ice formation primarily by supercooling and preventing nucleation. There is evidence that tissues of some herbaceous plants are capable of surviving by freezing avoidance during the active growth period (Sakai and Larcher, 1987).

Cold acclimation

In many herbaceous plants the freezing stress resistance increases if the plants are exposed to a brief period (days to few weeks) of chilling (Levitt, 1980). This increase is known as acclimation or hardening. This acclimation is reversed upon exposure to warm temperature, thus resulting in deacclimation. In many overwintering plants, acclimation and deacclimation are seasonal. Plants have the ability to increase both freezing tolerance and freezing avoidance by acclimation.

As will be clear from the discussion presented in this chapter, all three components of freezing stress - namely freezing tolerance, freezing avoidance and capacity to acclimate - are heritable traits. Some examples given here suggest that genetic variability exists for these three traits and separate selection is feasible.

FREEZE-THAW PROCESS IN NATURE

During typical spring and fall frosts, air temperatures usually drop at the rate of 1-2°C h^{-1} (Levitt, 1980; Steffen *et al.*, 1989). Examination of climatological data for this century, recorded at Madison, WI, supports this conclusion (Table 1). The maximal air cooling rate below 0°C was less than 2°C h^{-1}. This was true even during the extreme drops in temperature and during the spring and fall frosts. A continuous record of air temperature for

Table 1 Air cooling rates below 0°C during the last spring minimum or first fall minimum of -4.4°C or below from 1975 through 1985, inclusive, in Madison, WI. Also included are two of the greatest historic declines in air temperature over a 24-h period in Wisconsin.

Season	Date (prefrost)*	Day/night temperature range (°C)†	Maximal cooling rate below 0°C (°C h^{-1})‡
Historic	11-11-11	21.1 to -10.0	1.7
	11-11-40	10.0 to -12.2	1.8
Spring	05-08-77	25.0 to -3.9	1.1
	04-20-83	13.3 to -3.3	1.9
	04-06-84	15.0 to -5.6	1.9
Fall	10-13-79	8.3 to -5.6	1.4
	10-22-81	7.2 to -5.6	1.4
	10-20-82	4.4 to -5.6	1.7
	1975-1985	Mean ± SD	1.2 ± 0.36
		Range	0.6-1.9

Source: Steffen *et al.* (1989).

*Date which includes the highest daily temperature preceding the frost.

†The range includes the maximum daily temperature preceding the frost to the minimum temperature reached during the frost.

‡Temperature data were obtained from the National Weather Service in which hourly readings had been rounded to the nearest Fahrenheit degree. In order to minimize rounding influences, three consecutive hourly temperatures in the freezing range (covering 2 h) were regressed to determine the midpoint freezing rate.

a typical spring frost is shown in Figure 1. Again, the maximum cooling rate was about 1°C h⁻¹. Thus, the experimental tests to simulate the impact of freeze-thaw stress on plants should use cooling rates of 1-2°C h⁻¹.

Figure 1 A typical spring frost in central Wisconsin. A) Hourly air temperature collected by a Wisconsin Automated Agricultural Weather Station on the University of Wisconsin Agricultural Experiment Station at Hancock, WI. Data were collected on 60-s intervals and reported as a mean at the end of an hourly period ending on the hour. B) Air cooling rates. Rates were calculated from two consecutive hourly readings and reported here as a midpoint cooling rate on the half hour (source: Steffen *et al.*, 1989).

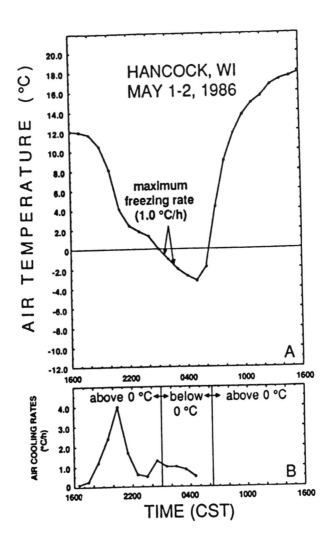

However, several recent studies investigating the mechanism of freezing injury have utilized cooling rates of 6-60°C h⁻¹. Such fast cooling rates can result in non-equilibrium freezing (supercooling), and thus rapid intracellular ice formation upon nucleation. Such a freezing event is usually lethal (Levitt, 1980; Sakai and Larcher, 1987). In a recent study we showed that simply increasing the cooling rate from 1.0 to 2.9°C h⁻¹ meant the difference between cell survival and cell death (Steffen *et al.*, 1989). Results presented above underscore the importance of utilizing a realistic freeze-thaw protocol for studying mechanisms of freezing injury, for evaluating freezing tolerance and for screening germplasm for freezing tolerance.

Ice initiation in the tissue requires nucleation, and plant tissue is usually nucleated by the time tissue temperatures drop to between -1 and -3°C (Levitt, 1980; Ashworth *et al.*, 1985). Slow cooling (1-2°C h⁻¹) results in ice initiation in the extracellular water. This is because extracellular water has a lower solute concentration than the intracellular water and because ice nucleators such as dust and bacteria are prevalent in the extracellular environment. As the air temperature drops further, ice grows in the extracellular space, withdrawing water from the cell and resulting in collapse of the cell walls (Figure 2). Water moves from the inside to the outside of the cell because the vapor pressure of extracellular ice is lower than vapor pressure

Figure 2 A diagrammatic description of the process of dehydration at the cellular level by freezing, water and osmotic stresses (source: Palta, 1990).

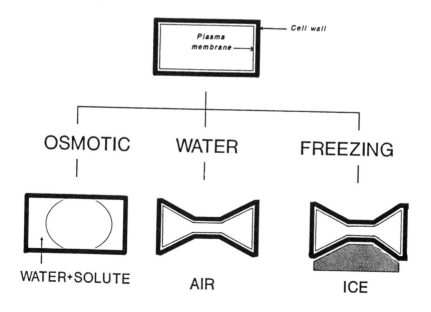

of vacuolar and cytoplasmic water at the same temperature. This behavior of the cell during freezing stress is well documented (see Levitt, 1980; Sakai and Larcher, 1987). Recent direct microscopic examination of frozen tissue also supports this conclusion (Pearce, 1988).

A common feature among several environmental stresses is dehydration. The process of dehydration and stresses that result from it are similar for both water and extracellular freezing stresses (Figure 2). However, for osmotic stress this dehydration and accompanied stresses are very different (Figure 2). This is important because several recent studies have used protoplasts to investigate the mechanism of freezing injury (Dowgert and Steponkus, 1984; Steponkus and Weist, 1978; Steponkus, 1984). Because protoplasts are bathed in an isotonic external solution, the freezing stress is essentially an osmotic stress in this system. Use of protoplasts thus eliminates the role of the cell wall in freezing stress. This is an important consideration because freezing stress imposes mechanical stress due to the collapse of the cell walls which is absent in osmotic stress (Figure 2). In support of this criticism, a large difference in the freezing tolerance of intact cells and their protoplasts has been reported (Tao *et al.*, 1983). Released protoplasts had LT_{50} of -21.5°C as compared to -12°C for intact cells suspended in the same medium. These results highlight the importance of using intact tissues and cells for the study of the mechanism of freezing injury.

MEMBRANE PERTURBATION AS AN EARLY MANIFESTATION OF FREEZING INJURY

The visual symptoms of freeze-thaw, stress-induced injury are often a water soaked appearance, leakage of solutes and loss of turgor. These symptoms suggest that the cell membrane is a site of injury and was recognized almost 80 years ago (Maximov, 1912). However, the nature of the injury to the cell membrane at a more fundamental level has been investigated only recently. This has been, in part, due to the assumption that freezing injury results in a complete loss of membrane semipermeability or the membrane ruptures. Contrary to this, leakage of solute occurs from injured, yet living, cells (Palta *et al.* 1977a,b; Palta and Li, 1978a, 1980). In spite of enhanced ion leakage and water soaked appearance following freeze-thaw injury, the cells were able to plasmolyze and exhibit protoplasmic streaming (Table 2). Furthermore, the injured, yet alive, cells could be vitally stained with fluorescein diacetate (Arora and Palta, 1988). In another study no ultrastructural perturbation of freeze-thaw injured potato leaves was found when examined with an electron microscope (Palta *et al.*, 1982; Palta and Li, 1978a,b). All of the organelles, including chloroplasts and mitochondria, appeared normal in spite of the injury. These results show that increased

Table 2 Effect of freezing stress on ion efflux, infiltration of the tissue, cell viability and the permeability of cell membranes to water. Onion bulbs were transferred directly to freezing chambers maintained at -4±0.5°C and -11±5°C. Observations were made immediately after thawing.

Freezing temp (°C)	Infiltration* of scale tissue (%)	Conductivity effusate ($\mu S g^{-1}$)	Cell viability	Water permeability constant† ($\mu m s^{-1}$)
-11	100	152.0	All alive	2.06±0.26
-4	50	83.2	All alive	1.97±0.38
+3 (control)	0	66.7	All alive	1.76±0.14

Source: Palta *et al.* (1977b).

*Estimated visually.

†Measured with radiotracer technique using inner epidermal cell layer.

ion leakage from the cells following freeze-thaw injury does not necessarily result from membrane rupture.

Recently, the impact of a freeze-thaw cycle on three important cellular functions - namely photosynthesis, respiration and membrane permeability - was investigated (Steffen *et al.*, 1989). The purpose of this study was to discover the initial site of perturbation by freezing injury by investigating the sequential development of injury to these three cellular functions. Photosynthesis was found to be much more sensitive than respiration (Figure 3). The first change observed in cellular function was a small but significant increase in ion leakage at incipient injury. This level of ion leakage represented a two- to threefold increase over the control values (Figure 3 inset) at -2°C which is non-lethal reversible injury. Increased ion leakage suggests perturbation of the transport properties of the cell membranes that would result in an altered cellular environment. Alteration in cytoplasmic environment could impair organelle functions such as photosynthesis (Figure 3). These studies along with other published results (for review, see Palta, 1989) have established that early events in freeze-thaw injury are subtle rather than cataclysmic.

From the results summarized above, it can be concluded that enhanced ion efflux following incipient freezing injury is not necessarily due to membrane rupture. The freeze-injured, yet alive, cells provide an excellent material for the study of the nature of alteration to the cell membranes that

Figure 3 Differential sensitivity of photosynthesis, respiration, and cellular membranes (measured as % ion leakage) to extracellular freezing stress in leaf tissue of *Solanum commersonii* L. potato. Excised leaflets were frozen to desired temperature at a cooling rate of 1°C h⁻¹ with ice nucleation at -0.5°C. Following slow thaw, various cellular functions were measured on the same leaflet (source: Steffen *et al.*, 1989).

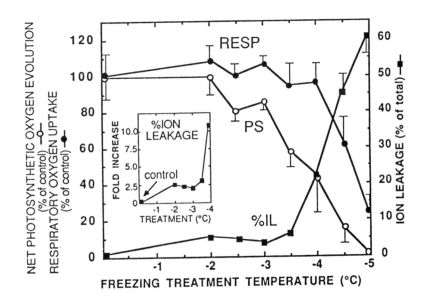

could account for increased ion efflux. We studied transport properties of the cell membranes to water and non-electrolyte (Palta *et al.*, 1977a; Palta and Li, 1980). Although there was more than a twofold increase in ion efflux following freeze-thaw treatment (-11°C), no significant change in the water permeability could be detected (Table 2). Because water is supposed to diffuse through the lipid portion of the membrane, these results suggest that membrane lipid transport properties remain unaltered following injury. Furthermore, no change in the passive permeability to non-electrolytes such as urea and methyl urea was detected following freeze-thaw injury (Palta and Li, 1980). These results further strengthen our earlier findings that physical state (which determines the transport properties) of the membrane lipids remains essentially unaltered following incipient freeze-thaw injury. We analyzed the chemical composition of the effusate. The major cation that leaks out of the cell was found to be K^+ (Palta *et al.*, 1977a). The transport of K^+ is known to be coupled to the activity of membrane pumps, the H^+-ATPases (Sze, 1985). A direct measurement of K^+ transport across the cell membranes following injury revealed that a specific alteration in the K^+

transport across the cell membranes resulted from freeze-thaw injury (Palta and Li, 1980).

POST-THAW RECOVERY

One aspect of freeze-thaw stress is the ability of the injured tissue to recover following stress and, depending upon the degree of initial injury, a freeze injured tissue is able to repair the damage (Palta *et al.*, 1977b). This repair is accompanied by reabsorption of leaked ions (primarily K^+) and a complete disappearance of the water infiltration (soaking) of the tissue (Palta *et al.*, 1977b,c). The study of the recovery process has become a powerful tool to understand the mechanism of injury because during the repair process one can study the initial perturbations by freezing stress which are being reversed. The recovery was first reported for onion bulb tissue over a decade ago (Palta *et al.*, 1977b). Since then, we have observed the recovery of freezing injury in potato leaves, celery (*Apium graveolens* L.) petiole, cabbage (*Brassica* sp.) leaves, green bean (*Phaseolus vulgaris* L.) fruits and carrot (*Daucus carota* L.) roots. More recently we have demonstrated that a recovery in the visual appearance of the tissue is accompanied by a reduction (return to the normal) in the ion leakage rate from the cells (Table 3). These

Table 3 Ion leakage rate (ion efflux h^{-1}) and visual observations on water soaking following reversible freeze-thaw injury and recovery of onion bulb scale tissue. One-half of freeze-injured and respective unfrozen control tissue was analyzed immediately after a complete thaw and the other half was analyzed after complete recovery. All the values are mean of four separate experiments $\pm$ SE. Each experiment included three separate measurements.

| Treatment | Extent of visual water soaking* | | Relative ion leakage rate† (efflux h^{-1}) | |
	After thaw (%)	After recovery	After thaw (%)	After recovery
Control (unfrozen)	0	0	100	100
Freeze-thaw	30-40	0-5	126±2.5	101.9±5.4

Source: Arora and Palta (1991).

*Values for freeze-injured scale tissue are within $\pm 5\%$.

†Ion leakage rate (efflux h^{-1}) was expressed as percentage of ions leaked in 1 h with respect to total internal ions during that hour. Ion leakage rate for control and freeze-injured tissue, right after thawing, were 26% $\pm$ 0.9 and 33.1% $\pm$ 1.6, respectively.

results show that a return to the normal condition is associated with the recovery of the membrane transport properties.

THE MEMBRANE ATPase HYPOTHESIS

From early experiments with onion bulb scale cells we proposed that a sublethal freezing stress results in an alteration of the transport properties of the cell membranes (Palta *et al.* 1977a,b; Palta and Li, 1980). This is an early manifestation of injury resulting from slow freeze-thaw stress. These alterations result from functional and or structural changes in the membrane associated H^+-ATPases as a result of the stress; how these changes are brought about is not clearly understood. Palta and Li (1980) suggested that these may result from (1) direct denaturation of membrane transport proteins or (2) changes in lipid-protein interaction.

In recent years it has become clear that the lipid bilayer is a major factor in determining the function of the integral membrane proteins (Carruthers and Melchoir, 1986). Various features of the membrane bilayer lipids that can drastically alter the activity of transport proteins include lipid head groups, lipid acyl chain length, saturation/unsaturation, lipid backbone, and bilayer fluidity. Alteration in the functions of H^+-ATPases following a sublethal freeze-thaw injury could be brought about by such changes in the membrane lipids associated with these proteins. However, conformational alterations of membrane proteins could also explain this injury. Recently, Arora and Palta (1986, 1988) have implicated a role for Ca^{2+} in freezing injury. Intracellular calcium is thought to be a "second messenger" which can influence many metabolic functions through its effect on calmodulin (Poovaiah and Reddy, 1987). Changes in the functions of H^+-ATPases could be brought about by changes in the intracellular concentration of Ca^{2+}. Further work at the membrane and cellular level would be necessary to elucidate the molecular sequence of events during freezing injury.

The membrane ATPase hypothesis was proposed more than 10 years ago by Palta [see Palta (1989) for details]. Since then, several independent groups have provided evidence in support of this hypothesis - e.g.;

(1) Following freezing injury plasma membrane ATPase activity decreased before any effect on organelle ATPase activity was found (Jian *et al.*, 1982).

(2) In purified plasma membrane preparations from pine needles the ATPase activity was totally gone after lethal freeze-thaw stress (Hellergren *et al.*, 1985).

(3) A decrease in total amount of plasma membrane ATPase activity was found following lethal and non-lethal freezing injury (Uemura and Yoshida, 1985). At the same time an increased inhibition by *N,N*-dichlorohexyl carbondiimide (DCCD) of the ATPase activity was observed following injury (Uemura and Yoshida, 1985).

(4) In winter wheat cells the earliest manifestations of icing stress was found to be inhibition of ^{86}Rb uptake indicating alteration in the functions of ion transport system (Pomeroy *et al.*, 1983).

Recently we have provided further evidence in support of this hypothesis by studying changes in the functions of plasma membrane ATPase following freezing injury and recovery from this injury (Iswari and Palta, 1989; Arora and Palta, 1991). In one study, using purified plasma membranes prepared from the leaf tissue, we found a progressive decrease in specific activity of the plasma membrane ATPase as the injury by freeze-thaw stress increased. This decrease appeared to be selective because the activity of several marker enzymes was unaltered even at the level of substantial injury (Table 4). Furthermore, at a very slight level of injury an activation in the activity of the plasma membrane ATPase was found (Table 4). This level of injury is totally reversible, and we have found no changes in ultrastructural or tissue respiration and photosynthesis rates at this injury level (Palta *et al.*, 1982; Palta and Li, 1978b). The increased ATPase activity can help explain the complete recovery following slight injury. Activation of this ATPase will help in reuptake of leaked K^+ which is then followed by reabsorption of

Table 4 Effect of degrees of freeze-thaw stress in *Solanum commersonii* leaflets on the specific activity of plasma membrane ATPase, CCO and CCR. Data calculated from Iswari and Palta (1989).

Category of damage	Visual water soaking (% area)	Ion leakage (% of total)	Specific activity of ATPase (%)		Specific activity (%)	
			Mic	PM	CCR	CCO
Control	0	9±2	100	100	100	100
Slight	10-20	14±6	216	189	105	102
	10-20	27±6	200	216	--	--
Substantial	70-90	76±9	26	24	98	108
	70-90	--	45	68	--	--
Total	100	85±5	ND	ND	62	127
	100	--	19	27	--	--

Specific activity of ATPase in control was about 0.06 μmol mg^{-1}min^{-1} in microsomes and 0.3 μmol mg^{-1}min^{-1} in plasma membrane fraction. Cytochrome c oxidase (CCO) activity in control was 3.7 nmol mg^{-1}min^{-1}, cytochrome c reductase (CCR) activity in control was 30 nmol mg^{-1}min^{-1}. ND = not detectable. Mic = Microsomal membranes; PM = plasma membrane.

water by the cells leading to the disappearance of water soaking of the tissue.

We have also studied the plasma membrane ATPase activity and ion leakage rates immediately following freeze-thaw stress and after complete recovery in onion bulb scale tissue (Arora and Palta, 1991). In injured tissue (30-40% water soaking), plasma membrane ATPase activity was reduced by about 30%, and this was paralleled by about 25% higher ion leakage rate (Table 5). As the water soaking of tissues disappeared during recovery, the plasma membrane ATPase activity and ion leakage rate returned to about the same level as the respective controls. Furthermore, in this study we also demonstrated that inhibiting the activity of the plasma membrane ATPase by vanadate prevented the recovery process.

Table 5 Vanadate sensitive ATPase activity in microsomal membrane fraction and ion leakage rate of onion scale tissue following reversible freeze-thaw injury and recovery. Standard assays were used to measure the activity of vanadate sensitive ATPase (in the presence of 100 μM vanadate) in the microsomes isolated from the same tissue which was used for ion leakage rate measurements.

Expt. no.	Treatment	Ion leakage rate measured* (% h^{-1})		Vanadate-sensitive ATPase activity (μmol Pi 50 g^{-1} 35 min^{-1})	
		After thaw	After recovery	After thaw	After recovery
1	Control (unfrozen)	25.5 ± 0.4	22.4 ± 1.4	49.2(65.2)†	44.2(78.3)
	Freeze-thaw	31.3 ± 2.1	21.3 ± 4.0	25.0(41.1)	35.5(61.1)
2	Control (unfrozen)	24.1 ± 1.1	20.0 ± 1.5	50.2(73.1)	51.3(86.0)
	Freeze-thaw	30.7 ± 1.7	22.3 ± 0.3	38.9(53.5)	56.2(84.8)

Source: Arora and Palta (1991).

*Ion leakage rate (efflux h^{-1}) was expressed as percentage of total internal ions.

†Values in parentheses are total activity in the absence of vanadate. The values for ion leakage rate and ATPase activity are from two representative experiments and are average of triplicates and duplicates, respectively.

ROLE OF MEMBRANE/CYTOSOLIC CALCIUM AS A MEDIATOR OF STRESS RESPONSE

Recent evidence implicates free (cytosolic) Ca^{2+} as a major metabolic and developmental controller in plants (Hepler and Wayne, 1985; Poovaiah, 1985; Poovaiah and Reddy, 1987). Because calcium concentration in the cytoplasm is very low, small changes in the absolute amount of calcium can create a 10- to 100-fold difference in concentration without upsetting the ionic balance of the cell. This feature makes calcium an excellent second messenger. The combination of a relatively large ionic radius and low free energy of hydration contribute, in part, to the high degree of selectivity of calcium compared to other cations for its role as a second messenger. Stimuli, such as hormones, light, gravity or other environmental conditions can cause an increase in the calcium concentration of the cytoplasm (Figure 4). A rise in the level of free of Ca^{2+} in the cytoplasm can bring about a cellular response by activating or deactivating enzymes. This regulation of enzymes has been demonstrated to be via protein kinase-induced phosphorylation of proteins. Evidence has also been presented for protein kinases to be regulated by either Ca^{2+}- or Ca^{2+}-activated calmodulin.

Figure 4 A flow diagram illustrating the pathway by which cellular calcium changes can occur which in turn result in cellular response to stress. Based on the scheme given by Poovaiah and Reddy (1987). PIP2 = phosphatidyl inositol 4,5 bisphosphate; DAG = diacylglycerol; IP3 = inositol 1,4,5 trisphosphate.

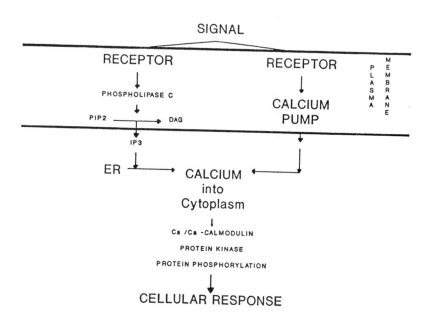

It is possible that plant response to low temperature stress can be brought about by a stress-induced change in the free Ca^{2+} in the cytoplasm. Our recent research contains evidence that suggests such a role of calcium in freeze-thaw stress induced injury and recovery (Arora and Palta, 1986,

1988, 1989). Using a chlorotetracycline (CTC) probe for membrane calcium, we studied changes in the membrane associated calcium following a freeze-thaw stress. A bright fluorescence was observed in control unfrozen cells (Figure 5). Both reversibly (non-lethal) and irreversibly (lethal) injured cells exhibited significant reduction in Ca^{2+}-CTC fluorescence (Figure 5). The fluorescent signal was quantified using image analysis. A linear inverse relationship was found between the extent of freezing injury (ion leakage) and the membrane associated Ca^{2+} (Figure 6). The fact that the Ca^{2+}-CTC signal was reduced (even in the cells that were able to completely repair the damage) suggests that perturbation of membrane associated Ca^{2+} takes place in the early stages of freezing injury. Such perturbation could help mediate recovery during the post-thaw period.

If injury leads to perturbation of certain cellular functions such as functioning of PM-ATPase, the recovery should require restoration of these cellular functions back to normal. Calcium-mediated changes in the enzyme functions, such as illustrated in Figure 4, could possibly restore this normal functioning. As mentioned above, recovery requires activity of the H^{+}-ATPase to pump back the leaked ions. We have found activation of the plasma membrane ATPase following reversible injury in potato leaves (Table 4). We have also found that recovery of the functions (activity) of the plasma membrane ATPase during the post-thaw recovery of freeze-injured onion bulb tissue (Arora and Palta, 1991). Recently, it has also been demonstrated that a Ca^{2+} requiring protein kinase is able to modulate plasma membrane associated ATPase (Schaller and Sussman, 1988). Thus, it is possible that recovery, in part, is brought about by changes in cytosolic Ca^{2+} which activate a protein kinase (Figure 4). This protein kinase, in turn, could activate plasma membrane ATPase leading to reuptake of leaked ions and thus recovery.

Loss of Ca^{2+} occurs during the early stages of injury and we have suggested this loss is linked to the initiation and the progression of injury (Arora and Palta, 1988). These results suggest that it may be possible to increase freezing tolerance by increasing the level of membrane associated Ca^{2+}. Our preliminary results support this suggestion (Arora and Palta, 1989). We found that when membrane Ca^{2+} was artificially lowered by bathing the cells in ethylene glycol tetra-acetic acid (EGTA), the cells became more sensitive to freezing stress. In another study we found that raising Ca^{2+} level, in a normally deficient potato tuber tissue, resulted in an increase in the freezing tolerance. Thus, it appears that it should be possible to increase freezing tolerance by increasing Ca^{2+} level of the otherwise Ca^{2+}-deficient tissue.

It is also tempting to suggest that a low temperature acclimation response could be mediated by perturbation of cellular calcium. Many herbaceous plant species acclimate when exposed to chilling temperatures for a period of about 2 weeks. During this period freezing tolerance can increase by

Figure 5 Photomicrographs of adaxial epidermal cells of onion bulb (both unfrozen control and freeze-thaw stressed) treated with FDA (fluorescence diacetate) and CTC (chlorotetracycline) staining solutions. a, Control cells showing fluorescence from FDA staining; b, freeze-thaw stressed (-11.5°C) cells showing fluorescence from FDA staining; c, control cells exhibiting Ca^{2+}-CTC fluorescence; d, bright field picture of (c); e, freeze-thaw stressed (-8.5°C) cells showing Ca^{2+}-CTC fluorescence; f, bright field picture of (e); g, freeze-thaw stressed (-11.5°C) cells showing Ca^{2+}-CTC fluorescence; h, bright field picture of (g). p, protoplasmic surface (plasma membrane); cw, cell wall. Magnification in (a) and (b) was x 180 and in (c) to (h) was x 500 (source: Arora and Palta, 1988).

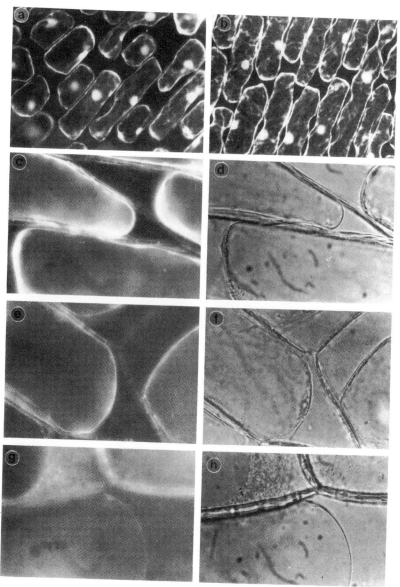

Figure 6 Relationship between membrane damage (expressed as % ion leakage) and membrane associated calcium (measured as Ca^{2+}-CTC fluorescence signal). Onion bulb scale tissue was subjected to slow freeze-thaw stress. Following thawing, the inner epidermal cells were stained with CTC (chlorotetracycline) fluorescent dye. The signal from this stain was quantified by image analysis (source: Arora and Palta, 1988).

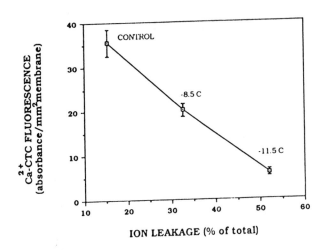

several fold. Low temperature acclimation is a complex reaction that involves gene induction. In the early stages of acclimation, ABA is known to increase and several new proteins are produced. It is possible that changes in cytosolic Ca^{2+} (by a scheme illustrated in Figure 4) can bring about a cascade of reactions that lead to acclimation. In support of such a suggestion, changes in the activity of Ca-dependent protein kinases have been reported during cold acclimation (Iswari *et al.*, 1990).

ALTERATION IN THE PROPERTIES OF MEMBRANE LIPIDS IN RESPONSE TO STRESS

The fluid mosaic model regards the biological membranes as a two-dimensional solution of oriented proteins in a fluid lipid bilayer. It has now been established that membrane integral proteins, such as transport "pumps" (ATPases), have an associated lipid component (annular lipid) that is essential for functioning of these proteins (for review, see Carruthers and Melchior, 1986). Features of lipids that can alter membrane protein functions include lipid head groups, lipid acyl chain length, lipid backbone and lipid fluidity (viscosity). Thus, a change in either the physical or chemical properties of lipids can bring about a change in the function of membrane proteins. A flow diagram illustrating such a response by freezing stress is shown in Figure 7.

Figure 7 A flow diagram illustrating membrane lipids as a site of membrane/ cellular response to stress. Changes in the physical state of membrane lipids from liquid crystalline (fluid) to solid gel can result in cascade of reactions leading to the cellular response (source: Palta, 1990).

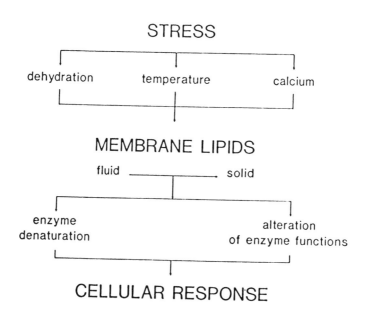

A change in the physical state of membrane lipids can be brought about by lowering temperature, by freeze induced dehydration and/or by a change in membrane associated calcium during freeze-thaw stress. Evidences in support of these possibilities have been presented in various plant tissues (Lyons *et al.*, 1979; Legge *et al.*, 1982; Kuiper, 1985). The influence of lowered temperature is obvious. The solidification temperatures of major membrane fatty acids range from -11 to +69.6°C depending upon the length of the fatty acid chain and upon the degree of unsaturation (Table 6). As the temperature is lowered, some of the membrane lipids will become solidified which could result in the denaturation of membrane associated enzymes such as plasma membrane-ATPase, thus causing injury to the plant.

Researchers have utilized two approaches to understand the role of lipids in the response of membrane to freezing stress. In one approach, an attempt has been made to correlate the differences in lipid composition or lipid

Table 6 Melting points (solidification temperatures) of common fatty acids found in plant membrane lipids.

No. carbon atoms	No. unsaturated bonds	Common name	Melting point (°C)
16	0	Palmitic	62.7
18	0	Stearic	69.6
18	1	Oleic	10.5
18	2	Linoleic	-5
18	3	Linolenic	-11

physical properties (phase transition temperatures, microviscosity) to the differences in freezing stress resistance of various plant species. However, very little effort has been made to compare genetically related plant species. In another approach, changes in these lipid properties have been investigated during cold acclimation. In these studies an attempt is made to determine lipid property changes that coincide with increase in freezing stress resistance during acclimation. One good way to study such changes is first to obtain purified membrane preparations.

We have studied lipid compositional changes in genetically related potato species (Palta and Meade, 1987, 1989). When lipids were extracted from whole leaf tissue, no significant changes following cold acclimation in the phospholipid fatty acid composition were found. However, in purified plasma membrane an increase in the phospholipid fatty acid 18:2 (linoleic acid) and a decrease in 16:0 (palmitic acid) was found following cold acclimation. There was, in general, also an increase in the unsaturation level of fatty acids suggesting an increase in membrane "fluidity" (decrease in microviscosity) after acclimation. These changes in lipid composition were accompanied by an increase in freezing tolerance and were found only in the potato species, *Solanum commersonii* L., that was able to cold acclimate (Palta and Meade, 1989). We have also found a similar consistent increase in 18:2 with increase in freezing tolerance of cranberry (*Oxycoccus marcocarpus* Pers.) leaves (Abdallah and Palta, 1989) and pine (*Pinus* sp.) needles (Sutinen and Palta, 1989) during fall and winter.

Recently we have found an increase in the specific activity of the plasma membrane ATPase in parallel with increase in freezing tolerance following cold acclimation in leaf tissue of *S. commersonii* (Iswari and Palta, 1989; Palta and Meade, 1989). In this potato species an increase in 18:2 and a decrease in 16:0 coincided with this increase in ATPase activity (Palta and Meade, 1989). Studies by Palmgren *et al.* (1988) suggest that an increase in plasma membrane ATPase activity *in vitro* can be brought about by an increase in 18:2 fatty acid. This finding, combined with our results stated above, suggests that membrane lipid changes can modulate plasma membrane ATPase during cellular response to low temperature stress.

ROLE OF LIGHT IN LOW TEMPERATURE STRESS

The primary role of photosynthesis is to trap energy and convert it into forms which can be utilized in biosynthesis. Light energy flux through the chloroplast can be divided into three processes - namely, light energy capture, light energy transduction and light energy utilization. Light energy capture consists of absorption of light by chlorophyll and other pigments. Transduction results in conversion of this energy into stable chemical forms. Utilization consists of making use of this chemical energy to fix CO_2 and transport sugar from the chloroplast. Both transduction and utilization are temperature-dependent as these processes involve a number of enzymatic reactions. By contrast, the light energy capture is essentially temperature-independent.

During cold acclimation, as the temperature is lowered, the ability of the plant to process and utilize the captured light energy is substantially reduced. This condition results in an overabundance of trapped light energy. Injury to the photosynthetic apparatus can occur whenever light energy trapped within the chloroplasts exceeds the capacity of the plant to process (thus dissipate) this light energy (Powles, 1984). The injury is brought about by photooxidation. The damaging impact of excess light energy at low temperature has been studied in several plant species (Osmond, 1981; Oquist, 1983; Ogren *et al.*, 1984; Steffen and Palta, 1987, 1989).

Since the trapped light energy can be excessive under low temperature acclimation, it would appear that a plant that is able to cold acclimate must possess the ability to prevent light dependent injury to photosynthetic apparatus. Several lines of evidence support this proposal:

(1) We have found that a cold-sensitive potato species that is unable to acclimate suffers much greater photosynthetic inhibition at low temperatures and moderate light level than a cold-tolerant species that is able to cold acclimate (Steffen and Palta, 1986).

(2) In a related study we found that growth and development of a potato plant at lower temperature results in an increased capacity to tolerate high light stress at chilling temperatures, as a plant acclimates to lower temperatures (Steffen and Palta, 1989).

(3) In our early studies on the leaf anatomy in relation to freezing tolerance (Palta and Li, 1979), we found that tolerant species had two palisade layers compared to only one layer in sensitive species (Figure 8). In the same study a significant positive correlation was found for palisade thickness and frost-killing temperature among 24 potato

Figure 8 Photomicrographs of leaf cross-section of (a) tolerant wild (*S. commersonii*) and (b) sensitive cultivated (*S. tuberosum*) potato species (source: Palta and Li, 1979).

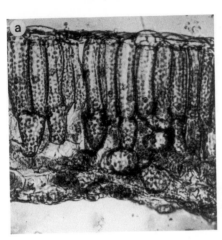 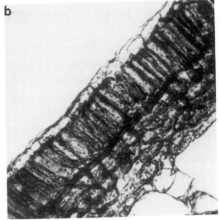

species (Figure 9). We suggested that multiple and thicker palisade cells could be used as a screening tool (Palta and Li, 1979). Later, Estrada (1982) reported multiple palisade (up to three) in all of the hardy clones. Significance of these anatomical differences perhaps lies, in part, in the ability of these species to protect themselves from photooxidation by filtering out the light in the multiple palisade layers and thereby reducing the effective amount of light reaching the leaf interior.

(4) Many evergreens, such as cranberries and rhododendrons that over-winter in the temperate regions, develop pigmentation and lose chlorophyll during cold acclimation in fall. Conversely, the leaves of these plants lose pigmentation and gain in chlorophyll contents during deacclimation in spring. This decrease in chlorophyll and increase in pigments in response to low temperatures could be related to the ability of these plants to prevent photooxidation at low temperatures. Reduced chlorophyll content would mean reduced amount of light energy captured by the plant. An increase in the other pigments would mean that less light energy is absorbed by the chlorophyll. In support of this idea, Estrada (1982) reported that, other factors being equal, pigmented potato plants were injured less by frost.

Cold acclimation is an energy-requiring process (Levitt 1980). This energy comes either from store reserves or from photosynthesis. Thus, ability of the plant to carry out net positive photosynthesis is important in cold acclimation.

Figure 9 Relationship between palisade thickness and frost killing temperature of several *Solanum* species. Palisade layer thickness = palisade/total thickness excluding upper and lower epidermis. The equation of regression line is y = 35.20 + 5.75 x with $\gamma = 0.59**$ (source: Palta and Li, 1979).

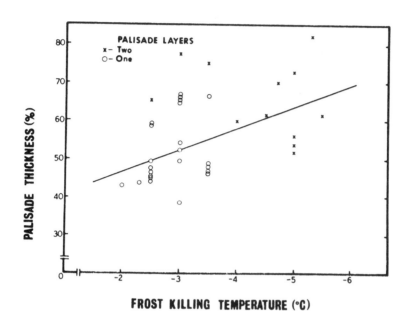

INHERITANCE OF FREEZING TOLERANCE AND COLD ACCLIMATION

The mode of inheritance of freezing stress resistance has been investigated in several plant species. The majority of the research has been conducted with wheat (Gullord, 1975; Gullord *et al.*, 1975; Sutka, 1981, 1984; Sutka and Veisz, 1988; Brule-Babel and Fowler, 1988; Limin and Fowler, 1989) and other cereals (Amirshashi and Patterson, 1956; Jenkins, 1969; Pfahler, 1966; Brule-Babel and Fowler, 1989). It has been generally concluded that inheritance of freezing stress resistance in cereals involves multiple genes with both additive and non-additive effects (Marshall, 1982; Stushnoff *et al.*, 1984; Blum, 1988; Thomashow, 1990). In some studies freezing resistance has been shown to be partially dominant, whereas in others freezing sensitivity has been shown to be partially dominant. These apparently conflicting results have been, in part, explained in terms of the freezing protocols used. At severe freezing stress, freezing sensitivity was dominant whereas at moderate freezing stress, freezing tolerance was dominant (Sutka and Veisz, 1988).

Breeding for improved resistance to low temperature stress has been difficult and met with very little success in the past. The reasons for this have been pointed out in recent reviews to be lack of effective selection criteria, complexity of the genetic control, and the narrow range of genetic variability in gene pools of many crop species (Marshall, 1982; Stushnoff *et al.*, 1984). In these reviews it has been concluded that the field survival remains the best screening method. While that may be the case, field selection has many inherent problems. "Winter survival" encompasses multiple facets of plants' ability to survive biotic and abiotic stresses. Variation in snow cover, midwinter thaw periods, water and nutrient status, and disease infestation contribute to the seasonal variation in winter survival. Ideally, one hopes for a test winter that is severe enough to kill most of sensitive cultivars, cause variable degrees of injury to the intermediate cultivars, and no injury to the most resistant cultivars. However, such test winters are rare.

We are utilizing potato species to understand genetic mechanisms of freezing tolerance and cold acclimation. Several non-cultivated tuber-bearing potato species are freezing tolerant and are able to acclimate in response to cool day/night temperatures (5/2°C), while the commonly cultivated species *Solanum tuberosum* is freezing sensitive and fails to cold-acclimate. Based on their response to cold temperatures, potato species have been classified into four groups (Table 7) - freezing tolerant and able to cold-acclimate, freezing tolerant but unable to acclimate, freezing sensitive but able to acclimate, and freezing sensitive and unable to acclimate. This classification indicates that freezing tolerance and acclimation ability are independent traits.

Recently, we have investigated the mode of inheritance of freezing tolerance and of cold acclimation in potato species (Stone *et al.*, 1991). In this study, we demonstrated that freezing tolerance (in unacclimated state) and capacity to cold acclimate were genetically distinct traits. We utilized *S. commersonii* (freezing tolerance of -3.9°C and cold acclimate in 10 days down to -7.2°C) and *S. cardiophyllum* Lindl. (freezing tolerance of -1.6°C and unable to cold acclimate) as diploid parents. The progeny, developed from the backcrosses of the F_1s with the parents, were screened for both the freezing tolerance and cold acclimation ability. Both these parameters segregated independently. In relatively few individuals examined we were able to recover both the parental types in backcross progeny for cold acclimation ability and nearly both the parental types for freezing tolerance. From these studies we concluded (1) freezing tolerance and cold acclimation must have separate genetic mechanisms, (2) freezing tolerance and cold acclimation appear to be quantitatively inherited, (3) both traits can be explained by a simple additive dominance model - lack of acclimation appears to be somewhat dominant, and (4) few major genes must control freezing tolerance and cold acclimation in these potato species. Thus, selection for

Table 7 Classification of tuber-bearing potato species in terms of their freezing tolerance and capacity to acclimate.

Categories	Species (examples)	Killing temp (°C) Before treatment*	After treatment†
Group I: Frost resistant and able to cold harden	S. acaule (Oka 3885)‡	-6.0	-9.0
	S. commersonii (Oka 5040)	-4.5	-11.5
Group II: Frost resistant but unable to cold harden	S. sanctae-rosae (Oka 5697)	-5.5	-5.5
	S. megistacrolobum (Oka 3914)	-5.0	-5.0
Group III: Frost sensitive but able to cold harden	S. oplocense (Oka 4500)	-3.0	-8.0
	S. polytricho;1 (PI 184773)	-3.0	-6.5
Group IV: Frost sensitive and unable to cold harden	S. tuberosum	-3.0	-3.0
	S. stenotomum (PI 195188)	-3.0	-3.0

Source: Li *et al.* (1979).

*Plants were grown in a regime of 20/15°C day/night, 14 h.

†Plants were grown in a regime of 2°C day/night, 14 h, for 20 days.

‡Identification number at Potato Introduction Station, Sturgeon Bay, WI.

these two traits in a breeding program should not be difficult, at least at the diploid level.

To date, no systematic attempt has been made to separate various components of winter survival in wheat and other cereals. Our results with potato suggest that it should be possible to screen separately for freezing tolerance and capacity to acclimate. During early fall the survival would primarily be determined by the basic freezing tolerance (unacclimated state) because acclimation requires several weeks of chilling (sometimes freezing) temperatures (Levitt, 1980). However, during the midwinter period, survival would be primarily dependent on capacity to acclimate. Furthermore, an extended winter thaw can deacclimate the plant. Thus, during a very cold spell following winter thaw, the freezing tolerance (unacclimated state) would be expected to play an important role for the survival. As mentioned above, winter survival encompasses multiple facets of the plant's ability to survive biotic and abiotic stresses. From the results discussed here, it appears that some progress in improving winter survival can be made by selecting both for freezing tolerance and capacity to acclimate. This can be only achieved by selection under controlled conditions by using rapid screening tools. Our studies with potatoes suggest development of these screens should be possible. We measure ion leakage following a simulated freeze-thaw stress to various temperatures (Palta *et al.*, 1981). Using this test we were able to screen 20 lines every day.

SELECTION FOR FREEZING TOLERANCE AND FREEZING AVOIDANCE IN THE SAME PLANT

Various parts of the same plant can survive at freezing temperatures by separate mechanisms. For example, in overwintering woody plants, the wood and the buds are known to survive winter by supercooling whereas the bark tissue actually freezes and survives by tolerance (Sakai and Larcher, 1987). Most studies on herbaceous plants have not attempted to separate these mechanisms in various plant parts. Undoubtedly, some of the crown tissue in alfalfa (*Medicago sativa* L.) and winter cereals have the potential to supercool, whereas the roots of these plants must freeze and thus survive by tolerance. The question should be asked whether root tolerance and crown survival in winter cereals are controlled by separate genes. If they are, then these traits should be selected separately and then combined.

Recent studies with carrot freezing stress resistance show that both freezing tolerance and avoidance are important for successful breeding for frost survival (G. Simon and J. Palta: unpublished data). Carrots are grown in southern France and harvested during the winter months for fresh market. Frost during this period can lead to two undesirable outcomes - (1) injury to the leaf tissue which when left in the field causes rot which invariably enters the root (because carrots are commercially harvested by machine which pulls on the foliage, frost damage thus prevents mechanical harvest); (2) following freeze-thaw stress, the roots develop surface cracks that enlarge, thus deteriorating fresh market quality of the carrots.

We found that the injury to leaf tissue was related to the freezing tolerance of carrot, whereas the crack in the root tissue was related to the formation of ice in the root. We were able to show that avoidance of ice formation is important for preventing root cracks. Using the ion leakage test, we were able to show that considerable variations in the freezing tolerance existed among the carrot germplasm. Carrot lines varied considerably for the position of the growing tip (root tip) in relation to soil. In some carrot lines the root tops were actually up to 5 cm above the ground, whereas in other lines root tops were about 2 cm below the soil surface. The carrots with root tops below the soil surface will be best at avoiding freezing of the root, thus avoid formation of cracks. Freezing avoidance of the roots was also negatively correlated with water content and positively correlated with root sugar content. From these studies, we concluded that, for successful cultivation of carrots in frost prone areas, the carrots should be selected for freezing tolerance of the foliage and freezing avoidance of the roots. These two characteristics segregate independently; thus, by precise tests, it should be possible to select separately for freezing tolerance and avoidance which would have been impossible to do by field selection.

SCREENING TECHNIQUES

Survival under natural conditions is the ultimate test of freezing resistance of plant material. However, field trials are often inconclusive due either to complete kill or lack of it (Fowler and Gusta, 1979). Furthermore, high variability in stress level often makes it difficult to identify small, but important, differences among genotypes (Fowler, 1979). For these and other reasons, selection under field conditions has resulted in only limited success (Stushnoff *et al.*, 1984).

Starting in the early 1900s, researchers have sought various morphological, anatomical, physiological and biochemical characteristics related to freezing stress resistance and cold acclimation (see review by Levitt, 1980). Based on this information, throughout this period there has been a continuous search for rapid, reliable and accurate screening methods. Major emphasis has been to relate screening methods to winter survival in the field. For example, Fowler *et al.* (1981) evaluated 34 biochemical, physiological and morphological characters of winter wheat for predicting winter survival and found several to be useful screening tools. However, the complexity of winter survival would suggest no single test would accurately predict winter survival over many seasons. This is likely due to different plant parts utilizing different mechanisms of survival. The most critical part of the plant to be protected for winter survival could vary from season to season. Perhaps selection should be separate for each critical part of the plant.

From our results (presented in the previous section), it is clear that freezing tolerance, freezing avoidance and cold acclimation abilities can be selected individually. In some germplasm, all of these mechanisms could be important for survival at freezing temperatures. It would be prudent to design screening techniques for specific mechanisms. Some parameters that need further research and have the potential for useful rapid screen are presented below.

Leaf anatomy and pigmentation

Ability to handle light under cold acclimation conditions appears to be an important aspect of cold acclimation. Thus, morphological or anatomical factors that reduce incident light and/or facilitate filtering of light entering the interior of leaf tissue could be useful screening tools. In agreement with this, we found multiple and thicker palisade layers to be associated with freezing tolerance in potato species. This character was used as a screening tool by Estrada (1982). In addition, production of pigments during cold acclimation also needs to be further investigated. There is some evidence in the research reported by Estrada (1982) that, other factors being equal, pigmented plants are more protected against frost injury than the non-pigmented plants. Also, in the seedlings of *Pinus contorta* Dougl., a strong negative correlation between frost damage and the anthocyanin color was found at the population level (Jonsson *et al.*, 1981).

Post-thaw recovery

Research summarized in this chapter shows clearly that one aspect of freeze-thaw stress is the ability of the plant to recover. This aspect of freeze-thaw needs further investigation. Undoubtedly, freezing tolerance (in part) is determined by the ability of the plant to recover following stress. Is this a separate genetic trait? We do not know. In some cases, visual observation on the disappearance of the water soaking during the post-thaw period (Palta *et al.*, 1977c) could be used as a simple and rapid tool to screen for recovery.

Electrolyte leakage: LT_{50}

The LT_{50} is determined by assessing injury to the tissue following a freeze-thaw stress under controlled conditions. In many studies only one freezing temperature is used. With alfalfa roots we found that freezing to one temperature, e.g., $-8°C$ (a standard temperature for assessing freezing tolerance), did not maximize differences among various cultivars (Sulc *et al.*, 1991). Electrolyte leakage following a freeze-thaw stress appears to be a sensitive, rapid and reliable technique for measuring relative freezing tolerance. This method, which was first developed by Dexter (1932), is useful in assessing early stages of freeze-thaw injury. Recently we have used this method to estimate LT_{50} for the backcross progenies of tolerant and sensitive potato species (Stone *et al.*, 1991). The LT_{50} value was estimated by extrapolating from the midpoint of the maximum and minimum ion leakage value (Figure 10). Using this procedure, we were able to screen about 20 genotypes per day.

Figure 10 Typical data from an *in vitro* ion leakage for estimating LT$_{50}$. Excised leaflets of potato were frozen at 1°C h^{-1} with ice nucleation at -1°C and thawed over ice.

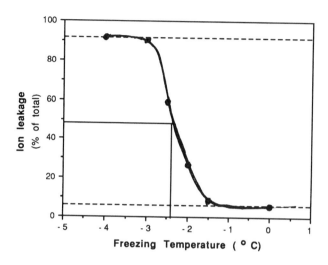

SUMMARY AND CONCLUSIONS

(1) Three components of freezing stress resistance - namely, freezing tolerance, freezing avoidance and capacity to acclimate - are heritable traits. There seems to be genetic variability for these traits present in the germplasm of some plant species. Thus, breeding to improve freezing stress resistance can be made by individually selecting for these traits and recombining them to get the desired genotype.

(2) In nature, cooling rates experienced by plants are generally low (1-2°C h^{-1}) during a frost episode. This results in ice formation in the extracellular space causing dehydration of the cell. Freezing tolerance thus largely depends upon the ability of the plant cell to tolerate dehydration.

(3) The plasma membrane plays a central role in freezing tolerance and cold acclimation. Alteration in the functions of plasma membrane ATPase is one of the earliest manifestations of freeze-thaw stress injury. Our results provide evidence that these alterations could be mediated by perturbation of cellular calcium and/or by changes in the membrane lipid properties (Figure 11).

(4) During cold acclimation there are key changes in the composition of plasma membrane lipids.

(5) The ability of the plant to cold acclimate, in part, is related to the ability of the plant to carry out net positive photosynthesis and to the

Figure 11 Sequence of events leading to injury or recovery following stress induced perturbations in membrane enzymes/membrane lipids/cellular calcium (source: Palta, 1990).

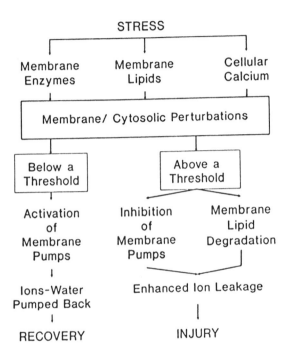

(6) Based on our understanding of the mechanisms of freezing tolerance, freezing avoidance and cold acclimation, it should be possible to design screening techniques that will be rapid, reliable and able to distinguish small, but important, changes in these traits.

ability of the plant to dissipate excess light energy and thereby prevent photo-oxidation.

REFERENCES

Abdallah, A. Y.; Palta, J. P. (1989) Specific changes in membrane polar lipid fatty acid composition coincide with initiation of fruit ripening and changes in freezing stress resistance of leaves of cranberry. *Plant Physiology* 89 (Supplement):100.

Amirshashi, M. C.; Patterson, F. L. (1956) Cold resistance of parent varieties, F_2 populations and F_3 lines of 20 oat crosses. *Agronomy Journal* 48:184-188.

Arora, R.; Palta, J. P. (1986) Protoplasmic swelling as a symptom of freezing injury in onion bulb cells. Its simulation in extracellular KCl and prevention by calcium. *Plant Physiology* 82:625-629.

Arora, R.; Palta, J. P. (1988) *In vivo* perturbation of membrane-associated calcium by freeze-thaw stress in onion bulb cells. Simulation of this perturbation in extracellular KCl and alleviation by calcium. *Plant Physiology* 87:622-628.

Arora, R.; Palta, J. P. (1989) Perturbation of membrane calcium as a molecular mechanism of freezing injury. In: Cherry, J. H. (ed.), *Environmental Stress in Plants*. NATO ASI series, Vol. G19. Springer Verlag Inc., New York, pp. 281-290.

Arora, R.; Palta, J. P. (1991) A loss in the plasma membrane ATPase activity and its recovery coincides with incipient freeze-thaw injury and post-thaw recovery in onion bulb scale tissue. *Plant Physiology* 95:846-852.

Ashworth, E. N.; Davis, G. A.; Anderson, J. A. (1985) Factors affecting ice nucleation in plant tissues. *Plant Physiology* 79:1033-1037.

Blum, A. (1988) *Plant breeding for stress environments*. CRC Press, Boca Raton, FL, pp. 79-127.

Boyer, J. S. (1982) Plant productivity and environment. *Science (Washington, DC)* 218:443-448.

Brule-Babel, A. L.; Fowler, D. B. (1988) Genetic control of cold hardiness and vernalization requirement in winter wheat. *Crop Science* 28:879-884.

Brule-Babel, A. L.; Fowler, D. B. (1989) Use of controlled environments for winter cereal cold hardiness evaluation: Controlled freeze tests and tissue water content as prediction tests. *Canadian Journal of Plant Science* 69:355-366.

Burke, M. J.; Gusta, L. V.; Quamme, H. A.; Weiser, C. J.; Li, P. H. (1976) Freezing injury in plants. *Annual Review of Plant Physiology* 27:507-528.

Carruthers, A.; Melchoir, D. L. (1986) How lipid bilayer affect membrane protein activity. *Trends in Biochemical Sciences* 11:331-335.

Dexter, S. T. (1932) Studies of the hardiness of plants: a modification of the Newton pressure method for small samples. *Plant Physiology* 7:721-726.

Dowgert, M. F.; Steponkus, P. L. (1984) Behavior of the plasma membrane of isolated protoplasts during a freeze-thaw cycle. *Plant Physiology* 75:1139.

Estrada, N. R. (1982) Breeding wild and primitive potato species to obtain frost resistant cultivated varieties. In: Li, P. H.; Sakai, A. (eds.), *Plant cold hardiness and freezing stress: Mechanisms and crop implications*. Vol 2. Academic Press, New York, pp. 615-634.

Fowler, D. B. (1979) Selection for winter hardiness in wheat. II. Variation within field trials. *Crop Science* 19:773-775.

Fowler, D. B.; Gusta, L. V. (1979) Selection for winterhardiness in wheat. I. Identification of genotypic variability. *Crop Science* 19:769-772.

Fowler, D. B.; Gusta, L. V.; Tyler, N. J. (1981) Selection for winter hardiness in wheat. III. Screening methods. *Crop Science* 21:896-901.

Gullord, M. (1975) Genetics of freezing hardiness in winter wheat (*Triticum aestivum* L.). Unpublished Ph.D. dissertation, Michigan State University, East Lansing.

Gullord, M.; Olien, C. R.; Everson, E. H. (1975) Evaluation freezing hardiness in winter wheat. *Crop Science* 15:153-157.

Hellergren, J.; Widell, S.; Lundborg, T. (1985) ATPase in relation to freezing in purified plasma membrane vesicles from nonacclimated seedlings of *Pinus sylvestris*. *Acta Horticulturae* 168:161-166.

Hepler, P. K.; Wayne, R. O. (1985) Calcium and plant development. *Annual Review of Plant Physiology* 36:397-439.

Iswari, S.; Palta, J. P. (1989) Plasma membrane ATPase as a site of functional alteration during cold acclimation and freezing injury. In: Li, P. H. (ed.), *Low Temperature Stress Physiology in Crops*. CRC Press, Inc., Boca Raton, FL, pp. 123-137.

Iswari, S.; Palta, J. P. (1989) Plasma membrane ATPase activity following reversible and irreversible freezing injury. *Plant Physiology* 90:1088-1095.

Iswari, S.; Weiss, L. S.; Bertics, P. J.; Palta, J. P. (1990) Plasma membrane bound protein kinase in cold acclimating and nonacclimating *Solanum* species. *Plant Physiology* 93(Suppl.):84 (Abstr. 491).

Jenkins, G. (1969) Transgressive segregation for frost resistance in hexaploid oats (*Avena* spp.). *Journal of Agricultural Sciences* 73:477-482.

Jian, C. L.; Sun, L. H.; Dong, H. Z.; Sun, D. L. (1982) Changes in ATPase activity during freezing injury and cold hardening. In: Li, P. H.; Sakai, A. (eds.), *Plant cold hardiness and freezing stress: Mechanisms and crop implications*. Vol. 2. Academic Press, New York, NY, p. 243.

Jonsson, A.; Eriksson, G.; Dormling, I.; Ifver, J. (1981) Studies on frost hardiness of *Pinus contorta* Dougl. seedlings grown in climate chambers. *Studia Forestalia Suecica* No. 157.

Kuiper, P. J. C. (1985) Environmental changes and lipid metabolism of higher plants. *Physiologia Plantarum* 64:118-122.

Legge, R. L.; Thompson, J. E.; Baker, J. E.; Lieberman, M. (1982) The effect of calcium on the fluidity and phase properties of microsomal membranes isolated from postclimateric golden delicious apples. *Plant and Cell Physiology* 23:161-169.

Levitt, J. (1980) *Responses of plants to environmental stresses. Vol. 1: Chilling, freezing, and high temperature stresses*. Academic Press, New York, NY.

Li, P. H. (1987) *Plant cold hardiness. Plant biology*. Vol. 5. Alan R. Li, New York.

Li, P. H.; Palta, J. P. (1978) Frost hardening and freezing stress in tuber-bearing *Solanum* species. In: Li, P. H.; Sakai, A. (eds.), *Plant and cold hardiness and freezing stress: Mechanisms and crop implications*. Vol. 2. Academic Press, New York, NY, pp. 49-71.

Li, P. H.; Palta, J. P.; Chen, H. H. (1979) Freezing stress in potato. In: Li, P. H.; Sakai, A. (eds.), *Low temperature stress in crop plants: The role of the membrane*. Academic Press, New York, NY, pp. 291-303.

Li, P. H.; Sakai, A. (1978) *Plant cold hardiness and freezing stress: Mechanisms and crop implications*. Vol. 1. Academic Press, New York.

Li, P. H.; Sakai, A. (1982) *Plant cold hardiness and freezing stress: Mechanisms and crop implications*. Vol. 2. Academic Press, New York.

Limin, A. E.; Fowler, D. B. (1989) The influence of cell size and chromosome dosage on cold-hardiness expression in the Triticeae. *Genome* 32:667-671.

Lyons, J. M.; Graham, D.; Raison, J. K. (eds.) (1979) *Low temperature stress in crop plants*. Academic Press, New York.

Marshall, H. G. (1982) Breeding for tolerance to heat and cold. In: Christiansen, M. N.; Lewis, C. F. (eds.), *Breeding plants for less favorable environments*. John Wiley & Sons, New York, pp. 47-69.

Maximov, N. A. (1912) Chemische Schutzmittel der Pflanzen gegen erfrieren. *Berichte der Deutschen Botanischen Gesellschaft* 30:52.

Ogren, E.; Oquist, G.; Halgren, J. E. (1984) Photoinhibition of photosynthesis in *Lemana gibba* as induced by the interaction between light and temperature. I. Photosynthesis *in vivo*. *Physiology Plant* 62:181-186.

Oquist, G. (1983) Effect of low temperature on photosynthesis. *Plant and Cell Environment* 6:281-300.

Osmond, C. B. (1981) Photorespiration and photoinhibition: some implications for the energetics of photosynthesis. *Biochimica Biophysica Acta* 639:77-98.

Palmgren, M. G.; Sommarin, M.; Ulvskov, P.; Jorgensen, P. L. (1988) Modulation of plasma membrane H^+-ATPase from oat roots by lysophosphatidycholine, free fatty acids and phospholipase A_2. *Physiologia Plantarum* 74:11-19.

Palta, J. P. (1989) Plasma membrane ATPase as a key site of perturbation in response to freeze-thaw stress. *Current Topics Plant Biochemistry and Physiology* 8:41-68.

Palta, J. P. (1990) Stress interactions at the cellular and membrane levels. *HortScience* 25:1377-1381.

Palta, J. P.; Chen, H. H.; Li, P. H. (1981) Relationships between heat and frost resistance of several potato species: Effect of cold adaptation on heat resistance. *Botanical Gazette* 142:311-315.

Palta, J. P.; Jensen, K. G.; Li, P. H. (1982) Cell membrane alterations following a slow freeze-thaw cycle: Ion leakage, injury and recovery. In: Li, P. H.; Sakai, A. (eds.), *Plant cold hardiness and freezing stress. Mechanisms and crop implications.* Academic Press, New York, pp. 221-242.

Palta, J. P.; Levitt, J.; Stadelmann, E. J. (1977a) Freezing injury in onion bulb cells. I. Evaluation of the conductivity method and analysis of ion and sugar efflux from injured cells. *Plant Physiology* 60:393-397.

Palta, J. P.; Levitt, J.; Stadelmann, E. J. (1977b) Freezing injury in onion bulb cells. II. Post thawing injury or recovery. *Plant Physiology* 60:398-401.

Palta, J. P.; Levitt, J.; Stadelmann, E. J. (1977c) Freezing tolerance of onion bulbs and significance of freeze-induced tissue infiltration. *Cryobiology* 14:614-619.

Palta, J. P.; Li, P. H. (1978a) Cell membrane properties in relation to freezing injury. In: Li, P. H.; Sakai, A. (eds.), *Plant cold hardiness and freezing stress: Mechanisms and crop implications.* Academic Press, New York, pp. 93-115.

Palta, J. P.; Li, P. H. (1978b) Examination of ultrastructural freeze-injury in the leaf cells of tender and hardy potato species. *HortScience* 13:387.

Palta, J. P.; Li, P. H. (1979) Frost hardiness in relation to leaf anatomy and natural distribution of several *Solanum* species. *Crop Science* 19:665-670.

Palta, J. P.; Li, P. H. (1980) Alterations in membrane transport properties by freezing injury in herbaceous plants. Evidence against rupture theory. *Physiologia Plantarum* 50:169-175.

Palta, J. P.; Meade, L. S. (1987) Fatty acid composition changes after cold acclimation: Comparison of leaf, callus and purified membranes. *Plant Physiology* 83(supplement):71.

Palta, J. P.; Meade, L. S. (1989) During cold acclimation of potato species, an increase in 18:2 and a decrease in 16:0 in plasma membrane phospholipid coincides with an increase in freezing stress resistance. *Plant Physiology* 89(Supplement):89.

Pearce, R. S. (1988) Extracellular ice and cell shape in frost-stressed cereal leaves: A low temperature scanning-electron-microscopy study. *Planta* 175:313-324.

Pfahler, P. L. (1966) Small grain improvement by breeding and selection. Report of the Florida Agriculture Experiment Station, p. 52.

Pomeroy, K. M.; Pihakaski, J. S.; Andrews, C. J. (1983) Membrane properties of isolated winter wheat cells in relation to icing stress. *Plant Physiology* 72:535.

Poovaiah, B. W. (1985) Role of calcium and calmodulin in plant growth and development. *HortScience* 20:347-352.

Poovaiah, B. W.; Reddy, A. S. N. (1987) Calcium messenger systems in plants. *CRC Critical Reviews in Plant Sciences* 6:47-102.

Powles, S. B. (1984) Photoinhibition of photosynthesis induced by visible light. *Annual Review of Plant Physiology* 35:15-44.

Sakai, A.; Larcher, W. (1987) *Frost survival of plants: Responses and adaptations to freezing stress.* Springer-Verlag, Berlin and New York.

Schaller, G. E.; Sussman, M. R. (1988) Phosphonylation of the plasma membrane H^+-ATPase of oat roots by a calcium-stimulated protein kinase. *Planta* 173:509-518.

Steffen, K. L.; Arora, R.; Palta, J. P. (1989) Sensitivity of photosynthesis and respiration to a freeze-thaw stress: Role of realistic freeze-thaw protocol. *Plant Physiology* 89:1372-1379.

Steffen, K. L.; Palta, J. P. (1986) Effect of light on photosynthetic capacity during cold acclimation in a cold-sensitive and a cold-tolerant potato species. *Physiologia Plantarum* 66:353-359.

Steffen, K. L.; Palta, J. P. (1987) Photosynthesis as a key process in plant response to low temperature: Alteration during low temperature acclimation and impairment during incipient freeze-thaw injury. In: Li, P. H. (ed.), *Plant cold hardiness.* Alan R. Liss, Inc., New York, pp. 67-99.

Steffen, K. L.; Palta, J. P. (1989) Growth and development temperature influences level of tolerance to high light stress. *Plant Physiology* 91:1558-1561.

Steponkus, P. L. (1984) Role of the plasma membrane in freezing injury and cold acclimation. *Annual Review of Plant Physiology* 35:543.

Steponkus, P. L.; Weist, S. C. (1978) Plasma membrane alterations following cold acclimation and freezing. In: Li, P. H.; Sakai, A. (eds.), *Plant cold hardiness and freezing stress: Mechanisms and crop implications.* Academic Press, New York, p. 75.

Stone, J. M.; Palta, J. P.; Bamberg, J. B.; Weiss, L. S. (1991) Freezing tolerance and capacity to acclimate are conferred by different genes in *Solanum* species. *Agronomy Abstracts* 1991:200.

Stushnoff, C.; Fowler, D. B.; Bruele-Babel, A. (1984) Breeding and selection for resistance to low temperature. In: Voss, P. B. (ed.), *Plant breeding - A contemporary basis.* Pergamon Press, Elmsford, pp. 115-136.

Sulc, M. R.; Albrecht, K. A.; Palta, J. P.; Duke, J. H. (1991) Effects of different freezing temperatures on leakage of intracellular substances from alfalfa roots. *Crop Science:*(in press).

Sutinen, M. K.; Palta, J. P. (1989) Seasonal changes in phospholipid fatty acid composition coincide with changes in freezing stress resistance in pine needles. *Plant Physiology* 89(Supplement):28.

Sutka, J. (1981) Genetic studies of frost resistance in wheat. *Theoretical Applied Genetics* 59:145-152.

Sutka, J. (1984) A ten-parental diallel analysis of frost resistance in winter wheat. *Zeitschrift fur Pflanzenzuechtung* 93:147-157.

Sutka, J.; Veisz, O. (1988) Reversal of dominance in a gene on chromosome 5A controlling frost resistance in wheat. *Genome* 30:313-317.

Sze, H. (1985) H^+-translocating ATPases: Advances using membrane vesicles. *Annual Review Plant Physiology* 36:175.

Tao, D.; Li, P. H.; Carter, J. V. (1983) Role of the cell wall in freezing tolerance of cultured potato cells and their protoplasts. *Physiologia Plantarum* 58:527-532.

Thomashow, M. F. (1990) Molecular genetics of cold acclimation in higher plants. In: Scandalios, J. G. (ed.), *Advances in genetics.* Vol. 28. *Genomic responses to environmental stress.* Academic Press, New York, pp. 99-125.

Uemura, M.; Yoshida, S. (1985) Studies on freezing injury in plant cells. *Plant Physiology* 80:187-195.

Chapter 11
Breeding plants for enhanced beneficial interactions with soil microorganisms

Fredrick A. Bliss

Department of Pomology, University of California, Davis, CA 95616

INTRODUCTION

Plant growth is often enhanced by the beneficial effects of soil microorganisms. In fact, growth without extensive interaction is virtually impossible, especially in the field. In addition to plant-microbe interactions, those among microbes can also affect plants by changing the microbial population dynamics and by producing synergistic effects (Sarig *et al.*, 1986). Specific effects of soil microbes on plant vigor, growth and economic yield are difficult to determine because a dynamic situation exists, and it is hard to observe activity in the rhizosphere.

Probably there is as much variability for root as for shoot traits. Judging by the extensive genetic variability among plants for reaction to aerial and soil-borne pathogens, a similar range of variation for root response to beneficial soil microorganisms might exist. However, much less is known about such host variability because of limited interest and technical difficulties associated with studying subterranean systems.

Breeding plants to enhance beneficial interactions with soil microorganisms requires a thorough understanding of a complex, interdependent system. It is not surprising that there have been few attempts to alter the host plant genetically to improve performance over the "wild-type" situation.

TYPES OF BENEFICIAL INTERACTIONS

Beneficial plant-soil microbe interactions include those that (1) enhance growth by providing more mineral elements or growth-promoting hormones and (2) provide protection of the host plant against antagonistic or pathogenic microorganisms. Fixation of atmospheric N_2 in legumes infected by rhizobia or bradyrhizobia is the best understood plant-microbe relationship. However, because of the importance of non-legumes - especially cereal grains and forage grasses - and the fact that they require large amounts of N for high yields, biological nitrogen fixation (BNF) by grasses, through either rhizosphere bacteria or development of a symbiosis with rhizobia, is of considerable interest.

Non-symbiotic BNF by bacteria that associate with the roots, including *Azotobacter*, *Klebsiella* and *Azospirillum*, contributes to soil fertility in some tropical and temperate settings (Van Berkum and Bohlool, 1980). Sugar cane production systems have been reported in which no fertilizer N was added and, despite large N removal in the cane, there was no apparent decline in soil N content (Dart, 1986). Miranda *et al.* (1990) cited studies of tropical pasture grasses where BNF contributed between 20 and 40% of the total N accumulation, and up to 10 kg N $ha^{-1}month^{-1}$. Associative N_2-fixing bacteria can produce enhanced plant growth, but the actual amount of N_2 fixed has been difficult to determine. Responses such as increased root growth and branching may lead to greater nutrient uptake and more N accumulation, irrespective of greater fixation potential.

Response to inoculation with associative nitrogen-fixing organisms has been equivocal, and their recovery from the roots in sufficient numbers to support large amounts of fixation has been difficult. There have been few attempts to identify host plant genotypic differences that affect associative fixation (e.g., Urquiaga *et al.*, 1988). However, Miranda *et al.* (1990) have reported variation among ecotypes of *Panicum maximum* Jacq. for associative BNF based on greenhouse trials.

BNF by nodulated non-legumes in association with the *Frankia* endophyte is found primarily in woody plants. The use of fast-growing trees for wood production and intercropping is attractive, and selection of plants capable of enhanced N_2 fixation and rapid growth is desirable. It is difficult to accurately measure N_2 fixation by trees; however, differences among three clones of swamp oak, *Casuarina equisetifolia* Forst., were found using ^{15}N-labeled fertilizer (Sougoufara *et al.*, 1990).

The use of P-scavenging endomycorrhizae (e.g., *Glomus*) should be viewed differently than N_2 fixation because, instead of synthesizing a new compound (i.e., organic N), they aid in extracting P from the soil and providing more to the plant. The vesicular arbuscular (VA) endomycorrhizae represent the most widespread type of fungal infection in plant roots (Gianinazzi-Pearson *et al.*, 1991) and are potentially most valuable where soils are high in total P but low in available P. For legumes growing on low-P soils, the tripartite interaction between the mycorrhizae, N_2-fixing bacteria and the host plant should be considered, since fixation is limited by low P, and a positive effect often occurs when more P is assimilated by the plant. Despite the wide distribution and importance of these fungi, knowledge about the complex interactions with plants is not yet sufficient to implement plant breeding programs for improving the host plant. The first report of non-mycorrhizal plant mutants, in pea (*Pisum sativum* L.) and faba bean (*Vicia faba* L.), appeared only recently (Duc *et al.*, 1989).

Plant stress from nitrogen and phosphorus deficiencies is widespread because of the large demands for these elements during plant growth and the extensive cropping areas with soils low in available N and P. It is hoped that proper management of N_2-fixing bacteria and P-scavenging endomycorrhizae in concert with responsive host plant cultivars will lead to decreased fertilizer use in the future (Pacovsky *et al.*, 1986).

In addition to the beneficial effects provided to plants from increased availability of N and P, and possible indirect effects from stimulated root growth, some microbes act as antagonists to harmful microorganisms. There are many microbial interactions, but only a few examples of successful biocontrol of plant disease and productivity enhancement (Handelsman and Parke, 1989). Biocontrol using microbes is now at the stage of identifying useful organisms and formulating effective inoculum for physical application; a situation comparable to that for legumes and rhizobia several decades ago. Selection of crop cultivars to support beneficial biocontrol agents has yet to be undertaken.

Only for a few legume *Rhizobium* systems is there sufficient knowledge and enough economic interest to support host plant breeding programs. Even there, only a few efforts to select host plants for increased BNF have been reported. When there is interest in improving the host plant to provide greater support for other useful microbes, breeding approaches similar to those used in legumes for increased BNF by *Rhizobium* should prove useful.

IMPROVING PLANTS FOR BIOLOGICAL NITROGEN FIXATION

In the U.S.A., corn (*Zea mays* L.), cotton (*Gossypium hirsutum* L.), wheat (*Triticum aestivum* L.), and soybean [*Glycine max* (L.) Merr.] receive the most N fertilizer, with the estimated total use on those crops in 1985 being about 6 million metric tons (Tauer, 1989). Although the unit cost of N has remained low because of low energy prices and improved technology for manufacture, transport and application, costs will rise as energy becomes more scarce and costly. Tauer (1989) concluded that BNF technologies potentially have a high value to society. Based on modeling studies, he stated that "Increasing the efficiency of legumes to fix N_2 may have an annual U.S. benefit of \$1,067 million while decreasing N fertilization by 1,547 thousand metric tons. Total elimination of nitrogen fertilization of the major crops has an annual U.S. benefit of \$4,484 million."

Improved BNF is not an end unto itself. In practice, the benefits are measured as increased herbage (i.e., forages), fruit [i.e., snap beans (*Phaseolus vulgaris* L.), peas], or seed and protein [i.e., soybeans, peanuts (*Arachis hypogaea* L.)] yield. To more effectively achieve increased protein productivity in grain crops, Cregan and van Berkum (1984) suggested that physiological and biochemical traits reflecting N metabolism - such as N_2 fixation, N accumulation, N remobilization and nitrogen harvest index (NHI) - in addition to seed yield, should be included in plant selection schemes.

Two approaches for plant improvement to increase BNF in legumes can be followed. One is to identify and characterize important traits that affect

N_2 fixation, then devise ways to overcome existing limitations and improve the processes essential to enhanced fixation (Mytton, 1983). The altered traits can be incorporated into an adapted host cultivar. Another approach is to select legume plants for increased N_2 fixation, either using amount of fixed N_2 or related traits as selection criteria. The improved populations that result can be used as cultivars and as contrasting genotypes for studying important plant traits affecting fixation.

Selection for increased dinitrogen fixation

Effective selection depends on the following: (1) Choice of traits as selection criteria that can be measured precisely and economically, while allowing for discrimination between superior and inferior selection units; (2) variability in legume germplasm and heritability of differences for either plant N derived from the atmosphere (Ndfa) or traits indicative of fixation potential; (3) identification of genetically diverse parents that are also suitable agronomically; (4) choice of selection units (i.e., individual plants or families) that facilitate precise quantification of selection criteria and allow production of progeny from selected plants; and (5) use of a breeding procedure (e.g., mass selection, family selection) that provides maximum genetic gain per year for increased N_2 fixation and recombination with essential agronomic traits.

Selection criteria and measurement of plant response

Effective selection for increased N_2 fixation requires that the plant be measured accurately and precisely, and if large populations are studied, rapidly and economically. This requires a method for distinguishing Ndfa from N from the soil (Ndfs) and from fertilizer (Ndff). Amount of total N is often determined by the Kjeldahl method; but use of near-infrared reflectance spectroscopy (NIRS) is also a rapid, accurate and satisfactory method (Mundel and Schaalji, 1988; Miranda and Bliss, 1991).

The N difference method can be used when plants are grown where the exact amount of N is known and they extract similar amounts of N from the growth medium. Ndfa is estimated by comparing the total N in fixing plants with that in suitable non-fixing plants grown under the same conditions. When unknown levels of soil N are present and plants extract different amounts of soil N, ^{15}N dilution methods should be used to obtain accurate estimates of Ndfa. If proper precautions are followed for using non-fixing standard crops, application of labeled fertilizer, plant sampling and sample analysis, total plant N can be partitioned to the three sources - atmospheric (Ndfa), soil (Ndfs), and fertilizer (Ndff). Choice of a suitable non-fixing plant for both the difference and ^{15}N-dilution methods is important to obtain valid estimates of N_2 fixation (Henson and Heichel, 1984; Boddey et al.,

1990). Non-legumes {e.g., cereals, sorghum [*Sorghum bicolor* (L.) Moench]} and non-nodulating, near-isogenic lines of legumes (i.e., non-nodulating soybean) have been used. When plant Ndfa is the selection criterion, values can be expressed as total N (weight plant^{-1}) or proportion (%) of plant N from fixation (% Ndfa). Another decision is whether to select for total plant biomass and tissue N concentration (e.g., fresh and dry weight of forage legume herbage) or for seed yield, total seed N and N (protein) concentration of grain legumes. When N-dilution methods are used, apportionment of fixed N$_2$ to the grain as well as total N$_2$ fixed can be included as selection criteria.

The use of ^{15}N-labeled fertilizer for research is relatively recent. Although it is the preferred method for accurately estimating Ndfa in field settings, other methods have been used because of simplicity and lower cost. Alternative methods should be chosen for their effectiveness in improving the trait of choice (i.e., increased plant Ndfa) rather than simplicity and low cost.

Measures of acetylene (C$_2$H$_2$) reduction (AR) activity by excavated roots as an estimate of nitrogenase activity (NA) have been used to detect relative differences in N$_2$ fixation potential among different plant genotypes. Values presented as specific nodule activity (C$_2$H$_2$ reduced nodule mass^{-1}time^{-1}) or plant activity (C$_2$H$_2$ reduced root^{-1}time^{-1}). AR should not be used to estimate the amount of N$_2$ fixed because of difficulty in relating activity to actual fixation and in recovering all of a plant's roots during excavation. The substantial contribution of fixation by lateral root nodules to total plant N, especially during later vegetative growth stages and seed fill, further limits the utility of AR values estimated from only crown root nodules at a single sampling date. Other traits used as indirect measures of fixation and fixation potential include nodulation characteristics such as nodule number and dry weight plant^{-1}, visual nodulation scores, early nodulation and delayed nodule senescence. For plants grown on minimal-N media, plant dry weight, intensity of green color (Graham and Temple, 1984), plant and seed N, and seed yield (Miranda and Bliss, 1991) have been used as indirect measures of N$_2$ fixation. The effectiveness of direct versus indirect selection can be compared on the basis of expected and realized gain from using each criterion. Other efficiencies that accrue because some traits are easier and cheaper to evaluate, especially in large populations, should also be taken into account.

Heritability and gain from selection

Although direct measures of Ndfa are desirable and ^{15}N-labeled fertilizer is available, there are only a few examples of ^{15}N-dilution methods being used to assess variability in the germplasm and for estimating Ndfa as the primary selection criterion. Sometimes, either the ^{15}N-dilution method or the difference method has been used to verify progress from indirect selection based on other traits. Implicit in the expectation that indirect selection will be effective is the assumption that secondary traits are positively correlated genetically with the amount of Ndfa. Despite the presence of substantial genetic variability for BNF in most grain and forage legumes, and the feeling that improvement of most legumes is needed, there are few examples of actual selection programs.

Variability for BNF and related traits has been reported in forage legumes. The studies in red clover (*Trifolium pratense* L.) and subterranean clover (*T. subterraneum* L.) were the first to demonstrate the feasibility of improving N_2 fixation by plant breeding (Nutman, 1984). Seetin and Barnes (1977) reported heritable variation for AR among alfalfa (*Medicago sativa* L.) genotypes. In alfalfa populations grown in the greenhouse and subjected to two generations of bidirectional selection, there was a positive, direct response for the traits selected - shoot dry weight, nodule mass score, fibrous root score and nitrogenase activity (NA) measured by acetylene reduction (AR) - and positive correlated responses for the other traits (Viands *et al.*, 1981). Estimates of realized heritability generally ranged from low to intermediate values (Table 1). Unselected populations, selected populations and population hybrids were characterized for herbage yield and N_2 fixation using ^{15}N dilution. The results showed gain from selection for increased N_2 fixation, although the level of performance depended on field N levels (Viands *et al.*, 1981; Barnes *et al.*, 1984).

The comprehensive studies of selection for increased BNF in alfalfa not only demonstrated progress from selection, but also provided valuable

Table 1 Effect of two cycles of bidirectional selection for four characteristics in the *MnNC* and *MnPL* alfalfa gene pools (adapted from Viands *et al.*, 1981).

| | Realized heritability | | | |
| | MnNC selection cycle | | MnPL selection cycle | |
Characteristic	*1*	*2*	*1*	*2*
High nitrogenase activity	0.25	0.25	0.01	0.06
Low nitrogenase activity	0.29	0.16	0.08	0.10
High top dry weight	0.21	0.32	0.14	0.24
Low top dry weight	0.06	0.22	0.00	0.28
High nodule mass score	0.29	0.35	0.17	0.45
Low nodule mass score	0.05	0.27	--	--
High fibrous root score	0.29	0.75	0.27	0.62
Low fibrous root score	0.02	0.33	0.26	0.30

materials in which to investigate other important questions regarding C and N partitioning. Barnes *et al.* (1984) discuss the steps involved in a breeding program where, although the primary aim is to improve BNF, other important traits must be considered if new cultivars are to be useful. Sometimes improvement of a single trait does not lead to expected or hoped for consequences; increased N_2 fixation did not always result in more herbage yield, and there was an association between higher fixation potential and increased susceptibility to bacterial wilt of alfalfa in some populations (Barnes *et al.*, 1984).

Among grain legumes, the highest levels of N_2 fixation appear to be in the faba bean. Using ^{15}N-dilution methods, Brunner and Zapata (1984) found that total Ndfa was in excess of 200 kg ha^{-1} and % Ndfa was consistently greater than 85% in mutant lines that had been selected for improved yield and yield-related traits. Duc *et al.* (1988) compared cultivars at two locations and two years and found a wide range of values for Ndfa. The mean % Ndfa was from 60 to 75% and the mean total Ndfa was up to 162 kg ha^{-1}, with individual values as high as 250 kg N ha^{-1} (Table 2). These high levels of N_2 fixation would be difficult to improve, except if certain varieties are low for fixation but have outstanding agronomic traits.

Table 2 Characteristics of faba bean seed from two field locations and 2 years that relate yield, quantity of biologically fixed dinitrogen and soil-derived nitrogen (for 1981 n, number of genotypes = 16; 1982, n = 10) (adapted from Duc *et al.*, 1988).

| | Location 1 | | | Location 2 | | |
| | | N source | | | N source | |
Year	Seed yield (kg ha^{-1})	Fixation (%)	Soil (kg ha^{-1})	Seed yield (kg ha^{-1})	Fixation (%)	Soil (kg ha^{-1})
1981	4180	75	162	47	4132	70	137	54
1982	3323	61	112	66	1470	61	48	26

In contrast to faba bean, common bean is generally considered to be a poor fixer. However, considerable genetic variation for N_2 fixation and related traits has been found (e.g., Westermann *et al.*, 1981; Rennie and Kemp, 1983; Pereira *et al.*, 1989). The Mexican common bean landrace cultivar Puebla 152 (black-seeded), identified by Graham and Rosas (1977) as having high fixation potential, was used by McFerson (1983) in crosses with adapted cultivars to produce populations of inbred backcross lines.

Heritability estimates for AR ranged from 0.25 to 0.71. From segregating populations of BC_2S_3 lines grown on low-N soils, he made selections for high AR score, seed yield and plant type.   Evaluation of BC_2S_4 lines using the ^{15}N dilution method that selected lines fixed more N_2 and had higher yields than 'Sanilac' the low-fixing commercial parent, but did not fix as much N_2 as the donor parent Puebla 152 (Table 3) (St. Clair et al., 1988). Selected BC_2S_5 lines were intercrossed to produce new populations of recombinant inbred F_3 families which were evaluated in the field for N_2 fixation (St. Clair and Bliss, 1991).   The best F_3 families fixed amounts similar to the donor parent.

Table 3 Estimates of N_2 fixation, N yield per plant and % Ndfa for common bean inbred backcross lines and parents at the R9 growth stage, Hancock, WI, 1985 (adapted from St. Clair et al., 1988).

	\multicolumn{5}{c}{Amount of nitrogen applied}				
	\multicolumn{2}{c}{10 kg ha^{-1}}		\multicolumn{2}{c}{62 kg ha^{-1}}	
	N_2 fixed		Seed yield	N_2 fixed		Seed yield
Line	(kg ha^{-1})	(%)	(kg ha^{-1})	(kg ha^{-1})	(%)	(kg ha^{-1})
Puebla 152	72	63	3267	75	49	3920
24-55	32	37	1742	31	31	2395
24-17	55	48	2831	48	39	3158
24-21	43	41	2178	24	21	2505
24-48	36	41	1851	15	16	1960
24-65	45	45	2505	26	24	3050
Sanilac	8	10	1742	2	3	2178

Selection in other inbred backcross populations from the cross of Puebla 152 with tropical black-seeded parents has produced progeny lines with traits similar to the recurrent parent and higher N_2 fixation potential.   Five black-seeded germplasm lines from ICA Pijao x Puebla 152 that are potential new cultivars for Brazil have been released (Bliss et al., 1989).

The positive correlations among amount and percentage of seed Ndfa, amount and percentage of total seed N and seed yield (Ronis et al., 1985; St. Clair et al., 1988; Duc et al., 1988) suggest additional approaches to obtaining estimates of fixation. Using the difference method with non-nodulating soybean as the non-fixing control, Miranda and Bliss, (1991) reported realized heritability of total seed N in common bean ranging from 0.18 to 0.50 in two populations of F_3 lines.   Positive gain from selection was found for both total seed N and seed yield. The increased total plant and seed N were indicative of increased N_2 fixation. Increased nodule number, nodule mass and nodulation score have been reported to be positively correlated with increased N_2 fixation in several species (e.g., common bean) (Rosas and Bliss, 1986; Burias and Planchon, 1990). In common bean, recurrent mass selection for increased nodule number on 21- to 28-day-old seedlings was practiced in segregating populations grown in a controlled environment (Pereira and Bliss, unpublished data). Plants selected for high nodule number were transplanted, then intercrossed to produce progeny for another cycle of selection. There was a 50% increase in nodule number per plant, with a positively correlated response for nodule mass per plant, and a negative response for individual nodule size. Seeds from the two cycles of selected plants along with the original 10 parents were planted in the field. Lines from the C_2 cycle averaged more total plant N indicating higher N_2 fixation, and some lines had yields higher than the parents. This method of indirect selection for increased fixation is quite efficient since three cycles of mass selection per year can be completed in controlled growth facilities.

In peanut germplasm, there is variability for traits related to N_2 fixation, but improvement of host cultivars is complicated because the greatest differences occur among different botanical varieties (Elkan et al., 1980). Improvement of Spanish-type plants was attempted by crossing with Virginia-type parents, followed by estimation of variability and selection in segregating populations. There was significant genetic variation for nodulation traits and AR among F_2-derived F_5 and F_6 families (Arrendell et al., 1985; 1986).

Arrendell et al. (1989) estimated the variance components and heritability for AR among F_5 lines resulting from a cross of Virginia-type x Spanish-type peanut parents. The five lines with the highest and five with the lowest mean AR values were selected and evaluated in the field for nodulation traits, shoot N, fruit weight (yield) and vegetative, fruit and seed traits for which the parents differ. Heritability estimates for AR at three growth stages were 0.63, 0.67 and 0.60, and the selected F_6 lines differed significantly for AR. The lines selected for high AR had significantly higher values for nodule rating, number and weight, plant weight and fruit weight, but not specific nodule activity and total plant N. Nodule rating was positively correlated with other traits, and because it is easy, cheap and quick to evaluate, it can be used as an efficient early selection criterion.

Genetic variability for BNF in soybean has been reported among and within maturity groups (Kvien et al., 1981; Talbott et al., 1985; Betts and Herridge, 1987; Neuhausen et al., 1988). The traits recorded included N_2 fixed per plant measured by the difference method, nodule traits and scores, AR, and ureide content of xylem sap. Ronis et al. (1985) used the ^{15}N-dilution method to obtain estimates (i.e., 0.35-0.85) by analyzing P_1, P_2, and F_2 soybean populations for mass and percentage of fixed N in the seed, mass and percentage total N in the seed, seed yield and plant weight. Those estimates were based on single plant measures and did not account for potential genotype x environmental interactions.

Three F_2 populations of soybean plants were grown in soil-less culture in aerated tanks by Burias and Planchon (1990). Selections based on AR values were made among F_2 plants and seed yield of F_4-derived lines was evaluated under field conditions. Seed yield was positively correlated with AR, nodule volume and nodule dry weight, leading the authors to state that, "improvement of nodulation and N_2 fixation abilities might thus constitute a way of selection for increasing yield in parallel with the search for the best host microsymbiont combinations." Greder et al. (1986) reported heritable variation for nodule mass in three populations of F_3-derived F_5 and F_6 lines grown at three locations. Based on the positive correlation between greater nodule mass and seed yield, they suggested that selection for nodule mass should be effective and that increased yield, presumably affected by increased N_2 fixation, would be expected.

Despite the evidence in soybean that variability for N_2 fixation is heritable and that fixation and yield are positively correlated, there are few published reports of either selection for increased N_2 fixation or verification of gain from selection. This may be due to the general belief that soybeans already fix high levels of N_2. Also, in the U.S.A., soybeans are grown on relatively N-rich soils and in rotation with crops receiving large amounts of fertilizer N. Although they often are good fixers, more N is removed through a high seed yield than is returned to the soil. Most NHIs of soybean are between 0.65 and 0.75. Proportions of crop N derived from fixation vary widely (i.e., 0.13-0.85), rarely exceed 0.75 and usually are around 0.50. For the soil N pool to remain undepleted, the proportion of total crop N derived from fixation must exceed the NHI (i.e., as much N should be fixed by the crop as is removed) (Betts and Herridge, 1987). A biological yield in soybean of 3360 kg ha^{-1} total dry matter contains about 134 kg N of which 0.65-0.75 is in the seed (Ronis et al., 1985).

These values can provide goals for increasing total Ndfa and % Ndfa through plant breeding, improved inoculant application and better production management. In general, the studies of other grain legumes indicate genetic parameter estimates for BNF similar to those described above, e.g., cowpea [Vigna unguiculata (L.) Walp.] (Zary et al., 1978), garden pea (Hobbs and Mahon, 1982), chickpea (Cicer arietinum L.) (Jaiswal and Singh, 1990).

Host plant control of infection and nodulation

Susceptibility to infection and spectrum of nodulation response

The susceptibility of the host is defined by the infectivity (pathogenicity) of the interacting strain (Graham, 1981). In order to define clearly the scope of nodulation response of the host, different strains of the microsymbiont must be available.

Nodulation mutants were first discovered under field conditions in the presence of strain mixtures, and others have been found by screening putative mutants against specific rhizobial strains. Mutant alleles giving rise to "non-nodulating" phenotypes have been found in alfalfa, red clover, white clover (*T. repens* L.), soybean, chickpea, garden pea, cowpea, common bean and perhaps other legumes (e.g., Holl and LaRue, 1976). Four qualitatively inherited genes, *Rj1, Rj2, Rj3* and *Rj4*, controlling nodulation response have been described in soybean, and allelism tests among *Rj1, Rj2* and *Rj4* have shown them to be distinct and located at different loci (Devine and O'Neill, 1989).

As with other traits, centers of genetic diversity are a likely source of variation for genes controlling nodulation response. Devine and Breithaupt (1981) and Devine (1987) found variation in nodulation response to two bradyrhizobial strains among cultivated soybean and *G. soja* (Sieb. and Zucc.) considered to be the wild progenitor species. The frequency of ineffective nodulation was related to geographic origin of the host plants.

The cultivated pea and wild pea (*P. sativum* L. ecotype *fulvum*) show distinctly different nodulation patterns with rhizobial strains collected from different geographical areas. Three host genes were found in primitive and wild pea plants from the Middle East which confer resistance to nodulation by rhizobial strains of cultivated pea from Europe. One gene controlled temperature sensitive nodulation, another conferred resistance to a wide spectrum of European races and the third gave specific resistance to a single European strain (Lie, 1984).

The plant genes that control nodulation response can be utilized in several ways. Non-nodulating phenotypes are helpful when estimating Ndfa using either the difference method or ^{15}N-dilution method. It has been especially difficult to control the microsymbiont nodulating a legume cultivar when there is a competitive indigenous bacterial population. Even though superior natural or engineered strains are available, there is still no economically feasible method of establishing only the desired microsymbiont with an improved host cultivar (Devine and O'Neill, 1989). A strategy suggested by Devine and Breithaupt (1980) and Devine (1984) has been to accumulate genes that exclude the undesirable strains yet favor nodulation by desired ones. This strategy entails considerable work when developing a new legume cultivar. Also, there is the possibility that should inoculation fail for any reason, no fixation will result since the indigenous strains cannot nodulate the plants. Strain exclusion may be a risky strategy where soils are N-deficient when there is no guarantee that good inoculant will be available and applied properly, and it is not feasible to apply fertilizer N as a backup alternative.

Another strategy is selection for host genotypes that are susceptible but show differential preference for invasion by more efficient bacterial strains. Variability in invasion of white clover by rhizobia was reported by Jones and Hardarson (1979); strain preference differences were heritable with inheritance being additive (Hardarson and Jones, 1979). Barnes *et al.* (1984) found that host plant selection for physiological and morphological traits associated with N_2 fixation also modified host-*Rhizobium* compatibility.

Wild progenitors and relatives are often used as sources of variability for a wide range of traits for plant improvement. Because genes controlling nodulation are segregating in those populations, care should be taken to avoid incorporating genes that dramatically change the spectrum of nodulation either through random incorporation during crossing and selection, or because of linkage between a desirable trait and nodulation response genes.

Increased nodulation and modified nodule traits

In the absence of limiting physical factors and the presence of abundant effective rhizobia, nodulation density and pattern are controlled by plant factors. The sparseness (an exception being peanut) and uneven distribution of nodules are postulated to result from inhibition in zones adjacent to existing nodules.

It is often difficult to establish a strong direct link between amount of nodulation and amount of Ndfa in the plant, especially in the field. Infection by less-effective strains may produce large numbers of nodules but variable amounts of Ndfa. Nodule number, produced even by effective strains, is not always a good predictor of N_2 fixation. There is a strong negative correlation between nodule number and individual weight, but nodule mass (weight) usually is positively correlated with nodule number and gives a better prediction of total N_2 fixation or fixation potential. Graham (1981) stated that nodule weight can be used to define host susceptibility in quantitative terms, and that it should be possible to increase the genes for susceptibility.

Mutagenesis has been used to produce supernodulating plants from standard genotypes in pea, bean, chickpea and soybean. Usually, expression is controlled by one or relatively few genes. For most supernodulating mutants there has been little increase in total N_2 fixation - e.g., mutant soybean plants (Gresshoff *et al.*, 1991). Although it has been postulated that increased energy cost of excessive nodule development and maintenance offsets greater N_2 fixation potential and results in impaired plant growth, definitive results await the availability of near-isogenic lines for comparisons. In contrast to most studies of non-nodulating mutants using near-isogenic lines produced by backcrossing, those of supernodulating mutants have utilized the raw mutants.

Park and Buttery (1988) produced a supernodulating mutant of common bean. When grown in the absence of combined N, nodule numbers were greater, but individual nodule size was smaller and nodule dry weight per plant was similar in the mutant and standards lines (Buttery *et al.*, 1990). They found that plant dry weight of the mutant plants was less than that of the parent both with and without added N, indicating that perhaps other factors associated with growth and vigor were affected in addition to nodulation. Under field conditions the super nodulating mutant produced less grain yield. While the authors' conclusion that "... the relatively greater mass of nodules in the supernodulator drains the plants' limited resources and further retards growth" (Buttery *et al.*, 1990) may be correct, other effects related to the mutation cannot be excluded until comparisons are made with near-isogenic lines. Perhaps the super nodulating mutants can be used for breeding by crossing with standard lines, followed by selection of plants for intermediate nodule numbers, normal plant growth and high yield. Incorporation of an intermediate level of nodulation may be optimum depending on the ability of each plant genotype to supply adequate photosynthate.

Another method of producing lines with greater nodulation is to select for increased nodule number among plants in genetically variable natural populations. Selection for increased AR in F_2 soybean plants produced a correlated response for increased nodulation and there was a positive correlation with seed yield (Table 4) (Burias and Planchon, 1990). Positive selection response also was obtained in common bean, resulting in lines with a 50% increase in nodule number and 40% increase in nodule weight (Pereira and Bliss, unpublished results). In contrast to the negative correlation between super nodulation and plant growth seen for many of the induced mutants, higher yields were associated with increased nodulation and greater fixation by superior selections from breeding populations.

Nodulation and fixation in the presence of mineral N

Most legumes show a positive response to fertilizer N unless soil N is quite high, there is residual fertilizer N present, or factors other than N deficiency are limiting. Response to rhizobial inoculation is often variable, but usually biomass and seed yields are less than from plants receiving adequate fertilizer N. When substantial mineral N is present, the proportion of total Ndfa is often reduced. Without high levels of fixation, the increased uptake of nitrate results in depletion of soil N; therefore, fertilizer N must be added to maintain productivity and the use of a legume in the crop rotation is inefficient (Hansen *et al.*, 1989). The inhibitory effects of nitrate on N_2 fixation occur during several phases of symbiosis including root hair infection, nodule growth and development, level of nitrogenase activity and nodule longevity (Harper and Gibson, 1984). The microsymbiont is also affected by the presence of NO_3^-. One approach is to identify NO_3^--tolerant

Table 4 Correlations in three soybean populations of N_2 fixation, nodule volume, and nodule dry weight of F_2 plants in controlled conditions with yield of derived F_4 lines in the field (for each cross) (adapted from Burias and Planchon, 1990).

Yield of F_4 lines	No. observations	N_2 fixation	Nodule volume	Nodule dry weight
		Traits of F_2 plants		
Weber x Maple Arrow	30	0.433**	0.580**	0.474**
Weber x Jiling 14	30	0.394*	0.488*	0.396*
Weber x Kingsoy	26	0.597**	0.459*	0.522*

*,**Significant at the 0.05 and 0.01 probability levels, respectively.

rhizobial strains, but their use may be minimized by barriers to inoculant establishment when large indigenous populations are present. The use of host genotypes tolerant to high NO_3^- conditions may encounter no such barriers and would be a feasible approach to improving fixation in the presence of inorganic N (Betts and Herridge, 1987).

Supernodulating induced mutants have been studied also for response to high NO_3^-. In soybean, common bean and pea, some mutants show enhanced ability to nodulate and fix more N_2 under high NO_3^- conditions than their normal counterparts. Genetic analyses of progeny from crosses between induced mutants and natural variants tolerant to high NO_3^- should provide further information about the genetic control of nodulation.

N_2 fixation is affected differentially by NO_3^- in different legume species (Harper and Gibson, 1984), and among different genotypes within species. Hardarson *et al.* (1984) and Gibson and Harper (1985) identified soybean genotypes that showed differential NO_3^- tolerance. Betts and Herridge (1987), using ureide content of the xylem sap as an index of fixation, found a substantial number of soybeans to be NO_3^- tolerant. More than 50% of the tolerant lines were from Korea, whereas only 5% of some 470 lines of other origin were NO_3^- tolerant.

There is enough natural variation to warrant selection for increased fixation under high NO_3^- levels, while the utility of induced mutants is yet to be fully determined. The criteria for tolerance to high levels of mineral N include ability to nodulate effectively, produce large amounts of nodule mass and maintain high nitrogenase activity per unit nodule weight. Since there is often a compensation effect (i.e., when more combined N is used, N_2 fixation is suppressed), plant response must be measured at two or more levels of N availability including growth at minimal N. The objective is to obtain plants capable of unrestricted N_2 fixation and growth in the absence of NO_3^- and near-maximum levels of N_2 fixation in the presence of high levels of NO_3^- (Betts and Herridge, 1987).

St. Clair *et al.* (1988) reported differences among common bean breeding lines for N yield, N_2 fixed and % Ndfa based on field trials in which N_2 fixation was estimated using ^{15}N-depleted and ^{15}N-enriched $(NH_4)_2SO_4$. While most lines fixed less N_2 (mg plant^{-1}) when higher levels of $(NH_4)_2SO_4$ were applied, the high-fixing parent, Puebla 152, and one resulting progeny line, 24-55, fixed similar amounts at both high (61.25 kg N ha^{-1}) and low (10 kg N ha $^{-1}$) application levels, suggesting that they may be tolerant to high NO_3^- (Table 3). Also in common bean, Park and Buttery (1989) reported a wide range in variation of response to high NO_3^-. One of the promising lines for high nitrate tolerance was a strain of Puebla 152.

Plant molecular genetics

Each partner in symbiosis contains genes that are expressed only in the symbiotic state, but there is little information at the molecular level about specific plant genes for increased fixation. Nodulins are proteins present in nodules but not roots, with the so-called "early nodulin" genes being involved in infection and nodule formation. Plant genes that encode compounds that induce *Rhizobium* nod genes may be important in breeding host plants with superior BNF potential. Among those compounds are the flavonoids which are widely distributed in plants and have been implicated in bacterial infection and nodule induction (Hirsch *et al.*, 1991).

Plant lectins occur in legumes and are postulated to have diverse functions including a role in symbiosis involving rhizobial attachment to root hairs and determination of host plant specificity (Lugtenberg *et al.*, 1991). Variability in nodulation does not correspond with genetic differences in seed lectin expression in legumes, but related lectin genes that have evolved different functions still could be involved in symbiosis.

Construction of a restriction fragment length polymorphism linkage map is underway with the intention of identifying and characterizing genes affecting fixation (Gresshoff *et al.*, 1991). Marker-assisted selection for host plant improvement should be possible when sufficient information has been accumulated.

Plant traits that affect BNF

In the legume-*Rhizobium* symbiosis, the plant roots provide sites for nodule establishment, the shoot provides oxidizable carbon substrates to the nodule as an energy source, and reduced N is exported from the nodules to the shoot for protein synthesis in the host (Atkins, 1984). The plant characteristics that influence fixation, include photosynthetic capacity, amount of H_2

evolution, transport and assimilation of fixed N, as well as nodulation properties such as early nodulation, nodule persistence and duration of active fixation (Vincent, 1980). The extent to which these traits can be altered genetically depends on the amount of variability and the changes in other important agronomic traits produced by selection.

Plant growth, photosynthesis and nitrogen transport

There is extensive information about photosynthesis in legumes, yet with a few exceptions, that knowledge has not contributed greatly to developing new, high-fixing cultivars. The interrelationship between photosynthesis and N_2 fixation is complex because more available N stimulates more photosynthesis which in turn can support more fixation. In addition to total photosynthetic capacity, the coordination of photosynthate production and N_2 fixation with peak demands of the maturing plant are important. In this chapter, no attempt has been made to review photosynthesis, but rather to cite instances where genotypic-based differences in plant traits affect N_2 fixation and populations selected for differences in fixation have been used to investigate processes associated with photosynthate distribution.

Studies of N nutrition of legumes often have been conducted assuming that growth is similar regardless of N source, despite observations that growth patterns and yield response are often very different. Studies of N_2 fixation conducted using pot experiments with plants growing in controlled environments may be necessary to provide control over certain variables, but extrapolation of results to field conditions should be done carefully. While some relationships between plant traits and N_2 fixation suggest possible limiting factors, cause and effect is difficult to establish because of confounding among factors. From a practical standpoint, trait modification is important only after they are incorporated into an adapted cultivar and if there is an increase in plant Ndfa without a dramatic change in other desirable traits (e.g., delayed maturity, undesirable plant type, reduced yield etc.).

Carbohydrate availability to the nodules limits N_2 fixation. Plant type, size and amount of leaf canopy which are influenced by such factors as N supply, flowering time, total leaf area, and leaf area duration often affect fixation. In peanut, Wynne et al. (1982) found that 70-75% of the variation in nodulation and fixation among eight cultivars could be attributed to differences in leaf area duration. Indeterminate, climbing plants of common bean usually fix more N_2 than determinate, bush plants (Graham, 1981). A greater proportion of the non-structural carbohydrates was transferred to the nodules in climbing plants than in plants with other growth habits. There are also genetic differences for ability to apportion and mobilize carbohydrates (Adams et al., 1978; Wolyn et al., 1991). Greater fixation often is associated with delayed flowering and seed maturation because developing pods may compete for photosynthate. The effects of delayed leaf senescence have been studied in soybean using lines that were either male sterile or which were male fertile but with a delayed leaf senescence (DLS) phenotype (Phillips *et al.*, 1983). Although results have varied, in neither case has there been a marked increase in N_2 fixed, and the yields of plants with DLS were less than or similar to normal counterparts.

Flowering is genetically controlled, but also affected by many environmental factors. Late flowering is of limited utility when it is associated with late maturity, since early maturing varieties are sought for many production areas to minimize pest attacks, avoid adverse weather and fit better into crop rotations.

There are differences among rhizobial strains for hydrogen evolution which affects the efficiency of N_2 fixation; the so-called *Hup*$^+$ strains having hydrogenase activity are more efficient than *Hup*$^-$ strains which lack activity. Host plant cultivar effects on H_2 evolution by nodulated pea were seen when nodules resulted from a *Hup*$^+$ but not a *Hup*$^-$ strain of *R. leguminosarum* (Bedmar *et al.*, 1983).

Nitrogen-fixing plants can be classified as amide or ureide exporters based on the composition of the xylem fluid. The absolute amounts and relative proportion of these primary and secondary nitrogenous solutes vary with developmental and environmental factors (Schubert, 1986). The N exported from the nodules is utilized in metabolism and for production of seed storage proteins.

The abundance of ureides in xylem sap and in plant tissues has been used as an indicator and a quantitative assay of N_2 fixation in soybean (McClure *et al.*, 1980; Herridge, 1982). It has been suggested that rate of translocation of N compounds is directly related to plant performance of common bean (Thomas and Sprent, 1984) and soybean (Thomas *et al.*, 1983). There is little evidence for genetic variability for differences in concentration of xylem sap or translocation rate and the extent to which this is a factor limiting N_2 fixation. Differences in rate and amount of ureide transport were related to presumed levels of N_2 fixation in common bean (Thomas *et al.*, 1984).

Increasing N_2 fixation by selecting plants with greater nodule mass has raised questions about photosynthate partitioning to competing organs. In both forage and grain legumes the carbon economy of the nodules is closely related to the CO_2 assimilation of the leaf canopy, and in grain legumes allocation to reproductive sinks. Cralle *et al.* (1987) measured photosynthate partitioning in alfalfa populations selected for increased N_2 fixation. In plants that differed by 40% in nodule mass, they found no difference in photosynthate partitioning and concluded, "...that selection for alfalfa for N_2 fixation capability by selection for nodule mass may maintain or increase yields without changes in carbon allocation". Among faba bean genotypes that varied for both yield and Ndfa, higher yielding genotypes also fixed more N_2 (Duc *et al.*, 1988). They state that, "presumably dinitrogen fixation regulated the yield, or some common factor such as limit in carbon source regulated both yield and fixation". Clearly high N_2 fixation *per se* does not necessarily limit plant growth and seed yield even though there are high energy requirements for fixation.

Root characteristics and nodulation pattern

The genetic variability in size and form of root systems is difficult to characterize precisely because the environment around the plants and the medium in which they are grown may alter the form and function. Thus, studies of plants grown in liquid culture and in soil-containing pots provide only approximations of what may be occurring under less constrained field conditions. The difficulties in directly studying roots greatly limit our ability to understand which characteristics are important for enhanced BNF and the extent to which they are heritable.

For forage and grain legumes, considerable fixation occurs in nodules on roots in the upper soil profile. Because number of nodulation sites is important, extensive root branching and lateral root development, rather than a single deep tap root or a few primary roots probably favors N_2 fixation. In alfalfa, Viands *et al.* (1981) found positive correlations between fibrous root mass and other indirect measures of fixation potential. In different subpopulations there was a positive selection response for fibrous root mass, and realized heritability for high fibrous root score was 0.75.

Nodulation of seedlings occurs along the primary root; and, especially when inoculant is applied to the seed or into the row at sowing, much of the nodulation is in the crown area. This conspicuous concentration of nodules and the difficulty in excavating entire root systems of field-grown plants has led to a presumed importance of nodules in the central root zone. Most estimates of fixation using AR have been on root samples from the central root core. Those sampling strategies have contributed to the mistaken belief that N_2 fixation in grain legumes always declines rapidly at the onset of seed fill and that nodulation of lateral roots was inconsequential.

In some grain legumes, considerable fixation occurs after the onset of pod fill. Harper (1974) reported that fixation in soybean peaked during pod fill some 3 weeks after full bloom. In common bean, there was genotypic variability among breeding lines for proportion of total N accumulation during pod fill. Wolyn *et al.* (1989) found that nodules on lateral roots were active during pod fill and contributed substantial N to the large demands of the maturing seeds. In soybean, nodulation pattern differed with location of inoculant, and nodules formed on the lower part of the root system fixed more N_2 than the crown nodules (Hardarson *et al.*, 1989). There too, the later formed nodules contributed substantially to total Ndfa. The difficulties in studying roots in undisturbed field settings will continue to confound efforts to breed for better rooting habits and to incorporate selection for root traits into a program to increase N_2 fixation.

CONCURRENT IMPROVEMENT OF PLANTS AND MICROSYMBIONTS

The closely coordinated functions of the symbiotic participants suggest that changes in either partner may affect the performance of the other. Even with this knowledge at hand, most attempts to improve either the microbe or host plant have been done separately. Too often the work has not produced greater N_2 fixation, which in retrospect is not surprising. Grain legume production is often under suboptimal conditions where there is stress on both symbionts. Moisture stress, usually insufficient but sometimes excessive, is common. Salinity is becoming more prevalent especially where irrigation is poorly managed. Wide ranges in soil pH with direct and indirect affects on plant growth and microbial populations are often present in legume cropping areas. In most cases there is genetic variability in both symbionts for response to these conditions. Maximum benefit from BNF will result when both participants show optimal expression under the conditions in which production occurs.

REFERENCES

Adams, M. W.; Wiersma, J. V.; Salazar, J. (1978) Differences in starch accumulation among dry bean cultivars. *Crop Science* 18:155-157.

Arrendell, S.; Wynne, J. C.; Elkan, G. H.; Isleib, T. G. (1985) Variation for nitrogen fixation among progenies of a Virginia x Spanish peanut cross. *Crop Science* 25:865-869.

Arrendell, S.; Wynne, J. C.; Elkan, G. H.; Schneeweis, T. J. (1986) Bidirectional selection for nitrogenase activity and shoot dry weight among late generation progenies of a virginia x spanish peanut cross. *Peanut Science* 13:86-89.

Arrendell, S.; Wynne, J. C.; Rawlings, J. O. (1989) Genetic variability and selection for acetylene reduction in peanut. *Crop Science* 29:1387-1392.

Atkins, C. A. (1984) Efficiencies and inefficiencies in the legume/*Rhizobium* symbiosis - A review. *Plant and Soil* 82:273-284.

Barnes, D. K.; Heichel, G. H.; Vance, C. P.; Ellis, W. R. (1984) A multiple trait breeding program for improving the symbiosis for N_2 fixation between *Medicago sativa* L. and *Rhizobium meliloti*. *Plant and Soil* 82:303-314.

Bedmar, E. J.; Edie, S. A.; Phillips, D. A. (1983) Host plant cultivar effects on hydrogen evolution by *Rhizobium leguminosarum*. *Plant Physiology* 72:1011-1015.

Betts, J. H.; Herridge, D. F. (1987) Isolation of soybean lines capable of nodulation and nitrogen fixation under high levels of nitrate supply. *Crop Science* 27:1156-1161.

Bliss, F. A.; Pereira, P. A. A.; Araujo, R. S.; Henson, R. A.; Kmiecik, K. A.; McFerson, J. R., Teixera, M. G.; da Silva, C. C. (1989) Registration of five high nitrogen fixing common bean germplasm lines. *Crop Science* 29:240-241.

Boddey, R. M.; Urquiaga, S.; Neves, M. C. P. (1990) Quantification of the contribution of N_2 fixation to field-grown legumes - A strategy for the practical application of the ^{15}N isotope dilution technique. *Soil Biology Biochemistry* 22:649-655.

Brunner, H.; Zapata, F. (1984) Quantitative assessment of symbiotic nitrogen fixation in diverse mutant lines of field bean (*Vicia faba minor*). *Plant and Soil* 82:407-413.

Burias, N.; Planchon, C. (1990) Increasing soybean productivity through selection for nitrogen fixation. *Agronomy Journal* 82:1031-1034.

Buttery, B. R.; Park, S. J.; Dhanvantari, B. N. (1990) Effects of combined nitrogen, *Rhizobium* strain and substrate on a supernodulating mutant of *Phaseolus vulgaris* L. *Canadian Journal of Plant Science* 70:955-963.

Cralle, H. T.; Heichel, G. H.; Barnes, D. K. (1987) Photosynthate partitioning in plants of alfalfa population selected for high and low nodule mass. *Crop Science* 27:96-100.

Cregan, P. B.; van Berkum, P. (1984) Genetics of nitrogen metabolism and physiological/biochemical selection for increased grain crop productivity. *Theoretical and Applied Genetics* 67:97-111.

Dart, P. J. (1986) Nitrogen fixation associated with non-legumes in agriculture. *Plant and Soil* 90:303-334.

Devine, T. E. (1984) Genetics and breeding of nitrogen fixation. In: Alexander, M. (ed.), *Biological nitrogen fixation*. Plenum, New York, pp. 127-154.

Devine, T. E. (1987) A comparison of rhizobial strain compatibilities of *Glycine max* and its progenitor species *Glycine soja*. *Crop Science* 27:635-639.

Devine, T. E.; Breithaupt, B. H. (1980) Significance of incompatibility reactions of *Rhizobium japonicum* strains with soybean host genotypes. *Crop Science* 20:279-281.

Devine, T. E.; Breithaupt, B. H. (1981) Frequencies of nodulation response alleles Rj2 and Rj4 in soybean plant introduction and breeding lines. USDA Technical Bulletin 1628, U.S. Government Printing Office, Washington, DC.

Devine, T. E.; O'Neill, J. J. (1989) Genetic allelism of nodulation response genes Rj1, Rj2, and Rj4 in soybean. *Crop Science* 29:1347-1350.

Duc, G.; Mariotti, A.; Amarger, N. (1988) Measurements of genetic variability for symbiotic dinitrogen fixation in field-grown faba bean (*Vicia faba* L.) using a low level ^{15}N-tracer technique. *Plant and Soil* 106:269-276.

Duc, G.; Trouvelot, A.; Gianinazzi-Pearson, V.; Gianinazzi, S. (1989) First report of non-mycorrhizal plant mutants (*myc⁻*) obtained in pea (*Pisum sativum* L.) and fababean (*Vicia faba* L.). *Plant Science* 60:215-222.

Elkan, G. H.; Wynne, J. C.; Schneeweis, T. J.; Isleib, T. G. (1980) Nodulation and nitrogenase activity of peanuts inoculated with single strain isolates of *Rhizobium*. *Peanut Science* 7:95-97.

Gianinazzi-Pearson, V. S.; Gianinazzi, J. P.; Guillemin, J. P.; Trouvelot, A.; Duc, G. (1991) Genetic and cellular analysis of resistance to vesicular arbuscular (VA) mycorrhizal fungi in pea mutants. In: Hennecke, H.; Verma, D. P. S. (eds.), *Advances in molecular genetics of plant-microbe interactions*. Kluwer Academic Publishers, Dordrecht, The Netherlands, pp. 336-342.

Gibson, A. H.; Harper, J. E. (1985) Nitrate effect on nodulation of soybean by *Bradyrhizobium japonicum*. *Crop Science* 25:497-501.

Graham, P. H. (1981) Some problems of nodulation and symbiotic nitrogen fixation in *Phaseolus vulgaris* L.: A review. *Field Crops Research* 4:93-112.

Graham, P. H.; Rosas, J. C. (1977) Growth and development of indeterminant bush and climbing cultivars of *Phaseolus vulgaris* L. inoculated with *Rhizobium*. *Journal of Agricultural Science (Cambridge)* 88:503-508.

Graham, P. H.; Temple, S. R. (1984) Selection for improved nitrogen fixation in *Glycine max* (L.) Merr. and *Phaseolus vulgaris* L. *Plant and Soil* 82:315-327.

Graham, R. A. (1981) Demonstration of constant ranking for *Rhizobium* pathogenicity and host susceptibility. *Tropical Agriculture (Trinidad)* 58:319-323.

Greder, R. R.; Orf, J. H.; Lambert, J. W. (1986) Heritabilities and associations of nodule mass and recovery of *Bradyrhizobium japonicum* serogroup USDA 110 in soybean. *Crop Science* 26:33-37.

Gresshoff, P. M.; Landau-Ellis, D.; Funke, R.; Sayavedra-Soto, L.; Caetano-Anolles, G. (1991) Plant genetic control of nodulation in legumes. In: Hennecke, H.; Verma, D. P. S. (eds.), *Advances in molecular genetics of plant-microbe interactions*. Kluwer Academic Publishers, Dordrecht, The Netherlands, pp. 331-335.

Handelsman, J.; Parke, J. L. (1989) Mechanisms of biocontrol of soilborne plant pathogens. In: Kosuge, T.; Nester, E. W. (eds.), *Plant-microbe interactions*. Vol. 3. McGraw-Hill, New York, pp. 27-61.

Hansen, A. P., Peoples, M. B.; Greshoff, P. M.; Atkins, C. A.; Pate, J. S.; Carroll, B. J. (1989) Symbiotic performance of supernodulating soybean [*Glycine max* (L.) Merr.] mutants during development on different nitrogen regimes. *Journal of Experimental Botany* 40:715-724.

Hardarson, G.; Golbs, M.; Danso, S. K. A. (1989) Nitrogen fixation in soybean (*Glycine max* L. Merrill) as affected by nodulation patterns. *Soil Biological Biochemistry* 21:783-787.

Hardarson, G.; Jones, D. G. (1979) The inheritance of preference for strains of *Rhizobium trifolii* by white clover (*Trifolium repens*). *Annals of Applied Biology* 92:329-333.

Hardarson, G.; Zapata, F.; Danso, S. K. A. (1984) Effect of plant genotypes and nitrogen fertilizer on symbiotic nitrogen fixation by soybean cultivars. *Plant and Soil* 82:397-405.

Harper, J. E. (1974) Soil and symbiotic nitrogen requirements for optimum soybean production. *Crop Science* 14:255-260.

Harper, J. E.; Gibson, A. H. (1984) Differential nodulation tolerance to nitrate among legume species. *Crop Science* 24:797-801.

Henson, R. A.; Heichel, G. H. (1984) Denitrogen fixation of soybean and alfalfa: Comparison of the isotope dilution and difference methods. *Field Crops Research* 9:333-346.

Herridge, D. F. (1982) Relative abundance of ureides and nitrate in plant tissues of soybean as a quantitative assay of nitrogen fixation. *Plant Physiology* 70:1-6.

Hirsch, A. M.; Bochenek, B.; Lobler, M.; McKhann, H. I.; Reddy, A.; Li, H.-H.; Ong, M.; Wong, J. (1991) Patterns of nodule development and nodulin gene expression in alfalfa and Afghanistan pea. In: Hennecke, H.; Verma, D. P. S. (eds.), *Advances in molecular genetics of plant-microbe interactions*. Kluwer Academic Publisher, Dordrecht, The Netherlands, pp. 317-324.

Hobbs, S. L. A.; Mahon, J. D. (1982) Heritability of N_2 (C_2H_2) fixation rates and related characters in peas (*Pisum sativum* L.). *Canadian Journal of Plant Science* 62:265-276.

Holl, F. B.; LaRue, T. A. (1976) Genetics of legume plant hosts. In: Newton, W. E.; Nyman, C. J. (eds.), *Proceedings of the 1st international symposium on nitrogen fixation*. Washington State University Press, Pullman, pp. 391-399.

Jaiswal, H. K.; Singh, R. K. (1990) Breeding for increased nitrogen fixing ability among wild and cultivated species of chickpea. *Annals of Applied Biology* 117:415-419.

Jones, D. G.; Hardarson, G. (1979) Variation within and between white clover varieties in their preference for strains of *Rhizobium trifolii*. *Annals of Applied Biology* 92:221-228.

Kvien, C. S.; Ham, G. E.; Lambert, J. W. (1981) Recovery of introduced *Rhizobium japonicum* strains by soybean genotypes. *Agronomy Journal* 73:900-905.

Lie, T. A. (1984) Host genes in *Pisum sativum* L. conferring resistance to European *Rhizobium leguminosarum* strains. *Plant and Soil* 82:415-425.

Lugtenberg, B. J. J.; Diaz, C.; Smit, G.; DePater, S.; Wijne, J. W. (1991) Roles of lectin in the *Rhizobium* symbiosis. In: Hennecke, H.; Verma, D. P. S. (eds.), *Advances in molecular genetics of plant-microbe interactions*. Kluwer Academic Publishers, Dordrecht, The Netherlands, pp. 174-181.

McClure, P. R.; Israel, D. W.; Volk, R. J. (1980) Evaluation of the relative ureide content of xylem sap as an indicator of N_2 fixation. *Plant Physiology* 66:720-725.

McFerson, J. R. (1983) Genetic and breeding studies of dinitrogen fixation in common bean, *Phaseolus vulgaris* L. Ph.D. Thesis, University of Wisconsin, Madison (Dissertation Abstracts International DA 8313188, Ann Arbor, MI).

Miranda, B. D.; Bliss, F. A. (1991) Selection for increased nitrogen accumulation in common bean: Implications for improving dinitrogen fixation and seed yield. *Plant Breeding* 106:301-311.

Miranda, C. H. B.; Urquiaga, S.; Boddey, R. M. (1990) Selection of ecotypes of *Panicum maximum* for associated biological nitrogen fixation using the ^{15}N-isotope dilution technique. *Soil Biology Biochemistry* 22:657-663.

Mundel, H. H.; Schaalji, G. B. (1988) Use of near infrared reflectance spectroscopy to screen soybean lines for plant nitrogen. *Crop Science* 28:157-162.

Mytton, L. R. (1983) Host plant selection and breeding for improved symbiotic efficiency. In: Jones, D. E.; Davies, E. R. (eds.), *Temperate legumes: Physiology, genetics and nodulaton*. Putnam Adv. Publishing Program, London, United Kingdom, pp. 373-393.

Neuhausen, S. L.; Graham, P. H.; Orf, J. H. (1988) Genetic variation for dinitrogen fixation in soybean of maturity group 00 and 0. *Crop Science* 28:769-772.

Nutman, P. S. (1984) Improving nitrogen fixation in legumes by plant breeding; the relevance of host selection experiments in red clover (*Trifolium pratense* L.) and subterranean clover (*T. subterraneum* L.). *Plant and Soil* 82:285-301.

Pacovsky, R. S.; Paul, E. A.; Bethlenfalvy, G. J. (1986) Response of mycorrhizal and P fertilized soybeans to nodulation by *Bradyrhizobium* or ammonium nitrate. *Crop Science* 26:145-150.

Park, S. J.; Buttery, B. R. (1988) Nodulation mutants of white bean (*Phaseolus vulgaris* L.) induced by ethyl-methane sulphonate. *Canadian Journal of Plant Science* 68:199-202.

Park, S. J.; Buttery, B. R. (1989) Identification and characterization of common bean (*Phaseolus vulgaris* L.) lines well-nodulated in the presence of high nitrate. *Plant and Soil* 119:237-244.

Pereira, P. A. A.; Burris, R. H.; Bliss, F. A. (1989) ^{15}N-determined dinitrogen fixation potential of genetically diverse bean lines (*Phaseolus vulgaris* L.). *Plant and Soil* 120:171-179.

Phillips, D. A.; Pierce, R. O.; Edie, S. A.; Foster, K. W.; Knowles, P. F. (1983) Delayed leaf senescence in soybean. *Crop Science* 24:518-522.

Rennie, R. J.; Kemp, G. A. (1983) N_2 fixation in field beans (*Phaseolus vulgaris* L.) quantified by ^{15}N isotope dilution. II. Effect of cultivars of beans. *Agronomy Journal* 75:645-649.

Ronis, D. H.; Samons, D. J.; Kenworthy, W. J.; Meisinger, J. J. (1985) Heritability of total and fixed N content of the seed in two soybean populations. *Crop Science* 25:1-5.

Rosas, J. C.; Bliss, F. A. (1986) Host plant traits associated with estimates of nodulation and nitrogen fixation in common bean. *HortScience* 21:287-289.

Sarig, S.; Kapulnik, Y.; Okon, Y. (1986) Effect of *Azospirillum* inoculation on nitrogen fixation and growth in several winter legumes. *Plant and Soil* 90:335-342.

Schubert, K. R. (1986) Products of nitrogen fixation in higher plants: Synthesis, transport and metabolism. *Annual Review of Plant Physiology* 37:539-574.

Seetin, M. W.; Barnes, D. K. (1977) Variation among alfalfa genotypes for rate of acetylene reduction. *Crop Science* 17:783-787.

Sougoufara, B.; Danso, S. K. A.; Diem, H. G.; Dommergues, Y. R. (1990) Estimating N_2 fixation and N derived from soil by *Casuarina equisetifolia* using labelled ^{15}N fertilizer: Some problems and solutions. *Soil Biological Biochemistry* 22:695-701.

St. Clair, D. A.; Bliss, F. A. (1991) Intra population recombination for ^{15}N-determined dinitrogen fixation ability in common bean. *Plant Breeding* 106:215-225.

St. Clair, D. A.; Wolyn, D. J.; DuBois, J.; Burris, R. H.; Bliss, F. A. (1988) Field comparison of dinitrogen fixation determined with nitrogen-15-depleted and nitrogen-15-enriched ammonium sulfate in selected inbred backcross lines of common bean. *Crop Science* 28:773-778.

Talbott, H. J.; Kenworthy, W. J.; Legg, J. O.; Douglass, L. W. (1985) Soil-nitrogen accumulation in nodulated and non-nodulated soybeans: A verification of the difference method by a ^{15}N technique. *Field Crops Research* 11:55-67.

Tauer, L. W. (1989) Economic impact of future biological nitrogen fixation technologies on United States agriculture. *Plant and Soil* 119:261-270.

Thomas, R. J.; Jokinen, K.; Schrader, L. E. (1983) Effect of *Rhizobium japonicum* mutants with enhanced N_2 fixation activity of N transport and photosynthesis of soybeans during vegetative growth. *Crop Science* 23:453-456.

Thomas, R. J.; Sprent, J. I. (1984) The effects of temperature on vegetative and early productive growth of a cold tolerant and cold sensitive line of *Phaseolus vulgaris*. II. Nodular uricase, allantoinase, xylem transport of N and assimilation in shoot tissues. *Annals of Botany (London)* 53:589-597.

Thomas, R. J.; McFerson, J. R.; Schrader, L. E.; Bliss, F. A. (1984) Composition of bleeding sap nitrogen from lines of field-grown (*Phaseolus vulgaris* L.). *Plant and Soil* 79:77-88.

Urquiaga, S.; Caballero, S. S.; Botteon, P. de T. L.; Boddey, R. M. (1988) Selection of sugar cane cultivars for associated biological nitrogen fixation using ^{15}N-labelled soil. In: Skinner, F. A.; Boddey, R. M.; Fendrick, I. (eds.), *Nitrogen fixation with non-legumes*. Kluwer Academic Publ., Dordrecht, pp. 311-319.

van Berkum, P.; Bohlool, B. B. (1980) Evaluation of nitrogen fixation by bacteria in association with roots of tropical grasses. *Microbiological Review* 44:491-517.

Viands, D. R.; Barnes, D. K.; Heichel, G. H. (1981) Nitrogen fixation in alfalfa-responses to bidirectional selection for associated characteristics. USDA Technical Bulletin 1643, U.S. Government Printing Office, Washington, DC.

Vincent, J. M. (1980) Factors controlling the legume-*Rhizobium* symbiosis. In: Newton, W. E.; Orme-Johnson, W. H. (eds.), *Nitrogen fixation*. Vol. 2. University Park Press, Baltimore, MD, pp. 103-129.

Westermann, D. T.; Kleinkopf, G. E.; Perter, L. K.; Leggett, G. E. (1981) Nitrogen sources for bean seed production. *Agronomy Journal* 73:660-664.

Wolyn, D. J.; Attewell, J.; Ludden, P. W.; Bliss, F. A. (1989) Indirect measures of N_2 fixation in common bean (*Phaseolus vulgaris* L.) under field conditions: The role of lateral root nodules. *Plant and Soil* 113:181-187.

Wolyn, D. J.; St. Clair, D. A.; DuBois, J.; Rosas, J. C.; Burris, R. H.; Bliss, F. A. (1991) Distribution of nitrogen in common bean (*Phaseolus vulgaris* L.) genotypes selected for differences in nitrogen fixation ability. *Plant and Soil* (in press).

Wynne, J. C.; Ball, S. T.; Elkan, G. H.; Isleib, T. G.; Schneeweis, T. J. (1982) Host and host plant factors affecting nitrogen fixation in peanut. In: Graham, P. H.; Harris, S. C. (eds.), *Biological nitrogen fixation technology for tropical agriculture*. Centro Internacional de Agricultura Tropical, Cali, Colombia, pp. 67-75.

Zary, K. W.; Miller, J. C., Jr.; Weaver, R. W.; Barnes, L. W. (1978) Intraspecific variability for nitrogen fixation in southern pea [*Vigna unguiculata* (L.) Walp.]. *Journal of the American Society of Horticultural Science* 103:806-808.

Discussion

T. E. Carter, Jr., Moderator

You described the drought resistance of *L. pennelii* as a species that has very hairy leaves. Could the hairy leaves be the mechanism of drought resistance?

I don't know, but this does seem a plausible explanation. Hairy leaves increase the boundary layer resistance to air movement at the leaf surface, thus restricting water loss.   [J. S. Boyer]

In tomato, you cited that genetic differences in water use efficiency were large under conditions of plentiful moisture but small under moisture stress. If genetic differences in water use efficiency are expressed only when moisture is not limiting, how is this useful in development of a drought-tolerant tomato?

This is a good point. In this example of water use efficiency, the application would be to production systems where most of the crop's water is supplied through irrigation. Genotypes with enhanced water use efficiency would require less irrigation and thus be cheaper to produce.   [J. S. Boyer]

Is osmotic adjustment more important as (1) a survival mechanism so that plants may capitalize upon rain when it finally comes, or as (2) a mechanism to maintain growth and productivity during a stress period?

Osmotic adjustment helps the plant in both situations but may be more important in maintaining growth during the stress. To examine the role of osmotic adjustment properly, we need to develop genotypes (within a species) that differ only for this trait. Such material has been available in only one case, that of Morgan cited in my chapter.   [J. S. Boyer]

Are there other plant strategies for tolerating soil stress in addition to deep rooting?

Yes. Deep rooting seems to play a large role for many success stories in drought resistance breeding. Other mechanisms may be important as well, such as (1) maintenance of viable leaf tissue also known as the stay-green trait, (2) persistent reproductive development during drought, and (3) high osmoregulatory capacity to maintain growth. Osmoregulation may be good because it doesn't cost much, energetically, in comparison with production of a large root system.

[J. S. Boyer]

CIMMYT physiologists think that there may be little variation in water use efficiency in tropical maize under conditions of severe water deficit. The only observed difference between tolerant and susceptible genotypes has been in the number of days from anthesis to silking. Tolerant genotypes have a short anthesis-silk interval because of early ear/shoot initiation, making pollination less affected by drought. Would you comment on the relative importance of anthesis/silking events in comparison to water use efficiency?

Pollination, a major event in corn reproduction, is especially sensitive to drought. When drought occurs at flowering, coordination of anthesis and silking events is probably a more important factor than water use efficiency. But one must also pay attention to later stages of seed development. I speculate that drought resistance may be found to be a significant factor during the seed filling period in tropical maize even though it may not be so critical at flowering.

[J. S. Boyer]

Is it possible to select for "general stress" tolerance that would simultaneously improve tolerance to drought, aluminum, salt, adverse temperature, and low soil nutrients? In other words, are there mechanisms in common?

Perhaps. A mechanism of tolerance described by several speakers today is osmotic adjustment. Osmotic adjustment has been shown to be involved in drought, heat, cold, and salt tolerance. Also, vigorous rooting probably affords some protection to both chemical barriers in the soil and to drought.

[J. S. Boyer]

Are there other plant strategies for tolerating soil stress in addition to deep rooting?

Deep rooting is an obvious strategy, but mechanisms vary considerably. Probably there are several traits a breeder may consider as selection criterion in a breeding program. Direct selection for high performance under stress is perhaps the best way to initiate a breeding program. Specific selection for tolerance mechanisms could follow later.

[J. Dvořák]

You mentioned that halophytes accumulate high levels of salt to overcome the osmotic stress from salt build-ups in the soil. In contrast, the salt-tolerant non-halophytes produce organic osmotica as a protection against salt injury. Which mechanism is less costly for a plant and, hence, more efficient for a plant breeder to incorporate into agronomic crops?

The path taken by the halophytes seems less expensive metabolically, but is a constitutive trait. In contrast, osmotic adjustment by non-halophytes is usually inducible and, thus, has a cost only under stress conditions.

[J. Dvořák]

Could salt includers (i.e., halophytes) be used to desalinate soils with high salt levels?

This is an important question which we are working on in California. The major problem is not salt in the soil - we can leach that out - but what do you do with the drainage water? Salt includers may help in mopping up this problem. [J. Dvořák]

Have you looked for the presence of ice-nucleating bacteria in your studies? What role do they play in cold tolerance?

No. Ice-nucleating bacteria are only one of several factors involved in ice formation - dust plays a major role, for example. However, some plants isolate themselves from the bacteria by means of a waxy leaf surface.

[J. P. Palta]

What is the role of the plant hormone ABA in freezing stress tolerance? What is the role of stress proteins?

ABA seems to be involved in acclimation to temperature stress. It has been used successfully to acclimate callus tissue to freezing stress, but there has been no success at the whole plant level as yet. Stress proteins are induced by freezing stress. Some are like the heat-shock or drought-stress proteins, but others are specific to cold shock. The roles of these proteins are not known at present. [J. P. Palta]

Will the current emphasis on sustainable agriculture and reduction of nitrate contamination of ground water provide new impetus for genetic improvement of biological nitrogen fixation?

I believe so, especially for sandy soils where leaching effects may be great. In snap beans, we may be able to reduce starter nitrogen fertilization by one-half and obtain the rest of the nitrogen requirements for the crop from nitrogen fixation. [F. A. Bliss]

Has breeder selection for high yield in well-fertilized plots inadvertently selected against positive responses to soil beneficials?

Definite data have not been reported on this, but some work with beans would suggest the answer is yes. When high yielding lines are selected under good fertility conditions, nodulation and nitrogen fixation are generally, but not always, low.

[F. A. Bliss]

PART THREE

MODIFICATION OF PLANTS TO TOLERATE STRESSES DUE TO DISEASE AND INSECTS

Chapter 12
Selecting components of partial resistance

J. E. Parlevliet

Plant Breeding Department, Agricultural University, Wageningen, The Netherlands

INTRODUCTION

Plants are exposed to a wide range of potential parasites. In order to restrict the damage from such parasites, plants employ a wide range of defense mechanisms. These can be classified as avoidance, resistance and tolerance mechanisms (Parlevliet, 1981b). With avoidance the parasitic contact is reduced (thorns, unpleasant smells or tastes). Resistance and tolerance operate after parasitic contact has been established. With resistance the growth and development of the parasite is reduced. It is mostly of a chemical nature. With tolerance there is no interference with the growth and development of the parasite, but the damage resulting from the parasite's activities is reduced.

The defense mechanisms used by breeders against pathogens nearly always appear to belong to the category of resistance. And it is within resistance that the problems of race-specificity and loss of effectiveness occur. Because avoidance does not seem to be employed frequently by plants against pathogens (mostly it is used against animal parasites and herbivores), while tolerance to pathogens is very difficult to assess and may not occur frequently either (Parlevliet, 1981a,c), breeders have to use resistance mechanisms to protect crops from pathogens.

Resistance in crops to most pathogens varies greatly in intensity; from almost imperceptible (only a slight reduction in the growth of the pathogen) to complete (no growth of the pathogen). If host genotypes show a continuous range of variation in resistance, from no resistance (extremely susceptible) to good resistance (low susceptibility) one could speak of quantitative or partial resistance.

Loss of resistance due to adaptation of the pathogen population is a very frequent experience. Resistance neutralized by new races of the pathogen is invariably simply inherited and often of a very high level. Partial resistance (when based on few to several genes) has never been shown to erode or break down. This is the reason why partial resistance is often assumed to be considerably more durable than the monogenic race-specific resistance so often used by plant breeders because of its ease of manipulation.

PARTIAL RESISTANCE

Assessment of partial resistance

As mentioned above, partial resistance (PR) is here equated to quantitative resistance, characterized by a continuous variation ranging from hardly any resistance to fair levels. Because the resistance is meant to be used in the field under certain commercial conditions, PR is defined in this chapter as the quantitative resistance, which reduces the pathogen growth in the field under the normal growing conditions. In order to select for PR, one must be able to measure PR. An accurate and proper assessment of PR is often not easy to obtain. A very common reason is the fact that in a large number of pathosystems the PR has to be evaluated indirectly. The pathogen itself and its reduction in growth due to the PR are not visible. The assessment is done on the basis of the effects of the pathogen in the host.

With a number of pathogens it is possible to assess the amount of pathogen directly (powdery mildews) or nearly so (rusts) by scoring the percentage tissue covered with the pathogen. For a large number of pathogens - most hemi-biotrophic and necrotrophic fungi, for instance - the pathogen itself is not visible; but the symptoms of its presence are more or less easily discernible and restricted to the parts of the host tissue invaded. Discoloration of the invaded and immediately adjacent tissue is the most general indication of the pathogen. The area of the tissue covered with the pathogen or showing the symptoms of its presence through discoloration tends to be a fairly good to good measure of resistance.

In a number of cases there are true disease symptoms which are characteristic of the pathogen and more or less systemically expressed by the whole plant - for example, wilting caused by vascular pathogens or leaf rolling, mottling, stunting, etc. caused by viruses. These symptoms tend to be rather unreliable for assessing resistance because the relationship between the amount of pathogen present and the severity of symptoms is often rather poor (Parlevliet, 1989).

Another problem is that PR cannot be assessed in absolute terms. It has to be assessed relative to the performance of well-known standards. Differences in the level of pathogen or disease at a given assessment date are then taken as representing differences in PR (Table 1). But other factors may affect these differences interfering with the proper evaluation. Interplot interference, earliness and height of the plants, inoculum pressure and moment of assessment may be the cause of a wrong assessment of the PR present.

Assessment in small plots may suggest much smaller differences in PR than really available as in the case of barley (*Hordeum vulgare* L.) for the pathogen *Puccinia hordei* Otth. (Table 1), or even give a wrong impression, as is the case with barley for *Erysiphe graminis* DC. ex Mèrat f. sp. *hordei*

Table 1 Percentage leaf area covered with barley leaf rust urediosori of four barley cultivars in 2 years in plots isolated from one another (no interplot interference) and in adjacent, single row plots (after Parlevliet and van Ommeren, 1975).

Cultivar	Isolated plots		Single rows	
	1973	*1974*	*1973*	*1974*
L94	29.0	8.2	17.0	25.0
Sultan	8.2	0.42	8.2	14.9
Zephyr	5.5	0.39	6.0	10.5
Rika	1.3	0.07	4.6	6.0
Julia	0.16	0.02	2.5	4.6
Vada	0.01	0.01	1.0	1.4

when evaluation occurs in single row plots (Table 2) (Nørgaard Knudsen *et al.*, 1986).

Earliness and tallness interfere greatly in the assessment of PR of wheat (*Triticum aestivum* L.) to septoria glume (*Leptosphaeria nodorum* E. Müller) and leaf blotch [(*Mycosphaerella graminicola* (Fückel) Schröt.] (Tavella, 1978; Rosielle and Brown, 1980; Shaner *et al.*, 1975). Inoculum pressure is also a factor to take into account. Often very high inoculum densities are pursued to prevent escapes. This, at the same time, tends to reduce or even prevent the expression of small differences in PR. The

Table 2 Partial resistance of some spring barley cultivars to powdery mildew, expressed as the area under the disease progress curve relative to that of Golden Promise (= 100) in isolated plots, adjacent plots of 1.4 m² and adjacent single row plots (after Nørgaard Knudsen *et al.*, 1986).

Cultivar	Plot situation and type		
	Isolated 11 m²	*Adjacent* 1.4 m²	*Adjacent single row*
Golden Promise	100	100	100
Pallas	57	67	106
Proctor	46	49	85
CI 9670	36	34	40
Hannchen	28	36	82
Gloire du Velay	24	26	71
Nottingham Longear	19	22	58

moment of assessment should be chosen to enable a maximum expression of differences in PR. Too early or too late an assessment can reduce the visible differences considerably.

Components of partial resistance

Partial resistance is expressed as a reduced amount of tissue of the host invaded or affected or as a reduced concentration of the pathogen (viruses) in the host tissues compared with that of a highly susceptible standard. A reduction in the tissue area invaded/affected can be the result of fewer and/or smaller lesions. A reduced concentration of the pathogen may lead to less intensive symptoms.

The PR as measured in the field is the result of a series of events. The propagules of the pathogen have to reach the proper host tissue and may infect this tissue under certain conditions. The chances of successful infections are determined by these events. Differences in host genotypes may affect the chance of infection resulting in differences in **infection frequency** or **density**. Once inside, the host colonization of the host tissue occurs. Host genotypes may provide different environments leading to different rates of colonization and so to differences in **lesion size** or in **concentration** of the pathogen. After some time the affected host tissue starts to produce pathogen propagules for further spread. A high **propagule production** per unit of host tissue affected over a period of time is essential for most pathogens in order to reach sufficient new host tissue in time. These three steps or phases could be seen as the three components, together determining the disease build-up in the field.

This is a highly generalized description, too general to apply in each pathosystem. Depending on the pathosystem and on the way of assessment, different sets of components can be defined.

In the case of non- or partly systemic leaf pathogens such as barley leaf rust (*P. hordei*) and wheat leaf rust (*P. recondita* Rob. ex Desm. f. sp. *tritici*), infection frequency, spore production per urediosorus per day, and the infectious period in days together determine the total amount of spores produced and are seemingly a good measure of PR. Johnson and Taylor (1976) indeed concluded that the total spore production would provide an accurate method of measuring the pathogenicity of the pathogen and the resistance of the host, the resistance being the sum of the effects of all components. This, however, is not the case for two reasons. The spore production, except at very low pathogen densities, is determined by the host tissue area itself and not by the tissue area invaded because of the strong interference between spore production per lesion and the lesion density (Mehta and Zadoks, 1970; Teng and Close, 1978). The second reason is that spores produced very early after the incubation period are of much greater importance in furthering the epidemic than those formed later on the same lesion. A short latency period, in case of polycyclic pathogens, is essential for the pathogen, a long one for good PR. Thus in many pathogens latency period is a component of great importance, while the infectious period is much less so.

Spore production, whether it is per lesion or per tissue area is difficult to measure. Lesion size, generally highly correlated with spore production (Parlevliet, 1979) is much easier to evaluate and is therefore used an an indirect measure of spore production. Infection frequency or infection density, latency period and lesion size could be used as components to assess PR against this group of pathogens.

In this group of pathosystems a quite different set of components is distinguished through histological analysis (Zadoks, 1972; Zadoks and Schein, 1979). In this approach the percentage of successful units is determined at each of a number of successive steps such as germination, penetration, formation of infective hyphae, formation of haustoria (failure of it would result in early abortion), formation of sporulating tissue (failure of it would result in late abortion) (Heath, 1974; Niks and Dekens, 1987). Since histological assessments are too laborious to be done on a large scale on many entries, this component type of approach is not likely to play a role as an instrument to select PR. It is, therefore, not elaborated on in this chapter.

In many other, partly or fully, systemic pathogens the propagules produced do not always become visible. This reduces the components to be studied to infection frequency, measured as lesion number per plant or plant part or as incidence, and lesion size or disease severity. In the case of partially systemic pathogens, infection frequency and lesion size can be used; in the case of completely systematic pathogens, incidence and severity may be the parameters. An example of the former is *Xanthomonas campestris* pv. *oryzae* (Ishiyama) Dye of the bacterial blight pathogen of rice (*Oryza sativa* L.). When the inoculation is done by the clipping or pinprick method, the usual procedure, the infection frequency as a component is by-passed and only lesion size remains (Koch *et al.*, 1991). In the case of field or partial resistance in potatoes to various potato viruses, the incidence, representing the chance of a plant becoming infected, is one component. The rate of multiplication within the host assessed through the severity of the symptoms might be used as the second component.

For many pathosystems it is possible to recognize one component or a set of components that represent the PR in the field in a satisfactory way. For *P. hordei* in barley, the latency period assessed in the adult plant stage measures the PR extremely well (Parlevliet and van Ommeren, 1975; Parlevliet *et al.*, 1985). In this pathosystem there is no need to evaluate the other possible components such as infection frequency, urediosorus size, spore production and infectious period. In the rice-bacterial blight pathosystem, the association between PR and lesion size (lesion length) is quite good, but one misses a useful part of PR - the part provided by a reduced infection frequency if one uses the clipping inoculation method (Koch *et al.*, 1991).

For the potato leaf roll virus in *Solanum tuberosum* L., the incidence reflects the resistance to infection, a lower incidence indicating a higher resistance. It actually is associated strongly with the rate of multiplication in the host (Barker and Harrison, 1985) making incidence a proper component to evaluate PR in this pathosystem.

Components of PR can be used with success in a numer of pathosystems to select for PR. Whether it will be done by breeders is a different question. This depends on various aspects such as - (1) the heritability of PR itself; (2) the heritability of the component(s) during the screening in the laboratory, greenhouse or field; and (3) the ratio in costs between selection for PR directly or indirectly for the components.

Selection for PR through its components is a kind of indirect selection and has its advantages and disadvantages as discussed below.

USE OF COMPONENTS TO SELECT PR; POSSIBILITIES AND RESTRICTIONS

There are basically three reasons why one might use the indirect approach of selecting PR.

1. **Indirect selection** for PR through one or more components because the PR is difficult to evaluate (low heritability), while one or more of the components can be assessed with much greater accuracy (high heritability).

2. **Accumulate genes for PR** - Through selection of two or more components of PR one is certain that at least a few genes for PR are accumulated. And these genes have quite different effects, resulting in a more complex type of resistance, which is expected to be more durable. In leaf pathogens, one might bring together PR genes thereby reducing the infection frequency, increasing the latency period and reducing the spore production. This, of course, implicitly assumes that the components are under the control of different genes.

3. **Producing superior parental material** - Even if selection through component selection would be too costly for selecting PR, one might contemplate the accumulation of high levels of PR through component selection in material to be used as parents in breeding programs.

Indirect selection; is it worthwhile?

Indirect selection is applied only when the trait to be improved is difficult to assess and another trait exists which has a considerably higher heritability, is highly correlated to the desired trait, and can be measured with the same or even lower costs. Selection for low tannin field beans (*Phaseolus vulgaris* L.), for instance, can be done by selecting for white-flowered genotypes as the trait "white flowers" is highly correlated with "low tannins" and is much cheaper to assess.

So for indirect selection of PR, at least five variables have to be evaluated before the right choice can be made. The breeder, therefore, should know the following: (1) The amount of variation shown by the component [this is not always sufficient (Parlevliet, 1979)], (2) the heritability of PR in the field, (3) the heritability of the chosen component, (4) the correlation between PR and the chosen component, and (5) the costs of assessing entries for PR in the field and of assessing them for the chosen component.

Each of these five aspects may vary greatly with the pathosystem as the following examples illustrate. A general answer therefore is not possible. The only generalization one can make is that in most pathosystems insufficient data are available to make the right choice. A few examples may illustrate some of the problems one may meet. The examples are barley (*P. hordei*), potato [*Phytophthora infestans* (Mont.) de Bary], rice (*X. campestris* pv. *oryzae*), and potato (some potato viruses).

Barley leaf rust in barley

In this pathosystem the PR and its components in both the seedling stage and in the adult plant stage have been studied in detail as well as the selection response in the field for PR or for one or more components in the greenhouse.

PR in the field has a high heritability, whether the selection is carried out in large or small plots (Parlevliet and Van Ommeren, 1975; Parlevliet *et al.*, 1980). Only when a high level of PR has been reached does the discrimination between entries become much more difficult, especially in small adjacent plots as the level of rust remains very low in all entries. A real need for the use of indirect selection does not exist in this pathosystem; in addition, there is a second reason why breeders are not likely to use the indirect approach in this pathosystem. All components have to be assessed in the greenhouse in monocyclic tests. Of the components (infection frequency, latency period, pustule size, spore production per pustule, and infection period), latency period is by far the best component to use. Infection frequency, spore production - whether it is per pustule or per leaf - and infectious period are far more difficult to assess with a reasonable degree of accuracy and are less well correlated with PR than latency period (Parlevliet and Kuiper, 1977; Neervoort and Parlevliet, 1978; Parlevliet, 1979). This is also the case for pustule size (Parlevliet, unpublished data).

Selection in the seedling stage for increased latency period and decreased infection frequency (fewer pustules per leaf) was very effective in selecting entries with partial resistance, but so was selection for PR in the field using single plants and especially using small plots (Table 3) (Parlevliet *et al.*, 1980). Both selection procedures, the direct and the indirect one, are about

Table 3 Relative response to selection (the maximum possible response is set at 100) in a barley population for partial resistance to *Puccinia hordei* after seedling selection in the greenhouse for two components and after selection for partial resistance in the field when the plot consists of a single plant or of two short rows at two selection intensities (after Parlevliet *et al.*, 1980).

	Selection intensity	
Selection procedure	*5%*	*25%*
Seedling screening on basis of:		
Long latency period	61	60
Low infection frequency	55	35
PR in the field:		
Single plants	55	52
Small plots	86	69

equal in effectiveness. Thus, other factors will decide which approach is taken. The advantage of the seedling test is its relative simplicity and reliability as far as exposure to the barley leaf rust is concerned. Its disadvantages are that it requires an increased organizational complexity, it may be more expensive than field testing and, with seedling screening, it is not possible to exploit all PR as only part (a considerable part, though) of the PR is expressed in this stage (Parlevliet, 1975; Parlevliet and Kuiper, 1977). The disadvantage of the field test is the unpredictability of the occurrence of the pathogen which, however, can be greatly improved upon through introducing the pathogen via spreader rows or spreader plants.

Selection in the adult plant stage for an increased latency period is even more effective than selection in the seedling stage (Parlevliet and Kuiper, 1985; Parlevliet *et al.*, 1985). By selecting for an increased latency period, the PR in the field was increased some hundred fold (Table 4). But such a greenhouse screening is rather expensive and, for that reason, probably uninteresting for a breeder. Only when very high levels of PR are required should a greenhouse test be considered because field selection among entries with high levels of PR (higher than that of "Vada" in Table 1) becomes rather difficult (the leaf rust levels remain very low). But such high levels are not required in western Europe (Habgood and Clifford, 1981). Selection for PR through increased latency periods in the adult plant stage does not harbor the danger that only part of the PR can be selected for as in the seedling stage. This is because the genes for latency period are the same genes as those resulting in a reduced infection frequency and reduced pustule size and so reduced spore production (Parlevliet, 1986; Arntzen and Parlevliet,

Table 4 Latency period (LP) in the young flag leaf stage relative to that of L94 (which carries no partial resistance at all and is set at 100) and the number of urediosori per tiller in plots isolated from each other to prevent interplot interference of eight barley cultivars and lines exposed to *Puccinia hordei* (after Parlevliet *et al.*, 1985).

Cultivar/line	LP	Sori per tiller
Akka	113	5000
Berac	147	750
Vada	185	100
42-1-9	212	30
26-6-16	246	7
17-5-9	281	1
17-5-16	281	0.4
26-6-11	291	1

1986; Parlevliet *et al.*, 1985). The components of PR are associated with one another through pleiotropy.

The conclusion is that, although indirect selection is very effective and could be done without serious problems in this pathosystem, there is no real need for it because the PR is quite easy to select for in the field (Parlevliet and van Ommeren, 1988).

Late blight in potato

This pathogen attacks foliage as well as tubers. Apart from at least 13 major race-specific resistance genes, PR (field resistance) has been recognized and studied extensively. The resistance factors in the tubers and foliage are to a large extent independent of one another (Swiezynski, 1990) (Table 5). Cultivars with the same 5 rating for PR in the foliage vary in PR in the tubers from very susceptible (3 rating) to very resistant (9 rating). This means that PR must be selected independently of tissues as if two different traits are involved. The PR in the foliage is considered to be more important than PR in tubers as it determines the rate of epidemic development and influences the rate of tuber infection later in the season.

Another problem in this pathosystem is the strong association between PR and lateness (Table 6). Very early cultivars with a high level of PR have never been observed.

PR of the foliage is assessed by evaluating the percentage foliage affected using a pictorial or a written descriptive key. As for the tubers, the incidence is assessed after harvesting. The association of PR with lateness

Table 5 Resistance values of foliage and tubers, and the maturity values of six potato cultivars without known major resistance genes to *Phytophthora infestans* (after Anonymous, 1987, 1989).

Cultivar	Resistance value*		Maturity*
	Foliage	Tuber	
Meerlander	6	8	7
Spunta	6	5	7
Romano	5	9	7
Monalisa	5	6	7 1/2
Eigenheimer	5	3	7
Climax	3	8	7 1/2

*Values on a scale of 1 to 10, 1 being extremely susceptible or extremely late, a 10 being completely resistant (seemingly immune) or extremely early.

Table 6 Mean values for resistance of foliage and tubers of four maturity groups of potato cultivars without known major resistance genes to *Phytophthora infestans* (after Anonymous, 1989).

Maturity*		Number of	Resistance value*	
Mean	Range	cultivars	Foliage	Tubers
10	8.5-9.5	8.8	3.9	5.5
19	7 1/2-8	7.7	4.1	6.9
20	5.5-7	6.3	4.8	6.6
16	2.5-5	4.5	6.0	7.2

*Values on a scale of 1 to 10, 1 being extremely susceptible or extremely late, a 10 being completely resistant (seemingly immune) or extremely early.

interferes considerably with a proper assessment as the differences in maturity are so large in potato. One should compare the PR primarily between entries within maturity classes. This is not possible in the very early generations of the breeding program because the maturity class of an entry is not or insufficiently known.

The PR in the foliage is based on fewer lesions that grow slower and sporulate less than in susceptible genotypes. Actually, the same components as with the barley leaf rust in barley can be discerned - infection frequency, latency period, lesion size and spore production. These components are clearly associated, although infection frequency behaves more independently than the other components (Van der Zaag, 1959; Umaerus, 1970; Umaerus and Lihnell, 1976). The PR is most clearly expressed in the young adult plant stage (at flowering) and very young and very old plants are the most susceptible. This, together with the association of PR with lateness, makes a reliable and accurate assessment of PR quite difficult, especially among seedlings or first year clones. The problem of the association with maturity remains if one or more components are used to select for PR. If one would decide to use indirect selection, the component which could be assessed with the greatest accuracy is probably lesion size. Lesion number always carries a high coefficient of variation, whereas latency period and spore production are strongly dependent on the environmental conditions. Due to the somewhat independent variation of lesion number and other components, selection for one component would mean that only part of the PR present is exploited. It is not yet clear whether component selection is more cost-effective than selection for PR in the field, as breeders did not give PR a high priority in the past decades because very effective fungicides were available. It can be concluded that selection for PR, either directly or indirectly, is not easy, and it is not clear which approach is the most advisable in a breeding program.

Bacterial leaf blight in rice

This pathosystem is quite different from the former two systems. Assessment of PR in the field is very difficult because it is almost impossible to expose rice entries uniformly to the pathogen in a natural way (Alvarez *et al.*, 1989; Koch, 1989; Reddy, 1989). PR in the field can be considered to have a very low heritability and indirect selection might solve this problem. The pathogen enters the host tissue and leaves through wounds or natural openings such as hydathodes (Horino, 1984). After entering the leaf, the bacteria spread through the xylem vessels of the vascular system, killing the tissue and causing stripe type lesions. The number and length of stripes are considered to reflect the level of susceptibility of the entry; they are often assessed together as the percentage diseased leaf area (DLA) (Koch, 1989). Because a reliable field test for the assessment of PR under more or less natural conditions (a polycyclic test started from spreader plants or rows) does not exist, the screening for bacterial leaf blight resistance uses a field test, which is in fact a monocyclic, indirect selection test for the most important component, lesion length. The selection is not directed at PR, but at major-gene resistance (at least 14 are identified), a resistance shown to be elusive (Ogawa and Khush, 1989). In this test, the entries are clip-inoculated, which enables the inoculum to enter directly through the wounds. After 2 to 3 weeks the DLA is measured. This test is quite efficient for the selection of entries carrying an effective major gene for resistance, as the difference in lesion length (and, thus, DLA) between susceptible and resistant entries is great. If, however, PR is the objective, this test is insufficient for two reasons. First, DLA assessment is inadequate to measure the PR because only a part of the PR is selected. Although rice entries differ greatly in length of their leaves, the lesion length (reflecting the component of PR affecting the spread of the pathogen in the host tissue) is independent of the leaf length. Entries with long leaves can therefore have a relatively low DLA when the lesions are actually fairly long, while entries with short leaves tend to score high for DLA even if their PR is fairly good (Koch and Parlevliet, 1990). Second, the clip-inoculation method bypasses the other component, the resistance to entering and establishing in the host tissue. This component can be assessed by a spray inoculation test. Differences in the number of lesions reflect differences in PR for this second component. Koch *et al.* (1991) showed that resistances to entering and spreading are at least partially independent components, and that the spray inoculation test selected entries with a good level of PR that would have escaped the attention in the clip-inoculation test (cultivars IR36 and IR40 in Table 7). However, the spray-inoculation test has a considerably higher coefficient of variation and is therefore less suitable to use as the only screening test. In order to select entries that carry PR based on both components, Koch *et al.* (1991) advised screening entries first with the more accurate clip-inoculation test for reduced lesion length. Selected entries would then be spray-inoculated to select entries which show reduced lesion length and number. In this pathosystem indirect selection has to be practiced because there does not exist a reasonably reliable and accurate field test for PR.

Table 7 Mean lesion length following clip (C) inoculation and mean lesion number of nine cultivars following spray (S) inoculation with *Xanthomonas campestris* pv. *oryzae* (Koch *et al.*, 1991).

Cultivar	C-inoculation lesion length (cm)	S-inoculation lesions per leaf
TN1	18.9	1.6
Biplab	16.1	3.2
IR 36	14.7	0.9
BR51-282-8	13.0	1.6
IR40	12.0	0.7
IR32	11.7	1.5
Cisadane	10.5	2.0
IR54	9.1	0.9
IR48	7.3	2.3

Potato viruses in potato

The most important viruses in potato in western Europe are potato leaf roll virus and potato virus Y, of which several distinct strains are known, potato virus X and potato virus A. Major resistance genes to all of the viruses except leaf roll virus are present in the cultivated potato. These major genes restrict the disease to a local lesion reaction, i.e. the virus is localized around the infection point and a small necrotic fleck may be visible. Against all these viruses there is another type of resistance, which shows a quantitative pattern and is often thought to be polygenic (Wiersema, 1972; Ross, 1986) but sufficiently detailed genetic analyses have not been carried out yet. This resistance, also a PR, is variously called field resistance or resistance to infection, as the incidence of infected plants varies greatly after exposure to the virus (Table 8). Cultivar differences are considerable for leaf roll to very large for virus Y^n (Table 9). Both the PR and the major gene resistance are specific for each virus (Table 9). The major gene resistance is also strain-specific (Ross, 1978, 1986) but for the PR there is no reported strain specificity.

With virus diseases it is difficult to discern components within the PR because the assessment itself is far from straightforward. One might consider the incidence and the severity of the disease as two separate components. If these were representing the resistance to infection and the resistance to the spread and multiplication of the virus, respectively, one could consider them as two components. Unfortunately, this is probably not the case. The incidence to leaf roll, for instance, is clearly associated with the virus concentration in the host. The infected plants of cultivars with a reduced incidence have a much lower virus concentration than the infected plants of very susceptible cultivars (Barker and Harrison, 1985). Similar observations have been reported for virus X (Kurzinger and Schenk, 1988). At the same time, the association between the severity of the symptoms and

Table 8 Percentage of virus Y^n-symptomatic plants in screening trials of potato cultivars with varying PR resistance levels (after Van der Woude, 1985).

Resistance level of cultivars*		*Year*			
	1973	*1975*	*1977*	*1979*	*Mean*
3	76	88	98	62	81
5	40	52	89	25	52
8	5	8	29	3	11

*Values on a scale of 1 to 10, 1 being extremely susceptible or extremely late, a 10 being completely resistant (seemingly immune) or extremely early.

Table 9 PR values of eight potato cultivars for four potato viruses (Anonymous, 1987, 1989).

Cultivar	Leaf roll	A	Virus* X	Y^n
Arkula	8	8	8	9
Doré	7 1/2	2	6	2
Eigenheimer	4	6	4	5 1/2
Elles	5 1/2	R	9	6 1/2
Timate	6	7	R	9 1/2
Mansour	5 1/2	3	R	7
Astarte	4	R	R	8 1/2
Corine	5	R	7	R

*Values on a scale of 1 to 10, 1 being extremely susceptible or extremely late, a 10 being completely resistant (seemingly immune) or extremely early. R means a major resistance gene present.

the level of susceptibility (incidence and virus concentration of infected plants) is not very good. Barker and Harrison (1985) observed that genotype G746 had a fairly high resistance rating for leaf roll (reduced incidence and virus concentration), but the infected plants had symptoms similar in severity to those of very susceptible genotypes. The other resistant genotypes had mild symptoms compared with susceptible ones. Cuperus *et al*. (1987) assessed the incidence and severity of symptom expression of over 30 potato cultivars for virus M. The association between incidence and severity was rather poor.

In the field, the breeder would screen for a reduced incidence and/or reduced symptom expression. This is only useful if the breeder is able to expose the entries uniformly to the virus concerned without interference of other viruses, but this is not easy. It is possible to obtain a reasonably uniform exposure by planting the entries in two plant plots between spreader rows. The spreader rows consist of a cultivar susceptible for and infected with the respective virus (Wiersema, 1972; Ross, 1986), and each entry should be replicated. If the selection is done on the basis of reduced symptoms, then three problems arise. First, highly susceptible genotypes still may be selected because of an insufficient association between symptom expression and PR. Second, if selection against symptom expression is successful, then one may have selected highly resistant genotypes as well as susceptible genotypes (symptomless carriers). The third problem is that once fairly resistant genotypes have been identified which show few symptoms; then the removal of virus-infected plants during maintenance and multiplication may become problematic.

Selection for a reduced incidence might therefore be a better and more advisable breeding approach. The ideal cultivar would, of course, have a very high level of PR, but with clear symptoms if infected. Once a breeder chooses reduced incidence as a selection objective, then a fair number of plants of each entry in the field is necessary because incidence cannot be measured on the basis of single plant plots and results are very inaccurate when only few plants are tested. This is the reason why selection for PR for viruses in potatoes is not pursued vigorously even though sufficient genetic variation exists in the cultivated potato. Further, field tests are rather laborious and expensive. But as long as cheap, fast and reliable methods to measure virus concentrations in individual potato plants are not available, such field tests form the only approach possible for selecting partially resistant genotypes.

Another possibility is assessing resistance to infection of clones by evaluating their seedling progenies. If selfing is possible, then the selfed progenies could be compared. Jones (1978) reported that the incidence of leaf roll in clones (cultivars) could be assessed through exposing a sufficient number of rooted cuttings of each clone to aphid viruses in the greenhouse. Cultivar differences were associated with the disease incidence observed in seedling progenies obtained by open pollination. Such a seedling test in a greenhouse might form a practical alternative for the elaborate field test of the cultivars/entries themselves, especially if the seedlings were obtained through selfing. Clones that do not flower or which are male sterile create problems in such a test, however.

The conclusion is that selection of PR to viruses is very difficult in potatoes and that selection for components of PR is not practical because of the difficulty in discerning clear components of resistance.

From the above four examples it is clear that no generalizations can be made about selecting PR. The recognizability of PR in the field varies from very good to very poor. For many pathogens, especially the leaf pathogens with no or little systemic growth, it is possible to discern several components. However, for other pathogens this is more difficult or nearly impossible. Only in a restricted number of pathosystems can one component of resistance be identified which is strongly correlated with and easier to assess than PR in the field. However, even then it is not always possible to use that component because of the extensive efforts and costs involved in identifying the desirable genotypes.

USE OF COMPONENTS TO ACCUMULATE PR; POSSIBILITIES AND RESTRICTIONS

The ultimate goal of selection is for resistance that is durable. Partial resistance is often considered to be durable because of its complex inheritance (Van der Plank, 1968). Although it is convenient to classify the endless

variation in resistance in two types - (1) the race-specific, non-durable, qualitative, monogenic type and (2) the race-non-specific, durable, quantitative,
polygenic type - this does not represent the reality in the field. PR, although
no guarantee for durability, is a better bet than simply inherited complete
resistance, especially in those host-pathogen systems where the durability of
monogenic resistances is very low (Parlevliet, 1990).

Resistances which are based on a few to several genes do seem considerably more durable than monogenic resistances. As yet, no case of erosion of
partial resistance has been reported when based on more than a single gene
(Parlevliet, 1990).

PR based on at least a few genes could be obtained if the PR is the accumulated result of an improvement of two or more of its components, provided the genes controlling these components are not identical. Such an approach will very probably result in a durable resistance. But in order to
materialize this idea into a screening procedure, a number of requirements
must be met such as (1) at least two components of the PR should vary independently of each other or, if they are correlated with one another, the
association should be relatively slight; (2) it should be possible to assess
each component independently and reasonably accurately without great costs
(i.e., a practical screening procedure for each component should be possible); and (3) PR in the field has a low heritability.

Independency of components

Associated variation of the components of PR is a general phenomenon. The
association in the barley *P. hordei* pathosystem is very strong. In other
cereal-rust pathosystems the components also vary in association with each
other (Parlevliet, 1979). However, the association seems less complete.
Especially the infection frequency appears less strongly associated with the
other components than in the case of barley leaf rust in barley (Ohm and
Shaner, 1976; Jacobs, 1989).

The observations in the potato-*Phytophthora infestans*, wheat-*M. graminicola* and wheat-*Leptosphaeria nodorum* pathosystems are similar. The
components vary in association with each other, whereby infection frequency
shows the least association to the other components (Van der Zaag, 1959;
Umaerus, 1970; Umaerus and Lihnell, 1976; Shearer, 1978; Griffiths and
Jones, 1987; Cunfer *et al.*, 1988). Cunfer *et al.* (1988) did not include
infection frequency in their studies, but assumed from the high association
between the other components that a single or interrelated mechanism controls the expression of resistance in wheat to glume blotch. Griffith and
Jones (1987) came to a slightly different conclusion by observing the same
pathosystem. The associations they found were not as high as those of
Cunfer *et al.* (1988), and they suggested a selection index based on several

components where spore production and incubation period should have the highest weight for the selection of PR.

Independency of resistance components to a single pathogen does not seem to occur often. The best chance of independency is between the components expressed as frequency of infection (lesion number, incidence) and as size of the infection (lesion size). Instead of lesion size, latency period or incubation period can be used when these are easier to assess.

Assessment of the independent components

When components of resistance are inherited independently, they most often appear to be infection frequency (expressed as lesion number, on one hand) and any of the components that indicate the rate of spread in the host tissue, such as incubation and latency period, lesion size and spore production, on the other hand.

The assessment of infection frequencies (i.e., lesion numbers) nearly always goes together with a high coefficient of variation and at least a series of assessments are needed for an accurate evaluation of the entries (Parlevliet and Kuiper, 1977). The same can be said of spore production. Only incubation and latency period can be assessed with a reasonable accuracy in a simple test in many, but not all, pathosystems.

A simple, fairly inexpensive screening procedure based on the stepwise selection of entries toward improvement for two components of resistance, therefore, is not likely to be developed with the present knowledge and facilities except in incidental cases.

Heritability of PR in the field

The ease by which PR can be recognized varies greatly from one pathosystem to the next as the four examples discussed illustrated. PR can be easily confounded with late plant maturity (septoria leaf and glume blotches, Fusarium head blight) and plant height (septoria leaf and glume blotches), underestimated (barley leaf rust), assessed incorrectly (barley powdery mildew) in small plots, not measurable at all in the field using a polycyclic test (bacterial blight of rice), not measured on a small number of plants (potato viruses), or hidden behind race-specific major genes (wheat rusts, powdery mildew in barley, downy mildew in lettuce).

If the heritability is high, then there is no need to use a selection procedure based on individual components. Most, if not all, genetic variation for PR present in the population to be improved is available for accumulation. Only in the case of a relatively low heritability is the alternative (component selection) of interest. In this case, the component with the highest heritability will be selected when the selection is done for PR in the field. The other component(s), with a lower heritability, may not even be selected at all. By component selection this can be prevented.

Component selection with the aim of accumulating PR genes can not be more than an incidental case if the three requirements discussed above are considered jointly. Independency of components is not common and, if it occurs, at least one of the components (lesion number) requires a rather laborious and costly assessment. Only when PR itself is very difficult to evaluate does component selection come into consideration. The bacterial blight pathosystem in rice is one of the very few good examples. Here, only selection through components will be effective and, in order to ensure a higher durability, the two step screening procedure leading to a joint reduction in lesion length and lesion number is highly advisable.

INDIRECT SELECTION OF COMPONENTS

The components of PR in most pathosystems can be considered as quantitative traits. When such a component is controlled by several genes, then indirect selection of the trait might be possible by using the polymorphism that is found in the sequence of base pairs making up the DNA. This polymorphism can be made visible through restriction enzymes that cleave the DNA at specific sites producing DNA fragments of different lengths, the restriction-fragment length polymorphism (RFLP). Each RFLP can be seen as a marker and, once a sufficient number of RFLP markers are identified which are well spread over the genome, it would be advantageous to investigate linkage of RFLP markers with components of resistance. Selection with seedlings or young plants using RFLP markers would result in the selection for the component in question (Tanksley *et al.*, 1989). This gives the advantage of not requiring the pathogen when selecting for the component of resistance.

One problem, which is often overlooked when suggesting this approach, is that in order to establish the linkage between the component of PR and certain RFLP markers, one must be able to measure the component with reasonable accuracy. If the heritability of the component is too low, then it is not possible to obtain an accurate estimate of the association with the RFLP markers. This is, unfortunately, rather often the case.

Indirect selection through RFLP markers could also be done with PR itself. Here the same possibilities and restrictions apply. Too low a heritability may interfere severely with the possibilities to apply this approach. In such cases, though, one of its components may have a sufficiently high heritability to use the RFLP markers selection approach with some success.

On the other hand, if the PR itself or one of the components has a very high heritability, as in the case of barley against barley leaf rust, the RFLP approach is of no use because direct selection is sufficiently effective. Use of RFLP marker selection will be especially effective in those pathosystems where (1) the heritability of PR is relatively low, (2) the heritability of one of its components is fairly high, and (3) the costs of assessing this component are fairly high.

One other fairly common situation where RFLP marker selection for PR or one of its components can be useful is when race-specific major genes are present in high frequencies. This interferes very strongly with the proper assessment of PR (Parlevliet, 1989). Once RFLP markers strongly linked with PR or its components have been identified, then the major race-specific genes do not obstruct selection anymore.

SELECTION OF PARENTAL MATERIAL

In this chapter it was shown that it is unlikely that component selection will be applied in breeding for PR, except for a few pathosystems. Component selection, as well as the indirect RFLP marker selection, are normally accompanied by increased costs, thus preventing their use in many breeding programs.

In cases where higher investments are worthwhile, the more laborious and costly indirect selection procedures (component as well as RFLP marker selection) would be worthwhile. Development of superior parental material is such a case. Accumulation of high levels of PR in parental material can have a very strong spin-off through the material generated from it. This is especially useful if the PR has a low heritability. Investment in good parental material, especially when done on a wide basis (more than a single good parent), may carry high dividends.

However, in order to develop that superior parental material, suitable screening procedures for the PR and its components must be available. In many pathosystems the knowledge needed for selecting them is still inadequate and a considerable amount of research is required in those pathosystems.

CONCLUSION

Component selection, meant to enhance the breeding for PR, is not generally recommended. For each pathosystem a separate decision has to be taken, a decision based on the heritability of PR itself, on the heritability of the component or components to be used, on the magnitude of the association between PR and the components and on the costs of component selection relative to the costs of selection for PR. It is expected that only in a restricted number of pathosystems will component selection improve the selection for PR.

REFERENCES

Alvarez, A. M.; Teng, P. S.; Benedict, A. A. (1989) Methods for epidemiological research on bacterial blight of rice. In: *Bacterial blight of rice*. Proceedings International Workshop IRRI and AGCD, Belgium. Los Baños, 14-18 March, 1988. IRRI, Manila, Philippines, pp. 99-110.

Anonymous (1987) *62e Beschrijvende Rassenlijst voor Landbouwgewassen 1987*. (62nd descriptive cultivar list for arable crops 1987). Leiter-Nypels, Maastricht, The Netherlands, pp. 248-253.

Anonymous (1989) *64e Beschrijvende Rassenlijst voor Landbouwgewassen 1989*. (64th descriptive cultivar list for arable crops 1989). Leiter-Nypels, Maastricht, The Netherlands, pp. 254-259.

Arntzen, F. K.; Parlevliet, J. E. (1986) Development of barley leaf rust, *Puccinia hordei*, infection in barley. II. Importance of early events at the site of penetration for partial resistance. *Euphytica* 35:961-968.

Barker, H.; Harrison, B. D. (1985) Restricted multiplication of potato leaf roll virus in resistant potato genotypes. *Annals of Applied Biology* 107:205-212.

Cunfer, B. M.; Stooksbury, D. E.; Johnson, J. W. (1988) Components of partial resistance to *Leptosphaeria nodorum* among seven soft red winter wheats. *Euphytica* 37:129-140.

Cuperus, C.; de Bokx, J. A.; Piron, P. G. M. (1987) De vatbaarheid van Nederlandse rassen voor aardappelvirus M (the susceptibility of Dutch cultivars for potato virus M). *Gewasbescherming* 18(1):19-25.

Griffiths, H. M.; Jones, D. G. (1987) Components of partial resistance as criteria for identifying resistance. *Annals of Applied Biology* 110:603-610.

Habgood, R. M.; Clifford, B. C. (1981) Breeding barley for disease resistance: The essence of compromise. In: Jenkyn, J. F.; Plumb, R. T. (eds.), *Strategies for the control of cereal diseases*. Blackwell Scientific Publications, Oxford, pp. 15-25.

Heath, M. C. (1974) Light and electron microscope studies of the interactions of host and non-host plants with cowpea rust-*Uromyces phaseoli* var. *vignae*. *Physiological Plant Pathology* 4:403-414.

Horino, O. (1984) Ultrastructure of water pores in *Leersia japonica* Mikino and *Oryza sativa* L.: Its correlation with the resistance to hydathodal invasion of *Xanthomonas campestris* pv. *oryzae*. *Annual Phytopathologists Society of Japan* 50:72-76.

Jacobs, Th. (1989) The occurrence of cell wall appositions in flag leaves of spring wheats, susceptible and partially resistant to wheat leaf rust. *Journal Phytopathology* 127:239-249.

Johnson, R.; Taylor, A. J. (1976) Spore yield of pathogens in investigations of the race-specificity of host resistance. *Annual Review of Phytopathology* 14:97-119.

Jones, R. A. C. (1978) Progress in leaf roll virus resistance. In: *Development in the control of potato virus diseases*. Proceedings of a Planning Conference, Lima, 14-18 Nov. 1977, CIP, Lima, Peru, pp. 15-26.

Koch, M. F. (1989) Methods for assessing resistance to bacterial blight. In: *Bacterial blight of rice*. Proceedings of an International Workshop, IRRI and AGCD, Belgium, Los Baños, 14-18 March 1988. IRRI, Manila, Philippines, pp. 111-123.

Koch, M. F.; Parlevliet, J. E. (1990) Assessment of quantitative resistance of rice cultivars to *Xanthomonas campestris* pv *oryzae*: A comparison of assessment methods. *Euphytica* 51:185-189.

Koch, M. F.; Parlevliet, J. E.; Mew, T. W. (1991) Assessment of quantitative resistance of rice cultivars to *Xanthomonas campestris* pv *oryzae*: A comparison of inoculation methods. *Euphytica* 54:169-175.

Kurzinger, B.; Schenk, G. (1988) Beitrag zur Bestimmung der relativen Resistenz gegen Potato Virus X (PVX) an In-Vitro Pflanzen der Kartoffel. *Potato Research* 31:49-53.

Mehta, Y. R.; Zadoks, J. C. (1970) Uredospore production and sporulation period of *Puccinia recondita* f. sp. *triticina* on primary leaves of wheat. *Netherlands Journal of Plant Pathology* 76:267-276.

Neervoort, W. J.; Parlevliet, J. E. (1978) Partial resistance of barley to leaf rust, *Puccinia hordei*. V. Analysis of the components of partial resistance in eight barley cultivars. *Euphytica* 27:33-39.

Niks, R. E.; Dekens, R. G. (1987) Histological studies on the infection of triticale, wheat and rye by *Puccinia recondita* f. sp. *tritici* and *P. recondita* f. sp. *recondita*. *Euphytica* 36:275-285.

Nørgaard Knudsen, J. C.; Dalsgaard, H. H.; Jorgensen, J. H. (1986) Field assessment of partial resistance to powdery mildew in spring barley. *Euphytica* 35:233-243.

Ogawa, T.; Khush, G. S. (1989) Major genes for resistance to bacterial blight in rice. In: *Bacterial blight of rice*. Proceedings of an International Workshop IRRI and AGCD, Belgium, Los Baños, 14-18 March 1988. IRRI, Manila, Philippines, pp. 177-192.

Ohm, H. W.; Shaner, G. E. (1976) Three components of slow leafrust at different growth stages in wheat. *Phytopathology* 66:1356-1360.

Parlevliet, J. E. (1975) Partial resistance of barley to leaf rust, *Puccinia hordei*. I. Effect of cultivar and development stage on latent period. *Euphytica* 24:21-27.

Parlevliet, J. E. (1979) Components of resistance that reduce the rate of epidemic development. *Annual Review of Phytopathology* 17:203-222.

Parlevliet, J. E. (1981a) Disease resistance in plants and its consequences for breeding. In: Frey, K. J. (ed.), *Plant breeding II*. Iowa State University Press, Ames, pp. 309-364.

Parlevliet, J. E. (1981b) Race non-specific resistance. In: Jenkyn, J. F.; Plumb, R. T. (eds.), *Strategies for the control of cereal disease*. Blackwell Scientific Publications, Oxford, pp. 47-54.

Parlevliet, J. E. (1981c) Crop loss assessment as an aid in the screening for resistance and tolerance. In: Chiarappa, L. (ed.), *Crop loss assessment methods*. Supplement 3. FAO and Commonwealth Agricultural Bureaux, London, pp. 111-114.

Parlevliet, J. E. (1986) Pleiotropic association of infection frequency and latent period of two barley cultivars partially resistant to barley leaf rust. *Euphytica* 35:267-272.

Parlevliet, J. E. (1989) Identification and evaluation of quantitative resistance. In: Leonard, K. J.; Fry, W. E. (eds.), *Plant disease epidemiology, Vol. 2, Genetics, resistance and management*. McGraw-Hill Publ., New York, pp. 215- 248.

Parlevliet, J. E. (1990) Breeding for durable resistance to pathogens. In: Tanner, D. G.; van Ginkel, M.; Mwangi, W. (eds.), *Proceedings Sixth Regional Wheat Work shop for Eastern, Central and Southern Africa*. Addis Ababa, Ethiopia, 2-6 Oct., 1989. CIMMYT, Mexico, D.F., pp. 14-27.

Parlevliet, J. E.; Kuiper, H. J. (1977) Partial resistance of barley to leaf rust, *Puccinia hordei*. IV. Effect of cultivar and development stage on infection frequency. *Euphytica* 26:249-255.

Parlevliet, J. E.; Kuiper, H. J. (1985) Accumulating polygenes for partial resistance in barley to barley leaf rust, *Puccinia hordei*. I. Selection for increased latent periods. *Euphytica* 34:7-13.

Parlevliet, J. E.; Leijn, M.; van Ommeren, A. (1985) Accumulating polygenes for partial resistance in barley to barley leaf rust, *Puccinia hordei*. II. Field evaluation. *Euphytica* 34:15-20.

Parlevliet, J. E.; Lindhout, W. H.; van Ommeren, A.; Kuiper, H. J. (1980) Level of partial resistance to leaf rust, *Puccinia hordei* in west European barley and how to select for it. *Euphytica* 29:1-8.

Parlevliet, J. E.; van Ommeren, A. (1975) Partial resistance of barley to leaf rust, *Puccinia hordei*. II. Relationship between field trials, micro plot tests and latent period. *Euphytica* 24:293-303.

Parlevliet, J. E.; van Ommeren, A. (1988) Accumulation of partial resistance in barley to barley leaf rust and powdery mildew through recurrent selection against susceptibility. *Euphytica* 37:261-274.

Reddy, A. P. K. (1989) Bacterial blight: Crop loss assessment and disease management. In: *Bacterial blight of rice*. Proceedings of an International Workshop IRRI and AGCD, Belgium, Los Baños, 14-18 March 1988. IRRI, Manila, Philippines, pp. 79-88.

Rosielle, A. A.; Brown, A. G. P. (1980) Selection for resistance to *Septoria nodorum* in wheat. *Euphytica* 29:337-346.

Ross, H. (1978) Methods for breeding virus resistant potatoes. In: *Development in the control of potato virus diseases*. Proceedings Planning Conference, Lima, 14-18 Nov. 1977, CIP, Lima, Peru, pp. 93-114.

Ross, H. (1986) Potato breeding. Problems and perspectives. *Advances in Plant Breeding*. Suppl. J. Plant Breeding 13. Paul Parey, Berlin.

Shaner, G.; Finney, R. E.; Patterson, F. L. (1975) Expression of resistance in wheat to septoria leaf blotch. *Phytopathology* 65:761-766.

Shearer, B. L. (1978) Inoculum density-host response relationships of spring wheat cultivars to infection by *Septoria tritici*. *Netherlands Journal of Plant Pathology* 84:1-12.

Swiezynski, K. M. (1990) *Resistance to Phytophthora infestans in the potato*. Institute Potato Research, Bonin, Poland.

Tanksley, S. D.; Young, N. D.; Paterson, A. H.; Bonierbale, M. W. (1989) RFLP mapping in plant breeding: New tools for an old science. *Biotechnology* 7:257-264.

Tavella, C. M. (1978) Date of heading and plant height of wheat varieties, as related to septoria leaf blotch damage. *Euphytica* 27:577-580.

Teng, P. S.; Close, R. C. (1978) Effect of temperature and uredinium density on urediniospore production, latent period, and infectious period of *Puccinia hordei* Otth. *New Zealand Journal of Agricultural Research* 21:287-296.

Umaerus, V. (1970) Studies on field resistance to *Phytophthora infestans*. 5. Mechanisms of resistance and application to potato breeding. *Zeitschrift fuer Pflanzenzuechtung* 63:1-23.

Umaerus, V.; Lihnell, D. (1976) A laboratory method for measuring the degree of attack by *Phytophthora infestans*. *Potato Research* 19:91-107.

Van der Plank, J. E. (1968) *Disease resistance in plants*. Academic Press, London.

Van der Woude, K. (1985) Het kweken en het onderzoek van nieuwe aardappelrassen (he breeding and the testing of new potato cultivars). *De Pootaardappelwereld* 7(1):5-16.

Van der Zaag, D. D. (1959) Some observations on breeding for resistance to *Phytophthora infestans*. *European Potato Journal* 2:278-87.

Wiersema, H. T. (1972) Breeding for resistance. In: de Bokx, J. A. (ed.), *Viruses of potatoes and seed-potato production*. PUDOC, Wageningen, The Netherlands, pp. 174-187.

Zadoks, J. C. (1972) Modern concepts of disease resistance in cereals. In: Lupton, F. G. H.; Jenkins, G.; Johnson, R. (eds.), *The way ahead in plant breeding*. Proceedings of the 6th Congress Eucarpia, Cambridge, 29 June-2 July 1971. PBI, Cambridge, UK, pp. 89-98.

Zadoks, J. C.; Schein, R. D. (1979) *Epidemiology and plant disease management*. Oxford University Press, New York, Oxford.

Chapter 13
Selecting rice for simply inherited resistances

Gurdev S. Khush

Division of Plant Breeding, Genetics and Biochemistry, International Rice Research Institute, P. O. Box 933, Manila, Philippines

During this century, plant breeders have developed numerous disease- and insect-resistant cultivars of crop plants by incorporating major genes for resistance. Large-scale adoption of these cultivars has resulted in major increases in crop productivity. I shall discuss the use of major genes in developing resistant cultivars, taking rice (*Oryza sativa* L.) as an example.

Rice is a host to numerous diseases and insects. The year-round warm and humid climate in the tropics and subtropics, where more than 80% of the rice crop is grown, is also conducive to the development of diseases and insects. After the introduction of cultivars with improved plant type, farmers started using improved cultural practices, such as better water and weed control, higher rates of fertilizers and higher plant populations per unit area. The development of irrigation facilities and the availability of early maturing, photoperiod-insensitive cultivars have enabled the farmers in tropical Asia to grow successive rice crops throughout the year in large areas. A limited number of improved cultivars have replaced numerous tall traditional ones.

The reduced genetic variability, improved cultural practices, and continuous rice cropping - necessary for increased rice production - have increased the genetic vulnerability of the crop. The incidence of diseases and insects harmful to the rice crop has increased in recent years. Very little research has been done on chemical control of diseases in the tropics. Chemical control of high insect populations for prolonged periods under the tropical climate - where insect generations overlap throughout the year - is very expensive. Social and economic conditions in the tropics present other obstacles to the chemical control of rice diseases and insects.

Host resistance is the major strategy being followed by the international and national rice improvement programs for the control of diseases and insects. A large number of pathologists, entomologists, and breeders are working together throughout the rice-growing world to develop cultivars with multiple resistances to major diseases and insects. Screening techniques have been developed, germplasm collections have been evaluated, donors for resistance have been identified, inheritance of resistance has been studied, and genes for resistance have been incorporated into improved cultivars now widely grown in major rice-growing countries.

SELECTING FOR DISEASE RESISTANCE

About 60 rice diseases have been recorded (Ou, 1972); 37 of them are caused by fungi. Blast, caused by *Pyricularia grisea* Sacci, sheath blight, and brown spot are of wide occurrence. Of these, blast has received major attention in the host resistance programs. Of the bacterial diseases, bacterial blight [causal bacterium commonly referred to as *Xanthomonas campestris* pv *oryzae* (Xco)] is the most important. Of the 12 known virus diseases of rice, tungro and grassy stunt are the most serious in the tropics.

Blast

The blast disease causes serious losses and is the most important disease of rice. It occurs in all rice-growing areas of the world and is most serious on upland rice and in the regions where low temperatures prevail at the time of flowering. The disease is a highly variable organism. The differences in the pathogenicity of the fungus strains were first recorded by Sasaki (1922, 1923). He reported that rice cultivars resistant to one strain were severely infected by another. Pathogenic variability has been reported in the U.S.A. (Latterel *et al.*, 1960), India (Padmanabhan, 1965), China (Chiu *et al.*, 1965), Korea (Ahn and Chung, 1962), and Philippines (Bandong and Ou, 1966). Ou and Jennings (1969) reported that more than 100 races of blast exist in the Philippines.

Several methods for evaluating resistance to blast have been developed. The uniform blast nursery, where test entries are planted in upland seedbeds surrounded by a susceptible check, is most commonly used for screening germplasm and breeding materials. This test allows quick evaluation of the blast reaction of a large number of rice cultivars to a number of races present in the locality. Thus, entries with a broad spectrum of resistance can be selected. Most of the rice improvement programs have identified a large number of donors for blast resistance through uniform blast nursery tests and have utilized them for developing blast-resistant cultivars.

In genetic studies, pure cultures of known pathogenicity are used for inoculating parents and segregating populations. Seedlings of test entries are sprayed with a spore suspension of a specific race in the greenhouse chamber. High humidity maintained in the inoculation chamber helps the germination of spores and disease development. Genetic studies using pure cultures were initiated by Niizeki (1960) and continued by Kiyosawa and co-workers in Japan. They used seven fungus strains of varying pathogenicity for genetic studies of several domestic and introduced rice cultivars. As summarized by Kiyosawa (1972, 1974), 10 loci for blast resistance were identified. They were designated as (1) *Pi-1*, (2) *Pi-b*, (3) *Pi-f*, (4) *Pi-i*, (5) *Pi-k*, (6) *Pi-m*, (7) *Pi-s*, (8) *Pi-t*, (9) *Pi-ta*, and (10) *Pi-z*. There are multiple alleles at some loci. For example, at the *Pi-k* locus originally identified by

Yamasaki and Kiyosawa (1966) in "Kanto 51", at least three other distinct alleles, e.g., *Pi-k*^s (Kiyosawa 1969a), *Pi-k*^p (Kiyosawa 1969b), and *Pi-k*^h (Kiyosawa and Murty, 1969) have been detected. Similarly *Pi-ta* and *Pi-ta*² are two distinct alleles at the *Pi-ta* locus (Kiyosawa, 1966, 1967, 1969b). Table 1 lists the known genes for blast resistance.

Table 1 Genes for resistance to blast in rice identified with the use of Japanese isolates of blast fungus.

Gene locus	Allele	Type variety
Pi-a	*Pi-a*	Aichi Asahi
Pi-b	*Pi-b*	BL8
Pi-f	*Pi-f*	St1
Pi-i	*Pi-i*	Shikare-shiroke
Pi-k	*Pi-k*	Kanto 51
		Pi-k^hK3
		Pi-k^PK2
		Pi-k^sShin 2
Pi-m	*Pi-m*	Minehikari
Pi-s	*Pi-s*	65A15
Pi-t	*Pi-t*	K59
Pi-ta	*Pi-ta*	Yashiro-mochi
		*Pi-ta*²Pi4
Pi-z	*Pi-z*	Fukunishiki
	Pi-z^t	Toride 1

More recently, Mackill and Bonman (1991) developed a set of near-isogenic lines (NILs) in order to systematically study the genetics of resistance to blast in tropical rice cultivars. Major genes from four donor parents ("Lac 23", "5173", "Pai-kan-tao", and "Tetep") were transferred into the genetic background of the susceptible cultivar CO 39 by backcrossing. After six backcrosses, plants were selfed for three generations to obtain homozygous resistant lines. Twenty NILs were inoculated with a diverse set of Philippine blast isolates and were classified into six groups based on their reactions to the isolates. One NIL from each group was used to make all possible crosses among resistance groups. Susceptible recombinants were recovered in all crosses, except the cross representing groups III and VI. The absence of susceptible recombinants in this cross indicates that resistance in these two groups is controlled by two very tightly linked or allelic genes. All the other crosses segregated in a ratio of 15 resistant to 1 susceptible indicating that independent dominant genes confer resistance in the four groups. These genes were tentatively designated as *Pi-1* to *Pi-4*. The allelic relationships of these genes with the genes identified in Japan, however, need to be determined.

Bacterial blight

Bacterial blight occurs in almost all the rice-growing countries except those in Europe. The disease symptoms usually appear at the flowering stage in the field. Lesions start at the margins of the leaf blades; they enlarge and form a wavy margin. As the disease advances, the lesions cover the entire leaf blade and may even advance into the leaf sheath. Under heavy disease pressure all the leaves may be killed, and severe yield losses may result. In the tropical areas, disease can attack the crop at the seedling stage, and some tillers or the entire plant may die. If the attack occurs at seedling stage, the disease is referred to as kresek.

Variation in the pathogenicity of the disease organism has been recognized in several countries. At least six races are known in the Philippines (Mew, 1987) and five in Japan (Horino *et al.*, 1981). Similar variations have been reported in several other countries (Yamamoto *et al.*, 1977; Vera Cruz and Mew, 1989).

Several techniques have been developed for assessing cultivar resistance to bacterial blight. The clipping technique (Kauffman *et al.*, 1973) is most widely used. It involves clipping the upper one-third of the leaves with scissors. The scissor blades are dipped in a bacterial inoculum before clipping. The cut ends of the leaves are inoculated and the lesions develop rapidly. Scoring is done 14 to 18 days after inoculation. Using this technique, large volumes of germplasm have been evaluated for resistance and many donors have been identified.

Studies on the genetics of resistance were initiated in Japan by Sakaguchi (1967). He analyzed 27 cultivars and identified two dominant genes for resistance. These genes, designated as *Xa-1* and *Xa-2* were found to be linked with a recombination value of 2-16%. From linkage analysis with marker genes, *Xa-1* and *Xa-2* were located on chromosome 4. Ezuka *et al.* (1975) identified another dominant gene in "Wase Aikoku" and designated it as *Xa-w*. This designation was later changed to *Xa-3* to conform to the rules of gene symbolization in rice (Petpisit *et al.*, 1977).

Genetic studies on resistance to bacterial blight were initiated at the International Rice Research Institute (IRRI) in 1969. Philippine races of the bacterium were used in these studies. From genetic analysis of a large number of cultivars over a 20-year period, nine additional genes have been identified.

A dominant gene *Xa-4* and a recessive gene *xa-5* were identified by Petpisit *et al.* (1977). These two genes segregate independently of each other. Sidhu and Khush (1978) identified another dominant gene designated as *Xa-6*. It was found to be linked with *Xa-4*. Sidhu *et al.* (1978) identified two additional genes designated as *Xa-7* and *xa-8*. These two genes were found to be independent of *Xa-4*, *xa-5*, and *Xa-6*. On the basis of genetic analysis of a large number of resistant cultivars, Singh *et al.* (1983) identified a recessive gene designated as *xa-9*.

Up to the early 1980s, genetic studies for resistance to bacterial blight at IRRI were carried out using Philippine race 1 of the pathogen. However, Mew *et al.* (1982), after an evaluation of a large number of isolates of the pathogen from the Philippines and using a set of differential cultivars, identified four races of bacterial blight. Among the differentials used, "CAS 209" was resistant to race 2 only. Yoshimura *et al.* (1983) carried out genetic analysis of CAS 209 using race 2 and found a dominant gene for resistance. Designated as *Xa-10*, it was found to be linked with *Xa-4*, with a recombination value of 27.4%.

Rice breeding line IR944-102-2-3 was analyzed by Ogawa and Yamamoto (1986), using Japanese races IA, II, and IIIA. It was found to have two dominant genes; one was allelic to *Xa-3* and the other was designated as *Xa-11*. *Xa-11* does not convey resistance to Philippine races of the pathogen. The well-known rice cultivar IR8 has *Xa-11* for resistance. A dominant gene, *Xa-kg*, for resistance to Japanese race V was identified in rice cultivars Kogyoku and Java 14. It was redesignated as *Xa-12* to conform to the rules of gene symbolization in rice (Ogawa, 1987). A recessive gene for resistance to Philippine races 4 and 6 was identified by Ogawa *et al.* (1987) and designated as *xa-13*. Taura *et al.* (1987) identified a dominant gene for resistance to race 5 and designated it as *Xa-14*. In a re-examination of earlier results using four Philippine races, Ogawa *et al.* (1990a,b) showed that *Xa-3*, *Xa-6*, and *xa-9* are allelic; gene symbols *Xa-6* and *xa-9* are thus redundant.

Thus, 12 distinct genes for resistance to Philippine races of bacterial blight are known. The differential reactions of these genes to Philippine races are shown in Table 2. These genes have been transferred to the genetic background of "IR24" by backcrossing. These NILs (Ogawa *et al.*, 1988) are being used as differentials to identify races of bacterial blight in different countries.

To date, more than 1000 resistant cultivars from many Asian countries have been analyzed. These can be classified into groups depending upon the gene(s) for resistance they possess (Ogawa *et al.*, 1991). Group designation is based upon the name of the cultivar in which the gene was first discovered.

Table 2 Major genes for resistance to bacterial blight in rice and their reaction to Philippine races of the pathogen.

Gene	Chromosome location	Reaction to Philippine races of bacterial blight*					
		1	*2*	*3*	*4*	*5*	*6*
Xa-1	4	S	S	S	S	S	S
Xa-2	4	S	S	S	S	S	S
Xa-3	11	R	R	R	R	R	S
Xa-4	11	R	S	S	MR	R	S
xa-5	5	R	R	R	MR	R	S
Xa-7	-	R	R	R	S	R	S
xa-8	-	R	R	R	R	R	S
Xa-10	11	S	R	S	S	R	S
Xa-11	-	S	S	S	S	S	S
Xa-12	4	S	S	S	S	S	S
xa-13	5	S	S	S	R	R	R
Xa-14	-	S	S	S	S	R	S

*R = resistant; MR = moderately resistant; S = susceptible.

Different cultivar groups, their resistance genes, and their reactions to Philippine races are shown in Table 3.

Of all the known genes for resistance, *Xa-4* has been most extensively utilized in developing resistant cultivars. It has been incorporated into most of the IR cultivars. Numerous other breeding lines developed at IRRI with *Xa-4* have been released in other countries. Many improved breeding lines with *xa-5*, *Xa-7*, and *xa-13* are now available and some will become commercial cultivars (Khush *et al.*, 1989).

Tungro

Of all the virus diseases of rice, tungro is the most serious. It is known by different names in different countries. The disease is caused by co-infection of two virus particles, the bacilliform virus and the spherical virus. It is transmitted by the green leafhopper *Nephotettix virescens* Distant. Two strains of virus, S and M strains, were identified at IRRI on the basis of differential chlorotic symptoms on "FK 135" and "Acheh" (Rivera and Ou,

Table 3 Rice cultivar groups based on the genes for resistance to bacterial blight and their reaction to Philippine races of the pathogen.

Cultivar group	Resistance gene	Reaction to Philippine races of bacterial blight*					
		1	2	3	4	5	6
Java 14	*Xa-3*	R	R	R	R	R	S
TKM 6	*Xa-4*	R	S	S	MR	R	S
DZ 192	*xa-5*	R	R	R	MR	R	S
CAS 209	*Xa-10*	S	R	S	S	R	S
BJ1	*xa-5 + xa-13*	R	R	R	R	R	R
Mond Ba	*Xa-4 + Xa-10*	R	R	S	MR	R	S
DV 85	*xa-5 + Xa-7*	R	R	R	MR	R	S

*R = resistant; MR = moderately resistant; S = susceptible.

1967). Very little information is available on the strain variation from country to country.

A large amount of germplasm has been evaluated for resistance to tungro at IRRI and by the various national rice improvement programs. However, very few good sources of resistance have been identified. Good protection against the virus is provided by resistance to the virus vector, the green leafhopper. If the level of resistance to the vector is high, the cultivar does not get infected by the virus. A field screening technique is employed to identify the germplasm that does not get infected. Test lines are planted interspersed with spreader rows of a susceptible cultivar. The susceptible cultivar is also planted around the borders of the field. To ensure good disease pressure, artificially infected seedlings of a susceptible cultivar are planted among the spreader rows. The planting time of the materials is arranged so that the rapid tillering phase of the test plants coincides with the maximum population density of the insect vector. In this way, the test materials are exposed to heavy disease pressure and the entries with the least infection are selected. Most of the IR cultivars released to date were selected through this screening technique. When inoculated artificially with viruliferous insects they became infected. Therefore, field tolerance for the virus is due to their resistance to the vector (Khush, 1977).

In recent years at IRRI, emphasis has been put on incorporating resistance to the virus into improved germplasm. Cultivars known to have resistance to the virus were crossed with an improved plant type line IR1561-228-3-3, which is highly susceptible to the virus as well as to the vector (IRRI,

1990). The F_1s were resistant, indicating the dominant nature of resistance. Four backcrosses were made using IR1561-228-3-3 as the recurrent parent. Improved plant type progenies from the fourth backcrosses have been selected, which are susceptible to the vector but resistant to the virus. The progenies are being evaluated as potential tungro-resistant cultivars. Since the resistance has been transferred through several backcrosses, it appears that major genes for resistance are involved. We are in the process of studying inheritance of resistance in these lines.

Grassy stunt

Grassy stunt was considered a minor disease before the early 1970s. However, with the increased incidence of brown planthopper (*Nilaparvata lugens* Stal) (BPH), which vectors the grassy stunt virus, incidence of this disease has also increased. As in tungro, resistance to the vector provides good protection against grassy stunt. A large volume of germplasm was evaluated for resistance to grassy stunt at IRRI in the early 1970s using an artificial mass screening technique. Several cultivars with field tolerance were identified. One accession of *O. nivara* Sharma and Shastry, a wild relative of cultivated rice, was found to be highly resistant (Ling *et al.*, 1970). Genetic analysis revealed that a single dominant gene designated as *Gs* confers resistance in *O. nivara* (Khush and Ling, 1974). This gene segregates independently of *Bph-1*, the dominant gene for resistance to the vector (BPH) of this disease.

Oryza nivara has many undesirable traits such as weak stems, a spreading growth habit, shattering panicles, and long awns. Therefore, the dominant gene for grassy stunt resistance was transferred to cultivated rice by a backcrossing program using IR24 as a recurrent parent. Grassy stunt-resistant lines from the fourth backcross were used as donor parents and several high-yielding cultivars with resistance to grassy stunt such as IR28, IR30, IR32, IR34, IR36 and IR42 were developed (Khush *et al.*, 1977). The first of these grassy stunt-resistant cultivars were released in 1974. These were widely grown in Asia during the 1970s and 1980s. With large-scale adoption of these cultivars, grassy stunt ceased to be a problem in farmers' fields. A new strain of the disease called grassy stunt 2 appeared in the Philippines in 1985 (Hibino *et al.*, 1985). The *Gs* gene from *O. nivara* does not convey resistance to this new strain and, to date, good donors for resistance to this strain have not been identified. However, grassy stunt is still not a serious problem as most of the cultivars grown in Asia are resistant to the vector, the BPH.

SELECTING FOR INSECT RESISTANCE

About 100 species of insects infest and feed on the rice crop. Twenty of them are of major economic importance. Considerable yield losses are caused by different species of stem borers, planthoppers and leafhoppers, gall midge, and several other insects every year in most rice-growing countries. Prior to the 1960s, chemicals were the only known means to control insects in rice; however, clear-cut cases of host resistance to several important insect species were demonstrated at IRRI and elsewhere, and programs to develop insect-resistant cultivars were initiated.

Brown planthopper

BPH is the most destructive insect pest of rice. It occurs in all the rice-growing areas of Asia. Cases of hopperburn and serious yield losses caused by this insect have been reported from many countries during the 1970s and 1980s.

Differences in cultivar resistance to BPH under controlled (greenhouse) conditions were first demonstrated at IRRI in 1967 (IRRI, 1968; Pathak *et al.*, 1969). A mass screening technique was developed for evaluating germplasm for resistance to the leafhoppers and planthoppers. Seeds of test cultivars are planted in rows about 5 cm apart in 60 x 45 x 10-cm seedboxes. The 45-cm rows are divided in the middle. Thus, there are 24 rows in one seedbox. Twenty rows are planted to the test cultivars and two rows each to resistant and susceptible checks. The seedlings are infested at the one leaf stage (7-8 days after sowing) with second- or third-instar nymphs of the hoppers reared in cages on a susceptible cultivar. Reactions are recorded when seedlings of the susceptible check are killed, generally 8-9 days after infestation.

With this technique, more than 40,000 accessions from the germplasm collections have been evaluated for resistance to BPH and approximately 800 resistant genotypes have been identified. Initially, the germplasm was screened with biotype 1 of the insect. Biotypes 2 and 3 were identified in the mid 1970s and entries resistant to biotype 1 were then screened for reactions to biotype 2 and 3. The resistant entries could be classified into three categories; (1) resistant to biotypes 1 and 2, (2) resistant to biotypes 1 and 3, and (3) resistant to all three biotypes. Another biotype of BPH, biotype 4, occurs on the Indian subcontinent. Germplasm for resistance to this insect was screened at the Directorate of Rice Research in Hyderabad, India, and the Bangladesh Rice Research Institute (BRRI), Joydebpur, Bangladesh, and donors for resistance to biotype 4 were identified (Kalode *et al.*, 1978).

More than 100 resistant cultivars have been genetically analyzed. Athwal *et al.* (1971) identified dominant and recessive genes for resistance in "Mudgo" and "ASD7", respectively. These genes were designated as *Bph-1* and *bph-2*. They are closely linked and no recombination between them has been observed. Lakshminarayana and Khush (1977) identified two additional genes that were designated as *Bph-3* and *bph-4*, which are closely linked with each other. However, *Bph-1* and *bph-2* on one hand and *Bph-3* and *bph-4* on the other segregate independently of each other. On the basis of trisomic analysis, Ikeda and Kaneda (1982) located *Bph-3* and *bph-4* on chromosome 10. Ikeda and Kaneda (1983) located *bph-2* on chromosome 4 by virtue of its linkage with *d-2*, which is a well-known marker of linkage group 4.

Khush *et al.* (1985) carried out genetic analysis of "ARC 10550", which is resistant to BPH populations in Bangladesh and India (Biotype 4) but susceptible to biotypes 1, 2, and 3. It was found to have a single recessive gene, *bph-5*, which segregates independently of *Bph-1*, *bph-2*, *Bph-3*, and *bph-4*.

Several additional cultivars resistant to biotype 4 but susceptible to biotypes 1, 2, and 3 were genetically analyzed by Kabir and Khush (1988). Two new genes, one dominant and one recessive were identified. These were designated *Bph-6* and *bph-7*, respectively. Ikeda (1985) and Nemoto *et al.* (1989) identified two more genes - one recessive and the other dominant - and these were designated as *bph-8* and *Bph-9*, respectively. Thus, of the nine genes known for resistance to BPH, five are recessive and four are dominant. Their reactions to the four biotypes are shown in Table 4.

Table 4 Genes for resistance to brown planthopper in rice and their reaction to different biotypes.

Gene	Chromosome location	Reaction to biotypes*			
		1	2	3	4
Bph-1	4	R	S	R	S
bph-2	4	R	R	S	S
Bph-3	10	R	R	R	R
bph-4	10	R	R	R	R
bph-5	-	S	S	S	R
Bph-6	-	S	S	S	R
bph-7	-	S	S	S	R
bph-8	-	R	R	R	-
Bph-9	-	R	R	R	-

*R = resistant; S = susceptible.

Recently, additional genes for resistance to BPH have been transferred from wild species *Oryza officinalis* Well ex Watt and *O. australiensis* Domin into cultivated rice (Jena and Khush, 1990). The allelic relationships of these genes with known genes for resistance are being investigated.

The first BPH-resistant cultivar IR26 was released by IRRI in 1973 and had *Bph-1* conditioning resistance. It was rapidly adopted by farmers in the Philippines, Indonesia, and Vietnam. However, after 3 years of large-scale cultivation it became susceptible because of the development of the new biotype 2. "IR36", with the *bph-2* gene for resistance, was then released in 1976 (Khush, 1979) and it replaced IR26 within 1 year. It became the most widely planted cultivar of rice by the late 1970s and it was cultivated on 11 million ha of ricelands by the early 1980s. IR36 and several other cultivars with *bph-2*, such as IR42, IR50, IR54, and IR58, have been widely grown in many Asian countries for the past 15 years and BPH has not caused major damage to them. A new biotype capable of damaging cultivars with *bph-2* did appear in small areas of North Sumatra, Indonesia, and in Mindanao area of the Philippines in 1983, but it has not become widespread. Cultivars with *Bph-3*, such as IR56, IR60, IR62, IR68, IR70, IR72, and IR74, have been released as a precaution against the spread of this new biotype. Following the sequential release strategy (Khush, 1977), three genes for resistance have been utilized and we still have a stockpile of six genes. New genes are also being transferred from the wild species.

Whitebacked planthopper

The whitebacked planthopper, *Sogatella furcifera* Horvath, occurs in all the rice-growing areas of Asia and does moderate damage to the crop. Serious outbreaks of the insect have occurred only rarely (Khush, 1977). With the mass screening technique described in the last section, germplasm collections have been evaluated for resistance at IRRI, India, China, and Korea, and donors for resistance have been identified.

More than 100 resistant cultivars have been genetically analyzed, and five genes for resistance have been identified (Khush 1984). A single dominant gene, designated as *Wbph-1*, was found to convey resistance in "N22" (Sidhu *et al.*, 1979). Angeles *et al.* (1981) identified *Wbph-2* in "ARC 10239". An additional dominant gene *Wbph-3* and a recessive gene *wbph-4* were identified by Hernandez and Khush (1981). Wu and Khush (1985) analyzed 15 cultivars and discovered *Wbph-5* in "N'Diang Marie".

Several breeding lines with different genes for resistance have been developed at IRRI and by national rice improvement programs and are undergoing further evaluation.

Green leafhopper

The green leafhopper is distributed throughout Asia, but it is a more serious pest in the tropics and subtropics. A high population may reduce yields by direct feeding but the insect causes greater damage by vectoring the tungro virus.

A large number of varieties have been screened for resistance to green leafhopper at IRRI, India, China, and Bangladesh and many resistant donors have been identified (Heinrichs *et al.*, 1985). Athwal *et al.* (1971) carried out genetic analysis of three resistant cultivars and identified three independently segregating genes, which were designated *Glh-1*, *Glh-2*, and *Glh-3*. The inheritance of resistance was investigated in 13 additional cultivars by Siwi and Khush (1977) and a recessive gene *glh-4* and a dominant gene *Glh-5* were identified. Two dominant genes, *Glh-6* and *Glh-7*, were identified by Rezaul Karim and Pathak (1982) and a recessive gene *glh-8* by Ghani and Khush (1988). The known genes for resistance to green leafhopper are listed in Table 5.

Table 5 Genes conferring resistance to green leafhopper in rice.

Gene	Type variety	Reference
Glh-1	Pankhari 203	Athwal *et al.* (1971)
Glh-2	ASD 7	Athwal *et al.* (1971)
Glh-3	IR8	Athwal *et al.* (1971)
glh-4	Ptb 8	Siwi and Khush (1977)
Glh-5	ASD 8	Siwi and Khush (1977)
Glh-6	IR36	Rezaul Karim and Pathak (1982)
Glh-7	Modai Karuppan	Rezaul Karim and Pathak (1982)
glh-8	DV 85	Ghani and Khush (1988)

Most of the known genes for resistance have been incorporated into improved rice cultivars. All the IR cultivars with the exception of IR22 have at least one other gene for resistance. All the elite breeding lines which are currently distributed throughout Asia via international nurseries are resistant to green leafhopper.

Zigzag leafhopper

Zigzag leafhopper, *Recilia dorsalis* Motschulsky, occurs in the tropics and subtropics of Asia. However, it is a minor pest of rice. Several donors of resistance have been identified (Heinrichs *et al.*, 1985). Genetic analysis of three donors was carried out by Angeles *et al.* (1986) and three independently segregating genes designated as *Zlh-1*, *Zlh-2*, and *Zlh-3* were identified. These genes segregate independently of *Bph-1*, *bph-2*, *Bph-3*, and *bph-4*, the genes for BPH resistance. Since zigzag leafhopper is a minor pest, no serious efforts have been made to incorporate resistance into improved rice cultivars.

Gall midge

The rice gall midge, *Orseolia oryzae* Wood-Mason, is a very serious pest in several areas of Asia and Africa. It occurs in endemic areas in Indonesia, Malaysia, Thailand, Cambodia, Laos, Vietnam, Southern China, Burma, Bangladesh, Nepal, India, and Sri Lanka and has been reported in Sudan, Cameroons, and Nigeria. The female gall midge lays eggs on the leaves near the base of the plant. The newly hatched larvae creep down the leaf sheath and burrow into the interior of the shoot where they feed on the meristematic tissues. Their feeding stimulates the tillers to grow into tubular galls that resemble an onion leaf. The infested tissue does not develop panicles. Under heavy infestation, the crop is badly stunted and very few panicles are produced.

At least three biotypes of gall midge are known. They occur in India, Thailand, and Indonesia. Many donors for resistance have been identified on the basis of field screening (Heinrichs and Pathak, 1981). A greenhouse screening technique where plants are infested with artificially reared larvae has been used at the Directorate of Rice Research, Hyderabad, India.

Resistance to gall midge was postulated to be due to two genes in "W1263" and four genes in "Ptb 18" (Shastry *et al.*, 1972). In a more diagnostic study however, Satyanarayanaiah and Reddi (1972) showed that resistance in W1263 was governed by a single dominant gene. Chaudhary *et al.* (1986) studied the inheritance of resistance and identified two dominant genes designated *Gm-1* and *Gm-2*. Sahu *et al.* (1990) identified a recessive gene for resistance and designated it as *gm-3*.

Using the resistant donors, several improved cultivars with resistance to gall midge have been developed at IRRI and the national rice improvement programs in India, Sri Lanka, Thailand, and Indonesia. They have been grown for the last 15 years and the resistance has not broken down.

MULTIPLE RESISTANCE IN RICE

In most rice-growing countries, serious yield losses are caused by more than one disease or insect. One year there may be an epidemic of bacterial blight, the next year green leafhopper and tungro may cause serious damage and the following year an outbreak of BPH and grassy stunt might occur. To minimize losses from disease and insect attacks, cultivars with multiple resistances to most major diseases and insects are required. Four diseases - blast, bacterial blight, tungro and grassy stunt - and four insects - BPH, green leafhopper, stem borers, and gall midge - commonly occur in most countries of tropical and subtropical Asia. Development of improved germplasm with multiple resistance to these diseases and insects has been a major objective of the IRRI breeding program. Rapid progress was made in developing improved germplasm with resistance to blast, bacterial blight, tungro, grassy stunt, BPH, and green leafhopper. Moderate levels of resistance to stem borers have been incorporated in the improved cultivars. Because gall midge does not occur in the Philippines, it has not been possible to incorporate resistance to this insect in IR cultivars except in IR32, IR36, IR38 and IR42, which were screened collaboratively for gall midge resistance in India. As shown in Table 6, earlier IR cultivars were susceptible to most diseases and insects, but IR26 and later releases have multiple resistance.

MAJOR GENE VERSUS POLYGENIC RESISTANCE

We have utilized donors with major genes as well as polygenes for resistance. Major gene resistance to blast, for example, is ephemeral and breaks down in 2-3 years. Therefore, we have been selecting for polygenic or partial resistance to blast. In our screening program for blast resistance, we select only the progenies with a score of 3-6 on a scale of 1-9. Progenies with a score of 1 or 2 have high resistance conditioned by major genes and are discarded. Similarly progenies with a score of 7-9 have high susceptibility and are discarded. Fortunately many of the parents used in the breeding program have partial resistance. Thus, it is not necessary to employ special procedures to accumulate polygenes into the breeding lines. Just simple screening and selection for partial resistance has been effective for incorporating durable resistance to blast in the elite germplasm. IR36, for example, shows a blast score of 5-6 in the blast nursery and disease development is slow. It has been grown on millions of hectares of riceland every year for the last 15 years and has not suffered blast damage. IR50, on the other hand, had a high resistance level to blast (score 1) when released in 1979, but the resistance broke down within 2 years. Partial resistance to stem borers has also been selected. Major genes for resistance to stem borers have not been identified despite a thorough evaluation of germplasm

Table 6 Disease and insect reactions of IR cultivars of rice.

Cultivar	Blast	Bact. blight	Tungro	Grassy stunt	Green leafhop.	Brown planthop.	Stem borer	Gall midge
IR5	MR	S	S	S	R	S	MR	S
IR8	S	S	S	S	R	S	S	S
IR20	MR	R	S	S	R	S	MR	S
IR22	S	R	S	S	S	S	S	S
IR24	S	S	S	S	R	S	S	S
IR26	MR	R	MR	S	R	R	MR	A
IR28	R	R	R	R	R	R	MR	S
IR32	MR	R	R	R	R	R	MR	R
IR36	MR	R	R	R	R	R	MR	R
IR38	MR	R	R	R	R	R	MR	R
IR42	MR	R	R	R	R	R	MR	R
IR46	MR	R	R	R	R	R	MR	R
IR50	S	R	R	R	R	R	S	R
IR54	MR	R	R	R	R	R	MR	-
IR58	MR	R	R	R	R	R	S	-
IR60	MR	R	R	R	R	R	MR	-
IR62	MR	R	R	R	R	R	MR	-
IR64	MR	R	R	R	R	R	MR	-
IR66	MR	R	R	R	R	R	MR	-
IR68	MR	R	R	R	R	R	MR	-
IR72	MR	R	R	R	R	R	MR	-

*R = resistant; MR = moderately resistant; S = susceptible; - = not known.

collections (Khush, 1977; Chaudhary *et al.*, 1984). We have accumulated polygenes from several donors into IR cultivars. IR36 has the best level of resistance thus far.

For resistance to bacterial blight selection has primarily been for major gene resistance. The gene *Xa-4* has been incorporated into most of the IR cultivars since the first bacterial blight-resistant line was released in 1969. Since then, cultivars with *Xa-4* have been widely grown in the Philippines and elsewhere. *Xa-4* confers resistance to races 1 and 5, moderate resistance to race 4, but no resistance to races 2, 3 and 6. Races 2 and 3 are becoming predominant in the Philippines and we have developed germplasm with *Xa-5* and *Xa-7* which confer resistance to these races. Similarly, major genes for resistance to BPH and green leafhopper have been incorporated into the improved cultivars. Cultivars with *Bph-1* for resistance to BPH remained resistant for only about 3 years . However, cultivars with *bph-2* are still resistant after 15 years. Thus, the useful life of the different major genes may differ. Green leafhopper-resistant cultivars with different major genes have also been grown on a large scale for the last 20 years and not a single case of the breakdown of resistance has been recorded. These examples show that major gene resistance can be successfully employed in host resistance programs. The major advantage is that it can be incorporated into new cultivars in the short period of 3-4 years with time-tested breeding methods such as backcrossing and pedigree selection.

In the germplasm evaluation programs we have found parents with tolerance or low levels of resistance to each disease and insect pest studied. Such tolerance is presumably conditioned by polygenes and is perhaps synonymous with horizontal resistance. There are practical difficulties with incorporating this type of resistance into improved germplasm. The tolerant parents are landraces with very poor agronomic traits. In the process of selecting plants with better agronomic traits in crosses involving such parents, the tolerance levels are either lost or diluted. Moreover, it is difficult to identify the segregants with tolerance because the screening techniques generally used favor selecting genotypes with high levels of resistance.

Another difficulty is the length of time required for developing cultivars with polygenic resistance. At least 10-12 years is needed to accumulate polygenes from several parents to build sufficient levels of resistance. This is a rather long period in the face of recurring epidemics. When multiple resistance is required, as in the case of rice in the tropics and subtropics, it is next to impossible to develop cultivars with polygenically governed multiple resistance. It is difficult to accumulate polygenes for resistance to even one disease or insect, and nearly impossible to concurrently incorporate polygenes for resistance to each disease and insect pest into the same cultivar. The strategy adopted by rice breeders, e.g., the incorporation of polygenic resistance to some diseases (such as blast and stem borers) and major gene resistance to others (such as bacterial blight, grassy stunt, BPH and green leafhopper) has been very successful. This system may serve as a model for other crops.

REFERENCES

Ahn, C. J.; Chung, H. C. (1962) Studies on the physiologic races of rice blast fungus *Piricularia oryzae*. *Seoul University Journal of Biology and Agriculture Series B* 11:77-83.

Angeles, R. E.; Khush, G. S.; Heinrichs, E. A. (1981) New genes for resistance to whitebacked planthopper in rice. *Crop Science* 21:47-50.

Angeles, E. R.; Khush, G. S.; Heinrichs, E. A. (1986) Inheritance of resistance to planthoppers and leafhoppers in rice. In: *Rice genetics*. International Rice Research Institute, Manila, Philippines, pp. 537-550.

Athwal, D. S.; Pathak, M. D.; Bacalangco, E. H.; Pura, C. D. (1971) Genetics of resistance to brown planthoppers and green leafhoppers in *Oryza sativa* L. *Crop Science* 11:747-750.

Bandong, J. M.; Ou, S. H. (1966) The physiologic races of *Piricularia oryzae* Cav. in the Philippines. *Philippine Agriculturist* 49:655-667.

Chaudhary, B. P.; Srivastava, P. S.; Srivastava, M. N.; Khush, G. S. (1986) Inheritance of resistance to gall midge in some cultivars of rice. In: *Rice genetics*. International Rice Research Institute, Manila, Philippines, pp. 523-528.

Chaudhary, R. C.; Khush, G. S.; Heinrichs, E. A. (1984) Varietal resistance to rice stem borers in Asia. *Insect Science and Its Application* 5:447-463.

Chiu, R. J.; Chien, C. C.; Lin, S. Y. (1965) Physiologic races of *Piricularia oryzae* in Taiwan. In: *The rice blast disease*. Proceedings of a Symposium at IRRI, July 1963. The Johns Hopkins Press, Baltimore, Maryland, pp. 245-255.

Ezuka, A.; Horino, O.; Toriyama, K.; Shinoda, H.; Morinaka, T. (1975) Inheritance of resistance of rice variety Wase Aikoku 3 to *Xanthomonas oryzae*. *Bulletin Tokai-Kinki National Agricultural Experiment Station* 28:124-130.

Ghani, M. U.; Khush, G. S. (1988) A new gene for resistance to green leafhopper, *Nephotettix virescens* (Distant) in rice. *Journal of Genetics* 67:151-159.

Heinrichs, E. A.; Medrano, F. G.; Rapusas, H. R. (1985) *Genetic evaluation for insect resistance in rice*. International Rice Research Institute, Manila, Philippines.

Heinrichs, E. A.; Pathak, P. K. (1981) Resistance to the rice gall midge, *Orseolia oryzae* in rice. *Insect Science and Its Applications* 1:123-132.

Hernandez, J. E.; Khush, G. S. (1981) Genetics of resistance to whitebacked planthopper in some rice (*Oryza sativa* L.) varieties. *Oryza* 18:44-50.

Hibino, H.; Cabauatan, P. Q.; Omura, T.; Tsuchizaki, T. (1985) Rice grassy stunt virus strain causing tungro like symptoms in the Philippines. *Plant Disease* 69:538-541.

Horino, A.; Mew, T. W.; Khush, G. S.; Ezuka, A. (1981) Comparison of two differential systems for distinguishing pathogenic groups of *Xanthomonas campestris* pv. *Oryzae*. *Annals Phytopathological Society of Japan* 47:1-14.

Ikeda, R. (1985) Studies on the inheritance of resistance to the rice brown planthopper (*Nilaparvata lugens* Stal) and the breeding of resistant rice cultivars. *Bulletin National Agriculture Research Center* 3:1-54.

Ikeda, R.; Kaneda, C. (1982) Genetic relationships of brown planthopper resistance to dwarf disease and stripe disease in rice. *Japan Journal of Breeding* 32:177-185.

Ikeda, R.; Kaneda, C. (1983) Trisomic analysis of the gene *Bph-1* for resistance to the brown planthopper, *Nilaparvata lugens* Stal. in rice. *Japanese Journal of Breeding* 33:40-44.

International Rice Research Institute (IRRI) (1968) Annual report for 1967. Manila, Philippines.

International Rice Research Institute (IRRI) (1990) Annual report for 1989. Manila, Philippines.

Jena, K. K.; Khush, G. S. (1990) Introgression of genes from *Oryza officinalis* Well ex Watt to cultivated rice, *O. sativa* L. *Theoretical and Applied Genetics* 80:737-745.

Kabir, M. A.; Khush, G. S. (1988) Genetic analysis of resistance to brown planthopper in rice (*Oryza sativa* L.). *Plant Breeding* 100:54-58.

Kalode, M. B.; Krishna, T. S.; Gour, T. B. (1978) Studies on the patterns of resistance to brown planthopper (*Nilaparvata lugens*) in some rice varieties. *Proceedings of the Indian National Science Academy* B44:43-48.

Kaufmann, H.; Reddy, A. P. K.; Hsieh, S. P. Y.; Merca, S. D. (1973) An improved technique for evaluating resistance of rice varieties to *Xanthomonas oryzae*. *Plant Disease Reporter* 57:537-541.

Khush, G. S. (1977) Disease and insect resistance in rice. *Advances in Agronomy* 29:265-341.

Khush, G. S. (1979) Genetics of and breeding for resistance to brown planthopper. In: *Brown planthopper: Threat to rice production in Asia.* International Rice Research Institute, Manila, Philippines, pp. 321-332.

Khush, G. S. (1984) Breeding rice for resistance to insects. *Protection Ecology* 7:147-165.

Khush, G. S.; Ling, K. C. (1974) Inheritance of resistance to grassy stunt virus and its vector in rice. *Journal of Heredity* 65:134-136.

Khush, G. S.; Ling, K. C.; Aquino, R. C.; Aguiero, V. M. (1977) Breeding for resistance to grassy stunt in rice. Proceedings 3rd international Congress, SABRAO, Canberra, Australia. *Plant Breeding Papers* 1:3-9.

Khush, G. S.; Mackill, D. J.; Sidhu, G. S. (1989) Breeding rice for resistance to bacterial blight. In: *Bacterial blight of rice.* International Rice Research Institute, Manila, Philippines, pp. 207-217.

Khush, G. S.; Rezaul Karim, A. N. M.; Angeles, E. R. (1985) Genetics of resistance of rice cultivar ARC 10550 to Bangladesh brown planthopper biotype. *Journal of Genetics* 64:121-125.

Kiyosawa, S. (1966) Studies on inheritance of resistance of rice varieties to blast. 3. Inheritance of resistance of a rice variety Pi No. 1 to the blast fungus. *Japanese Journal of Breeding* 16:243-250.

Kiyosawa, S. (1967) Inheritance of resistance of the rice variety Pi No. 4 to blast. *Japanese Journal of Breeding* 17:165-172.

Kiyosawa, S. (1969a) Inheritance of resistance of rice varieties to a Philippine fungus strain of *Pyricularia oryzae. Japanese Journal of Breeding* 19:61-73.

Kiyosawa, S. (1969b) Inheritance of blast resistance in West Pakistani rice variety, Pusur. *Japanese Journal of Breeding* 19:121-128.

Kiyosawa, S. (1972) Genetics of blast resistance. In: *Rice breeding.* International Rice Research Institute, Manila, Philippines, pp. 203.226.

Kiyosawa, S. (1974) Studies on genetics and breeding of blast resistance in rice. Miscellaneous publications of the National Institute of Agricultural Sciences. Series D I. 58 pp. (In Japanese, English summary).

Kiyosawa, S.; Murty, V. V. S. (1969) The inheritance of blast resistance in Indian rice variety HR 22. *Japanese Journal of Breeding* 19:269-278.

Lakshminarayana, A.; Khush, G. S. (1977) New genes for resistance to the brown planthopper in rice. *Crop Science* 17:96-100.

Latterell, F. M.; Tullis, E. C.; Collier, J. W. (1960) Physiologic races of *Piricularia oryzae. Plant Disease Reporter* 44:679-683.

Ling, K. C.; Aquiero, V. M.; Lee, S. H. (1970) A mass screening method for testing resistance to grassy stunt disease of rice. *Plant Disease Reporter* 56:565-569.

Mackill, D. J.; Bonman, J. M. (1991) Inheritance of blast resistance in near-isogenic lines of rice. *Phytopathology* 91:(in press).

Mew, T. W. (1987) Current status and future prospects of research on bacterial blight of rice. *Annual Review of Phytopathology* 25:359-382.

Mew, T. W.; Vera, C. M.; Reyes, R. (1982) Interaction of *Xanthomonas campestris* pv. *Oryzae* and resistant rice cultivar. *Phytopathology* 72:786-789.

Nemoto, H.; Ikeda, R.; Kaneda, C. (1989) New genes for resistance to brown planthopper, *Nilaparvata lugens* Stal., in rice. *Japanese Journal of Breeding* 39:23-28.

Niizeki, H. (1960) On a gene for resistance to *Piricularia oryzae* in Japanese rice variety, Aichi Asahi. *Agriculture and Horticulture* 35:1321-1322.

Ogawa, T. (1987) Gene symbols for resistance to bacterial blight. *Rice Genetics Newsletter* 4:41-43.

Ogawa, T.; Busto, G. A.; Tabien, R. E.; Romero, G. O.; Endo, N.; Khush, G. (1991) Grouping of rice cultivars based on reaction pattern to Philippine races of bacterial blight. *Japanese Journal of Breeding* 41:41:109-119.

Ogawa, T., Lin, L.; Tabien, R. E.; Khush, G. S. (1987) A new recessive gene for resistance to bacterial blight of rice. *Rice Genetics Newsletter* 4:98-100.

Ogawa, T.; Yamamoto, T. (1986) Inheritance of resistance to bacterial blight in rice. In: *Rice genetics*. International Rice Research Institute, Manila, Philippines, pp. 471-480.

Ogawa, T.; Yamamoto, T.; Khush, G. S.; Mew, T. W. (1990a) Genetics of resistance in rice cultivars, Zenith and Cempo Selak to Philippine and Japanese races of bacterial blight pathogen. *Japanese Journal of Breeding* 40:183-192.

Ogawa, T.; Yamamoto, T.; Khush, G. S.; Mew, T. W. (1990b) Genetics of resistance in rice cultivar Sateng to Philippine and Japanese races of bacterial blight pathogen. *Japanese Journal of Breeding* 40:329-338.

Ogawa, T.; Yamamoto, T.; Khush, G. S.; Mew, T. W. (1991) Grouping of rice cultivars based on reaction pattern to Philippine races of bacterial blight. *Japanese Journal of Breeding* 41:109-119.

Ogawa, T.; Yamamoto, T.; Khush, G. S.; Mew, T. W.; Kaku, H. (1988) Near-isogenic lines as international differentials for resistance to bacterial blight of rice. *Rice Genetics Newsletter* 5:106-107.

Ou, S. H. (1972) *Rice diseases*. Commonwealth Mycological Institute, Kew, England, 368 pp.

Ou, S. H.; Jennings, P. R. (1969) Progress in the development of disease resistant rice. *Annual Review of Phytopathology* 7:383-410.

Padmanabhan, S. Y. (1965) Physiologic specialization of *Piricularia oryzae* Cav., the causal organism of blast disease of rice. *Current Science* 34:307-308.

Pathak, M. D.; Cheng, C. H.; Fortuno, M. E. (1969) Resistance of *Nephotettix impicticips* and *Nilaparvata lugens* in varieties of rice. *Nature (London)* 223:502-504.

Petpisit, V.; Khush, G. S.; Kauffman, H. E. (1977) Inheritance of resistance to bacterial blight in rice. *Crop Science* 17:551-554.

Rezaul Karim, A. N. M.; Pathak, M. D. (1982) New genes for resistance to green leafhopper, *Nephotettix virescens* (Distant) in rice, *Oryza sativa* L. *Crop Protection* 1:483-490.

Rivera, C. T.; Ou, S. H. (1967) Transmission studies of the two strains of rice tungro virus. *Plant Disease Reporter* 51:877-881.

Sahu, V. N.; Mishra, R.; Chaudhary, B. P.; Shrivastava, P. S.; Shrivastava, M. N. (1990) Inheritance of resistance to gall midge in rice. *Rice Genetics Newsletter* 7:118-121.

Sakaguchi, S. (1967) Linkage studies on the resistance to bacterial leaf blight, *Xanthomonas oryzae* (Uyeda et Ishiyama) Dowson, in rice. *Bulletin National Institute of Agricultural Sciences Series D* 16:1-18.

Sasaki, R. (1922) Existence of strains in rice blast fungus I. *Journal of Plant Protection (Japan)* 9:631-644.

Sasaki, R. (1923) Existence of strains in rice blast fungus II. *Journal of Plant Protection (Japan)* 10:1-10.

Satyanarayanaiah, K.; Reddi, M. V. (1972) Inheritance of resistance to insect gall midge (*Pachydiplosis oryzae* Wood Mason) in rice. *Andhra Agricultural Journal* 19:1-8.

Shastry, S. V. S.; Freeman, W. H.; Seshu, D. V.; Israel, P.; Roy, J. K. (1972) Host-plant resistance to rice gall midge. In: *Rice breeding*. International Rice Research Institute, Manila, Philippines, pp. 353-365.

Sidhu, G. S.; Khush, G. S. (1978) Dominance reversal of a bacterial blight resistance gene in some varieties of rice. *Phytopathology* 68:461-463.

Sidhu, G. S.; Khush, G. S.; Medrano, F. G. (1979) A dominant gene in rice for resistance to whitebacked planthopper and its relation to other plant characters. *Euphytica* 28:227-232.

Sidhu, G. S.; Khush, G. S.; Mew, T. W. (1978) Genetic analysis of bacterial blight resistance in seventy-four cultivars of rice, *Oryza sativa* L. *Theoretical and Applied Genetics* 53:105-111.

Singh, R. J.; Khush, G. S.; Mew, T. W. (1983) A new gene for resistance to bacterial blight in rice. *Crop Science*: 23:558-560.

Siwi, B. H.; Khush, G. S. (1977) New genes for resistance to the green leafhopper in rice. *Crop Science* 17:17-20.

Taura, S.; Ogawa, T.; Tabien, R. E.; Khush, G. S.; Yoshimura, A.; Omura, T. (1987) The specific reaction of Taichung Native 1 to Philippine races of the bacterial blight and inheritance of resistance to race 5 (PXO112). *Rice Genetics Newsletter* 4:101-102.

Vera Cruz, C. M.; Mew, T. W. (1989) How variable is *Xanthomonas campestris* pv. *oryzae*? In: *Bacterial blight of rice*. International Rice Research Institute, Manila, Philippines, pp. 153-166.

Wu, C. F.; Khush, G. S. (1985) A new dominant gene for resistance to whitebacked planthopper in rice. *Crop Science* 25:505-509.

Yamamoto, T.; Hartini, R. H.; Muhammad, M.; Nishizawa, T.; Tantera, D. M. (1977) Variation in *Xanthomonas orzyae* (Uyeda et Ishiyama) Dowson and resistance of rice varieties to the pathogen. *Contributions Central Research Institute of Agriculture, Bogor, Indonesia* 28:1-22.

Yamasaki, Y.; Kiyosawa, S. (1966) Studies on inheritance of resistance of rice varieties to blast. I. Inheritance of resistance of Japanese varieties to several strains of the fungus. *Bulletin National Institute of Agricultural Sciences Series D* 14:39-69.

Yoshimura, A.; Mew, T. W.; Khush, G. S.; Omura, T. (1983) Inheritance of resistance to bacterial in rice cultivar CAS 209. *Phytopathology* 73:1409-1412.

Chapter 14
Emerging breeding strategies for insect resistance

O. M. B. de Ponti[1] and C. Mollema[2]

[1]Nunhems Zaden BV, P. O. Box 4005, 6080 AA Haelen, The Netherlands and
[2]DLO-Centre for Plant Breeding and Reproduction Research (CPRO-DLO),
P. O. Box 16, NL-6700 AA Wageningen, The Netherlands

HOST PLANT RESISTANCE AS THE BASIC PRINCIPLE OF INTEGRATED PEST MANAGEMENT

Although crop losses due to pests and diseases are still significant and chemical pesticides certainly have great potential in reducing these losses, there is a growing awareness that crop protection should not rely on chemical control only but, rather, be based on an integration of protection strategies. An increasing number of developed and developing countries have adopted integrated pest management (IPM) as their crop protection philosophy and strategy. By doing so, they intend to improve the balance between economic, social, and environmental values while aiming at a sustainable, safe, and profitable agriculture. Obviously, this change is beneficial to all concerned - farmers, consumers, and the society as a whole.

IPM aims at promoting the natural, biological and cultural control potential of an agroecosystem as the first line of control. Chemical control is part of IPM, but is only added if non-chemical means perform insufficiently. It is essential to further develop and implement non-chemical alternatives, which fit into an overall IPM strategy and are adopted easily by farmers.

As the plant is the principal factor in agricultural production, it is obvious to exploit its genetic variability to prevent or reduce pest and disease incidence. From the very start of agriculture, farmers have been selecting resistant plants as seed sources for their next crop (de Ponti, 1985). The discovery by Mendel of the genetic basis of variation was the start of genetics as a scientific discipline and of plant breeding as a scientific "interdiscipline" and agricultural enterprise. This has facilitated the selection of improved genotypes with cultivar resistance since the beginning of this century.

Long before the discovery of chemical pesticides, cultivar resistance, therefore, formed the backbone of crop protection (Zadoks, 1991). However, the industrial development of resistant cultivars has also suffered from "scientific sophistication". It is common in many species to select for resistance to the major pests and diseases in seedling nurseries. Selection for agronomic characters such as yield is then performed in the field, generally under chemical protection. Selection under such a pesticide umbrella prevents selection for resistance to the minor pests and diseases and, as a consequence, resistance to these pests and diseases tends to erode (de Ponti, 1981). As deliberate selection for resistance to insects is less common than for resistance to diseases, maintenance of resistance to insects is at risk. Another problem created by the internationalization of plant breeding is that location-specific pest and disease problems tend to be neglected. This can result in unexpected outbreaks of secondary pests and diseases due to the introduction of new cultivars which have been bred elsewhere. To secure and improve the resistance to the entire complex of pests and diseases at a specific location, Buddenhagen and de Ponti (1983) advocated that breeding for resistance should be based on location-specific analyses of the entire pathosystem and that at least the final selection of advanced material should take place within that pathosystem without or with minimum chemical interference. As a consequence, cultivar trials should be performed without or with minimum chemical control. This is the only way to identify cultivars that best fit IPM strategies. It also makes farmers better aware of the potential of partial resistance.

The potential and profitability of partial resistance is often underestimated because, due to the cosmetic effects of chemical control, farmers became used to "clean" crops. With the adoption of IPM, farmers must learn to understand economic damage thresholds and to accept low levels of pest and disease incidence as non-economic. As a consequence, they will learn to appreciate the economic importance of partial resistance alone or in combination with other control strategies. Recently, Van Emden (1987, 1990) reviewed the often synergistic effects of partial resistance to insects and natural, biological, cultural, and chemical control. This is best illustrated by the observation that partial resistance often shows more clearly in the field than in greenhouse tests because of the additional effect of natural enemies in the field. Partial resistance to virus vectors can also reduce the incidence of virus diseases (Jones, 1990), although one should remain alert for an eventual opposite effect (Kennedy, 1976). In addition, partial resistance poses little selection pressure on the pathogen or insect population, resulting in durable resistance and, as part of an IPM strategy, in durable control. Cultivar resistance to pests and diseases, therefore, deserves continuous and increased attention as a basic principle of IPM to which other means of control can easily be added. One of the major advances of cultivar control is the ease of implementation with the seed, without additional knowledge and costs.

This paper focuses on breeding for resistance to insects (including other arthropods such as mites). We do not review the entire field, as several excellent reviews have recently been published (Singh, 1986; Fritzsche *et al.*, 1987; Wink, 1988; Smith, 1989; Campbell and Eikenbary, 1990). After a short analysis of the state of the art, we highlight recent developments which can improve screening and breeding for resistance to insects, including increased understanding of population dynamics and behavioral and molecular biology of insect-plant interactions.

BREEDING FOR RESISTANCE TO INSECTS: STATE OF THE ART

Investments in the development of disease-resistant cultivars are significantly larger worldwide than investments in insect-resistant cultivars. This probably is due to some degree of scientific isolation between entomologists and plant breeders. Pathologists and breeders are generally trained in botanical departments at a university, whereas entomologists are often trained in the zoology department. In addition, entomologists embraced the promise of pesticides to a greater degree than pathologists, who continued and strengthened their efforts on host plant resistance. As a consequence, entomologists earlier became aware of the negative side of the pesticide coin and started to develop IPM strategies. Additionally, when host plant resistance is utilized on a broad scale, it also has a negative side-effect in the repeated selection of new virulent races, strains, and biotypes. The lesson to be learned from this is that interdisciplinary cooperation is essential for the development of durable crop protection strategies including strategies for durable resistance. This should be based on a thorough understanding of the relevant pathosystems (Buddenhagen and de Ponti, 1983).

During the last ten years, breeding for resistance to insects has gained increasing interest in the U.S.A. and the International Agricultural Research Centers (IARCs). These efforts are based on public breeding efforts, and their successes have been commercialized by private breeding companies or handed over to national programs. In Europe, public breeding efforts on insect resistance developed at a slower pace, although interesting results have been reported (Ellis and Freuler, 1990). As a consequence, breeding companies in Europe still have a limited research effort in insect resistance.

Practical results in terms of insect-resistant cultivars have been reported with some of the main food crops - e.g., Hessian fly [*Mayetiola destructor* (Say)] and greenbug- [*Schizaphis graminum* (Rondani)] resistant wheat (*Triticum* spp.), brown planthopper- [*Nilaparvata lugens* (Stal)] resistant rice (*Oryza sativa* L.), and aphid-resistant alfalfa (*Medicago sativa* L.). According to Wiseman (1990), more than 200 cultivars with resistance to 50 insect species have been released and are in commercial production. Extensive research in crops like maize (*Zea mays* L.) and potato (*Solanum tuberosum* L.) will soon result in the release of resistant cultivars.

Terminology

Because resistance to pests and diseases has developed in different scientific schools - entomology and plant pathology, respectively - it is often suggested that the principles and approaches are different. This is mainly due to differences in terminology. The only essential difference is based on the mobility of insects, providing an additional modality of resistance. Terms for insect resistance were defined by the founder of insect resistance (Painter, 1951). His terminology has been used for many years. Presently, resistance to insects is classified in the two categories, antixenosis and antibiosis. Tolerance to insects is an additional modality enabling plants to withstand damage by insect attack. The three modalities are plant characters and are defined as follows:

Antixenosis is the complex of plant characters which prevents insects from approaching, landing, settling, feeding, or ovipositing. Antixenosis affects the behavior of the insects.

Antibiosis is the complex of plant characters which affects the reproduction of insects; it interferes with lifetable components such as longevity, oviposition rate, generation time, and preadult mortality. Antibiosis has an effect on the insect's physiology.

Tolerance is the complex of plant characters which allows the plant to endure some amount of insect attack without affecting its yield or the quality of its marketable product.

All three modalities have potential in protecting plants from insect attack and can be found in gene banks. In the following discussion we will focus on antixenosis and antibiosis, which best lend themselves to deliberate selection in specific laboratory and bench tests. Tolerance is an intriguing modality of plants, but is difficult to tackle with specific tests; it requires simultaneous observations on insect populations and yield potential of adult plants. The way to select for tolerance is to allow some degree of insect attack in the field when selecting for yield. Tolerance has no effect on insect population growth. A tolerant cultivar can, therefore, create problems in adjacent non-tolerant cultivars.

It is clear that breeding for insect resistance has great potential which should be exploited further. Prerequisites for a successful breeding program are (1) interdisciplinary cooperation, (2) availability of sources of resistance, either from the target or related species or from quite different organisms through molecular techniques, and (3) efficient and reliable test procedures based on an in-depth understanding of the specific insect-plant relation.

Durability

As with resistance to diseases, resistance to insects can be non-durable, particularly if the resistance is very high, its inheritance is simple, and if the resistant cultivars are grown on a large scale. This creates an environment for selection of new, virulent biotypes. Classical examples are the many biotypes of Hessian fly, brown planthopper, and alfalfa aphids.

Fortunately, resistance to insects tends to be partial and its inheritance polygenic, so that selection of biotypes breaking resistance is less likely when compared to resistance to diseases. Durability of resistance is now a key issue in breeding for resistance and strategies have been developed to promote it (Parlevliet, 1991). Most important is the limitation of selection pressure on the insect population through partial resistance and the deliberate deployment of monogenic resistance genes. Novel approaches to resistance breeding such as the breeding of transgenic plants carrying toxic genes from *Bacillus thuringiensis* (*Bt*) or the trypsin inhibitor gene from cowpea [*Vigna unguiculata* (L.) Walp] tend to focus on monogenic factors with a very high expression. As a consequence, experts in resistance breeding were not surprised to learn of the extremely fast adaptation of several insect species [*Heliothis zea* (Boddie) and *Leptinotarsa decemlineata* (Say)] to the *Bt* gene. The durability of the trypsin inhibitor gene is also at risk; Dick and Credland (1986) indicated that bruchids are able to adapt to the resistant accession from which this gene was isolated. Strategies for durable transgenic resistances will be discussed in more detail later in this chapter.

If resistance is combined with other crop protection strategies, then it is not necessary for the degree of resistance to be large. Rather, resistance should be partial and one should aim at a durable protection strategy instead of durable resistance alone.

NEW TRENDS IN BREEDING FOR RESISTANCE TO INSECTS

As mentioned before, breeding for resistance to insects is an interdisciplinary endeavour with contributions from a wide variety of scientific disciplines. Based on new scientific insights and tools in the field of population dynamics, insect physiology and molecular biology, new trends in breeding for insect resistance have developed recently. They concern (1) the use of population growth models for exploring resistance management strategies, (2) developing efficient test procedures, (3) a further exploitation of the resistance modality antixenosis, and (4) evaluating the potential of molecular biological techniques.

Population dynamic models for resistance management strategies

New technologies in computer science have contributed to developments in ecology to build insect population growth models. Simulation studies with these models explore and predict the relative effects of the different resistance modalities and individual life history components as potential resistance factors. Starting with a single crop and a single insect species, the complexity of these models can be increased by incorporating modules of more crops, more insects, their natural enemies, and all other control strategies, including insecticides. In this way, the entire patho- or pest system of a specific agricultural system and its pest management can be simulated. Such studies can substitute for vast amounts of experimentation and can select those experiments which are expected to provide useful information. It is obvious that simulation studies have to be validated by experiments.

The potential of this approach was demonstrated by simulation studies of the polyphagous insect *Heliothis zea* in the coastal plain of North Carolina, U.S.A. by Kennedy *et al.* (1987) using the HELSIM model of Stinner *et al.* (1974). This study indicated that a high level of a hypothetical antixenosis of corn whereby its attractiveness for oviposition by *H. zea* was reduced to 10% had little effect on the population of this insect on corn, but resulted in a dramatic calculated increase of the population on soybeans [*Glycine max* (L.) Merr.], tobacco (*Nicotiana tabacum* L.), and cotton (*Gossypium* spp.). Antixenosis of corn seems, therefore, to be completely counterproductive for this specific agroecosystem because of the compensating effect on the other crops. On the other hand, antibiosis, expressed in a hypothetical mortality of 50% of early or late instars, resulted in a dramatic reduction of the population on all crops, and 50% mortality of the late instars was more effective than that of the early instars. Based on these studies, it is clear that breeding for resistance in this specific relation should focus on antibiosis, and preferably on mortality of the late instars.

Hulspas-Jordaan and van Lenteren (1989) developed a population growth model of the greenhouse whitefly, *Trialeurodes vaporariorum* Westwood, on tomato (*Lycopersicon esculentum* Mill.). Sensitivity analyses indicated that of five life history factors, changes in the developmental period and fecundity have by far the greatest effect on population growth. A 10% increase of the developmental period led to a reduction of 43% in the total population size, and a 10% decrease of fecundity led to a 26% reduction. Developmental period and fecundity are, therefore, the most effective criteria for breeding greenhouse whitefly-resistant cultivars. Resistance tests should focus on these life history factors, as will be shown later.

Simulation models can also be used to investigate and predict the durability of specific resistance modalities and deployment strategies. In the same paper of Kennedy *et al.* (1987), it was indicated that antixenotic resistance of North Carolina field corn to *H. zea* will last longer than antibiotic resistance and that a combination of the two modalities will show the best durability. Gould (1986) calculated that a simultaneous release of two resistance genes to the Hessian fly in wheat will result in a better durability than a sequential release and that this durability will be further increased if 20% of the plants in each field contain no resistance genes. This confirms the potential of cultivar mixtures or multilines in the control of pests and diseases which tend to adapt to new resistance genes by a repeated selection for new biotypes.

Without exaggerating the potential of these simulation studies, they can contribute to our understanding of the quantitative aspects of insect-plant relationships and indicate which modalities and management strategies of resistance have the greatest potential. These simulation studies should also be used to assist breeders in exploring management strategies for transgenic resistances, as advocated by Gould (1988).

Population growth models for the development of selection procedures for antibiotic resistance

Breeding of resistant cultivars is a time-, labor- and money-intensive activity. Therefore, research is needed for the development of efficient selection procedures which can handle vast numbers of plants with minimum investments. A first screening of a wide variety of germplasm is generally carried out in a field with a heavy natural infestation of the target insect, a so-called hot spot. Potential sources of resistance are subsequently subjected to detailed investigations to confirm the resistance and to identify the various components of resistance. These are based on the life history factors of the insect, which are most affected by the resistant plant. A thorough understanding of these life history factors and their relative contribution to the population growth of the target insect is of key importance for the development of efficient and effective resistance tests. We call this the study of the biology of the resistance, which is primarily based on population growth theories and sensitivity analyses.

Over the last decade we have found the population model developed by Lewontin (1965) to be a useful instrument to compare population dynamics on different host plants and to select critical periods of the reproductive cycle to use as criteria for resistance. Lewontin's population growth model, often referred to as the Lewontin reproductive triangle (Figure 1), describes reproduction during the period of exponential population increase as a simple function, V, of age (x) where A, T, and W are the time reproduction starts, peaks and ends, respectively. The area, S, of the triangle represents total reproduction. From this reproductive triangle one can calculate r, the intrinsic rate of increase of the insect population. Thus, the triangle is a theoretical representation of reproduction over a single generation. To superimpose a triangle over varying data points from reproduction experiments, Romanow *et al.* (1991) described a method based on linear regression for determining the sides of the triangle. In this way, large data sets are compiled to a simple triangular visualization of the onset, peak, end, and total reproduction of a single generation (Figure 2). Another adaptation of the Lewontin's triangle concerns a shifting in the period of observation (i.e., from the emergence of parent to the emergence of offspring) instead of the common

Figure 1 The Lewontin reproductive triangle. O = emergence of parent, A = start of reproduction, T = peak reproduction, W = end of reproduction, and S (area) = total reproduction (Lewontin, 1965).

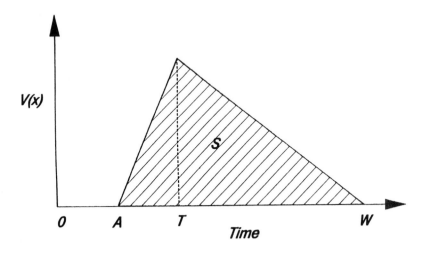

Figure 2 Lewontin reproductive triangles for greenhouse whiteflies on two tomato genotypes; the susceptible cultivar Allround and a resistant wild tomato accession of *L. hirsutum glabratum* (Romanow *et al.*, 1991).

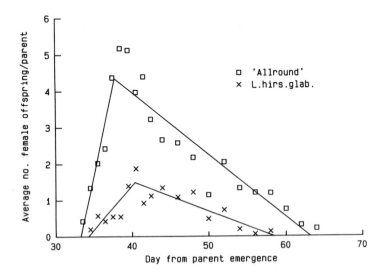

"egg-to-egg" period. In this way, survival of instars or nymphs and developmental time is automatically incorporated into the model.

Components of resistance of a set of genotypes can easily be investigated by comparing respective reproductive triangles and the intrinsic rates of increase based on the calculated values of A, T, W, and S. The reproductive triangle can also be used for simulation studies to predict the relative effects of hypothesized changes in one or more of the factors A, T, W, and S on r. Such simulation studies can predict the relative value of these factors as components of resistance and indicate for each factor separately what minimum change is needed to have an effect.

It is evident that large amounts of data must be collected to determine the calculated values of A, T, W, and S. Such laborious experiments in controlled conditions are justified for investigating the range of resistance in a selected set of resistant genotypes compared to a susceptible control. For screening large numbers of plants in segregating generations, this procedure is far too laborious. Knowing the potential variation among progeny from crosses between resistant and susceptible genotypes, one can select the shortest period of reproduction which represents a reliable estimate of the total reproduction. This truncation incorporates into one figure all reproduction or life history factors over a specific period and can be used for an efficient and effective routine resistance test.

We have used this approach to improve our understanding of the population dynamics of the greenhouse whitefly, on susceptible and resistant genotypes of tomato and of the two-spotted spider mite, *Tetranychus urticae* Koch, on susceptible and resistant genotypes of cucumber (*Cucumis sativus* L.).

With tomato and *T. vaporariorum*, Romanow *et al.* (1991) determined reproduction triangles for eight genotypes - the susceptible cultivar Allround, one resistant accession of the wild species *Lycopersicon hirsutum glabratum* C.H. Mull., and six breeding lines resulting from crosses between these two parental genotypes. In non-choice greenhouse experiments, most of these lines have repeatedly proven to match the resistant parent (de Ponti *et al.*, 1983; Table 1). Figure 2 represents the reproduction triangles of the parental genotypes. In Table 2 the intrinsic rate of increase, total female reproduction, and female reproduction until day 38 is summarized. These reproduction data demonstrate that the insect reproduction on the lines is not significantly different from the susceptible control. By comparing different sets of truncated data, we have found that female reproduction until day 38, which is close to the average day for peak reproduction (38.95), correlated reasonably well with r and total reproduction, whereby the differences from the susceptible control are sometimes significant. It is interesting to note that resistance in this case can be measured despite limiting the observation to a relative short period of early reproduction. Most interesting is that evidently a large amount of antibiotic resistance has been lost in

Table 1 Comparison of the number of empty pupal cases per tomato plant 45 and 90 days after infestation with 10 female greenhouse whiteflies in a non-choice greenhouse experiment. Figures followed by the same letter do not differ significantly at the 5% level (de Ponti *et al.*, 1983).

Material	No. empty pupal cases on	
	Day 45	Day 90
Tomato cv. Allround	403 a	5.737 a
L. hirsutum glabratum PI 251305	227 ab	735 b
Backcross line IVT 80347	247 ab	2.908 ab
Backcross line IVT 81131	156 b	1.298 b

Table 2 Reproduction data (number of empty pupae per female) of the greenhouse whitefly on tomato genotypes, as measured in clip-on leaf cages (Romanow *et al.*, 1991).

Material	No. empty pupal cases		r
	Total	Until day 38	
Tomato cv. Allround	130	22	0.0974
L. hirsutum glabratum	37	4	0.0666
Backcross line IVT 82216	95	14	0.0902
Backcross line IVT 82207	105	18	0.0922
Backcross line IVT 81440	118	14	0.0936

subsequent backcross generations. As most lines perform well in non-choice greenhouse tests, two conclusions have been drawn - (1) there is potential for an increase of antibiotic resistance and (2) *L. hirsutum glabratum* and the selected lines carry high levels of antixenotic resistance.

Based on these studies, a routine bench test for antibiotic resistance has been developed in which newly emerged adult female whiteflies are confined in clip-on leaf cages. On day 8, the leaf cages are removed and the females killed chemically. On day 38, empty pupal cases are counted as a measure for resistance.

Westerman and de Ponti (unpublished data) did a similar study for resistance of cucumber to the two-spotted spider mite, incorporating nine genotypes with various levels of resistance. The reproductive triangles of the susceptible control and the most resistant breeding line are presented in Figure 3. From these results we have developed the following bench test which is used routinely by various Dutch breeding companies. On the first true leaf of a young cucumber plant, 10 newly emerged adult female spider mites are introduced after the petiole has been smeared with sticky "Tanglefoot". After 13 days, when reproduction is about peaking, the number of female adult offspring is counted or the amount of leaf tissue damage assessed to estimate levels of resistance (Table 3). On the resistant genotypes, many of the introduced mites and their offspring got stuck in the "Tanglefoot", indicating the presence of antixenosis in addition to antibiosis. By excluding these stuck mites from the final count, this test combines nicely both antixenotic and antibiotic resistance.

Figure 3 Lewontin reproductive triangles for two-spotted spider mites on two cucumber genotypes; the susceptible line G6 and the resistant line IVT 78235.

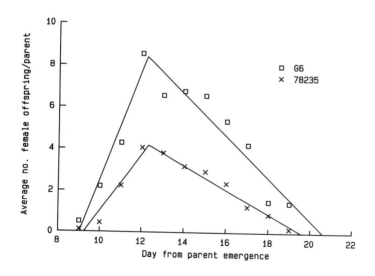

Antixenosis: The behavioral modality of resistance

Although in a previous example the effect of antixenotic resistance of corn to *H. zea* was negative, this is not the rule for all interactions. There is an

Table 3 Comparison of five cucumber genotypes for resistance to the two-spotted spider mite. Antixenosis was assessed by the percentage of mites migrated in 10 days (de Ponti, 1978) and antibiosis by the reproduction per female 13 days after introduction (unpublished data).

Material	Antixenosis	Antibiosis
Susceptible line G6	24	82
PI 220860	76	--
cv. Hybrid LGP	64	38
cv. Robin 50	45	40
Line IVT 78235	--	6

increasing interest in and understanding of antixenotic resistance, thanks to a better knowledge of the physiological and chemical factors that mediate the attractiveness of a plant for landing, settling, feeding, and oviposition by insects. Reduction of attractiveness can be very effective for insect control in the absence of alternative hosts, as it keeps the insects away from its food source and reproduction site. In this way, it stops the reproduction cycle at the first stage and the insects will use their limited energy by a continued search for a suitable host plant. As the physical contact with the rejected antixenotic host plant is relatively short, the chance for adaptation is expected to be smaller than for antibiotic resistance. Because there is little experience with antixenotic resistance in terms of antixenotic cultivars grown on a large scale, there is still little experimental evidence for this supposed durability. However, depending on the selection procedure used, many resistant cultivars have some amount of hidden antixenotic resistance which may contribute to the durability of the antibiotic resistance. In all cases studied we found interesting levels of antixenotic resistance, although the emphasis was on antibiotic resistance.

Several tomato lines resistant to *T. vaporariorum* performed much better in non-choice greenhouse tests than in the antibiosis bench tests. This is evident from comparing the results presented in Table 2 and Figure 2. One of the explanations is that these lines carry a substantial amount of antixenotic resistance which is easily expressed in greenhouse tests where the insects can freely move ("wasting their time"), but not in bench tests in which the insects are confined in small clip-on leaf cages.

In evaluating sources of resistance of cucumber to *T. urticae*, de Ponti (1978) tested both antibiosis and "acceptance" as a parameter of resistance.

"Non-acceptance" or antixenosis was measured by the percentage of female mites that were absent from the leaf 10 days after introduction. Large differences in antixenosis were found, often highly correlated with differences in antibiosis (Table 3).

In cucumber genotypes resistant to *T. urticae*, two types of antixenosis to western flower thrips (*Frankliniella occidentalis* Pergande) were distinguished (Mollema *et al.*, 1989). The first type resulted in reduced egglaying by thrips females, while the second type caused reduced feeding by the thrips larvae.

The first type of antixenosis was found in experiments in which spider mite-resistant and susceptible genotypes were exposed to thrips females for 5 h. Three days later the number of hatched eggs and the percentage of plants on which at least one egg had hatched were reduced on the resistant genotypes. The second type of antixenosis was demonstrated in experiments in which the behavior of young thrips larvae was observed on leaf discs from the same genotypes. The time spent on resting (feeding) or walking (searching) was recorded for periods of 15 minutes. Figure 4 shows that the percentage of time spent on walking is approximately four times greater on the spider mite-resistant genotypes. It will be clear that both types of antixenosis result in significantly less damage, as was shown in a greenhouse

Figure 4 The percentage of time spent on walking by young *F. occidentalis* larvae on spider mite-resistant and -susceptible cucumber genotypes; b (bitter, cucurbitacin containing) and n-b (non-bitter).

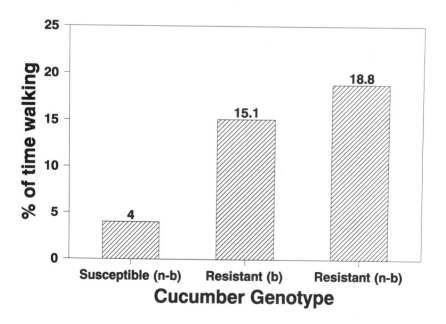

experiment (Figure 5). Even 8 weeks after inoculation, the spider mite-resistant genotypes were still green, while the susceptible plants were almost dead.

Figure 5 Damage symptoms on cucumber leaves 2 weeks after inoculation with *F. occidentalis* females and larvae. First two genotypes are bitter (cucurbitacin containing); last two genotypes are non-bitter. Bars indicated by the same letter do not differ significantly (Mollema *et al.*, 1989).

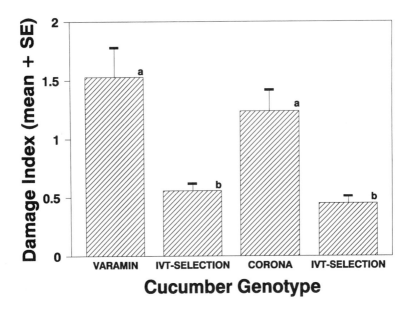

Molecular biological approaches and prospects

Breeding for resistance to insects often causes a dilemma for breeders. The demand for this type of resistance is great and increasing, but the bioassays are rather unpopular as the insects can be harmful to adjacent selections. This problem, however, may be solved by using techniques in which insect-bioassays are rarely involved. For this reason, molecular biology techniques are useful instruments and significant research in plant molecular biology concerns resistance to insects. Nevertheless, insect bioassays remain an important component in the validation of transgenic resistance. This is true for the very start as well as the end of a selection program for transgenic resistance.

The possibility of obtaining resistance to insects in transgenic plants was reported in 1987 (Hilder *et al.*, 1987; Vaeck *et al.*, 1987). Since then this strategy was further developed. Additional genes have shown expression in plants and transgenic plants have been field-tested.

Besides the transfer of genes from one organism to another, it is also possible to manipulate the site and magnitude of gene expression. Particularly in the context of resistance to insects, such tissue-specific expression is meaningful. Such resistance could be expressed only in tissues attacked by the insects (in many cases the leaves) and not in the plant parts harvested for consumption (e.g., fruits or seeds). If the host plant emits attractants of natural enemies as biological control agents, it is important to avoid misleading these natural enemies. The attractants should not be produced constitutively. In these situations, wound-inducible promoters enable the plants to express the gene products only after attack. This will be useful for genes affecting the release of infochemicals (e.g., attractants for natural enemies), an underexplored area of research.

Another application of molecular biology in breeding for resistance to insects is the use of DNA markers like restriction fragment length polymorphisms (see Tanksley *et al.*, 1989). These molecular markers are very useful, particularly in cases when resistance is inherited quantitatively. The advantage of this approach above conventional (phenotypic) methods is that the accuracy and efficiency of selection is greatly enhanced.

Transgenes for resistance to insects in theory

The most direct way to transfer resistance from one organism to another is when the resistance is caused by a proteinous compound. Some primary gene products like peptides and other protein-like structures can be extremely toxic to insects. In principle, these insecticidal compounds can be divided into (1) toxins, (2) inhibitors, (3) lectins, and (4) neuropeptides.

Nature provides a wide variety of proteinous toxins (Harris and Chapman, 1986). Some of them are known to be highly effective against insects. These toxins evolved in completely different organisms like bacteria, wasps, scorpions, or snakes. From this group, the bacterial toxins are studied most thoroughly (Falmagne and Alouf, 1986). From the beginning of this century, the entomopathogenic bacterium *Bacillus thuringiensis* has been used for control of insect pests. This bacterium produces an intracellular protein crystal (endotoxin) which damages the insect's midgut. Isolates of *B. thuringiensis* have been used for the production of bio-insecticides. An advantage of the *B. thuringiensis* endotoxin is the insect specificity. The expression of genes coding for this toxin in transgenic plants has been studied.

A second group of toxic proteins are the inhibitors of the insect's digestive enzymes. Such inhibitors are widespread in the plant kingdom (Richardson, 1990). Two major groups of inhibitors can be distinguished - proteinase inhibitors and amylase inhibitors. The first group of inhibitors is classified into several families, based on their primary structure (e.g.,

Bowman-Birk, Kunitz, Potato Inhibitor I, Potato Inhibitor II, etc.). For both groups the procedures for isolation and identification are available. There is significant variation in effectiveness among the proteinase inhibitors. This is partly due to the number of binding sites per inhibitor molecule. Recently, Broadway (1989) isolated trypsin and chemotrypsin inhibitors from cabbage which appeared to be very potent - a concentration of 0.1% was effective against some lepidopterans. The most well-known proteinase inhibitor is derived from cowpea. Cowpea Trypsin Inhibitor (CpTI) is from the Bowman-Birk family, showing two opposite binding sites per molecule (Hilder *et al.*, 1989). The inhibitor is effective against a large group of economically important lepidopteran and coleopteran pests.

The third group of insecticidal primary gene products (lectins) is also derived from plants. Lectins are proteins which cause agglutination of bloodcells. Therefore, they are known as "phyto-hemagglutinins." Another property of the lectins after oral intake is their binding to the epithelial cells of the insect's intestines. Consequently, the absorption of nutrients is obstructed with subsequent inhibition of growth (Liener, 1980; Jaffé, 1983).

The last group of proteinous compounds is the neuropeptides. These are rather small peptides (maximum 15 amino acids) which occur in very small quantities inside the insects. Neuropeptides have a hormonal function. As a consequence, they play a vital role in the overall regulation of the insect's growth and development. Small disturbances in hormonal balances can result in total disorientation and disfunctioning of the insect. In order to achieve lethality, or at least "loss of appetite" in insects, one must be sure that the hormones consumed are still active (Van Brunt, 1987). In this respect, the use of neuropeptides is not as simple as the previously discussed compounds. The most important problem with these substances thus far is their instability and rapid breakdown during digestion (Keeley and Hayes, 1987). A major advantage stemming from the fact that these hormones are small peptides is the possibility of developing synthetic genes. Application of neuropeptides is doubtfull if the problems indicated are not solved adequately.

If the resistance factor is not some primary gene product but a non-protein, transference of the resistance will be more complex. Many plant secondary metabolites are non-proteins. These compounds generally play an important role in the defense of plants to pests and diseases (Wink, 1988; Dawson *et al.*, 1989). The synthesis of plant secondary compounds depends upon enzymatic reactions. Because enzymes are proteins and thus primary gene products, genes coding for enzymes are feasible objects for transference. Many plant defense chemicals, however, are very complex molecules. Hence, many enzymes are involved in the synthesis of a single biologically active compound. If only one essential enzyme in the biochemical pathway of a particular compound is missing or disfunctioning, the plant will not be able to produce this compound. Such an omission can be the causal factor for susceptibility to insect attack. Addition of the key enzyme by genetic engineering may change this plant from susceptible to resistant. Another possibility is enhancement of the production of certain active compounds by manipulating the expression of the enzymes involved. This approach has been suggested in particular for the production of behavior-controlling chemicals (antixenosis factors like antifeedants, repellents and deterrents) by crop plants (Pickett, 1985). The very active antifeedant compound "polygodial" from marsh pepper (*Polygonum hydropiper* L.) could be synthesized by genetically modified crop plants.

Disturbance of the insect's hormonal balance can also be achieved by enzymes which break down the hormones (anti-hormones). Recently, Hammock *et al*. (1990) showed that caterpillars which were exposed to the anti-juvenile hormone (Juvenile Hormone Esterase, JHE) from genetically engineered baculoviruses cease feeding and growing.

Transgenes for resistance to insects in practice

It is obvious that molecular biological techniques are promising and have great advantages in breeding for resistance to insects. However, a few critical notes can be made.

A first comment concerns the bottlenecks in elementary components of genetic manipulation. Although appropriate transformation techniques have been developed, only a few genes for resistance to insects have been introduced in even fewer plant species. This results from the poor availability of (1) suitable resistance genes and (2) successful regeneration techniques for many of the economically important crops.

Because the search for suitable transgenes is rather expensive and time-consuming, it is expected that the supply of new genes will remain an important problem. This aspect may even play a greater role when genetic manipulation becomes more and more routine. The latter argument will also hold for the development of regeneration techniques in other than the usual plant species. In other words, who will wash the dishes in a high-tech environment?

A second comment considers the durability of the resistance obtained by genetic manipulation. In many of these cases the resistance is based on single dominant genes. As the selection pressures upon the insects are high, probably a fast evolution of virulent biotypes will occur. For *B. thuringiensis*, this has already been demonstrated (McGaughey, 1988). Also the durability of the resistance by introduction of CpTI genes may be limited. Dick and Credland (1986) have shown that the resistance in the cowpea cultivar from which this gene was isolated could be tolerated by some strains of bruchids. In order to avoid a rapid adaptation of insects to the resistance, molecular biologists are considering the transference of gene packages. By this technique, a couple of resistance genes with different actions could be introduced simultaneously in the host plant. The chance of adaptation to all types of resistance at the same time is negligible.

The successful introduction of resistance to insects in plants by means of genetic engineering was reported in 1987 and the work is still in progress. Most of the research concerns the expression of *B. thuringiensis* toxins (Vaeck *et al.*, 1987; Visser *et al.*, 1988). Several research groups are investigating this system and field experiments have already been conducted (Botterman and Leemans, 1989). The next success concerns the inhibitors of insect digestive enzymes. The CpTI gene (Hilder *et al.*, 1987), some Potato Inhibitor genes (Stiekema *et al.*, 1988; Willmitzer *et al.*, 1989), and a bean Amylase Inhibitor gene (Altabella and Chrispeels, 1990) have been expressed in other plant species. Finally, lectin genes from bean (*Phaseolus vulgaris* L.) have been introduced in tobacco (*N. tabacum*) (Voelker *et al.*, 1987). The resistance of most transformed plants has been evaluated using Lepidopteran pests. In many crops, however, these insects have no significant economic importance. It will be clear that in the near future the application of techniques to transform plants must be considered for major pests like spider mites, thrips, aphids, whiteflies, and leafminers.

COMBINED STUDIES OF HOST PLANT RESISTANCE AND BIOLOGICAL CONTROL: THE TRITROPHIC SYSTEM APPROACH

It is obvious that the use of resistant cultivars is a major element of IPM and that partial resistance can promote the effectiveness of endemic or introduced natural enemies. Price (1986) distinguished between intrinsic defense mechanisms of plants resulting in some degree of resistance and extrinsic defense mechanisms resulting in some degree of biological control of its herbivores by natural enemies. Based on a thorough understanding of the complex relations within a tritrophic system (plant-herbivore-natural enemy), it is possible to identify plant characteristics which influence directly the effectiveness of natural enemies. These plant characteristics are often morphological, interfering with the mobility of the natural enemy, but sometimes chemical, interfering with the communication of the natural enemy. As demonstrated by van Lenteren and de Ponti (1990) and by van Lenteren (1991), plant breeders can manipulate these characteristics to promote the effectiveness of natural enemies. This will be illustrated by two examples.

The hairiness of cucumber interfering with the biological control of the greenhouse whitefly by the parasitic wasp *Encarsia formosa* Gahan has been studied extensively. It appeared that on experimental hybrids, which have half the number of hairs, the walking speed of *E. formosa* was increased by 30% and, as a result, the time needed to find a whitefly larva was reduced by 30% (Li *et al.*, 1987; Table 4). Greenhouse experiments validated these results where average parasitism was significantly higher on the experimental hybrids (van Lenteren, 1991). Breeding companies are now developing these hybrids with reduced hairiness. In order to maximize cultivar control, they should combine this with resistance to the greenhouse whitefly. Preliminary data indicate a similar relation between the hairiness of gerbera (*Gerbera jamesonii* H. Bolus) and the effectiveness of *E. formosa* (van Lenteren and de Ponti, 1990).

Table 4 Influence of leaf hairiness of cucumber on the mobility and parasitization efficiency of *E. formosa* (van Lenteren and de Ponti, 1990).

Parameter	Cucumber genotype	Observation
Walking speed	Hairy	0.21 mm s^{-1}
	Glabrous	0.63 mm s^{-1}
	Hybrid	0.40 mm s^{-1}
Host-finding success	Hairy	56%
	Hybrid	77%
Host-finding time	Hairy	1564 s
	Hybrid	1111 s

Another example is the sticky pubescence of tomato interfering with the biological control of the two-spotted spider mite by the predatory mite *Phytoseiulus persimilis* Athias-Henriot. In order to find new prey, the predatory mites move upward to the next leaf along the stem. On standard tomato cultivars, these stems are extremely hairy and sticky, resulting in an entrapment of up to 61% and a consequent mortality of up to 73%. The wild tomato species *Lycopersicon peruvianum* (L.) Mill. has glabrous stems and, consequently, the entrapment and mortality is zero (Haren *et al.*, 1987). Instead of the wild species, it is also possible to use a *L. esculentum* mutant as a source of glabrousness.

Even more fascinating is the manipulation of existing or introduced chemical compounds of the plant which regulate the attraction and stimulation of natural enemies. Dicke (1988), Noldus (1989) and Türlings *et al.* (1990) investigated the effect of so-called infochemicals on the behavior and effectiveness of natural enemies. Further research may result in the identification and isolation of genes regulating specific infochemicals promoting the effectiveness of natural enemies. With molecular techniques, these genes may then be transferred to relevant plant species to accelerate the production and release of these infochemicals immediately after insect attack.

These infochemicals should not be produced constitutively and devaluate their activity by attracting natural enemies if the herbivorous insects are absent. The manipulation of morphological and chemical traits of plants in relation to the behavior and thus the effectiveness of natural enemies nicely illustrates the potential of interdisciplinary research in breeding for insect control.

Because of the evident tritrophic interrelations between plants, their insect pests and their natural enemies, development of host plant resistance and biological control should be addressed by teams of entomologists and plant breeders, experts in both host plant resistance, and biological control. This approach will result in new cultivars which are best suited for use in IPM.

CONCLUSIONS

In this paper, it is shown that insect resistance can play a key role in IPM. It is argued that the suitability of a plant as host for noxious insects on the one side, and for their natural enemies on the other, is a basic ingredient for integrated control; it keeps densities of noxious insects at low levels and it provides suitable sites for their natural enemies. In the context of durable protection and sustainable agriculture, it is not necessary to provide high levels of resistance. Partial resistance alone and in combination with other control measures will give good prospects for durable protection.

A crucial point in breeding for resistance to insects is the method by which the resistance can be assessed. Because resistance is defined as a partial or complete reduction of the insect's population growth rate, testing methods must be aimed at components which most affect this growth rate. Therefore, population dynamic models will be helpful instruments to select the most effective component in specific insect-plant combinations. Examples of antibiotic and antixenotic resistance illustrate that germplasm from genebanks often contains a wide variety of useful genes for resistance to insects.

Application of molecular biological techniques in breeding for resistance to insects has some attractive advantages. In particular, the selection with DNA markers and the use of tissue-specific or wound-inducible promoters seem very promising. Thus far, only genes with insecticidal primary gene products have been used in studies with transgenic resistance. The success of this approach fully depends on the supply of such genes and the way they are introduced. There is an evident risk that introduced resistance genes are insufficiently durable if their structure and action is simple. Lessons should be learned from the experience with durable resistance in "conventional" breeding for resistance.

REFERENCES

Altabella, T.; Chrispeels, M. J. (1990) Tobacco plants transformed with αai gene express an inhibitor of insect α-amylase in their seeds. *Plant Physiology* 93:805-810.

Botterman, J.; Leemans, J. (1989) Field testing of insect and herbicide resistant crops. *Vorträge Pflanzenzüchtung* 16:455-461.

Broadway, R. M. (1989) Tryptic inhibitory activity in wild and cultivated crucifers. *Phytochemistry* 28:755-758.

Buddenhagen, I. W.; Ponti, O. M. B. de (1983) Crop improvement to minimize future losses to diseases and pests in the tropics. *FAO Plant Protection Bulletin* 31:11-30.

Cambell, R. K.; Eikenbary, R. D. (eds.) (1990) *Aphid-plant genotype interactions*. Elsevier, Amsterdam.

Dawson, G. W.; Hallahan, D. L.; Mudd, A.; Patel, M. M.; Pickett, J. A.; Wadhams, L. J.; Wallsgrove, R. M. (1989) Secondary plant metabolites as targets for genetic modification of crop plants for pest resistance. *Pesticide Science* 27:191-201.

Dick, K. M.; Credland, P. F. (1986) Changes in the response of *Callosobruchus maculatus* (Coleoptera; Bruchidae) to a resistant variety of cowpea. *Journal of Stored Products Research* 22:227-233.

Dicke, M. (1988) Infochemicals in tritrophic interactions. Unpublished Ph.D. dissertation, Wageningen Agricultural University, The Netherlands.

Ellis, P. R.; Freuler, J. (eds.) (1990) Breeding for resistance to insects and mites. Bulletin IOBC/WPRS XIII/6.

Falmagne, P.; Alouf, J. E. (eds.) (1986) *Bacterial protein toxins*. Gustav Fischer Verlag, Stuttgart.

Fritzsche, R.; Decker, H.; Lehmann, W.; Karl, E.; Geialer, K. (1987) *Resistenz von Kulturpflanzen gegen tierische Schaderreger*. Springer-Verlag, Berlin.

Gould, F. (1986) Simulation models for predicting durability of insect resistant germ plasm: A deterministic diploid, two-locus model. *Environmental Entomology* 15:1-10.

Gould, F. (1988) Evolutionary biology and genetically engineered crops. *BioScience* 38:26-33.

Hammock, B. D.; Bonning, B. C.; Possee, R. D.; Hanzlik, T. N.; Maeda, S. (1990) Expression and effects of the juvenile hormone esterase in a baculovirus vector. *Nature (London)* 344:458-461.

Haren, R. J. F. van; Steenhuis, M. M.; Sabelis, M. W.; Ponti, O. M. B. de (1987) Tomato stem trichomes and dispersal success of *Phytoseiulus persimilis* relative to its prey *Tetranychus urticae*. *Experimental and Applied Acarology* 3:115-121.

Harris, J. B.; Chapman, D. A. (eds.) (1986) *Natural toxins; animal, plant and microbial*. Clarendon Press, Oxford.

Hilder, V. A.; Gatehouse, A. M. R.; Boulter, D. (1989) Genetic engineering of crops for insect resistance using genes of plant origin. In: Lycett, G. W.; Grierson, D. (eds.), *Genetic engineering of crop plants*. Butterworths, Borough Green, pp. 51-67.

Hilder, V. A.; Gatehouse, A. M. R.; Sheerman, S. E.; Barker, R. F.; Boulter, D. (1987) A novel mechanism of insect resistance engineered into tobacco. *Nature (London)* 330:160-163.

Hulspas-Jordaan, E.; Lenteren, J. C. van (1989) The parasite-host relationship between *Encarsia formosa* Gahan (Hymenoptera; Aphelinidae) and *Trialeurodes vaporariorum* (Westwood) (Homoptera; Aleyrodidae) XXX. Modelling population growth of greenhouse whitefly on tomato. *Wageningen Agricultural University Papers* 89.2:1-54.

Jaffé, W. G. (1983) Nutritional significance of lectins. In: Rechcigl, M. (ed.), *CRC Handbook of natural occurring food toxicants*. CRC Press Inc., Boca Raton, Florida.

Jones, A. T. (1990) Breeding for resistance to virus vectors. *Proceedings Brighton Crop Protection Conference* 3:935-939.

Keeley, L. L.; Hayes, T. K. (1987) Speculations on biotechnology applications for insect neuroendocrine research. *Insect Biochemistry* 17:639-651.

Kennedy, G. G. (1976) Host plant resistance and the spread of plant viruses. *Environmental Entomology* 5:827-832.

Kennedy, G. G.; Gould, F.; Ponti, O. M. B. de; Stinner, R. E. (1987) Ecological, agricultural, genetic and commercial considerations in the deployment of insect resistant germplasm. *Environmental Entomology* 16:327-338.

Lenteren, J. C. van. (1991) Biological control in a tritrophic system approach. In: Peters, D. C.; Webster, J. A. (eds.), *Aphid-plant interactions: Populations to molecules*. Proceedings of an International Symposium, 12-17 August 1990. Oklahoma State University Press, Stillwater, OK (in press).

Lenteren, J. C. van; Ponti, O. M. B. de (1990) Plant-leaf morphology, host plant resistance and biological control. In: Jermy, T.; Szentesi, A. (eds.), *Insect-plant rela tionships*. Proceedings of the 7th International Symposium, Budapest, Hungary, 3-8 July 1989. Junk, Rotterdam, pp. 365-386.

Lewontin, R. C. (1965) Selection for colonizing ability. In: Baker, H. G.; Stebbins, G. L. (eds.), *The genetics of colonizing species*. Academic Press, New York, pp. 77-94.

Li, Z. H.; Lamers, F.; Lenteren, J. C. van; Huisman, P. W. T.; Vianen, A. van; Ponti, O. M. B. de (1987) The parasite-host relationship between *Encarsia formosa* Gahan (Hymenoptera; Aphelinidae) and *Trialeurodes vaporariorum* (Westwood) (Homoptera; Aleyrodidae) XXV. Influence of leaf structure on the searching activity of *Encarsia formosa*. *Journal of Applied Entomology* 104:297-304.

Liener, I. E. (ed.) (1980) *Toxic constituents of plant foodstuffs (food science and tech nology, a series of monographs)*. Academic Press, New York.

McGaughey, W. H. (1988) Resistance of storage pests to *Bacillus thuringiensis*. *Proceedings International Congress of Entomology, Vancouver* XVIII:449.

Mollema, C.; Hoeven, W. van der; Steenhuis, M.; Groot, S. (1989) Resistance to the western flower thrips (*Frankliniella occidentalis*) in cucumber (*Cucumis sativus*). In: Thomas, C. E. (ed.), *Evaluation and enhancement of cucurbit germplasm*. Proceedings of Cucurbitaceae '89, Charleston, South Carolina, pp. 166-169.

Noldus, L. (1989) Chemical espionage by parasitic wasps. Unpublished Ph.D. dissertation, Wageningen Agricultural University.

Painter, R. H. (1951) *Insect resistance in crop plants*. Macmillan, New York.

Parlevliet, J. E. (1991) Selecting components of partial resistance. In: Stalker, H. T.; Murphy, J. P. (eds.), *Proceedings of a symposium on plant breeding in the 1990s*. CAB International, Wallingford, U.K., pp. 281-302.

Pickett, J. A. (1985) Production of behaviour-controling chemicals by crop plants. *Philosophical Transactions of the Royal Society of London series B* 310:235-239.

Ponti, O. M. B. de (1978) Resistance in *Cucumis sativus* L. to *Tetranychus urticae* Koch. 3. Search for sources of resistance. *Euphytica* 27:167-176.

Ponti, O. M. B. de (1981) Conserving the natural resistance to insects by a proper use of the pesticide umbrella. In: *Genetic resources and plant breeding for resistance to diseases, pests and abiotic environmental conditions*. IX. Eucarpia Congress, Leningrad, September 1980, pp. 236-248.

Ponti, O. M. B. de (1985) Keynote address. International study workshop on host plant resistance and its significance in pest management. *Insect Science and Application* 6:235-236.

Ponti, O. M. B. de; Steenhuis, M. M.; Elzinga, P. (1983) Partial resistance of tomato to the greenhouse whitefly (*Trialeurodes vaporariorum* Westw.) to promote its biological control. *Mededelingen Faculteit Landbouwwetenschappen Ryksuniversiteit Gent* 48/2:195-198.

Price, P. (1986) Ecological aspects of host plant resistance and biological control: Interactions among three trophic levels. In: Boethel, D. J.; Eikenbary, R. D. (eds.), *Interactions of plant resistance and parasitoids and predators of insects.* Wiley, New York, pp. 11-30.

Richardson, M. (1990) Seed storage proteins: The enzyme inhibitors. In: Rogers, L. (ed.), *Methods in plant biochemistry.* Vol. 2. Academic Press, New York.

Romanow, L. R.; Ponti, O. M. B. de; Mollema, C. (1991) Resistance in tomato to the greenhouse whitefly: Analysis of population dynamics. *Entomologia Experimentalis et Applicata* (in press).

Singh, D. P. (1986) *Breeding for resistance to diseases and insect pests.* Springer Verlag, Berlin.

Smith, C. M. (1989) *Plant resistance to insects; A fundamental approach.* John Wiley & Sons, Inc., New York.

Stiekema, W. J.; Heidekamp, F.; Dirkse, W. G.; Beckum, J. van; Haan, P. de; Bosch, C. Ten; Louwerse, J. D. (1988) Molecular cloning and analysis of four potato tuber mRNA's. *Plant Molecular Biology* 11:255-269.

Stinner, R. E.; Rabb, R. L.; Bradley, J. R., Jr. (1974) Population dynamics of *Heliothis zea* (Boddie) and *H. virescens* (F.) in North Carolina: A simulation model. *Environmental Entomology* 3:163-168.

Tanksley, S. D.; Young, N. D.; Paterson, A. H.; Bonierbale, M. W. (1989) RFLP mapping in plant breeding: New tools for an old science. *Bio/Technology* 7:257-263.

Türlings, T. C. J.; Tumlinson, J. H.; Lewis, W. J. (1990) Exploitation of herbivore-induced plant-odors by host-seeking parasitic wasps. *Science (Washington, DC)* 250:1251-1253.

Vaeck, M.; Reynaerts, A.; Höffe, H.; Jansens, S.; Beuckeleer, M. de; Dean, C.; Zabeau, M.; Montague, M. van; Leemans, J. (1987) Transgenic plants protected from insect attack. *Nature (London)* 328:33-37.

Van Brunt, J. (1987) Pheromones and neuropeptides for biorational insect control. *Bio/Technology* 5:31-36.

Van Emden, H. F. (1987) Cultural methods: The plant. In: Burn, A. J.; Coaker, T. H.; Jepson, P. C. (eds.), *Integrated pest management.* Academic Press, London, pp. 27-68.

Van Emden, H. F. (1990) The interaction of host plant resistance with other control measures. *Proceedings Brighton Crop Protection Conference* 3:939-949.

Visser, B.; Salm, T. van der; Brink, W. van den; Folkers, G. (1988) Genes from *Bacillus thuringiensis entomocidus* 60.5 coding for insect-specific toxins. *Molecular and General Genetics* 212:219-224.

Voelker, T.; Sturm, A.; Chrispeels, J. (1987) Differences in expression between two seed lectin alleles obtained from normal and lectin-deficient beans are maintained in transgenic tobacco. *EMBO Journal* 6:3571-3577.

Willmitzer, L.; Basner, A.; Frommer, W. B.; Höfgen, R.; Köster, M.; Liu, X. Y.; Mielchen, C.; Prat, S.; Recknagel, C.; Rocha-Sosa, M.; Sonnewald, U.; Stratman, M.; Vancanneyt, G. (1989) Expression of foreign genes in potato; promoters, RNA-stability and protein accumulation. *Vortrge Pflanzenzchtung* 16:423-439.

Wink, M. (1988) Plant breeding: Importance of plant secondary metabolites for protection against pathogens and herbivores, review. *Theoretical and Applied Genetics* 75:225-233.

Wiseman, B. R. (1990) Plant resistance: A logical component of sustainable agriculture. *Annual Plant Resistance to Insects Newsletter* 16:40.

Zadoks, J. C. (1991) A hundred and more years of plant protection in The Netherlands. *Netherlands Journal of Plant Protection* 97:3-24.

Discussion

Wanda W. Collins, Moderator

Dr. Allard stated that most resistance genes from wild species are not useful because of associated undesirable genes. Do you have a comment?

The traits we have transferred into cultivated rice, like the genes for resistance to blast or to brown planthopper, have not brought in undesirably linked traits with them. We are, perhaps, fortunate that we have not seen adverse or deleterious effects in the improved cultivars. This may not be true for other genes, but at least in these two cases we have not seen adverse effects. [G. S. Khush]

My curiosity was also piqued by that comment from Dr. Allard because I've worked nearly a third of a century on the transfer of genes for crown rust resistance in oats and have created many isolines. These genes have come from both the cultivated oat (*Avena sativa*) and from *A. sterilis* (the hexaploid progenitor of the oat). We have worked with about 35 genes that give vertical resistance to crown rust. We have found several cases where, under non-rust conditions, decreased yield occurred; an equal number had associated increased yield, but most had no effect. For oats, there are a few associated reductions in yield, vigor, or fitness. [K. J. Frey]

I did my Ph.D. thesis on a transfer of multiple disease resistance from *Triticum timopheevii* to common wheat. Several resistance genes were transferred, and I spent about 30 years trying to break the unfortunate deleterious effects associated with them. I succeeded in a couple of cases and didn't in others. Genes for stem rust resistance that came in are still being used, even though there is still, as far as I know, at least a 5% yield detraction in the absence of the pathogen. One of the main reasons I made the statement is because of the very good evidence we have from *Rhinchosporium*, a disease in barley. The barley-*Rhinchosporium* pathosystem is an excellent one for doing host-pathogen interactions. Twenty-nine different disease resistance loci have been identified and a series of races exist which have very specific interactions with the resistance genes. Isogenic lines have been tested in many years using many different races and in different locations. We find that in about two-thirds of those cases there is a negative effect. Some of them are neutral and some of them actually give a slight advantage. But, the main effects that we have had, particularly when we have had adequate data using really good isogenic lines, are negative. Obviously, as one of our previous speakers said, each disease system is a thing unto itself, and you have to test for each pathogen and plant genotype.   [R. W. Allard]

Do you see any advantage in using RFLPs to tag major genes for resistance when they are often relatively easy and cheap to screen effectively in the field?

If you can screen in the field on a large scale, as is the case with most major genes, there will not be any advantage in using RFLPs. However, when screening is difficult (e.g., for drought and quantitative traits such as partial resistance for blast), the RFLP markers are useful. Screening for major gene resistance under field conditions is very simple and cost-effective as opposed to expensive laboratory analyses. Also, there may be lack of the polymorphisms in the parents used in crosses because the parents we use in rice are often quite closely related.   [G. S. Khush]

Have any of the resistances to insects that you've described been overcome by new races?

Resistance to brown planthopper has broken down in 5-6 years, but for other insects resistance has not broken down.   [G. S. Khush]

How do you measure 24% antixenosis in a susceptible cultivar?

It indicates that even what we call "susceptible" is not 100% susceptible; e.g., not all spider mites accept the plant the same. On a susceptible cultivar, most of the spider mites will stay, but on an antixenotic cultivar they try to walk away and drop off the leaves.   [O. M. B. de Ponti]

How would you differentiate between partial resistance and complete resistance reactions at the molecular level? Where in the pathway does the difference occur?

The hypothesis is that at least several different signal systems are activated - on perception of microbial attack - and different batteries of genes with different defense functions are activated. It's conceivable that, if a particular gene-for-gene interaction triggers only a limited set of those signal pathways, you might have, say, the ability to effect partial resistance; whereas, if you triggered another signal pathway, you might activate all the defense responses and have complete resistance. I think there is a multiplicity of signal pathways (even within one kind of biological situation) and a multiplicity of plant responses, so partial activation of subsets might well account for partial resistance.   [C. Lamb]

Are you sure that the chemical substances which cause insect resistance are without risk for human nutrition?

We haven't found any evidence that the cultivars which are resistant to the insects are toxic to human beings or any mammals.   [G. S. Khush]

With resistance to spider mites and thrips on cucumber, we think that the type of chemicals which are in force are the volatile type which are interfering with the communication systems of the mites and the thrips. We don't think that those compounds have any toxic activity. In our whitefly work, we were interested in knowing potential dangers for consumption because we made crosses with the wild-type tomato species (*Lycopersicon hirsutum glabratum*), which is known to have a higher level of solanins in the leaves and fruits. After analyzing the amounts of solanins in tomato fruits, we did not find evidence that the high amount of solanins in the wild tomato was transferred to breeding lines. This means that the resistance in the tomato has to be based on other factors than solanins. But, of course, it is a very important question to make sure that the quality of the plants which have resistance to insects is not negatively influenced by the breeding for resistance. There have been some suggestions recently made that, by increasing the level of resistance *per se*, you are increasing the level of toxic compounds, but I think there is very little evidence for this. [O. M. B. de Ponti]

We also work on trypsin inhibitors for other reasons, but trypsin inhibitor is an antinutritional compound. When it's cooked in food, there appears to be no problem; but there are still areas of the world where raw vegetables or other raw products are consumed. I can see that, where trypsin inhibitors are increased to the level where they could control insects, there could possibly be a human or animal antinutritional effect.   [W. W. Collins]

By crossing with wild progenitors or wild relative species, you may introduce secondary substances in the cultivated crops and these may be sometimes toxic. In potatoes, for instance, it is clear that, by crossing with the wild material, quite high levels of solanin have been introduced and these are toxic to human beings. But that is not the resistance gene itself which is introduced, but linked genes which are transferred with the resistance gene.
[J. E. Parlevliet]

With your comment that major genes do not work with partial resistance genes, how and where would you recommend that partial resistance be used in a breeding program?

I did not say that major genes do not work together with partial resistance; I only said that in answer to the question whether major genes supported by partial resistance would last longer. That is not the case, but they do work as long as the pathogen population is not yet adapted. However, they seem to adapt as fast to major genes in the presence of high levels of partial resistance as they do in the presence of less partial resistance. Absolute susceptibility is actually extremely rare; major genes, if present, always occur together with some partial resistance.   [J. E. Parlevliet]

Your research group has been very effective in isolating and cloning genes in the phytoalexin pathway (defense genes, for example). Why has it been so difficult to clone and isolate resistance genes to date?

I think predictions would be that the kind of products encoded by resistance genes have very different functions from the products that are encoded by defense genes; that is, they are either active biochemically so that they are likely to be a receptor or an early component of a signal transduction pathway as compared to the regulator gene (the downstream genes) that I would see as defense genes in this instance. The defense genes are inducible and produce biochemically defined products which we can purify based on biochemical function; the gene can then be isolated using these protocols. There are a number of ways of getting to the genes, and they are also likely to be expressed when they are induced at reasonably high levels. On the other hand, if the disease resistance gene is going to be a receptor, there is no easy biochemical assay and it is very difficult even to detect binding activities. There is no easy biochemical assay which could then allow one to get to the gene through the protein, and there are not likely to be abundant gene products.   [C. Lamb]

Monsanto will soon release *Bt* plants with high antibiosis. Is this wise? Shouldn't they wait to combine antibiosis with antixenosis? Are they wasting a lot of effort for short-term gain?

I think it's a very good question but, in fact, it's an old story; I mean that, in terms of resistance management, there have been a lot of theories that use resistance genes in a wiser way than companies do at the moment. The problem is that, if a company has a very effective gene against a particular insect, then they are very much tempted to bring this product to market as soon as possible. I agree that it might be wiser to wait with this introduction and to first make the resistance more complex, either by introducing more antibiosis genes with different actions or by combining different modalities of resistance to releasing a line which tends to be less sensitive for adaptation by the insect. Breeders tend to incorporate single genes as soon as they are available and, in this way, cause a fast erosion of resistance genes for those pathogens and insects which are likely to develop new races or biotypes.   [O. M. B. de Ponti]

Are there strategies to combine partial resistance to several different diseases at the same time?

Nearly all partial resistance is pathogen species-specific; so, if you want to combine partial resistance to different pathogens, you have to select that separately (in the sense that you have to test with the different pathogens separately). Concerning combining resistances, in Table 9 of Chapter 12, where I showed partial resistance to the four virus diseases in potato, the first potato cultivar, Arkula, showed that you can have a high level of partial resistance to the four different pathogens in one cultivar.   [J. E. Parlevliet]

What are you going to do when you've used up all major genes?

I think if you continue to screen germplasm, then you will continue to find major genes. They can also be found in wild species.   [G. S. Khush]

To follow up on that a little more, when you have a number of resistance genes stockpiled to back up genes already deployed, do you release new genes only when the currently used genes break down? Have you considered alternative methods of gene deployment such as gene rotation?

I think it will be very wise to use the different genes in different areas rather than using the same gene everywhere that rice is grown. If you have a number of cultivars with different genes, then develop a deployment strategy where, depending upon the insect population and biotype, you use the specific gene against that biotype in that region.   [G. S. Khush]

Is it possible that many genes, such as those for disease resistance which appear to be quantitative in nature, are really single genes with regulatory functions such as producing different amounts of a functional protein in time and space?

I suppose you mean the resistance genes for quantitative resistance or partial resistance. It's conceivable, yes. But, I think until we can break down polygenic traits and understand them on an individual gene basis, either with respect to disease resistance or multigenic traits, it's going to be impossible to answer this question.   [C. Lamb]

For partial resistance to leafroll in potato there was a recent report that it is probably a single gene. Further, there are no indications that resistance is eroding. The quantitative effect is probably due to the tetraploid nature of potato which may give four or five different levels of expression, depending upon whether there is complete dominance or recessiveness.

[J. E. Parlevliet]

Given that antixenosis and antibiosis are more durable when combined, what is the effect of tolerance on the durability of antixenosis and antibiosis?

Tolerance is a totally different modality. The combination of antixenosis and antibiosis might be very beneficial in terms of durability. Tolerance is just the character of a plant that it is able to endure some amount of insects. It indicates that partial resistance in combination with tolerance might be very interesting because then you raise the level of endurance of a plant to insects.

[O. M. B. de Ponti]

Do you know what the effect of the "bitter" gene is in cucumber on the enemies of mites and thrips?

We have studied in our breeding lines the combination of resistance to spider mites and the predatory mite *Phytoseiulus persimilis*. This predatory mite is used commercially on a very large scale by about 80% of the growers in the Netherlands. If you introduce resistance to spider mites, it is important that resistance is not negatively interfering with the biological control system. Apparently the combination of this partial resistance and biological control was very positive. In fact, on the resistant cultivar, the predatory mites were able to destroy the population of spider mites in only a few weeks.

[O. M. B. de Ponti]

Does the presence of the resistance genes reduce yield in the absence of the pathogens and pests?

We don't have any evidence that cultivars with multiple resistances, if you grow them without disease or insect pressure, have lower yield. It is possible to combine improved plant type with resistance to several diseases and insects.

[G. S. Khush]

Is tungro virus only controlled by resistance to the green leafhopper?

Until now, most of the cultivars which have been released have vector resistance which is very effective in controlling the tungro virus; but the vectors do adapt on the resistant cultivars. Between 5 to 7 years after introduction, cultivars begin to show susceptibility. Very good sources of resistance to the virus exist which are now being combined with vector resistance.

The problem is that tungro is a complex disease which is caused by two different viruses. One is bacilliform and the other is a spherical particle. These two particles, combined together, cause the disease.   [G. S. Khush]

Would you please comment on the deleterious effects of antibiosis on the third trophic level, i.e., predators and parasites?

For integrated control it is very important that, if you increase the level of resistance, it is compatible with the effect of natural enemies, endemic or introduced. That is the reason I advocated in Chapter 14 that breeders and entomologists should work together both in breeding for resistance and in the development of biological control because, in some cases, breeding for resistance might have a negative effect on the vitality of a predator or parasite. It is important to check when you increase the level of resistance that it doesn't have a negative effect on the biological control system.

[O. M. B. de Ponti]

PART FOUR

CONTRIBUTIONS OF BIOTECHNOLOGY TO PLANT IMPROVEMENT

Chapter 15
RFLP analyses for manipulating agronomic traits in plants

T. G. Helentjaris
Department of Plant Science, University of Arizona, Tucson, AZ 85721

INTRODUCTION

Future directions for plant improvement

Just as agronomists are charged with enhancing environmental conditions in order to maximize crop productivity, the challenge for plant breeders is to consolidate positive genetic elements into the most favorable combinations. During the last 50 years these efforts have been very successful in many species, achieving dramatic increases in yield, about half of which can be attributed to genetic improvement (Duvick, 1984; and other articles in the same volume). Despite this impressive progress, acceptance of the status quo with regards to present methods for plant improvement is clearly impractical. As agronomic practices are altered by economic and environmental concerns to establish a truly sustainable agriculture system for this country, compromises resulting in real crop productivity must be anticipated and these decreases must be made up by concurrent genetic improvements. Many of the previous increases in crop productivity attributed to genetic improvements may in fact be due to changes in production systems. For instance, corn (*Zea mays* L.) yields have increased almost logarithmically since 1930 (see Figure 2-1 in Duvick, 1984), but it has been suggested that much of these improvements are due to the substitution of double-cross and single-cross hybrids for open-pollinated cultivars over this time period (Figure 2-2 in Duvick, 1984). It is not clear what new type of production scheme could replace the present systems to maintain the same rate of improvement and one could argue that it will require the implementation of new technologies such as marker-assisted selection or transformation to create novel genetic variability. With these considerations in mind, it is surprising that during the last 10 years some have argued that present methods of plant improvement are sufficient and that additional research in this area, particularly that characterized as "biotechnology", is not only unnecessary but also irrelevant. At no time previously has there ever been a conscious effort in any scientific field to cease research progress and to be content with the status quo, and agriculture should certainly be no exception with this regard.

Confounding the desire to move forward with the genetic improvement of many important crop species is the surprising lack of understanding of the basis for our previous advances. While the gains in corn represent almost a sixfold change in yield over the last 60 years, in general, we understand little about the actual genetic basis for these improvements; and this is a major impediment to future progress. After all, if we don't understand how the manipulation of chromosomal segments has resulted in the progress to date, how can we plan for future advances in any systematic fashion? This lack of understanding also leads many to comment that plant breeding at times is more art than science, an observation that reflects the inability of research to provide a usable scientific framework for progress in the field. In contrast to many areas of science where one can obtain a high level of practical expertise through examination of previous studies, in plant breeding one also requires a substantial level of personal hands-on experience over many years to achieve the same degree of intuition and proficiency.

This lack of basic knowledge of the factors underlying crop productivity is surprising, especially when compared to the tremendous increase in our knowledge of basic genetic principles accumulated over the last 30 years, particularly with the development of molecular genetics. While some might argue that this is primarily due to low research funding for this area of science, I believe that other reasons for this discrepancy should be considered. Much of our research to date has focused upon understanding the "basic" principles of genome structure and gene expression and has used "model systems" that are most tractable to existing methods of analysis. Consequently progress has been rapid in understanding these basic principles and in developing newer methods that specifically take advantage of the intrinsic characteristics of these model systems. The emphasis to date has clearly been to develop a basic understanding of how life forms function in general and then to assume that these principles apply broadly across all species and problems to be encountered later.

A basic problem with this strategy is that it has been much more difficult and expensive than anticipated to extrapolate the results of these efforts to other studies, particularly those of an applied nature. For instance, even so momentous a discovery that DNA is the basis for all heritability has yet to produce any practical impact upon classical plant breeding strategies. Essentially, the applied fields of genetics, plant and animal breeding, have missed out on the revolution of the last 30 years in molecular genetics and most of the scientific framework that breeders can productively use in their efforts has been provided by the fields of quantitative genetics and physiology. Without the eventual implementation of both "Mendelian" and molecular genetic approaches to augment these efforts, progress will continue to be slow due to a lack of a fundamental understanding of the underlying processes being manipulated and the application of novel strategies will be next to impossible. For instance, the utilization of genetic engineering is currently limited to the introduction of single genes which have been primarily isolated from non-plant species to address limited aspects of crop productivity. The traits of most agronomic importance, such as yield, maturity, etc. which are the results of endogenous plant gene function, will remain intractable until we develop a better understanding of their genetic and physiological basis. In sum, this predicament is not so much a failure of plant breeding but a failure of science in general to recognize that the eventual relevance of any research should also receive attention in order for society to benefit.

Molecular markers as a means for unifying basic and applied genetics

The apparent divergence in the fields of basic and applied genetics is a serious impediment to understanding progress to date and to determining how to develop future breeding programs. The development of the molecular marker concept offers a significant opportunity for applying a linkage or "Mendelian" genetic approach to agriculturally important species and for gaining access to the same types of tools that have facilitated the revolution in molecular genetics. Isozymes were first used in this context and many of the principles applicable to molecular markers were conceptualized during these earlier studies. Limitations on the numbers of informative loci within many species restricted their actual use, but initial results in many cases were quite promising and have served to maintain interest in these approaches. With the development of restriction fragment length polymorphisms (RFLPs) (Botstein *et al.*, 1980) a new class of DNA-based molecular markers has provided researchers with a tool that in many ways supersedes the limitations of isozymes, but also introduces new concerns in the form of cost and throughput. One should not be too concerned with these limitations at this time, however, as it is the exciting results with RFLPs themselves that have already generated the demands for higher usage. Newer types of markers, such as Random Amplified Polymorphic DNAs (RAPDs) (Williams *et al.*, 1990) or other Polymerase Chain Reaction (PCR)-based methods, that provide the same type of genotypic information as RFLPs but may solve the problems associated with increased demand, are currently being developed and tested. One can be assured that if molecular markers demonstrate a practical advantage in agriculture, then technical solutions will be identified to increase their usage.

Several intrinsic properties of molecular markers, in particular DNA-based markers, lend themselves to use in applied genetic programs. First, as any restriction fragment can potentially function as a marker locus and millions of fragments can be created by digestion of the genomes of higher eukaryotes, the practical number of marker loci is essentially unlimited. Hence, a resolution of the genome in different types of analyses is only limited by the amount of genetic recombination within the test population.

Second, as there is often an extremely high degree of informativeness* with these markers, they can usually be defined directly within the germplasm of interest. This effectively obviates the need for marker stocks, a characteristic of classical genetic approaches that has limited the application of markers to model species. Third, as the DNA itself is evaluated directly for sequence polymorphism and not by assaying for variation in a product of gene expression, environmental influences upon the expression of the polymorphism are removed as a source of confusion in the selection process. We can usually assume that the genomic DNA in an organism is relatively constant and that an assay of any tissue produced under many different conditions will still provide the same information. While we cannot ignore environmental influences upon the ultimate traits of interest, we can remove the environmental bias when assaying for the presence of the genes responsible for that trait. Genotypes can be selected in the off-season or in testing situations far removed from that required for field testing or production. We can also select for specific genotypes where expression of the phenotype is impossible, for instance, recessive genetic loci in the heterozygous state. Progeny from backcrossing programs can be screened at each generation for the presence of the desired genotype, even if the phenotype is prevented from being expressed due to the recessive nature of the underlying genes. Finally, small amounts of tissue can be assayed in a rapid fashion which is non-destructive to the original plant and can allow the direct selection of specific genotypes from amongst a complex group of individuals. Given these types of advantages, many investigators are enthusiastic about testing the use of molecular markers in many types of genetic analyses.

APPLICATION OF RFLPS TO SPECIFIC ASPECTS OF PLANT IMPROVEMENT

For several years, researchers have tested the utility of RFLPs with respect to specific aspects of plant breeding. Despite difficulties with respect to cost and throughput of RFLP analysis, some groups have obtained enough proficiency to address various questions with this stragey and generate preliminary information which bears upon the ultimate utility of this approach.

Use of RFLPs in understanding germplasm relationships

The primary tool of the plant breeder is the reservoir of natural variation found within the species of interest, not only in domesticated cultivars but sometimes also in its wild relatives. Understanding and management of that variability is the key to establishing an efficient program for crop

---

*By "informativeness", I am referring to the ability of a molecular marker to discriminate between chromosomal segments, i.e., to detect sequence polymorphisms.

improvement. Initial criteria usually involve study of morphological characteristics, and then later will rely upon careful pedigree records to establish the relationships amongst the available germplasm. Isozymes have previously been evaluated for their ability to establish relationships, and the results have been somewhat mixed, depending upon the species. In corn there are several informative isozyme loci with multiple alleles which can delineate a group of genotypes in a manner that accurately reflects our knowledge of their relationships (Stuber and Goodman, 1983). Unfortunately, there are not enough loci to cover the entire genome and some chromosome regions will necessarily go unsurveyed in any such analysis. The situation is much worse with other species in that there are generally fewer informative loci and fewer alleles at these loci. For instance with tomato (*Lycopsersicon esculentum* Mill.), few loci provide any discrimination within the domesticated species, but some resolution is observed in interspecific comparisons (Tanksley, 1983). The number of alleles detected at these loci is also very low (rarely greater than two), yielding a small amount of information.

With RFLPs the presence of unlimited numbers of loci and the high degree of informativeness at these loci is particularly useful in the delineation of relationships amongst germplasm. In corn we have found that there are large numbers of informative loci available in all but the closest comparisons of sister lines, and across Corn Belt germplasm the average number of alleles at these loci is greater than six (see review by Helentjaris, 1987; Walton and Helentjaris, 1987). Within a representative group of 96 inbreds, it was easy to select 35 RFLP loci that distinctly delineated these inbreds. Although separated by this type of analysis, a principal component analysis revealed groupings of materials that fit quite well with pedigree information. In fact, one can think of this analysis as a type of refined pedigree information where one not only knows the overall percentage of parental contributions, but also the contributions on a chromosomal region by region basis. In other species where comparable pedigree information does not exist, this type of analysis should allow an efficient survey of the available materials and their organization into groups that are reflective of their actual genetic relationships.

The utility of developing such information is also dependent upon how reflective it is of plant performance. Initial insights into this question were noted by S. Figdore in our group (Walton and Helentjaris, 1987) who showed that a plot of hybrid yield vs. the dissimilarity of the two parents used to create that hybrid, as measured by RFLPs, revealed an amazingly close relationship. From dissimilarities of 15% to about 60%, the relationship was almost linear, above 60% the relationship was less clear. Using this fairly simple and naive model, one could explain a very high percentage of hybrid yield in corn simply by knowing the RFLP patterns of inbred lines. Improving this model might involve developing weights for various loci or even for various allelic combinations at these loci. This observation was extended further by Smith and Smith (1989) who used more markers and inbred combinations to show they could account for almost 87% of the hybrid yield over a range of dissimilarities from less than 10 to greater than 90%. This is an amazing result in that such a high percentage of crop yield can be predicted by such a simple model based upon a laboratory analysis, far above what one could predict based upon pedigree information. In fact, the authors of the last study concluded that RFLP analysis was the best predictor of yield short of actual field analysis. It remains to be seen whether similar types of analyses in other crop species with less developed germplasm will also reveal such striking relationships between productivity and genotypic information developed by molecular markers.

The potential utility of this type of analysis is obvious. Knowledge of germplasm relationships can make a breeder more efficient by allowing him to sample the available germplasm in a more systematic fashion. Knowledge of how genomic dissimilarity relates to yield will also refine our knowledge of hybrid vigor and how heterotic groups are formed. It may allow us to even develop new heterotic groups in crops like corn or to make predictions about heterotic groups in crops such as wheat (*Triticum aestivum* L.) where we have not yet developed them. With the high level of informativeness, this type of analysis should also become the standard of the industry for cultivar identification, in combination with other types of characters. Better protection of proprietary materials and the investment in them by their developers will spur more efforts to create new cultivars. Finally, comparison of these types of genomic analyses with the extensive field data already collected by most breeding operations should allow the delineation of those areas of the genome which are most important for crop productivity. In contrast, RFLP analysis of segregating progeny from single crosses requires a large number of analyses to reveal this type of information for a very small selection of germplasm. It should be possible to compare data collected across a broader set of germplasm on both the genotypic and phenotypic levels and obtain insights as to the regions of the genome containing genes with the most significant impacts and even alternative alleles that should be substituted in those regions to obtain improved cultivars. Our initial investigations (unpublished data) have revealed that given the proper constraints on the materials examined (i.e. there must be a relatively high level of relatedness amongst the germplasm being analyzed), one can detect specific regions of the genome that, if altered, could result in hybrid improvement. That these same regions also correlated with those found in analysis of segregating progeny was encouraging. The advantages of this approach are that it requires a relatively small number of analyses, it takes advantage of existing field efforts, and the payoff is potentially quite high by directing breeding programs toward using the most productive materials and crosses. The materials a breeder chooses to emphasize in his program is the ultimate limitation of his progress and if this choice could utilize both intuition and genotypic information, the efficiency of that program could be significantly enhanced. Other areas for future research should involve investigation of how should species be sampled to determine a genotype where development of inbred materials is impractical, how should polyploids be dealt with in determining a "genotype", and what is the relationship in other species between parental dissimilarity and hybrid yield.

Use of marker loci to "map" single genes of agricultural importance

Although many breeders feel that single gene traits are usually easy to manipulate by conventional means, there are situations where indirect selection with marker loci might prove useful; e.g., if the gene product's expression is affected by the environment or if the trait is difficult to assay. Disease resistance evaluation can often be complicated by the environment such that the wrong weather can mean the loss of an entire year of selection. Chemical composition of sugars, oils, etc. may require extensive analyses that could be substituted for by genotypic analysis. Finally, because the gene can be detected even when its expression is not evident, many situations (recessive gene in a heterozygous state or selection in the off-season) can allow more efficient selection of desired genotypes.

Mapping of single genes is usually quite trivial with molecular markers. One simply constructs a segregating population and compares both genotypic and phenotypic data to detect a correlation between the two. The large number of marker loci and their detection of exact genotype are particularly attractive advantages of molecular markers in this application. While in most cases, one is interested in mapping a phenotype to a chromosomal location, at times it may be of interest to map a cloned and identified gene to a location in order to associate a phenotype with it. For instance a clone for an auxin-binding protein (abp) was obtained (Lobler and Hirsch, 1990) as it was believed to be important at some point in plant development. On the other hand, no known phenotype was associated with the disruption of this gene. By mapping it to the corn genome and comparing its location with mutant phenotypes also assigned to the same region, it might be possible to associate the two and learn something about the function of this gene.

The more common case is the assignment of a trait to a chromosomal region and this is now becoming almost routine. In the case of a Maize Dwarf Mosaic Virus (MDMV) resistance gene (McMullen and Louie, 1989), an association of this gene with kernel color was noticed and the invesitgators were able to narrow their mapping to a small region on chromosome 6. Even without this information an analysis of only 25-100 segregating individuals with 25-50 markers should allow the placement of any locus with high precision to a chromosomal arm. In most cases the rationale for such efforts would be the development of an indirect selection criterion (marker kit) to substitute for the actual analysis of the trait itself during a selection program. This is certainly feasible, but the economics of RFLP analysis currently preclude the wider application of this approach. Hopefully during the next 5 years, the development of alternative technologies, perhaps based upon PCR analysis, will result in a field station kit that personnel can use on site for these types of applications.

Analysis of quantitatively inherited traits of agricultural importance

Probably the most difficult problem facing breeders is the manipulation of metric traits with complex inheritance. Many strategies are available which rely upon the statistical analysis of field data to evaluate what has occurred on the genotypic level, but these inferences are often very imprecise as to the number of genes involved and their mode of action. This situation is even more frustrating for the physiologist or molecular biologist to address with their tools and for the most part has prevented in-depth study of these important but complex problems by other approaches. Molecular markers provide a mechanism for applying linkage genetic techniques to complex inheritance problems that almost reduces them to the level of studying single gene traits, although both the experimental design and phenotypic measurements are much more critical. Early studies using isozymes (Frei *et al.*, 1986) clearly showed that some chromosomal regions exhibited much more impact on quantitative traits than would have been predicted by chance. The limitations in numbers of informative loci clearly restricted these analyses from being exploited further.

In collaborative studies with C. W. Stuber's group at North Carolina State University (Edwards *et al.*, 1991), we utilized 114 molecular markers, including both isozymes and RFLPs, to analyze a set of F_2 plants produced from the cross, Co159 X Tx303. We were able to detect numerous major loci affecting several metric traits measured in this population. These factors were relatively localized along the chromosomes and not all regions displayed the same level of impact on these traits. Genetic factors affecting plant height were not found on all chromosomes nor were they of equal value (Figure 1). Interestingly, although this analysis represents a cross between one very tall and one very short inbred, factors contributing to plant height in a positive manner were contributed by both parents. Another interesting observation was that many of these traits could be broken down into components which revealed information about the genes underlying them. For instance, Table 1 shows that several loci are tabulated that exhibit an effect on plant height. Variation for plant height can occur either through changes in internode length or numbers of nodes. It can be seen that some genes affecting plant height primarily acted through changing the number of nodes (linked to marker loci, *NPI 205* and *NPI 43*) and others through altering the internode distance (linked to marker loci, *Adh2* and *Acp1*). Another interesting observation is seen in Table 1 where ear height

Figure 1 A schematic corn linkage map with the approximate positions of 114 marker loci used to analyze the genetic inheritance of height in F_2 plants generated from the cross Co159 x Tx303. Factors contributing positively to height by each parent are denoted, as is the relative impact of these loci by size of the oval in each region (Edwards *et al.*, 1991).

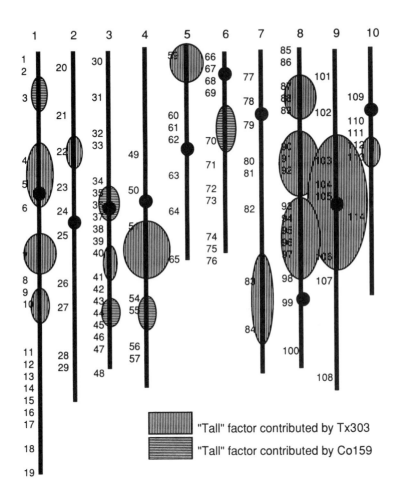

was also measured. A classical experiment in quantitative genetics with practical implications as well has been to attempt to shorten ear height to lessen lodging while maintaining a taller, higher yielding plant. One can clearly see that in this cross, such an attempt would eventually prove futile as most of the loci affecting ear height also affect plant height. Hence, in a single cross and analysis, one could predict the eventual result of several

Table 1 Chromosome origin and R^2 values of five marker loci that demonstrated correlations with plant height and related components in F_2 plants from the cross Co159 x Tx303 (Edwards *et al.*, 1991).

Marker designation	Chromosome origin	Ear height	Plant height	Internode distance	Number of nodes
NPI 205	1	13.5	8.1	--	16.7
Adh2	4	10.0	16.0	16.3	4.0
NPI 45	7	--*	5.7	5.9	--
NPI 43	8	12.3	14.4	4.9	17.9
Acp1	9	16.9	27.4	20.0	10.1

*Indicates non-significant correlation.

years of field analysis. Even without use of RFLPs as indirect selection criteria, molecular markers may provide important information to breeders as to which directions in their programs would prove most productive and aid them in prioritizing their field efforts.

Many of these types of analyses are in progress with different investigators and species and the results promise to change our thinking about these traits, if not also how we breed for their improvement. There has been particular concern expressed about the analyses of the data derived from these types of experiments. The development of maximum likelihood methods (Lander *et al.*, 1987) appears to offer real promise for analyzing Quantitative Trait Loci (QTLs), both in terms of the quality of data obtained and in setting uniform standards for its evaluation. I have been somewhat dismayed by the vigor of the arguments as to which method is best. In most of the experiments to date where highly structured populations have been used, it probably does not matter what type of analysis one uses. Simple analysis of variance (ANOVA) is usually sufficient to reveal the locations of major genes affecting these traits. This does not mean that these methods cannot be improved, but more concern should be directed towards experimental design considerations and what specific types of analysis actually reveal about the questions we ask. In my own experience, we have planned experiments, carried them out, analyzed the data, and then found out that other designs might have provided more insight into the problems we were attempting to address. There is a real need for a broader discussion of alternative types of designs which could incorporate genomic analysis and what the implications are for the data to be generated by them.

One concern often expressed is how useful this approach will be if the critical genes are different for every genetic background and environment.

Metric traits are known to be sensitive to environmental effects and one must also anticipate that the same genes cannot be acting in all genetic backgrounds, otherwise plant breeding would have already reached its zenith in obtaining the maximal yield in many species. Two interesting results from this type of analysis bear on these issues. In the first I was able to tabulate the data from 23 different molecular marker analyses of metric traits in corn, most of them unpublished. By simply plotting the approximate locations of the marker loci detecting significant correlations with various quantitative traits, I found that their locations were not random across the genome but that there were definite "hotspots". The locations of QTLs for yield are shown on chromosome 10 from many of these analyses where I have not attempted to control for either the genetic backgrounds used in the crosses or where they were analyzed (Figure 2). There is a definite concentration of influential loci near the centromere on chromosome 10. One must be cautious in this case, as the location here is near the centromere where recombination is reduced. This larger physical region could contain many genes and further efforts are needed to understand the numbers of functional elements within this region. Similar hotspots for yield were observed on chromosomes 1, 3, 4, and 9 and factors were also found to be similarly concentrated for plant height on several chromosomes. This observation is very encouraging as it suggests that the major loci for some of these traits may be more limited in number than originally suspected. How this should alter our approach to manipulating these traits should be considered further.

Similarly in attempting to analyze the effect of environment, I used data from a study conducted collaboratively with C. W. Stuber's group. S_4 lines from the cross Mo17 x B73 were test-crossed to the parents and analyzed in six environments ranging from North Carolina to the Midwest. Several marker loci revealed significant correlations with yield and height in this experiment (Table 2). Many of these major loci are also active across all or most of the environments tested. For instance, *BNL 15.45* and *NPI 296* were correlated with major effects upon both yield and height, *wx* and *Amp1* were located near genes affecting height, and factors near *Amp3* affected yield no matter which tester was used. These data do not show that the same lines would be most productive in all environments but simply state that the majority of yield in this population is due to factors that are important across all environments. A majority of variation in yield might be due to such factors but the remaining variation could still be due to environmentally sensitive loci such that yield rankings of the progeny across environments would vary significantly.

These results, demonstrating that major loci are active across both genetic backgrounds and environments, are important in that they reveal new insight into the genetic basis of important agricultural traits. Initially they should force us to re-examine many of the assumptions used in our current breeding strategies. Are there more effective strategies to take advantage of the

Figure 2 Marker loci on chromosome 10 summarized from 23 different studies
of corn yield are plotted by their approximate locations along the chromosome.
The location of these elements is distinctly non-random, reflecting a "hotspot"
near the centromere.

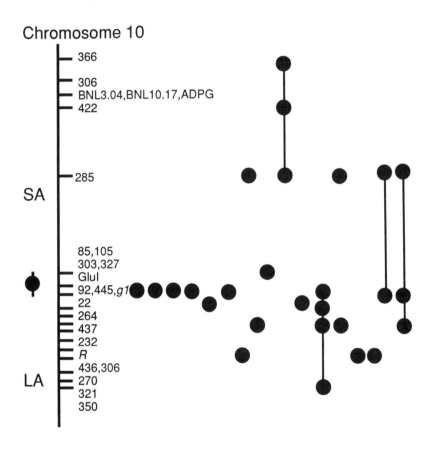

existence of these major genes? Should we move towards centralized development of materials with widespread field-testing? Should we use the data
gathered through the analysis of many different backgrounds and environments to determine if some backgrounds should receive more emphasis on
combining and fixing these major loci? These are questions we need to
address, particularly in the context of our past experiences. Perhaps it will
be possible to use molecular marker loci as indirect selection criteria to
follow these loci within breeding programs. The ability to remove the
environment as a confounding influence during the selection process could
be a powerful tool. Large-scale use of molecular markers will be thwarted,
however, until replacement technologies for genotypic screening can be

Table 2 R^2 values of marker loci affecting plant height and yield from analysis of S_4 lines derived from the cross Mo17 and B73 tabulated over six environments.

Trait	Locus	MO17 as tester						B73 as tester					
		#1	#3	#4	#5	#6	#7	#1	#3	#4	#5	#6	#7
Yield	Amp1	5.3*	2.7*				2.6*			4.1**	2.6*	6.1**	6.8**
Height				4.6**			4.3**	3.0*	3.5**	3.6**	6.8**	7.8**	7.4**
Yield	NPI 296	2.6*	7.2**	6.6**	7.5**	2.9*	3.3*	2.4*				4.3**	2.5*
Height		4.0**		5.3**	4.9**	4.0**	5.3**			3.5*			5.8**
Yield	BNL15.45	10.5**	6.2**	5.3**	4.1**		5.2**						
Height		7.7**	5.9**	5.3**	5.1**	6.3**	2.5*					3.5*	
Yield	Amp3	6.8**	9.8**	6.2**	6.3**	4.9**	2.8**	12.1**	4.7**	7.7**	10.2**	6.0**	8.8**
Height			3.0*			3.2*	4.8**						3.8**
Yield	wx	2.8*			10.3**		3.8**			3.7**		4.4**	6.0**
Height							3.1*	12.2**	8.5**	11.1**	8.0**	15.3**	4.4**

*, **Significant at $P = 0.01$ and $P = 0.001$, respectively.

perfected which are cheaper, easier, and faster than current field analysis. Even now, the quality of the information derived from such studies, if not the quantity, will continue to provide us with new insights into these complex problems.

CONCLUSIONS

The use of molecular markers to address many problems in plant breeding is now being widely explored. The progress to date has proven tremendously exciting and promises to continue. As we have seen here, RFLPs provide us with new insights into the relationships among germplasm of interest to the breeder. Besides just a systematics tool of basic interest, these types of analyses promise to streamline the use of complex sets of germplasm by allowing the breeder to sample them in a directed fashion. By relating genetic diversity to yield in hybrid crops, we may also anticipate the use of these results to refine our knowledge of heterotic groups and even aid in the design of new groups in many crops, particularly those where hybrids are not now extensively used. The use of high resolution finger-printing will allow the developer to protect his investment, an advance that will stimulate this process both publicly and privately. Use of molecular markers to tag single genes will become an important tool for selection once new technologies can be perfected that may be practical under field conditions. Finally, the use of molecular marker analysis to dissect complex traits promises to redefine our understanding of their actual genetic basis. By breaking these traits down into individual loci, we will begin to study these elements as we do any other gene. With new technologies, we may be able to select for improved combinations of genes on the genotypic level where the problem of environmental variation no longer confounds the process.

In the beginning of this chapter, I described one of the frustrations of this area as the inability to relate much of our other advances in basic sciences to the practical aspects of plant improvement. The most exciting facet of this work is the ability of molecular markers to facilitate comparison of data from many different types of analyses. For the first time we can combine Mendelian and molecular genetic approaches in plant breeding and begin to bring all of the advances in our basic studies to bear on this problem. The ability to understand complex traits in terms of individual gene actions opens up this area for study in ways which were previously impossible. Just a few years ago, no one would have considered that we would be able to use genetic engineering on our most complex but important traits such as yield, maturity, stress tolerance, etc. With our ability to detect those genes possessing the most influence upon these traits, we can begin to understand how individual genes function in translating genotype into phenotype and how that process is modulated by environmental influences. The recognition that many of these loci are important across both background and environment and that they are also pleiotropic in their action suggests that hormones may play a significant role in variation for crop productivity, a result that will surprise few physiologists. It should be possible within the next few years to obtain clones of some of these genes, perhaps from model species, such as *Arabidopsis thaliana*, and with advances in transformation technology, we should be able to insert altered copies of these genes into important cultivars. While initially it might be difficult to directly create improved cultivars through transformation of these single genes, we may be able to alter their function dramatically in ways that will create novel forms of plant variation. While much of this variation may prove unuseful, some may be modified within standard breeding programs to create new cultivars with improvement beyond that possible with existing alleles at these genes. In a sense this might work as a sort of directed mutagenesis program where only the genes contributing to variation in yield are targeted for alteration. Once improved alleles of single genes are developed, the insertion of these genes into many different backgrounds without extensive backcrossing programs could facilitate the rapid testing and development of new cultivars on an unprecedented scale, a result that would justify much of the initial enthusiasm for these techniques.

ACKNOWLEDGMENTS

Much of the work cited herein was extracted from published works and acknowledged as such. Most of the genotypic data were gathered at Ceres/ Native Plants Incorporated by Mark Walton, Scott Wright, Susan Neuhausen, and the technical staff. The phenotypic data were obtained either in collaboration with C. W. Stuber and his group at North Carolina State University (M. Edwards, R. Guffy) or with DeKalb-Pfizer Genetics. The author is grateful to all in participating in what are truly collaborative studies requiring the inputs of individuals with quite different backgrounds.

REFERENCES

Botstein, D.; White, R.; Skolnick, M.; Davis, R. W. (1980) Construction of a genetic linkage map in man using restriction fragment length polymorphisms. *American Journal of Human Genetics* 32:314-331.

Duvick, D. N. (1984) Genetic contributions to yield gains of U.S. hybrid maize, 1930 to 1980. In: Fehr, W. R. (ed.), *Genetic contributions to yield gains of five major crop plants.* Crop Science Society of America Special Publication 7, Madison, WI, pp. 15-47.

Edwards M. D.; Helentjaris, T.; Wright, S.; Stuber, C. W. (1991) Molecular-marker-facilitated investigations of quantitative trait loci in maize: IV. Analysis based on genome saturation with isozyme and restriction fragment length polymorphism markers. *Theoretical and Applied Genetics* (in press).

Frei, O. M.; Stuber, C. W.; Goodman, M. M. (1986) Yield manipulation from selection on allozyme genotypes in a composite of elite corn lines. *Crop Science* 26:917-921.

Helentjaris, T. (1987) A genetic linkage map for maize based upon RFLPs. *Trends in Genetics* 3:217-221.

Lander, E. S.; Green, P.; Abrahamson, J.; Barlow, A.; Daly, M. J.; Lincoln, S. E.; Newburg, L. (1987) Mapmaker: An interactive computer package for constructing primary genetic linkage maps of experimental and natural populations. *Genomics* 1:174-181.

Lobler, M.; Hirsch, A. M. (1990) RFLP mapping of the *abp1* locus in maize (*Zea mays* L.). *Plant Molecular Biology* 15:513-516.

McMullen, M.D.; Louie, R. (1989) The linkage of molecular markers to a gene controlling the symptom response in maize to Maize Dwarf Mosaic Virus. *Molecular Plant-Microbe Interactions* 2:309-314.

Smith, J. S. C.; Smith, O. S. (1989) The use of morphological, biochemical, and genetic characteristics to measure distance and to test for minimum distance between inbred lines of maize (*Zea mays* L.). Presented at the UPOV Workshop held in Versailles, France.

Stuber, C.W.; Goodman, M. M. (1983) Allozyme genotypes for popular and historically important inbred lines of corn, *Zea mays* L. USDA-ARS publication ARR-S-16, U. S. Government Printing Office, Washington, DC.

Tanksley, S. D. (1983) Introgression of genes from wild species. In: Tanksley, S. D.; Orton, T. J. (eds.), *Isozymes in plant genetics and breeding.* Elsevier, NY, pp. 331-337.

Walton, M.; Helentjaris, T. (1987) Application of restriction fragment length polymorphism (RFLP) technology to maize breeding. In: Wilkinson, D. (ed.), *Proceedings of the 42nd annual corn and sorghum research conference.* ASTA Publication, pp. 48-75.

Williams, J. G. K.; Kubelik, A. R.; Livak, K. J.; Rafalski, J. A.; Tingey, S. V. (1990) DNA polymorphisms amplified by arbitrary primers are useful as genetic markers. *Nucleic Acids Research* 18:6531-6535.

Chapter 16
Identification and isolation of agronomically important genes from plants

S. P. Briggs

Department of Biotechnology Research, Pioneer Hi-Bred International, Inc., Plant Breeding Division, P. O. Box 1004, Johnston, Iowa 50131

INTRODUCTION

For our purposes, an agronomic crop is one which contributes significantly to crop production. Genes which control maturity, yield, disease resistance, drought tolerance, nutrient assimilation, or nutritional content would be considered agronomic genes. Such genes, until recently, have been part of the natural variability within a species' gene pool. Progress in genetic engineering has removed this limitation, permitting plants to be transformed with genes from viruses and bacteria to confer disease resistance; genes from bacteria to confer male-sterility, herbicide resistance, and insect resistance; and genes from heterologous plant species to enhance nutritional content and to produce new pigments. We can expect to see genes from animals used as well, particularly transcription factors and antibodies. Generally, these are genes which confer novel traits for which little or no genetic potential is known within the germplasm of a given crop. Products with these new traits are moving rapidly to the marketplace. Heterologous organisms are certain to be a continuing source of genes for crop improvement. Nevertheless, with advances in gene mapping and cloning techniques, crops themselves will become a major source from which to isolate agronomic genes.

I am going to focus my discussion on crops as a source of genes. There are several reasons for seeking agronomic genes from crops rather than from microorganisms - most agronomic traits have no known corollary in microbes, the biochemical basis for agronomic traits is generally unknown, and plant-specific regulatory signals are not present in microbe genes. The isolation of crop genes has been retarded by the complexity of plant genomes and by the dearth of clearly defined agronomic genes. The complexity, or size, of most crop genomes exceeds that of the human genome. The effort required to isolate a human gene can easily run 5 years at a cost of several million dollars. It would be impossible to fund a plant gene isolation effort of this scale because of the small size of research grants for plant biology and because no known plant gene has such commercial value. Fortunately, plant geneticists have developed technologies which are far more efficient than those necessary for human gene isolation, permitting plant genes to be isolated at much less cost; these will be described in detail later. The advent of DNA restriction fragment length polymorphism (RFLP) maps is facilitating the clear definition of some agronomic genes. The development of gene isolation methods for use in plants and the placement of agronomic traits on a genetic map have set the stage for what should be a "golden age" of agronomy - over the next several years, many of the genes which contribute the most to crop production will be isolated and characterized.

IDENTIFICATION OF AGRONOMICALLY IMPORTANT GENES

Empirical

The only certain way to identify agronomic genes is to observe them segregating in a population. Often, simple Mendelian factors are used for crop improvement. For example, mutant alleles for several genes are known to cause desirable changes in the maize (*Zea mays* L.) endosperm (Table 1).

Table 1 Selected examples of mutant loci which cause beneficial phenotypes in the maize endosperm.

Gene	Trait
wx1	Blocks the conversion of amylose to amylopectin
ae1	Like *wx1* but less effective
o2	Increases the proportion of lysine
o7	Increases the proportion of lysine
fl2	Increases the proportion of lysine
sh2	Inhibits the conversion of sugar into starch
su1	Like *sh2* but half as effective
se1	Makes *su1* as effective as *sh2*
y1	Blocks carotenoid accumulation

Generally, however, agronomic genes are considered to be extraordinarily difficult to observe. Until recently, it has been possible only to estimate the number of genes involved in the expression of quantitative traits (i.e., traits which are expressed in a segregating population as a continuous range of phenotypes rather than as discrete classes). However, most agronomic traits are quantitative traits, and neither the genetic map location nor the proportion of the genetic variance for which a given gene accounts could be determined. Placing agronomic traits on the genetic map has not been possible because few mapped traits are segregating in breeding material. In contrast, RFLPs can be scored in any heterozygous population, regardless of phenotypes. The genotype of every plant in a population can be determined with certainty, limited only by the resolution of the RFLP map. When the genotype of a given plant is matched with its phenotype (e.g., an agronomic trait), and a population of such matches is statistically analyzed, patterns of inheritance can be distinguished which were previously obscure. Not only is a map location, or quantitative trait locus (QTL) determined, but also the proportion of the trait for which the QTL accounts can be assigned. The use of RFLPs overcomes another major impediment to mapping quantitative traits (i.e., environmental effects). The phenotypic differences resulting from alleles present in different elite inbreds or cultivars can be masked by environmental effects influencing the phenotypic expression of agronomic traits.

Quantitative trait loci have now been mapped for the most complex of traits, yield, as well as for maturity and a host of other agronomic traits. The resolution of QTL mapping is not yet good enough to permit firm conclusions to be drawn when a comparison is made to the standard genetic map. This is due in part to the fact that the standard map and the RFLP map(s) have not been fully integrated. However, the early results indicate that at least some QTLs may correspond to known genes on the standard map.

The first published reports which resolved quantitative traits into Mendelian factors were based on the tomato RFLP map. Association of insect resistance with RFLPs permitted introgression of resistance from exotic germplasm (Nienhuis *et al.*, 1987). Other traits such as soluble acids content (Osborn *et al.*, 1987; Paterson *et al.*, 1988) and water use efficiency (Martin *et al.*, 1989) have also been analyzed in tomato (*Lycopersicon esculentum* Mill.) using RFLPs, permitting QTLs for these traits to be identified. It appears that virtually any trait which can be measured is amenable to QTL mapping by RFLPs.

Rational

Another approach to identifying agronomically important genes is to make predictions, based on one's knowledge, and then to test a candidate gene in transgenic plants. This has been the most successful approach, partly because QTL mapping is so new. The most spectacular progress in rational gene identification has resulted from Roger Beachy's discovery that virus coat proteins can mimic cross-protection when expressed in transgenic plants [Beachy *et al.*, 1990; Hill *et al.*, 1991 (and references therein)]. This discovery has been extended to many viruses and to the use of other viral genes for similar purposes. Male sterility has been engineered by expressing RNAse in anthers (Mariani *et al.*, 1990). Similarly, *Bt*-toxin can protect plants from certain insects (Delannay *et al.*, 1989). Indeed, the whole gamut of prokaryotic, heterologous genes which have been put into plants is based upon a rational engineering approach.

Robertson's hypothesis

The most powerful method for gene identification may come from merging the QTL mapping with the detailed mutant maps which a few crops such as maize and tomato have available. For example, QTLs for height may coincide with dwarf genes in maize (Beavis and Grant, 1991; Helentjaris and Shattluck-Eidens, 1987; Jensen, 1989). Robertson put forth the hypothesis that QTLs are the same genes that geneticists study but, when two "wildtype" alleles are segregating rather than a mutant and a wild-type, the environment plays such a big role that the gene can only be detected as a QTL (Robertson, 1985). The observation that the achaete-scute complex is a QTL for bristle number in *Drosophila* supports Robertson's hypothesis (Mackay and Langley, 1990).

There are many Mendelian factors in maize which, if the genes could be cloned, could conceivably be used to engineer better plants (see Table 2). Most of these genes have not been evaluated as potential QTLs. It is likely that at least some of them play important roles in crop performance.

ISOLATION OF AGRONOMICALLY IMPORTANT GENES

Transposon tagging

Transposable elements

McClintock was the first to describe genes which move from one chromosome location to another. Such genes are now called transposable elements or transposons. The actions of transposable elements can be understood by knowing two basic properties. First, transposable elements are genes which encode proteins that are required for an element to excise from its old site and insert at a new site on the same or a different chromosome. Second, the nucleotide sequence which is cut is located at both ends of a transposable element. There may be many copies of an element scattered among the chromosomes and the proteins produced by any one of them can bind to every other copy, so that an element on one chromosome can cause the transposition of another element on a different chromosome. However, proteins produced by one element system cannot bind to or cut elements belonging to another system because the nucleotide sequences at their ends are different. The details of these processes are still being elucidated and the descriptions given here are significantly oversimplified. Consider the implications of the two properties. A mutation which destroys the ability of

Table 2 Selected examples of maize genes which are of potential use in genetic engineering.

Gene	Trait
an1	Gibberellin accumulation
aph1	Aphid resistance
bk2	Brittle stalk
bm3	Reduced lignin content
bx1	Corn borer and *H. turcicum* resistance
Cg11	*C. graminicola* resistance
d1	Gibberellin accumulation
d2	Gibberellin accumulation
d3	Gibberellin accumulation
d5	Gibberellin accumulation
D8	Gibberellin receptor (?)
dek	Many loci affecting kernel development
Ger	Glucoside earworm resistance
gl	Many loci affecting cuticle wax
hm1	Resistance to *C. carbonum*
hm2	Resistance to *C. carbonum*
ht1	Resistance to *H. turcicum*
ht2	Resistance to *H. turcicum*
ht3	Resistance to *H. turcicum*
Htn	Resistance to *H. turcicum*
ln1	Lower ratio of oleate to linoleate in kernel
loc1	Low oil content in kernel
lte1	Latente: Drought, heat, aluminum, frost resistance
Lte2	Epistatic to *lte1*
Mdm1	Maize dwarf mosaic virus resistance
mep1	Affects quantities of *Glb1* endosperm protein
Mer	Earworm resistance
mg1	Miniature germ
ms	Many loci affecting male sterility
Mv1	Resistance to maize mosaic virus I
o	Opaque: Many loci affecting kernel protein
Glb1&2	Globulin embryo protein
rhm1	Resistance to *H. maydis*
Rp1	Resistance to common rust
Rpp9	Resistance to southern rust
Zer	Earworm resistance

an element to produce its protein(s) can leave the sites at the end intact. Since there is no practical limit to the number of elements which can be present, another active (non-mutated) element can produce protein which will bind to and cause the mutated element to transpose (Figure 1 - activation of *Ds* by *Ac* protein). This establishes a dependence of the mutant element upon the wild-type for the ability to transpose. Two-element

Figure 1 Activation of *Ds* excision from R by *Ac*.

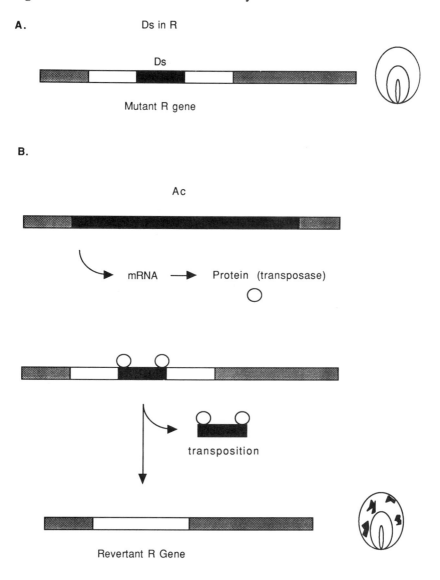

systems of this type are common. Most members of a transposable element system are of the defective type. Insertion of a transposable element into another gene, rather than into the vast stretches along the chromosome between genes, often destroys the ability of the disrupted gene to function. This is one of the most common causes of mutations. Occasionally, insertion into a gene will not be mutagenic because the element is spliced out of the RNA prior to translation into protein; this prevents the amino acid sequence of the protein from being disrupted, even though the DNA sequence is disrupted. These "non-mutagenic" insertions may be detected if the binding of transposase to the element within the disrupted gene suppresses the function of the gene, perhaps by derailing RNA polymerase from the DNA track. Thus, progeny which inherit an active element plus the mutant gene will display a mutant phenotype, but progeny which inherit only the mutant gene will appear as the wild-type.

Tagging strategies

The basic concepts in transposon tagging are very simple. The method is identical to a standard RFLP analysis except that transposable elements are used as probes and the population being examined is segregating for a mutation which arose as a result of a (putative) element insertion. DNA hybridization is only needed to visualize the transposons, to determine if one is co-segregating with the mutation. Thus, an investigation begins by obtaining a mutant, preferably from a population which is known to carry active transposable elements. DNA is isolated from several progeny in a population in which the mutant allele can be observed to segregate. All of the progeny which carry the mutant allele are run as a group of adjacent lanes on a Southern blot; the DNA is cut with an enzyme which cuts outside of the elements, so that they are each contained in a restriction fragment of different size (see Figure 2). Visual inspection of the lanes can reveal a band which is present in all the samples, suggesting that it represents an element which is co-segregating with the mutant allele. This can be confirmed by including siblings which did not inherit the mutant allele; they should lack the band in question.

Once a co-segregating band has been identified, it can be cloned and used as a probe to clone the wild-type allele (see Figure 3). However, independent confirmation is necessary to verify that the cloned band represents the gene of interest and is not simply linked to the gene. The co-segregating band seen in Figure 2, in fact, is not inserted into *Hm1*; it turns out to be approximately 6 cM proximal to *Hm1* (unpublished observations). Usually, the easiest means to verify an insertion is to obtain genetic revertants and see if the insertion has concomitantly excised. Proof of excision can be obtained by sequencing the insertion site to identify footprints. Revertants are easy to obtain with members of the *Spm/En* and *Ac* families but have

Figure 2 Co-segregation of a 5.7-kb *Mu1* fragment with the *hm1-1062* mutant allele. DNA was prepared from progeny of the cross *Hm*/hm* x *Hm/Hm*, cut with *Sst1*, fractionated on a gel, and Southern blotted. The blot was probed with the *Mu1* internal fragment.

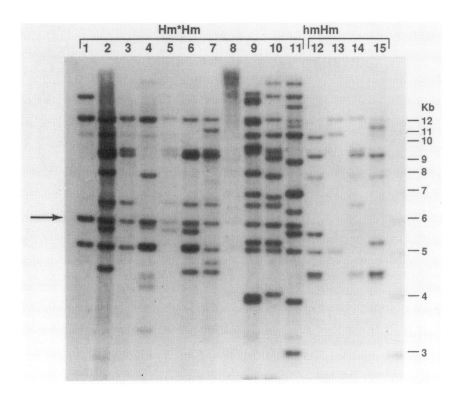

been difficult to obtain with *Mu* element insertions. The relative advantages of tagging are that it (1) can be used to clone any gene for which a phenotype can be reliably scored, (2) does not require a genetic map, (3) may be as efficient in complex genomes as in simple genomes, (4) has a good track record of success, (5) can be extended to heterologous organisms, and (6) is of relatively low cost, simple, and efficient.

T-DNA insertional mutagenesis

The integration of T-DNA into transgenic plants can create mutations similar to transposable elements. The principle and practice of isolating the disrupted genes is as described for element insertions. A large collection of such mutants has proven useful for isolating genes from *Arabidopsis*.

Figure 3 Steps to cloning the wild-type allele.

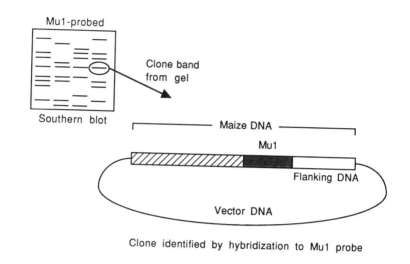

Clone identified by hybridization to Mu1 probe

The flanking DNA is used as a probe to clone
the wild-type allele from a separate library

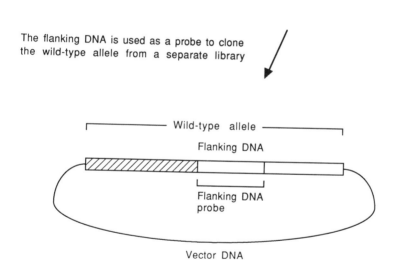

Map-based cloning

The goal of map-based cloning is to enable an investigator to clone any gene
by knowing only its position on the genetic map. This approach generally
requires a "chromosome walk" between RFLP markers which flank the
target gene (Figure 4). Obviously, the method is more practical for plants
with small genomes like *Arabidopsis* than for maize because the ratio of
kilobases of DNA to genetic map units is roughly proportional to the size of

Figure 4 Map-based cloning. X and Y are RFLP markers which flank Your Favorite Gene (YFG). * and # are probes made from one clone to enable the isolation of the next overlapping clone in the walk. Contig = an overlapping, contiguous array of ordered clones.

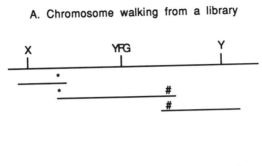

A. Chromosome walking from a library

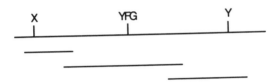

B. Chromosome walking across a contig

the genome. Cloning becomes much easier if the entire genome is already represented by a contiguous array of ordered clones (Figure 4).

Cloning by differential hybridization

The selective absence of a particular mRNA or region of DNA from a certain tissue, stage of development, or mutant can be used to clone the missing moiety. The method has been adapted for isolating genes based upon their predicted pattern of expression and is being developed for cloning genes which are represented as deletion mutations.

ISOLATION OF AN AGRONOMIC GENE FROM MAIZE

The role of *Hml* in disease resistance

There are races reported for the leaf blight and ear mold pathogen *Cochliobolus carbonum* (anamorph *Bipolaris zeicola* or *Helminthosporium carbonum*). Race 1 was originally misidentified as *H. maydis* (hence, the gene designation *hm*). The histology of pathogenesis is typical of

Helminthosporium - conidia germinate to form appressoria, infection hyphae ramify beneath the cuticle and invade underlying cells. Resistance is expressed in the first one or two cells invaded resulting in a chlorotic fleck; the fungus ceases to grow but remains viable. Compatible interactions lead to large, necrotic lesions which are oval and marked by concentric rings of dark pigment. Natural infection in the field is often so severe that the lesions coalesce, blighting the entire foliage. If ears are able to develop, they are typically covered by sooty spores (hence, the name carbonum). Race 1 can infect healthy corn only if it produces *HC*-toxin, a cyclic-tetrapeptide. The abilities of the fungus both to infect corn and to produce *HC*-toxin co-segregate as a single locus, *TOX2*. Tox⁻ isolates are arrested in their growth on susceptible genotypes of corn similar to Tox⁺ isolates on resistant genotypes. Exogenous application of *HC*-toxin permits the growth of Tox⁻ isolates to resume in susceptible but not resistant tissue. The role of *HC*-toxin, therefore, is to induce susceptibility (or compatibility), but it can do so only on certain genotypes of corn. Thus, *HC*-toxin accounts for host-specificity as well as pathogenicity. The receptor(s) for *HC*-toxin has not been identified nor is the mode of action known.

The genetic control of both susceptibility to race 1 and sensitivity to *HC*-toxin resides at the *hm1* locus on chromosome 1 in corn. Resistance is dominant or codominant and may be expressed at only some or all stages of plant development, depending upon which *Hm1* alleles are present. The role of *Hm1* is, thus, to prevent the action of *HC*-toxin. There is a second locus, *Hm2*, located on chromosome 9 which also confers resistance to race 1. It has not been determined whether or not *Hm2* affects sensitivity to *HC*-toxin. Resistance at either *Hm1* or *Hm2* is epistatic to susceptibility. The mechanism by which *Hm1* prevents the action of *HC*-toxin is not known; but there are three obvious hypotheses, each of which can be directly tested once the *Hm1* gene has been cloned. The *Hm1* gene may encode (1) an enzyme which directly inactivates *HC*-toxin, (2) a molecule which blocks the accessibility of *HC*-toxin to the toxin's receptor, or (3) a protein which duplicates the normal function of the toxin receptor but is not itself sensitive to *HC*-toxin. Of course, *Hm1* could also encode a gene which regulates one of these functions.

Robertson's Mutator

One of the most intensively studied transposable element systems in maize is Robertson's Mutator. This system is associated with an extremely high forward mutation rate and a very low reversion rate. Because of the high mutation rate and the fact that most of the members of the system have been molecularly characterized, Mutator is a good system for transposon tagging. The Mutator system consists of a master, or autonomous, element which may be present in one or several copies. In addition, at least eight slave, or non-autonomous, types of elements are known; each has unique internal DNA, but they share the 0.2-kb terminal inverted repeats. Each slave type can be present in many copies (greater than 50) or absent altogether, with all variations apparently possible. Transposition appears to result from the production of new copies of an element rather than from excision followed by reintegration at a new site, in contrast to the *Ac* and *Spm* elements. The large number of elements plus the fact that transposition, being a form of recombination, is targeted to genes may account for the high mutation rate.

Tagging *Hm1*

A restricted mutagenesis strategy was used to tag *Hm1* plants which carried both Robertson's Mutator system and *Hm1* (homozygous) were crossed with plants which lacked both traits. The F_1 progeny were all resistant to Race 1, as expected of heterozygotes, except for mutants which lost the function of the *Hm1* allele. While not every mutant was examined in detail to confirm that loss of resistance was caused by a mutation, the approximate frequency of mutation was 3×10^{-4}. Several mutants were examined, first by confirming the presence of the Mutator system to rule out contaminants and, second, by determining whether or not a *Mu* element could be associated with each mutation. As seen in Figure 2, *Mu* elements can be associated with a mutant gene by observing co-segregation of the gene (allele) and an element restriction fragment. While such an association does not prove that the mutation was caused by the element, it is a necessary first step toward establishing a causal relationship. By using the different *Mu* element probes and examining several different mutants, a *Mu1* element was found to be associated with the *hm1-656* allele and a *Mu3* element was associated with the *hm1-1369* allele. When these two elements plus their flanking DNA, were cloned and examined, it was revealed that the insertions were within a few hundred base-pairs of each other. Such a finding is not unexpected if the cloned elements caused the mutations. Using the flanking DNA as probes, other alleles could be examined or cloned. The *hm1-1790* allele, which behaves as a deletion (transmission is blocked through the pollen and reduced through the egg), was found to be missing the cloned sequences. The *hm1-1062* allele was found to have a small insertion near those of *hm1-656* and *hm1-1369*. Finally, the *hm1-1040* allele, which arose in a mutation screen involving the *Ac* element rather than Mutator, was found to contain an *Spm* element. The identification of insertions or deletions in the DNA of each allele provided strong circumstantial evidence that the cloned region contained at least part of the *Hm1* gene. Verification was obtained by searching for genetic revertants of the *hm1-1040* allele. Unlike Mutator, *Spm* elements excise from their locus as part of the transposition process. Excision frequently restores the function of the gene, resulting in a genetic revertant. Several revertants were obtained. In each case, the *Spm* element

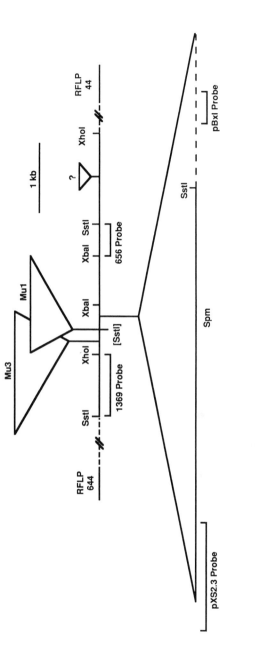

Figure 5 Restriction map of *Hm1* (B79), showing insertions in the *1369 (Mu3)*, *656 (Mu1)*, *1040 (Spm)*, and *1062* (unknown element) alleles; the *1790* allele is deleted in this region.

had excised from the cloned region. Similarly, revertant somatic sectors of mutant plants were associated with the excision of *Spm*. A summary map of the cloned region and sites of insertion is shown in Figure 5.

Utility

It may at first seem obvious that the utility of a cloned agronomic gene is to transfer the gene to cultivars which either lack the gene or are deficient for the respective trait which the gene affects. However, in most cases such applications may not be fruitful. For *Hm1*, transfer to susceptible lines has already been done by traditional methods, which will remain the practical means for intra-species gene transfer for some time. Inter-species transfer by molecular methods may occasionally be useful. The primary utility of cloned agronomic genes may be to elucidate key facets of plant cell biology, permitting a rational engineering approach to be conceived. For instance, in what cells is *Hm1* expressed, and at what stages of development? Such knowledge may be critical for directing the expression of heterologous genes to provide disease resistance. Because *Hm1* protects cultured cells as well as intact plants from *HC*-toxin, the gene may be useful as a selectable trans-formation marker. The promoter of *Hm1* may be useful for directing the expression of synthetic resistance genes (e.g., peptides which are toxic to the fungi) to protect all parts of the plant, including the ear. Alleles of *Hm1* are sensitive to the developmental stage of the plant (in contrast to the physiolog-ical age of the cells) and their promoters may be used for expressing genes only late in development, for example. Generally, functional gene duplica-tions are the result of structural duplications. Thus, *Hm1* may be useful as a probe to isolate *Hm2*. The *Hm1* clone can be used as a probe to determine whether or not other species possess and express the gene and, if so, to create anti-sense constructs for testing the function of the gene in those species. Likewise, ectopic expression of knock-out of *Hm1* in maize may reveal functions other than disease resistance. Perhaps a better understand-ing of *Hm1* will permit the engineering of new specificities.

REFERENCES

Beachy, R. N.; Loesch-Fries, S.; Tumer, N. (1990) Coat protein-mediated resistance against virus infection. *Annual Review of Phytopathology* 28:451-474.

Beavis, W. D.; Grant, D. (1991) Quantitative trait loci for plant height in four maize populations and their associations with qualitative genetic loci. *Theoretical and Applied Genetics* (in press).

Delannay, X.; LaVallee, B. J.; Proksch, R. K.; Fuchs, R. L.; Sims, S. R.; Greenplate, J. T.; Marrone, P. G.; Dodson, R. B.; Augustine, J. J.; Layton, J. G.; Fischhoff, D. A. (1989) Field performance of transgenic tomato plants expressing the *Bacillus thuringiensis* var. *kurstaki* insect control protein. *Bio/Technology* 7:1265-1269.

Helentjaris, T.; Shattluck-Eidens, D. (1987) A strategy for pinpointing and cloning major genes involved in quantitative traits. *Maize Genetic Cooperative Newsletter* 61:88-89.

Hill, K. K.; Jarvis-Eagan, N.; Halk, E. L.; Krahn, K. J.; Liao, L. W.; Mathewson, R. S.; Merlo, D. J.; Nelson, S. E.; Rashka, K. E.; Loesch-Fries, L. S. (1991) The development of virus-resistant alfalfa, *Medicago sativa* L. *Bio/Technology* 9:373-377.

Jensen, J. (1989) Estimation of recombination parameters between a quantitative trait locus (QTL) and two marker gene loci. *Theoretical and Applied Genetics* 78:613-618.

Mackay, T. F. C.; Langley, C. H. (1990) Molecular and phenotypic variation in the achaete-scute region of *Drosophila melanogaster*. *Nature (London)* 348:64-66.

Mariani, C.; DeBeuckeleer, M.; Truettner, J.; Leemans, J.; Goldberg, R. B. (1990) Induction of male sterility in plants by a chimaeric ribonuclease gene. *Nature (London)* 347:737-741.

Martin, B.; Nienhuis, J.; King, G.; Schaefer, A. (1989) Restriction fragment length polymorphisms associated with water use efficiency in tomato. *Science (Washington, DC)* 243:1725-1728.

Nienhuis, J.; Helentjaris, T.; Slocum, M.; Ruggero, B.; Schaefer, A. (1987) Restriction fragment length polymorphism analysis of loci associated with insect resistance in tomato. *Crop Science* 27:797-803.

Osborn, T. C.; Alexander, D. C.; Fobes, J. F. (1987) Identification of restriction fragment length polymorphisms linked to genes controlling soluble solids content in tomato fruit. *Theoretical and Applied Genetics* 73:350-356.

Paterson, A. H.; Lander, E. S.; Hewitt, J. D.; Peterson, S.; Lincoln, S. E.; Tanksley, S. D. (1988) Resolution of quantitative traits into Mendelian factors by using a complete linkage map of restriction fragment length polymorphisms. *Nature (London)* 335:721-726.

Robertson, D. S. (1985) A possible technique for isolating genomic DNA for quantitative traits in plants. *Journal of Theoretical Biology* 117:1-10.

Chapter 17
Novel approaches to the induction of genetic variation and plant breeding implications

R. L. Phillips, D. J. Plunkett and S. M. Kaeppler

Department of Agronomy and Plant Genetics and Plant Molecular Genetics Institute, University of Minnesota, St. Paul, MN 55108

The induction of genetic variation has been important for many genetic and breeding purposes. Useful mutants have been induced by a variety of chemical mutagens and ionizing/non-ionizing radiation. The "Mutation Breeding Newsletter" published by the Joint FAO/IAEA Division of Isotope and Radiation Applications of Atomic Energy for Food and Agricultural Development, International Atomic Energy Agency (IAEA), Vienna, Austria has documented hundreds of induced variants of interest to breeders in a wide variety of crop species.

The purpose of this chapter is to describe two seemingly unrelated mechanisms of altering the plant genome and the possible plant breeding implications. The first relates to the use of transposable elements to induce stable and unstable mutant forms. The second is to use the process of plant tissue culture as a mutagenic procedure. Because plant tissue culture causes alterations in DNA methylation, this mutagenic system may be unique and perhaps will generate some novel types of variants. Interestingly, the activity of transposable elements has been correlated with hypomethylation of certain DNA sequences (Schwartz and Dennis, 1986; Chomet *et al.*, 1987; Fedoroff *et al.*, 1989; Dennis and Brettell, 1990), and active transposable elements have been induced via the tissue culture process (Evola *et al.*, 1984; Peschke *et al.*, 1987, 1991; Peschke and Phillips, 1991). Although these two approaches are quite different, one might facilitate the other.

MODULATION OF SWEET CORN KERNEL DEVELOPMENT USING TRANSPOSABLE ELEMENTS

Several transposable element systems in maize (*Zea mays* L.) have been discovered. A well-studied system is Suppressor-Mutator (*Spm*) (McClintock, 1954), also known as Enhancer/Inhibitor (*En/I*) (Peterson, 1953, 1965). Suppressor-Mutator, originally named for (1) the element's ability when inserted into a gene to control that gene's level of expression, and (2) the element's ability to be inserted into and excised from a gene (or cause a defective element's excision), thereby altering the gene's product. This transposable element system contains two elements: the autonomous element (*Spm*) and the non-autonomous element (defective *Spm* or *dSpm*). The major difference between the *Spm* and *dSpm* elements is that the *Spm* element has an internal sequence that codes for a transposase, which has the ability to recognize the specific repeated DNA sequences at both ends of the *Spm/dSpm* elements and effect their transposition. The transposase acts to excise the elements allowing for their insertion in other genomic locations. The *dSpm* element affects gene expression in two different ways (Masson *et al.*, 1987) - (1) when the *dSpm* element is present within the gene, with no active *Spm* element present in the genome, it allows no expression of the gene, but when the active *Spm* element is present, gene expression is intermediate until excision restores the wild-type expression levels (*Spm*-dependent allele) and (2) when the *dSpm* element is present within the gene, with no active *Spm* element present in the genome, it may allow some level of expression of the gene, but when the active *Spm* element is present, gene expression is stopped until the *dSpm* element excises restoring wild-type expression levels (*Spm*-suppressible allele). In some cases, when the *dSpm* element is present within the gene, with no active *Spm* present in the genome, it allows no expression of the gene, and when the active *Spm* element is present, gene expression is still eliminated until excision restores wild-type expression.

The *Spm* transposable element has been found inserted into many genes, including specific genes that code for color of aleurone tissue. In these cases, when the plants are homozygous for *dSpm* at the locus, the *dSpm* element is stable and remains internal to the aleurone color gene, yielding a colorless phenotype. When crossed with a corresponding colorless aleurone line that is heterozygous for an active *Spm* element anywhere within the genome, a segregating population of kernels will result such that one half will have excision events occurring within the aleurone color gene during development, which gives colored spots on a colorless background (Peterson, 1961); the remaining half of the kernels will remain colorless due to the absence of an active *Spm*. Changes in the *Spm* or *dSpm* elements affecting the frequency and timing of the excision events can be selected so that the elements respond at specific developmental stages and at a specific frequency. These selections are known to be highly heritable (Peterson, 1961).

Studies using the *Spm/dSpm* system (Fedoroff and Baker, 1989), have shown that the transposition mechanism and requirements for an element's movement are similar to that of another well-studied system, Activator-Dissociator (*Ac/Ds*). These requirements include a gene internal to the element, encoding the transposase enzyme, and specific repeated DNA sequences that flank the element. *dSpm* elements, by definition, do not possess an active transposase gene. Dosage effects using the *Ac/Ds* system show a decreased frequency of transposition when more than one active Ac element is present within the genome. The *Spm/dSpm* system does not show a dosage effect.

The transposition phenomenon is believed to be primarily a mitotic event occurring at the time of DNA replication (Chen *et al.*, 1987). Transposition events also occur during meiosis, and are called germinal transpositions. Such transpositions can lead to a loss of an element from the gene within the germ cells giving a non-mutant progeny phenotype (Peterson, 1970). Elements causing high frequency, or developmentally early, somatic excision events, have additionally been shown to cause a higher frequency of germinal transpositions.

Kernel quality in maize breeding programs is a constant concern in the synthesis of today's hybrids. The same is true in sweet corn breeding programs. In addition to the mature kernel quality issues it shares with dent corn (e.g., pathogen-free, high germination), sweet corn must also maintain high quality kernel traits while immature. Selection pressures applied to sweet corn for achieving superior kernel characteristics at an immature stage have compromised the kernel's germination, seedling vigor and stand attributes. Several of the selected characteristics in sweet corn contribute to this problem. Thin pericarp tissue, an attribute involved in tenderness, can lead to increased pathogen invasion or mechanical damage at harvest. Smaller embryo size can lead to non-germination. A decrease in the concentration of starch in the kernel can result in insufficient energy reserves for the embryo and lead to poor germination and seedling emergence. A significant positive correlation exists between an increase in leachate (potassium, phosphate, sugars, amino acids, proteins and other electrolytes leaching from the kernel into the soil) during imbibition and high sucrose levels in sweet corn lines (Schmidt and Tracy, 1989). Whether this leachate increase is due to mechanical damage of the pericarp or other factors is unknown. Such kernel imperfections contribute to an increased likelihood of pathogen invasion leading to a decrease in germination frequency and/or seedling emergence.

One of the inherent properties of the maize endosperm mutant sugary (*su1*), the gene used in most of the commercial sweet corn acreage, is its short "harvesting window". This window is the time in which the corn can be harvested and still maintain high quality (Tsai and Glover, 1974). Through breeding, the harvest window has been expanded, but this characteristic can be greatly lengthened by the use of other endosperm mutations. The "super-sweet" varieties, brittle-1(*bt1*), and brittle-2(*bt2*), shrunken-2(*sh2*), in addition to a longer harvesting window, have a higher sucrose concentration at harvesting time than the *su1* varieties (Cameron and Teas, 1954; Tsai and Glover, 1974). Use of these genes emphasized, even more, the issues of poor germination and seedling vigor (Rowe and Garwood, 1978). Except for a few geographic areas, these genotypes have been considered "unacceptable" by the commercial growers. Producers have resorted to later planting dates to attain better germination and seedling vigor.   Being able to plant earlier in order to achieve a longer harvesting season would be very beneficial (M. Edwards, 1990, pers. comm.).   In each of the super-sweet mutants, the enzymatic pathway from sucrose to starch is altered leading to an increase in sucrose content along with a decrease in starch content.   In most mutants, the precise defect in starch biosynthesis has been determined (Shannon and Garwood, 1982).

The endosperm mutant *bt1* (Mangelsdorf, 1926), one of the high sucrose endosperm mutations, maintains a higher germination rate and improved seedling vigor when compared with other super-sweets (Styler *et al.*, 1980). The precise reaction which this mutation alters in the biosynthetic pathway to starch is unknown.   Starch analysis of mature *bt1* kernels showed higher content of amylose and amylopectin than non-mutant endosperm (Ninomya *et al.*, 1989).   This particular mutant is not used extensively in commercial production, but several breeders have used it in their breeding programs (Brewbaker, 1977).

We have been developing a procedure for the use of transposable elements controlling genes such as *bt1* and *sh2* in commercial sweet corn production.   The procedure facilitates the production of sweet corn kernels that are much less defective and germinate at a similar rate to normal corn, yet are the sweet corn type at the time of consumption.   The ultimate goal is using the *bt1* gene to obtain a "super-sweet" sweet corn line that has a germination/emergence frequency approaching that of normal dent corn but otherwise maintains the kernel quality of sweet corn lines at the immature eating stage.   The transposable element system, *Spm/dSpm*, is being used to re-establish the normal *Bt1* gene product after the approximate time of harvest for immature kernels.   This will allow the seeds harvested at maturity to contain higher amounts of starch than the *bt1* sweet corn line, yet maintain high sucrose concentrations at the eating stage.

An inbred *bt1* line that contains the *dSpm* element within the *Bt1* gene was identified by Phillips *et al.* (1986).   This line has been named "brittle mutable" (*bt-m*) and acts as a *Spm*-dependent allele.   An inbred *bt1* line possessing the standard *bt1* allele and an active *Spm* element within its genome has also been isolated.   This *Spm*, named *Spm-P*, is usually independent of *bt1* but on occasion was found linked with the *a2* aleurone color gene which is linked to *bt1* (Phillips *et al.*, 1986).

Crosses have been made using the *bt-m* inbred and the *bt1 Spm* inbred. In the presence of an active *Spm*, the *dSpm* apparently leaves the locus.   Because this event is controlled in time and frequency, a specific kernel type which is mosaic for normal and brittle tissue is produced (Figure 1).   This kernel type is much more like normal kernels than the standard *bt1* and, therefore, germinates better, especially under stress conditions.   Upon self-pollination or open-pollination in isolation, hybrids containing *bt-m*, *bt1*, and *Spm* should produce kernels that are either brittle or mosaic in phenotype.   If

Figure 1 Kernels displaying the brittle, mosaic, and normal phenotypes.

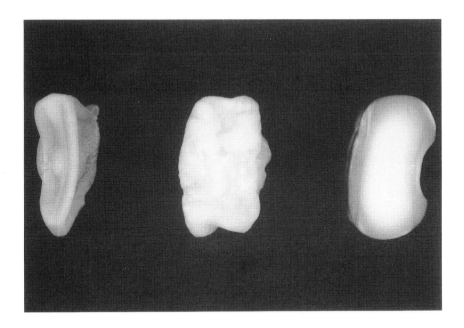

the timing of the mutation events induced by *Spm* is sufficiently late, the kernels may be indistinguishable from super-sweet *bt1* sweet corn at the time of harvest for consumption purposes. The idea described above and further outlined in Table 1 should apply to any sweet corn system but may be especially useful with *sh2* (super-sweet) which also has a germination problem. An *Ac*-controlled *sh2-m* is available from C. Hannah (University of Florida).

Crosses involving the initial versions of *bt-m* and *Spm*-P produced approximately 8% normal kernels in addition to the brittle and mosaic types in the F_2. This frequency of normal kernels reflects a 4% germinal transposition rate. The expected phenotypic frequencies, based on 4% germinal transposition, are 41.8% brittle, 50.4% mosaic, and 7.8% normal; the observed frequencies closely match expectations (Table 2). The normal kernels were self-pollinated to determine their genotypes; the genotypic frequencies matched our germinal transposition expectations (Table 3). These tests identified plants from normal kernels which possessed neither *bt-m* nor *bt1*. These reversions are likely *Bt1 dSpm* homozygotes; *Spm*-P may or may not be present. The *dSpm* is expected to be linked to *Bt1* in at least 25% of the cases (Peterson, 1970). Therefore, these stocks may be useful for producing new *bt* alleles since the *dSpm* would be expected occasionally to transpose back into *Bt1* in the presence of *Spm*-P.

Table 1 Genetic description of transposable element-controlled sweet corn.

Genotype	Description
bt-m	a) Brittle-mutable b) Possesses the standard brittle phenotype c) Represents the insertion of a defective *Spm* (*dSpm*) at the normal *Bt1* locus d) Stable in the absence of *Spm* e) Mutable in the presence of *Spm*
bt1	a) Brittle-1 b) Represents the standard brittle-1 phenotype c) Stable in the presence of *Spm* d) No active transposable elements present
Spm	a) Suppressor-mutator* b) Possesses no unique phenotype by itself c) Induces mutability in *bt-m*
bt-m/bt1 Spm	a) Hybrid of *bt-m* and *bt1 Spm* b) All kernels are mosaic due to the *dSpm* being induced to leave the *brittle-1* locus by the transacting *Spm* c) Time and frequency of the mutation events may be altered by changes in *dSpm* or *Spm*
bt-m/bt1 Spm ⊗	a) Self-pollinated (or open-pollinated in isolation) progeny kernels are either brittle or mosaic b) Independent segregation of *bt* and *Spm* expected to give 9 mosaic: 7 brittle in F_2 as follows:

♀ \ ♂	*bt-m*	*bt-m Spm*	*bt1*	*bt1 Spm*
bt-m	Brittle	Mosaic	Brittle	Mosaic
bt-m Spm	Mosaic	Mosaic	Mosaic	Mosaic
bt1	Brittle	Mosaic	Brittle	Brittle
bt1 Spm	Mosaic	Mosaic	Brittle	Brittle

*Spm-P (R. L. Phillips' version of *Spm*) is the element used in the University of Minnesota studies.

Table 2 Genetic expectation from *bt-m* x *bt1 Spm*-P cross considering germinal transposition (assuming *bt1* and *Spm*-P are not linked and 4% germinal transposition rate).

| Cross: | $\dfrac{bt\text{-}m;+}{bt\text{-}m;+}$ x | $\dfrac{bt1;Spm\text{-}P}{bt1;Spm\text{-}P}$ | | | | |

| F_1: | $\dfrac{bt\text{-}m;+}{bt1;Spm\text{-}P}$ | | | | | |

F_2:	*bt-m;+* 0.23	*bt-m;Spm-P* 0.23	*bt1;+* 0.25	*bt1;Spm-P* 0.25	*Bt1;+* 0.02	*Bt1;Spm-P* 0.02
bt-m;+ 0.23	Brittle	Mosaic	Brittle	Mosaic	Normal	Normal
bt-m;Spm-P 0.23	Mosaic	Mosaic	Mosaic	Mosaic	Normal	Normal
bt1;+ 0.25	Brittle	Mosaic	Brittle	Brittle	Normal	Normal
bt1;Spm-P 0.25	Mosaic	Mosaic	Brittle	Brittle	Normal	Normal
Bt1;+ 0.02	Normal	Normal	Normal	Normal	Normal	Normal
Bt1;Spm-P 0.02	Normal	Normal	Normal	Normal	Normal	Normal

Phenotypes of F_2 kernels

	Expected	Observed (n = 50 F_2 ears, approximately 400 kernels ear^{-1})
Brittle	41.8%	39.3%
Mosaic	50.4%	52.5%
Normal	7.8%	8.2%

Table 3 Expected and observed results from self-pollinating plants from phenotypically normal F_2 kernels derived from the cross described in Table 2.

Expected genotypes and frequencies	Segregation patterns	Expected genotypes and frequencies	Segregation patterns
$\underline{Bt1};+$ $bt\text{-}m;+$ 0.0092	3:1 Normal:Brittle		
$\underline{Bt1};+$ $bt1;+$ 0.0100	3:1 Normal:Brittle	$\underline{Bt1};+$ $Bt1;+$ 0.0004	All Normal
$\underline{Bt1};+$ $bt1;Spm\text{-}P$ 0.0100	3:1 Normal:Brittle	$\underline{Bt1};+$ $Bt1;Spm\text{-}P$ 0.0008	All Normal
$\underline{Bt1;Spm\text{-}P}$ $bt1;+$ 0.0100	3:1 Normal:Brittle		
$\underline{Bt1;Spm\text{-}P}$ $bt1;Spm\text{-}P$ 0.0100	3:1 Normal:Brittle	$\underline{Bt1;Spm\text{-}P}$ $Bt1;Spm\text{-}P$ 0.0004	All Normal
$\underline{Bt1};+$ $bt\text{-}m;Spm\text{-}P$ 0.0092	12:3:1 Normal:Mosaic:Brittle		
$\underline{Bt1;Spm\text{-}P}$ $bt\text{-}m;+$ 0.0092	12:3:1 Normal:Mosaic:Brittle		
$\underline{Bt1;Spm\text{-}P}$ $bt\text{-}m;Spm\text{-}P$ 0.0092	3:1 Normal:Mosaic		

Segregation patterns

	Expected	Observed (n = 59 F_3 ears, approximately 400 kernels ear^{-1})
3 Normal:1 Brittle	63%	67%
12 Normal:3 Mosaic:1 Brittle	23%	26%
3 Normal:1 Mosaic	12%	5%
All Normal	2%	2%

We are also selecting phenotypically distinct versions of mosaic kernels in crosses of the *bt-m* and *bt1 Spm*-P lines. Although the F_1 kernels are all mosaic, there is a range in phenotypes from "mostly-brittle" to "mostly-normal". The "mostly-brittle" phenotype reflects later and/or fewer transposition events while the "mostly-normal" phenotype reflects earlier and/or more transposition events. Some of these selections should be heritable for the modified expression of either *dSpm* or *Spm*-P and could be useful if sucrose levels and/or germination levels of the initial crosses are not adequate. Some of the "mostly-brittle" types may actually be new stable brittle alleles due to the *dSpm* not cleanly excising from the locus.

The breeding scheme (see Table 4; Figures 2 and 3) for hybrid sweet corn would be to first separately backcross *bt-m* and *bt-1 Spm* into different sweet corn lines. Six generations of backcrossing and one generation of selfing would likely be required. Hybrid seed would be produced by crossing the derived backcross-selfed lines. If the timing of the mutation events leading to mosaic kernels is too early and results in too much starch synthesis by the time of eating, then versions of *Spm* or *dSpm* with later activity or less frequent events might need to be selected via spontaneous changes in *Spm* or *dSpm*. These changes would initially be recognized by a variant mosaic mature kernel pattern. Preliminary germination/emergence data and sucrose determinations indicate that the concept being tested has merit.

VARIATIONS IN DNA METHYLATION INDUCED BY TISSUE CULTURE

Plant tissue culture is a mutagenic procedure. The question is often asked "Does tissue culture generate unique types of mutations?" Our results and the results of others show that, although classical types of mutations are frequent, novel types of variation are also occurring.

The mutagenic nature of tissue culture has long been recognized (reviewed by Orton, 1984; Larkin, 1987). Heritable variation has been seen among tissue culture regenerants in a wide range of species. The underlying mechanism of this variation is yet to be resolved. Several types of variation are seen at a high frequency among tissue culture regenerants from a number of species. Qualitative and quantitative trait variation is seen in nearly all studies in which plants are regenerated from tissue culture and analyzed [e.g., maize (Zehr *et al.*, 1987; Lee *et al.*, 1988), oats (*Avena sativa* L.) (Dahleen *et al.*, 1991), wheat (*Triticum aestivum* L.) (Maddock, 1986)]. Chromosomal aberrations are frequently found in plants; aneuploidy and polyploidy occur but at a much lower frequency than changes in chromosome structure [e.g., oats and maize (Benzion *et al.*, 1986)]. The chromosomal aberrations most frequently appear to have arisen from chromosome breakage events. Tissue culture-induced transposable element activation has been found among progeny of maize (Peschke *et al.*, 1987, 1991; Peschke

Table 4 Crossing scheme to develop the appropriate genetic stocks (stock 1 = homozygous *bt-m*; stock 2 = homozygous *bt-1 Spm*) for creating mosaic super-sweet kernels.

	Stock 1			Stock 2	
:---:	:---:		:---:	:---:	
Generation	Cross		Generation	Cross	
1	*bt-m* x elite sweet corn (ESC)		1	*bt1 Spm* x ESC	
2	Backcross F_1 to ESC (generates BC_1 seed); self same F_1		2	Backcross F_1 to ESC Test-cross F_1 x *bt-m*	
3	BC_1 x ESC (use at least 4 BC_1 plants) (use BC_1 where self-segregated for *bt-m*) Self same BC_1		3	BC_1 x ESC (use at least 4 BC_1 plants) (use BC_1 where test-cross segregated for mosaic kernels) Test-cross same BC_1 to *bt-m*	
4	BC_2 x ESC (use BC_2 where self-segregated for *bt-m*) Self same BC_2		4	BC_2 x ESC (use BC_2 where test-cross segregated for mosaic kernels) Test-cross same BC_2 to *bt-m*	
5-7	Repeat crosses to ESC and selfs		5-7	Repeat crosses to ESC and selfs	
8	Self BC_6 (provides homozygous *bt-m*)		8	Self BC_6 (provides homozygous *bt1 Spm*)	
			9	Self and test-cross *bt1* types to *bt-m* to identify homozygous *Spm*	

and Phillips, 1991) and alfalfa (*Medicago sativa* L.) (Groose and Bingham, 1984) regenerants. DNA methylation may be an important underlying factor in all of the above types of variation.

The most frequently methylated base in higher eukaryotic DNA is cytosine (5-methylcytosine). In plants, the sequence CpG is the most frequently methylated sequence although CpXpG can also be methylated, where X is C, A, or T (Gruenbaum *et al.*, 1981). DNA methylation is negatively correlated with gene expression in both plants and animals (reviewed by

Figure 2 Crossing scheme to obtain homozygous *brittle*-mutable seed.

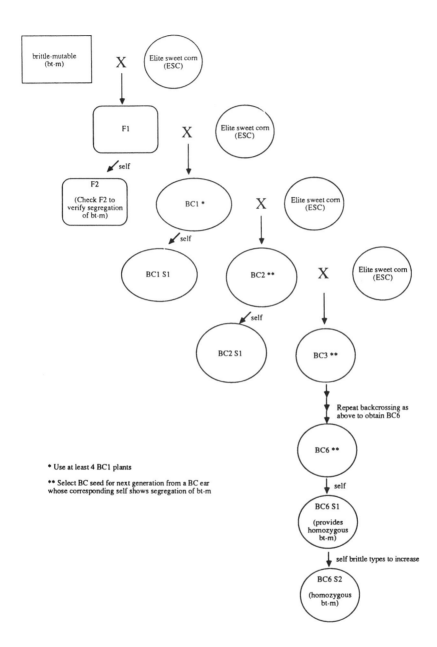

* Use at least 4 BC1 plants

** Select BC seed for next generation from a BC ear whose corresponding self shows segregation of bt-m

Figure 3 Crossing scheme to obtain homozygous *brittle-1 Spm* seed.

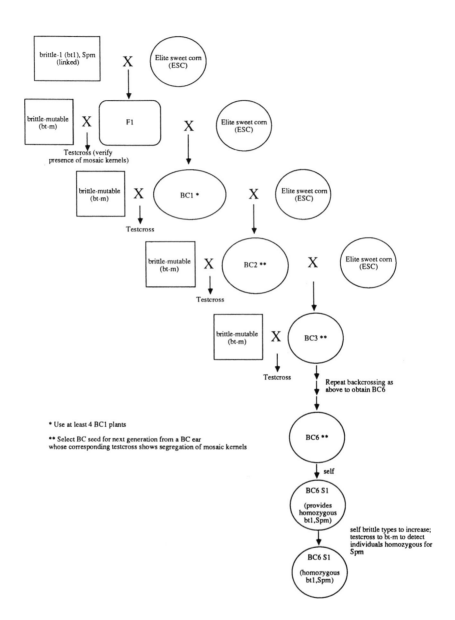

Holliday, 1987). DNA methylation has also been correlated with chromatin structure. For example, the highly heterochromatic, inactive X chromosome in humans is highly methylated (Monk, 1990). Also, DNase1 sensitive regions, which are mostly euchromatic, have significantly less cytosine methylation than total DNA (Klaas and Amasino, 1989).

We have been working under the hypothesis that changes in DNA methylation induced by tissue culture may (1) affect quantitative trait expression by changing the expression of specific loci or altering chromatin structure; (2) be involved in the activation of transposable elements, elements which may then cause qualitative variation; and (3) alter the structure of heterochromatic regions causing chromosome breakage events. We know that other types of variation (e.g., base changes) do occur in culture, but these may occur independently of DNA methylation alterations. We recognize that deamination of 5-methylcytosine results in thymine.

Brown (1989) reported that methylation variation occurs among maize regenerants, and Müller *et al.* (1990) have reported that methylation and sequence variation occur among progeny of rice (*Oryza sativa* L.) regenerants. We designed a study to examine the frequency and nature of variation in cytosine methylation among progenies of regenerated maize plants.

Twelve-day-old embryos from a selfed A188 plant were plated on modified MS media and induced to form callus cultures. Two of these embryo sources were grown as callus for approximately 7 months before plant regeneration. Regenerated plants were grown to maturity in a greenhouse and self-pollinated. Plants were grown from R_1 seed of those regenerants (R_0-derived R_1 families). Nine R_1 families from one embryo source (I) and 13 R_1 families from the second embryo source (J) were grown for 4 weeks in a greenhouse. The entire aboveground portion of five plants per family was harvested for evaluation by Southern analysis. The DNA was digested with the enzymes *HpaII* and *MspI*, run on an agarose gel and blotted to a nylon membrane. *HpaII* and *MspI* both recognize the sequence CCGG, although only *MspI* will cut the sequence if the internal C is methylated. The DNA was probed with 21 single copy probes which included three cDNA probes (alcohol dehydrogenase I, sucrose synthetase I and sucrose synthetase II) as well as 18 PstI genomic clones which were developed by B. Burr (Brookhaven National Laboratory) and D. Hoisington (University of Missouri-Columbia) for restriction fragment length polymorphism (RFLP) analysis.

In the analysis of results, the probes were separated into two classes (Table 5). In one class, the *HpaII* band was equal in size to the *MspI* band in the uncultured control. This indicated that the CCGG sequences in and/or around the probe sequence were not methylated at the internal C. Thus, these probes would not be expected to detect hypomethylation, only hypermethylation. We found no variant patterns using these probes (Table 5). In the other class, the *HpaII* band was larger than the *MspI* band in the

Table 5 DNA methylation changes among 22 R_1 families derived from two embryo sources (data presented as average per cent variant families probe^{-1}).*

Embryo	No. R_1 families	12 probes *HpaII = MspI*	9 probes *HpaII > MspI*
I	9	0.0	34.2
J	13	0.0	34.6

*Both explanted embryos derived from a single A188 ear. *HpaII = Msp I* indicates that the fragment sizes generated upon digestion with these enzymes are identical, thus only hypermethylation can be detected with these probes. *HpaII > MspI* indicates larger fragments occur with *HpaII* digestion, thus hypo- or hypermethylation is detectable.

uncultured control. In this case, at least one CCGG sequence in or around the probe sequence was methylated at the internal C. Thus, these probes would detect either hypo- or hypermethylation. Using these probes, 34% of the families showed variant methylation patterns (Table 5). Fifteen uncultured control plants were invariant for DNA methylation alterations with the probes tested.

Several important observations were made: (1) all observed changes were losses of methylation; (2) the frequency of methylation changes was high and regenerant-derived families within an embryo source often had different changes; and (3) 17% of the families with altered methylation patterns appeared to come from regenerated plants homozygous for the methylation change (Figure 4); 83% of the families were segregating for an altered pattern (Figure 5). Of those families segregating for a methylation change, 60% were segregating for two new bands. These results show methylation changes to be prevalant among progeny of tissue culture regenerants. The findings that all changes were losses of methylation and that some changes were homozygous were unexpected. Studies on the inheritance and cause of these changes are currently being conducted. The *MspI* digestion patterns were invariant across families for a particular probe. Thus, base sequence changes apparently did not occur. It was interesting that the results for the two sets of R_1 families were very similar.

There are several implications of these observations. First, DNA methylation changes may represent a type of mutagenesis unique to tissue culture. Second, these changes may be inherited faithfully upon selfing (Brown, 1989) but may be altered upon outcrossing, such as was shown for Mutator (*Mu*) elements in maize (Freeling, 1988). Others have found mutants that were not stable upon outcrossing. Oono (1985) described a dwarf mutant that was stable for eight generations of selfing but could never be recovered in selfed progeny of outcrosses.

Figure 4 DNA from R_1 plants derived from a selfed R_0 plant (lanes 1-5) demonstrating R_0 homozygosity for a methylation change. DNA digested with *HpaII* (A) and *MspI* (B) and probed with UMC54. Lane 6 is the non-cultured control (inbred A188).

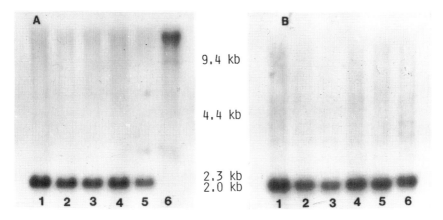

9.4 kb

4.4 kb

2.3 kb
2.0 kb

Figure 5 DNA from R_1 plants derived from a selfed R_0 plant (lanes 1-5) demonstrating R_0 heterozygosity for two new bands as compared to the non-cultured control. Lane 6 is the non-cultured control (inbred A188).

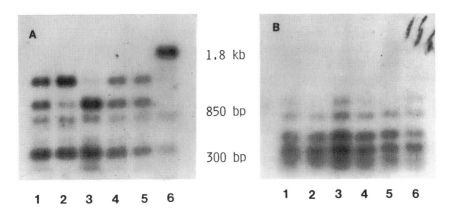

1.8 kb

850 bp

300 bp

Also, we are interested in the stability of quantitative variation. In a field evaluation, A634 and B73 R_2-derived R_4 families of maize were generally poorer in agronomic performance than the uncultured control. However, in crosses of the R_2-derived R_3 families to the uncultured control, the performance of the cross was often equal to the control. The crosses described here are all involving two lines of the same inbred, one representing the source inbred while the other was derived from the same inbred but had passed through tissue culture and three subsequent seed generations. The A634 data are displayed in Table 6; note that for yield and plant height the values for the crosses are generally near the A634 value whereas the self values are lower. In fact, heterosis over the higher parent (non-cultured A634) was sometimes observed. If the mutations induced by tissue culture at quantitative trait loci are not recessive, then additivity might be expected. The restoration of high yields after crossing leads us to suggest the possibility that the genetic defects in the tissue culture-derived lines may be corrected in progenies of crosses. Could this be because DNA methylation changes account for the reduced performance and, being epigenetic, might be modified in the crosses back to the source inbred? Fincher (reviewed in Phillips, 1989) also reported near normal performance of progenies in crosses of poor performing tissue culture-derived lines back to the source inbred. We are evaluating lines derived from these regenerant family x control crosses at several levels of inbreeding to test whether the genetic changes segregate in the progeny.

A final implication of methylation alteration is that this epigenetic variation may allow a plant to respond quickly, yet not permanently, to severe stresses such as imposed by tissue culture. It is interesting to speculate about what the signal might be that induces these changes. LoSchiavo *et al.* (1989) showed that the overall methylation of carrot (*Daucus carota* L.) suspension cultures could be altered by changing the auxin levels. Perhaps a hormone is involved. Imbalances of nutrients or other compounds could also play a role as the signal for change.

It is also interesting to speculate about whether these changes are equally frequent throughout the genome. The probes utilized in our studies detect sequences in eight of the 10 chromosomes. In an HPLC analysis of overall cytosine methylation in inbred A188, no differences were observed between non-cultured control plants, callus cultures, regenerated plants or progeny of regenerated plants (Amasino, Kaeppler, and Phillips, unpublished data). In our Southern analysis of single copy sequences, it was apparent that some probes detected much more variation than others. Perhaps methylation changes are somehow limited to certain subregions of the genome.

Methylation variation may be an underlying factor in many of the tissue culture-induced phenomena including chromosome breakage, quantitative trait variation, and transposable element activation (Phillips *et al.*, 1990). These changes are usually stable upon selfing. Epigenetic variation is not a classical type of mutation and may not result from treatments such as with ethyl methane sulfonate. Although certain tissue culture-derived qualitative mutants have been found to be useful (e.g., herbicide resistance), the importance and stability of the type of variation we describe here is yet to be determined.

Table 6 Comparisons of A634 regenerant-derived lines in selfs and in crosses with the non-cultured source inbred.*

Genotype	Yield (kg ha^{-1})	Genotype	Plant height (cm)
C7043	122.9	C7043	207.2
C7115	80.2	7061	174.2
C7063	73.5	C7061	171.7
C7049	73.0	C7057	170.0
7043	71.7	C7071	168.3
C7037	70.3	C7099	168.3
A634	70.0	7097	167.5
C7061	69.5	C7111	167.5
C7039	66.6	C7049	167.5
C7097	63.5	7071	165.0
7063	62.3	C7097	165.0
C7071	62.1	C7037	165.0
C7059	61.3	C7107	165.0
C7057	61.0	C7039	164.2
C7107	61.0	7099	163.3
C7111	59.1	C7063	163.3
C7105	58.8	C7105	163.3
C7099	54.8	**A634**	162.5
7049	54.3	C7115	162.5
7037	54.0	7107	161.7
7115	53.5	C7059	160.8
7107	52.6	7049	159.2
7097	50.6	7111	159.2
7039	49.0	7037	159.2
7111	47.8	7057	158.3
7071	47.6	7105	157.5
7105	41.2	7063	156.7
7061	38.7	7039	155.0
7057	35.6	7043	155.0
7099	32.3	7059	154.2
7059	24.8	7115	144.2
LSD =	9.3	LSD =	5.5

*C = cross of R_2-derived R_3 line with the non-cultured source inbred (A634); genotypes designated without a "C" are R_2-derived R_4 lines. Data represent means of two replications at each of three locations in 1990.

SUMMARY

Great strides have been made in our understanding of the genetics and cell biology of plants. Many of the basic discoveries now provide us with novel approaches to the induction of genetic variation. The plant breeding applications are becoming increasingly apparent.

We describe here, for the first time, a possible application of transposable elements to sweet corn production. The control of mutability afforded by a specific transposable element system is capitalized upon in the proposed approach. The genes used for the production of super-sweet sweet corn render the seed of poor quality as the result of the extreme collapse of the endosperm. The concept proposed is to utilize a transposable element-controlled super-sweet allele that is suppressed (mutant) at early stages of development such that high sucrose levels exist at the eating stage. However, at an appropriate developmental point, transposable element-induced mutability would create a mature kernel mosaic for mutant and normal tissue. Preliminary evidence with a brittle-mutable allele controlled by the Suppressor-Mutator (*Spm*) system indicates that the resultant mosaic kernels germinate at nearly normal rates. Sucrose analysis of these materials at different stages of maturity is underway.

Also reported here is evidence that the maize tissue culture process induces DNA methylation changes that can be detected several seed generations later. These changes are very frequent, represent loss of cytosine methylation compared to the controls, and are frequently homozygous in the original regenerated plant. The mechanism for this unexpected homozygosity is not understood at this time. The agronomic performance of lines derived from tissue culture of a maize inbred is usually reduced. Perhaps this should be expected based on previous reports of the frequent tissue culture-induced qualitative mutations, chromosome breakage, and transposable element activation. However, we were surprised to find that progenies of crosses of these derived lines with the inbred that served as the source of the tissue cultures usually had near normal agronomic performance. Future tests will determine whether the results can be explained in terms of DNA methylation alterations.

ACKNOWLEDGMENT

Support from the Green Giant Division of Grand Metropolitan Food Sector and the USDA Competitive Research Grants Office (Grant 88-37262-3919) is gratefully acknowledged.

REFERENCES

Benzion, G.; Phillips, R. L.; Rines, H. W. (1986) Case histories of genetic variability in vitro: Oats and maize. In: Vasil, I. K. (ed.), *Cell culture and somatic cell genetics of plants.* Vol. 3. Academic Press, New York, pp. 435-448.

Brewbaker, J. L. (1977) Hawaiian super-sweet #9 corn. *HortScience* 12:355-356.

Brown, P. T. H. (1989) DNA methylation in plants and its role in tissue culture. *Genome* 31:717-729.

Cameron, J. W.; Teas, H. J. (1954) Carbohydrate relationships in developing and mature endosperms of brittle and related maize genotypes. *American Journal of Botany* 14:50-55.

Chen, J.; Greenblatt, I. M.; Dellaporta, S. L. (1987) Transposition of Ac from the P locus of maize into unreplicated chromosomal sites. *Genetics* 117:109-116.

Chomet, P. S.; Wessler, S.; Dellaporta, S. L. (1987) Inactivation of the maize transposable element activator (Ac) is associated with its DNA modification. *EMBO Journal* 6:295-302.

Dahleen, L. S.; Stuthman, D. D.; Rines, H. W. (1991) Agronomic trait variation in oat lines derived from tissue culture. *Crop Science* 31:90-94.

Dennis, E. S.; Brettell, R. I. S. (1990) DNA methylation of maize transposable elements is correlated with activity. *Philosophical Transactions Royal Society of London B* 326:217-229.

Evola, S. V.; Burr, F. A.; Burr, B. (1984) The nature of tissue culture-induced mutations in maize. *Eleventh Annual Aharon Katzer-Katchalsky Conference*, Jerusalem, Israel, 8-13 January. (Abstr.).

Fedoroff, N. V.; Baker, B. (1989) Plant gene vectors and genetic transformation: The structure, function, and uses of maize transposable elements. In: Constabel, F.; Vasil, I. K. (eds.), *Cell culture and somatic cell genetics of plants*. Vol. 6. Academic Press, New York, pp. 101-133.

Fedoroff, N.; Banks, J.; Masson, P. (1989) Molecular genetic analysis of the maize suppressor-mutator element's epigenetic developmental regulatory mechanism. *Genome* 31:973-979.

Freeling, M. (1988) Mutagenesis using Robertson's mutator lines and consequent insertions at the *Adh1* gene in maize. In: Nelson, O. (ed.), *Plant transposable elements*. Proceedings of an International Symposium, Madison, WI., 22-26 Aug. 1987. Plenum Press, New York, pp. 279-288.

Groose, R. W.; Bingham, E. T. (1984) Variation in plants regenerated from tissue culture of tetraploid alfalfa heterozygous for several traits. *Crop Science* 24:655-658.

Gruenbaum, Y.; Naveh-Many, T.; Cedar, H. (1981) Sequence specificity of methylation in higher plant DNA. *Nature (London)* 292:860-862.

Holliday, R. (1987) The inheritance of epigenetic defects. *Science (Washington, DC)* 238:163-170.

Klaas, M.; Amasino, R. (1989) DNA methylation is reduced in DNase1-sensitive regions of plant chromatin. *Plant Physiology* 91:451-454.

Larkin, P. J. (1987) Somaclonal variation: History, method, and meaning. *Iowa State Journal of Research* 61:393-434.

Lee, M. L.; Geadelmann, J. L.; Phillips, R. L. (1988) Agronomic evaluation of inbred lines derived from tissue cultures of maize. *Theoretical and Applied Genetics* 75:841-849.

LoSchiavo, F.; Pitto, L.; Guiliano, G.; Torti, G.; Nuti-Ronchi, V.; Marazziti, D.; Vergara, R.; Orselli, S.; Terzi, M. (1989) DNA methylation of embryogenic carrot cell cultures and its variations as caused by mutation, differentiation, hormones and hypomethylating drugs. *Theoretical and Applied Genetics* 77:325-331.

Maddock, S. E. (1986) Field assessment of somaclonal variation in wheat. *Journal of Experimental Botany* 37:1065-1078.

Mangelsdorf, P. C. (1926) The genetic and morphology of some endosperm characters in maize. *University of Connecticut Agricultural Experiment Station Bulletin Number 279*, pp. 509-612.

Masson, P.; Surosky, R.; Kingsbury, J. A.; Fedoroff, N. V. (1987) Genetic and molecular analysis of the *Spm*-dependent a-m2 alleles of the maize a locus. *Genetics* 177:117-137.

McClintock, B. (1954) Mutations in maize and chromosomal aberrations in *Neurospora*. *Yearbook-Carnegie Institution of Washington* 53:254-260.

Monk, M. (1990) Changes in DNA methylation during mouse embryonic development in relation to X-chromosome activity and imprinting. *Philosophical Transactions Royal Society of London B* 326:299-312.

Müller, E.; Brown, P. T. H.; Hartke, S.; Lörz, H. (1990) DNA variation in tissue-culture-derived rice plants. *Theoretical and Applied Genetics* 80:673-679.

Ninomya, Y.; Okuno, K.; Glover, D. V.; Fuwa, H. (1989) Some properties of starches of sugary-1; brittle-1 (*Zea mays* L.). *Starch* 41:165-167.

Oono, K. (1985) Putative homozygous mutants in regenerated plants of rice. *Molecular and General Genetics* 198:377-384.

Orton, T. J. (1984) Genetic variation in somatic tissues: Method or madness? *Advanced Plant Pathology* 2:153-189.

Peschke, V. M.; Phillips, R. L. (1991) Activation of the maize transposable element *Suppressor-mutator (Spm)* in tissue culture. *Theoretical and Applied Genetics* 81:90-97.

Peschke, V. M.; Phillips, R. L.; Gengenbach, B. G. (1987) Discovery of transposable element activity among progeny of tissue culture-derived maize plants. *Science (Washington, DC)* 238:804-807.

Peschke, V. M.; Phillips, R. L.; Gengenbach, B. G. (1991) Genetic and molecular analysis of tissue culture-derived *Ac* elements. *Theoretical and Applied Genetics* 82:121-129.

Peterson, P. A. (1953) A mutable pale green locus in maize. *Genetics* 38:682-683.

Peterson, P. A. (1961) Mutable a_1 of the *En* system in maize. *Genetics* 46:759-771.

Peterson, P. A. (1965) A relationship between the *Spm* and *En* control systems in maize. *American Naturalist* 99:391-398.

Peterson, P. A. (1970) The *En* mutable system in maize. *Theoretical and Applied Genetics* 40:367-377.

Phillips, R. L. (1989) Somaclonal and gametoclonal variation. *Genome* 31:1119-1120.

Phillips, R. L.; Block, L. G.; Peschke, V. M.; Burnham, C. R. (1986) A suppressor-mutator transposable element system of independent origin. *Maize Genetics Cooperation News Letter* 60:115-117.

Phillips, R. L.; Kaeppler, S. M.; Peschke, V. M. (1990) Do we understand somaclonal variation? In: Nijkamp, H. J. J.; Van Der Plas, L. H. W.; Aartrijk, J. Van (eds.), *Progress in plant cellular and molecular biology.* Proceedings VIIth International Congress of Plant Tissue and Cell Culture, Kluwer Academic Publ., Dordrecht, pp. 131-141.

Rowe, D. E.; Garwood, D. L. (1978) Effects of four endosperm mutants on kernel vigor. *Crop Science* 18:709-712.

Schmidt, D. H.; Tracy, W. F. (1989) Duration of imbibition affects seed leachate conductivity in sweet corn. *HortScience* 24:346-347.

Schwartz, D.; Dennis, E. (1986) Transposase activity of the *Ac* controlling element in maize is regulated by its degree of methylation. *Molecular and General Genetics* 205:476-482.

Shannon, J. C.; Garwood, D. L. (1982) Genetics and physiology of starch. In: Whistle, R. L. *et al.* (eds.), *Starch: Chemistry and industry.* 2nd Ed. Academic Press, New York, pp. 26-28.

Styler, R. C.; Cantliffe, D. J.; Hannan, L. S. (1980) Differential seed and seedling vigor in *Shrunken-2* compared to three other genotypes of corn at various stages of development. *Journal of the American Society for Horticultural Science* 105:329-332.

Tsai, C. Y.; Glover, D. V. (1974) Effect of the *Brittle-1 Sugary-1* double mutant combination on carbohydrate and postharvest quality of sweet corn. *Crop Science* 14:808-810.

Zehr, B. E.; Williams, M. E.; Duncan, D. R.; Widholm, J. M. (1987) Somaclonal variation among the progeny of plants regenerated from callus cultures of seven inbred lines of maize. *Canadian Journal of Botany* 61:491-499.

Chapter 18
The value of model systems for the future plant breeder

Richard B. Flavell

John Innes Institute, AFRC Institute of Plant Science Research, Colney Lane, Norwich NR4 7UH, U.K.

INTRODUCTION

The initiators of this symposium wrote in the original invitations, "Cooperation and collaboration among productive specialists have arisen as the optimum strategy for meeting the continual needs for plant improvement. The plant breeder will remain the most critical element because he must evaluate and integrate the products devised by the team of specialists. However, the future plant breeder will rely more and more upon foundations built by specialists in physiology, molecular genetics, germplasm resources, plant pathology and entomology."

There is ample evidence that this is true and much is presented in this volume. I was asked to project the value of potential model systems for plant breeding. I want to start with some general considerations that are worth noting, I believe, as we think about the specialist teams referred to by the symposium organizers and about the origins of innovation for plant breeding. First we should note that the organizers said "the plant breeder will rely more and more upon foundations built by specialists in physiology, molecular genetics, germplasm resources", etc. They did not say **plant** physiology, **plant** molecular genetics, etc. This is because plant science through molecular biology and biochemistry benefits considerably from studies on the organisms of other kingdoms.

Second, if we were asked to answer the question, "What has made the biggest impact on plant science in the 1980s?" I suspect the answer would be *Escherichia coli* recombinant DNA techniques, *Agrobacterium tumefaciens* molecular genetics, plant tissue culture, plant regeneration and general molecular biology techniques from mammalian and prokaryotic biology. Would this have been predicted before 1975? Would *E. coli* plasmid and bacteriophage replication research have been considered vital for isolating genes from complex plant genomes? Surely not. Would the study of crown gall tumors and their causative bacterium have been considered a model and subsequently a vital system for gene transfer in plant improvement programs? I suspect not.

Third, the techniques of DNA/DNA hybridization to recognize genes in distant species, the making of antibodies against epitopes of conserved active sites in proteins and the ability to transfer genes between species and kingdoms, have told us in the last 10 years that a wide diversity of biological processes in organisms of all kingdoms are or will turn out to be model systems to provide knowledge and genes for the plant breeder of today and tomorrow. Therefore, as the symposium organizers imply, plant breeders must take note of progress in a very wide corridor of biological research. This is a huge bewildering challenge to an individual, small team, university department or a multinational industry. However, if we fail to do this adequately, we will miss some opportunities for quantum gains in plant performance through manipulation of small numbers of genes.

EXAMPLES WHERE ORGANISMS FROM OTHER KINGDOMS HAVE SERVED AS USEFUL MODELS

As noted above, the developments leading to the ability to produce transgenic plants have opened up the opportunity to use genes, or at least coding sequences, from other species and kingdoms to help solve plant breeding problems. Very shortly after the first transgenic tobacco (*Nicotiana tabacum* L.) and petunia (*Petunia* sp.) plants were produced, Monsanto and Dupont, for example, demonstrated that microorganisms could be convenient sources of genes to confer herbicide resistance in plants because the target enzymes that were inhibited by the herbicides were present in both kingdoms, were functionally equivalent and structurally related (Oxtoby and Hughes, 1990). The herbicide glyphosate (Monsanto) inhibits 5-enol-pyruvylshikimate-3-phosphate synthase, a step in aromatic amino acid biosynthesis and the sulphonyl urea herbicides (Dupont) inhibit acetolactate synthase, a step in lysine isoleucine and valine synthesis. These steps are conserved between the biosynthetic pathways of bacteria, yeast and higher plants. When some bacteria, yeast or plant cells are plated in the presence of the herbicide, occasional resistant strains can be isolated. Some of these carry equivalent amino acid changes in the relevant enzyme, causing it to be resistant to the herbicide but not reducing its catalytic efficiency to any detrimental effect (Mazur and Falco, 1989; Lee *et al.*, 1988). The case of isolating genes from bacteria and yeast, by transfer of whole genomic libraries into populations of recipient cells and discovering which DNA fragment carries the functional property, enabled the genes to be initially isolated more rapidly from these organisms than from higher plants. Subsequent isolation of the functionally homologous plant gene was possible, however, by exploiting the residual homology between plants and yeast. When the bacterial or yeast coding sequences for the herbicide-resistant enzymes were inserted into plant expression cassettes (with appropriate promoters and terminators) and transferred into a tobacco plant cell and plants regenerated, plants which were resistant to these herbicides were created (Oxtoby and Hughes, 1990). Another example of creating herbicide-resistant plants is the use of the

Streptomyces gene which confers self-resistance to its own antibiotic, phosphinothricin - a substance which also acts as a herbicide (trade name Basta) by inactivating glutamine synthase in bacteria and plants (De Block *et al.*, 1987).

These examples illustrate the value of bacteria and yeast as model systems to work out the mechanism of action of herbicides and to provide a convenient source of coding sequences to be used in plant improvement programs. They serve as models because of the conservation of metabolic pathways and the functional and structural characteristics of component enzymes. There will undoubtedly be many other examples emerging in the near future where microbial and animal genes are exploited in plants. The recent example of the use of bacterial ADP glucose pyrophosphorylase and yeast invertase to modify plant starch metabolism and simultaneously source-sink relationships and plant morphology (Stitt *et al.*, 1990; Schaewan *et al.*, 1990) is a case in point. As knowledge of the control of metabolic pathways in time and space during development emerges, many novel modifications to the plant phenotype will be created that may turn out to be extremely useful.

Another whole group of extremely valuable model systems are those concerned with disease control. Again, it is the opportunity of gene transfer into plants that has greatly enhanced the value of these model systems for plant breeding. The first example I wish to use is that of how lepidopteran larvae are controlled (killed) by strains of the bacteria *Bacillus thuringiensis*. This is an extremely interesting system about which much is now known. The bacteria produce a crystalline protein in spores which, when ingested by larvae is converted into a smaller protein that binds to receptors in the midgut and destroys the cells. The larvae subsequently die or survive only with difficulty. This biological system has now been exploited to provide insect resistance to plants because the coding sequence for the toxic protein has been incorporated as part of a new gene into many plant species and the plants show resistance to extensive damage by the larvae (Delannay *et al.*, 1989; Perlak *et al.*, 1990; Vaeck *et al.*, 1987). The second example refers to viruses as model systems for disease control. Unravelling the control steps of the life cycle of viruses is almost certainly going to lead to the design of new genes to combat virus diseases in plants (Wilson, 1989). Attenuation of symptoms of cucumber mosaic virus and tobacco ringspot virus infections has been known for many years to be associated with the presence of an extra viral RNA molecule. When these RNAs were expressed as messenger RNAs in transgenic plants, they conferred increased resistance to viral symptoms following attack by a virulent viral strain because the incoming virus recognized the novel plant mRNA and viral proliferation was consequently reduced (Harrison *et al.*, 1987; Gerlach *et al.*, 1987). Also, plants with considerably enhanced resistance to specific viruses have been created by the insertion of genes encoding viral coat protein into plants (Nelson *et al.*, 1988; van den Elzen *et al.*, 1989). The precise mechanism of how pre-existing coat protein in a plant cell inhibits proliferation of an incoming virus is unknown. However, the results are impressive. I predict that, in the medium term, plant breeders will routinely control pathogens and pests using novel genes inserted into plants where the gene has been designed from knowledge of the behavior of the pathogen or pest and the plant's response to them. Therefore, research into the biology of pathogens and pests should be carefully followed by plant breeders to maximize the efficiency of the recognition and exploitation of new disease resistance genes.

ARABIDOPSIS AS A MODEL PLANT SPECIES

While I have argued above that all sorts of organisms will provide valuable genes for plant breeders, undoubtedly, plants will continue to be a vital, rich source of genes and genetic variation for crop improvement via gene insertion. However, crop plants are not particularly amenable for a molecular genetics dissection of plant processes and, therefore, there is the need for a model plant species, or perhaps more than one, for primary research.

Recently, many laboratories and government funding programs in Europe and the U.S.A. have become committed to the development of *Arabidopsis thaliana* as a model for uncovering the molecular and developmental principles for plants. This is a very exciting and historically significant step. It has not been a hasty decision because it has been recognized by many for a long time that *Arabidopsis* has numerous ideal features for plant laboratory research (Meyerowitz, 1989; Somerville, 1989). What has precipitated the commitment is the realization that the new techniques of molecular genetics can be utilized so much more efficiently and in so many more laboratories worldwide in an organism with a short life cycle, with a small genome and that grows in little space (Meyerowitz, 1989). Furthermore, plant science needs one such organism to which many persons contribute so that the mass of information at the molecular, cellular, developmental and whole organism levels is sufficiently substantial that real practical and intellectual dissections of plant processes can take place in a manner that matches the sophistication in modern prokaryotic, fruitfly, yeast and mammalian research. Without such progress, plant science, including crop biology, will not progress at a competitive rate and will fail to attract the brightest research minds.

There has been debate on both sides of the Atlantic, to my knowledge, of whether the development of *Arabidopsis* science will turn out to be a distraction from and a nuisance to the development of our understanding of crop biology. There is no doubt in my mind that *Arabidopsis* research is part of the development of crop biology and, without it, crop improvement will not match expected potentials in the next century. This is for the simple reason that, with currently foreseeable technology, we will not be able to discover and manipulate in a precise enough way many of the processes worth manipulating further to improve crops if we have to rely on the information coming only from crop biology. Using *Arabidopsis* as a "bridge" organism, genetic manipulation of our crop plants will develop considerably faster. One fact to support this statement is that a biotechnologist wanting to manipulate genes and insert them in today's crops have few genes to exploit, especially of the kinds important to breeders that influence plant height, maturity, morphology and biomass potential. The recognition and molecular isolation of such genes will be done initially much more rapidly via *Arabidopsis*. The isolated genes or coding sequences will then be used directly from *Arabidopsis* or will be used to isolate the homologous genes from the crop species of interest. Some of our own programs in the Cambridge Laboratory (formerly the Plant Breeding Institute, Cambridge, U.K.) at the John Innes Centre, Norwich, U.K. serve as useful examples to emphasize this strategy and illustrate one of the reasons why we have embarked on substantial *Arabidopsis* research programs.

Research of several of my colleagues has for many years focused on the genes regulating height and flowering time in wheat (Gale and Youssefian, 1985; Law, 1986). These genes made a major contribution to the "Green Revolution" because they contributed much to increased wheat yields around the world. The dwarfing mutations at the *Rht-1*, *Rht-2* and *Rht-3* loci are homoeoallelic and map to chromosomes 4B or 4D. Several alleles at each locus have been identified. Each dwarfing allele confers on wheat insensitivity to exogenous gibberellic acid (GA) and a range of insensitivities is found among the alleles. This range is correlated with the degree of dwarfism. Substitution by alleles with more sensitivity increases height, reduces seed setting and tillering, and gives rise to larger grains. Although the semi-dwarfing alleles enhance yields in many environments, this is not so in all environments. Where the environments are more hostile to wheat production, for example, in the hotter Mediterranean countries, less sensitive alleles are more useful (Worland *et al.*, 1988).

Research has shown that breeders have not only selected the most useful dwarfing alleles for their environments but also co-selected alleles at other loci which interact with the GA insensitivity dwarfing alleles to produce, for example, taller dwarfs which display higher yields under appropriate environments. We have learned, therefore, that to optimize plant productivity in different locations it is desirable to have available a range of alleles at the primary locus as well as a range of alleles at other loci affecting the same phenotype. A similar conclusion has come from studying the effects on yields of various alleles affecting the vernalization and photoperiodic response in different environments (Worland *et al.*, 1988).

This research has taken a long time because it has been necessary to identify different alleles, make isogenic lines as well as, of course, carry out the field trials. One suspects that, once the importance of a dwarfing allele had been established, if it had been possible to manipulate a cloned gene *in vitro* to result in variation in its expression or structure and create variation in plant response, then optimization of alleles for different environments could have been developed much more quickly. How do these very important genes cause the phenotypic effect, and what are the details of the natural allelic variation? To answer these questions, cloned genes and transgenic plants are required. How do we go about cloning the genes from wheat? Strategies are emerging but none are straightforward or simple. It, therefore, is very attractive to be able to use *Arabidopsis* as a first source of the genes. The control of development and height by GA is almost certainly physiologically similar in different angiosperms and mutants are known in *Arabidopsis* that have similar phenotypes to the dwarfing, GA-insensitive, alleles in wheat (Kornneef *et al.*, 1987). Similarly, there are mutants known in *Arabidopsis* with phenotypes which respond to vernalization as do the wheat mutants (Martinez-Zapater and Somerville, 1990). The ease with which it is going to be possible to isolate genes and dissect the role of multiple loci in a phenotype from *Arabidopsis* versus crop plants in the near future surely means that further manipulation of important crop phenotypes with a plethora of alleles is going to be greatly helped by genes derived from *Arabidopsis* variants directly or by the homologous crop genes isolated by virtue of their DNA sequence or product homology to their *Arabidopsis* counterpart. By using *Arabidopsis* as a "bridge species" to guide understanding, to provide genes and as a rapid test organism, it will be possible to move more rapidly in crop improvement where transgenic approaches are possible.

INTERNATIONAL *ARABIDOPSIS* GENOME PROJECT

While many laboratories have initiated plant research using *Arabidopsis*, the National Science Foundation (NSF) in the U.S.A., the Agricultural and Food Research Council (AFRC) in the U.K., and the European Economic Community (EEC) (among others) have established special programs with the hope of stimulating more efficient progress by sharing expertise, know-how, and results. Furthermore, a group of scientists who have been especially involved in the research to develop *Arabidopsis* as a convenient source of genes have formed an international steering group to stimulate international collaboration and exchange of results. The group has set a number of goals to stimulate all potential participants around the world including the essential funding bodies to realize the opportunities that are close at hand. These goals include, within 5 years, the recognition of all genes that, when mutated, create a phenotypic difference; a complete library of overlapping ordered DNA fragments representing the complete genome; densely marked genetic maps of gene loci integrated with DNA sequence markers covering the whole genome; and by the year 2000 the complete nucleotide sequence of the genome. Progress has been remarkable over the past year or so, and I give below a brief overview [source material: NSF (1991) publication on "Multinational co-ordinated *Arabidopsis thaliana* genome research project"] to emphasize the extent of the new plant science base that is being created so rapidly by research on *Arabidopsis*.

New genes recognized by mutations

Over 200 new loci have been recognized over the past year (1989/90) including those influencing floral development, fatty acid biosynthesis, amino acid biosynthesis, root form and growth, mineral uptake, environmental sensitivities, photosynthesis, height, and other morphological features. Representing perhaps 4000 genetic loci on preliminary tests, 42,000 embryo-lethal and seedling lethal mutations were identified by G. Jurgen at the University of Munchen and by A. J. Muller at Gatersleben in Germany. In these mutant searches, over 3600 chlorophyll-deficient lines were also noted. Over 1000 new mutant lines have been produced from experiments in which T-DNAs have been inserted from *Agrobacterium* and, therefore, these genes may be directly isolated by cloning of the T-DNA. These mutants include all classes of visible mutations from wax coat, trichome and root hair abnormalities to homeotic and embryo-lethal developmental defects. In another series of experiments (e.g., C. Dean, G. Coupland and colleagues, Cambridge Laboratory, Norwich, U.K.), transposable elements from maize (*Zea mays* L.) and *Antirrhinum* have been inserted into *Arabidopsis* and shown to excise and integrate into a new location to cause mutations. These sorts of systems will be useful for tagging new gene loci and, hence, isolating directly the gene involved. Estimates of the genome size and knowledge of the structure of many *Arabidopsis* genes suggests that the species is unlikely to contain more than 25,000 genes. Therefore, perhaps 15-20% of all gene loci may now be known from the recent large mutant hunts.

Genetic and physical maps

In the latest published genetic map (Korneef, 1990), 86 genes are marked. These represent the position as of July 1989. Since then, at least another 30-40 have been mapped. Several restriction fragment length polymorphism (RFLP) maps have been published in the past 2 years (Chang *et al.*, 1988; Nam *et al.*, 1989), and a total of over 360 DNA fragments have been mapped. In addition, Dupont have established another detailed map with several hundred randomly amplified polymorphic DNA (RAPD) fragment markers. Efforts are now underway to integrate the different maps by classical mapping and computer-assisted manipulation of the primary mapping data.

Progress towards establishing an ordered overlapping set of DNA fragments for the whole genome is proceeding well by collaboration between many laboratories. In H. Goodman's laboratory (Harvard University), 20,000 cosmid clones have been aligned into 750 overlapping groups which represent approximately 91,000,000 base pairs of the estimated total of 100,000,000. Several libraries of much larger fragments have been cloned in yeast artificial chromosomes (e.g., Ward and Jen, 1990), and these are also being ordered to provide contiguous stretches of the genome. So far more than 120 such stretches comprising an estimated 28% of the genome have been placed on the genetic/RFLP map and some regions greater than 1,000,000 base pairs have been reconstructed. Therefore, there seems to be no reason why a complete ordered library of fragments will not be available for the plant science community in the relatively near future.

The combination of mapped loci controlling known phenotypic effects, DNA sequence markers, libraries of mapped large overlapping chromosomal segments, not to mention T-DNA or transposon tagged loci, means that there is the opportunity to isolate plant genes of known function on a scale undreamed of just a few years ago. This provides an unprecedented opportunity for plant science and, subsequently, plant breeding.

Resource bases, data bases, and communication

It is the intention that *Arabidopsis* laboratories will contribute to and exploit a rich, user-friendly, data base that contains complete DNA maps, sequence information, genetic maps, and sophisticated details of the genome. When this is working well, it will herald a new mode of plant research. Furthermore, it is expected that a large number of *Arabidopsis* (and other plant) genes will be recognized or isolated by reference to the rapidly growing catalog of yeast, *Drosophila*, bacterial and animal gene, and protein sequences. This further emphasizes that tremendous progress is coming from and will increasingly emerge from the use of interspecies comparisons via interactive computer data bases. Thus, these are of very great importance to the plant breeder in the future. Their special importance is that, although they contain information on organisms other than plants, nevertheless, information is vital for efficient plant genetics. The compilers of these data bases will greatly ease the load of the future plant breeder in trying to keep an eye on all potentially useful model systems.

International resource centers are being established and financed to hold and send out all the mutant seed stocks (e.g., Nottingham, U.K.), DNA clones and other essential resources to enable researchers anywhere in the world to benefit, efficiently, from the international progress on *Arabidopsis* genome research. These sorts of developments, coupled with electronic mail and other computer-linked communication networks between laboratories should set the stage and the standards for cooperation to enhance the rate of research progress and may provide a useful example for later equivalent developments in crop biology research.

The progress on exploitation of *Arabidopsis* as a model plant system has been already remarkably rapid and within a few years will truly revolutionize the genetic resource base for plant scientists and, as I have argued, for all plant breeders because of the functional, genetic, biochemical, and developmental evolutionary homologies within the plant kingdom.

CONCLUDING THOUGHTS

This symposium focuses on plant breeding for the 1990s. I expect transgenic cultivars in several crop species to be released in the 1990s, thereby giving unequivocal evidence of the value of research into the genetics of other organisms as very important steps in crop improvement. From the earliest times of plant breeding it has been interesting and important to consider what are the most likely genetic sources for "quantum leaps" in crop productivity. Now we must consider if these sources are likely to be organisms other than crop plants, i.e., bacteria, yeasts, animals, or model plant species. If the answer is "highly likely" (and I believe this to be the case), then those responsible for the crop productivity of the future need to consider very carefully whether sufficient resources are going into the exploitations of these non-crop species as sources of genes for plant breeders. The days when plant breeders can consider their species as the only source of genes for crop improvement are gone forever, and the repercussions of this are legion in industry, in the public sector, in government research planning, and in the public understanding of the nature of crop germplasm.

REFERENCES

Chang, C.; Bowman, J. L.; Dejshon, A. W.; Lauder, E. S.; Meyerowitz, F. M. (1988) Restriction fragment length polymorphisms linkage map for *Arabidopsis thaliana*. *Proceedings of the National Academy of Science* 85:6856.

De Block, M.; Botterman, J.; Vandewide, M.; Dockx, J., Thoen, C.; Gossele, V.; Morva, N. R.; Thompson, C.; Van Montagu, M.; Leemans, J. (1987) Engineering herbicide resistance in plants by expression of a detoxifying enzyme. *EMBO Journal* 6:2513-2518.

Delannay, X.; La Vallee, B. J.; Proksch, R. K.; Fuchs, R. L.; Sims, S. R.; Granplate, J. T.; Mamore, P. G.; Dodson, R. B.; Augustine, J. J.; Layton, J. G.; Fischoff, D. A. (1989) Field performance of transgenic tomato plants expressing the *Bacillus thuringiensis* var *kurstani* insect control protein. *Bio/Technology* 7:1265-1269.

Gale, M. D.; Youssefian, S. (1985) Dwarfing genes in wheat. In: Russell, G. E. (ed.), *Progress in plant breeding*. Butterworths, London, pp. 1-35.

Gerlach, W. L.; Llewellyn, D.; Hoseloff, J. (1987) Construction of a plant disease resistance gene from the satellite RNA of tobacco ringspot virus. *Nature (London)* 328:802-805.

Harrison, B. D.; Mayo, M. A.; Baulcombe, D. C. (1987) Virus resistance in transgenic plants that express cucumber mosaic virus satellite RNA. *Nature (London)* 328:799-802.

Kaniewski, W.; Lawson, C.; Sammons, B.; Ilalcy, L.; Hart, J.; Delannay, X.; Turner, N. E. (1990) Field resistance of transgenic russet burbank potato to effects of infection by potato virus. *Bio/Technology* 8:750-754.

Kornneef, M. (1990) *Genetic maps*. In: Brien, S. J. (ed.). Cold Spring Harbor Lab Press, Cold Spring Harbor, NY.

Kornneef, M.; Adamse, P.; Barendse, G. W. M.; Karssen, C. M.; Kendrick, R. E. (1987) The use of gibberellin and photomorphogenetic mutants for growth studies. In: Cosgrove, D. J.; Knievel, D. P. (eds.), *Physiology of cell expression during plant growth*. American Society of Plant Physiologists, Rockville, MD, pp. 172-179.

Law, C. N. (1986) The genetical control of day length response in wheat. In: Atherton, J. G. (ed.), *The manipulation of flowering*. Butterworths, London, pp. 225-240.

Lee, K. Y.; Townsend, J.; Tepperman, J.; Black, M.; Chui, C. F.; Mazur, B.; Dunsmuir, P.; Bedbrook, J. (1988) The molecular basis of sulfonylurea herbicide resistance in tobacco. *The EMBO Journal* 7:1241-1248.

Martinez-Zapater, J. M.; Somerville, C. R. (1990) Effect of light quality and vernalisation on late-flowering mutants of *Arabidopsis thaliana*. *Plant Physiology* 92:770-776.

Mazur, B. J.; Falco, S. C. (1989) The development of herbicide resistant crops. *Annual Review of Plant Physiology* 40:441-470.

Meyerowitz, E. M. (1989) *Arabidopsis*, a useful weed. *Cell* 56:263-269.

Nam, H. G.; Graudat, J.; den Boer, B.; Mooran, F.; Loos, W. D. B.; Hauge, B. M.; Goodman, H. M. (1989) Restriction fragment length polymorphism linkage map of *Arabidopsis thaliana*. *Plant Cell* 1:699.

Nelson, R. S.; McCormick, S. M.; Delannay, X.; Dube, P.; Layton, J.; Anderson, E. J.; Kaniewska, M.; Proksch, R. K.; Horsch, R. B.; Rogers, S. G.; Fraley, R. T.; Beachy, R. N. (1988) Virus tolerance, plant growth, and field performance of transgenic tomato plants expressing coat protein from tobacco mosaic virus. *Biotech* 6:403-409.

Oxtoby, E.; Hughes, M. A. (1990) Engineering herbicide tolerance into crops. *Tibtech* 1990:61-65.

Perlak, F. J.; Deaton, R. W.; Armstrong, T. A.; Fuchs, R. L.; Sims, S. R.; Greenplate, J. T.; Fischoff, D. A. (1990) Insect resistant cotton plants. *Bio/Technology* 8:939-943.

Schaewen, A. von; Stitt, M.; Schmidt, R.; Sonnewald, U.; Willmitzer, L. (1990) Expression of a yeast-derived invertase in the cell wall of tobacco and *Arabidopsis* plants leads to accumulation of carbohydrate and inhibition of photosynthesis and strongly influences growth and phenotype of transgenic tobacco plants. *The EMBO Journal* 9:3033-3044.

Somerville, C. (1989) *Arabidopsis* blooms. *Plant Cell* 1:1131-1135.

Stitt, M.; Schaewen, A. von; Willmitzer, L. (1990) "Sink" regulation of photosynthesis metabolism in transgenic tobacco plants expressing yeast invertase in their cell wall involves a decrease of the Calvin-cycle enzymes and an increase of glycolytic enzymes. *Planta* 183:40-50.

Vaeck, M.; Reynaerts, Q.; Hofte, H.; Jansens, S.; De Beuckelcer, M.; Dean, C.; Zabeau, M.; Van Montagu, M.; Leemans, J. (1987) Transgenic plants protected from insect attack. *Nature (London)* 328:33-37.

Van der Elzen, P. J. M.; Huisman, M. J.; Willink, D. P-L.; Jongedijk E.; Hockema, A.; Cornelissen, B. J. C. (1989) Engineering virus resistance in agricultural crops. *Plant Molecular Biology* 13:337-346.

Ward, E. R.; Jen, G. C. (1990) Isolation of single-copy sequence clones from a yeast artificial chromosome library of randomly sheared *Arabidopsis thaliana* DNA. *Plant Molecular Biology* 14:561-568.

Wilson, T. M. A. (1939) Plant viruses: A tool box for genetic engineering and crop protection. *BioEssays* 10:179-186.

Worland, A. J.; Law, C. N.; Petroric, S. (1988) Pleiotropic effects of the chromosome 2D genes Ppddd1, Rht8, and Yr16. In: *Proceedings 7th international wheat genetics symposium.* Institute of Plant Science Research, Cambridge, pp. 669-674.

Discussion

Paul H. Sisco, Moderator

In the figure from Smith and Smith regressing yield on genetic distance, I think the relationship is not heterosis but, rather, confounded effects - additive and non-additive variance. Do you have a comment?

I don't want to argue semantics as to what it is, but I'm impressed mostly by the relationship to yield. Whether we want to argue if that's heterosis or confounding effects, I don't know what to say about that. However, it's thought-provoking and should make us think about what is underlying hybrid vigor or heterosis. If it doesn't define heterosis itself, it probably is telling us something about the phenomenon. [T. G. Helentjaris]

Should plant breeders be selecting for the maximum divergence of marker alleles?

One of the things on the graph that Steve and Howie put together is the similarity within the extreme tail or the right end. As you look at their graph and at your graph from NPI, you will see that there probably is no relationship between dissimilarity and yield, that once you get above 60 or 70% dissimilarity, you just have a cloud of data points that you could put a line through at any point. I think that the plant breeders have known this for a long time because they don't produce hybrids that are related.

[Bill Beavis, Pioneer Hi-Bred International]

I wouldn't argue with that observation at all. Once you start picking out microregions of that curve, it's not as accurate as you want. Being an optimist though, I tend to look at the curve and say "What are the positive aspects? It's a naive model, and yet it works fairly well. So, why doesn't it work perfectly?" I think the idea is that we're going to have to go in and weight particular loci or give weight to particular allelic combinations and see if that straightens out the curve at the top. At the very least, we do a lot of wide crosses in plant breeding and it might direct us into particular parental combinations that we would want to explore in more detail. I know that in one case, when we tried to make improvements in certain inbreds, the marker data in almost every case suggested that, if we made it more heterozygous, it would result in improved performance of the hybrid.

[T. G. Helentjaris]

You asserted that recombination occurs primarily in functional genes. What evidence do you have to support this? What implications does this observation have for plant transformation?

They aren't my data, but data from several different laboratories. For the *R* *waxy*, *bronze*, and *a1* loci, where there are several different insertion alleles available, by looking at intra-allelic recombination you can determine the relationship between kilobase length and genetic length within the gene. When compared to the genome size divided by the length of the genetic map, it is seen that the frequency of recombination within a gene is approximately 100-fold greater than for the genome as a whole. What that says is that at least 99% of the recombination is occurring within genes. In regards to transformation, to the extent that transformation relies upon the normal recombination mechanism, it is going to direct insertion into genes. I think that the incredibly high frequency of mutants derived from DNA insertions supports the idea that DNA transformation is going into genes for the most part.
[S. P. Briggs]

Please outline a strategy for eliminating culture-induced "mutants" from transgenic populations. Is there evidence that culture-induced variation can remain cryptic in early tests but be expressed in later generations?

There are various lines of evidence that give us some routes to follow. One is that it is the callus phase that's mostly involved here; so, the longer your material is in the callus phase, the more likely you will have mutational events occurring. Obviously, one possibility to reduce the mutation frequency is to get into and out of culture as quickly as possible. The other approach would be not to go through a particular callus phase, but to do some type of micropropagation (or other more direct embryogenesis) where the embryogenesis is not rising after a long period of callus culture. Data from embryogenically derived materials from maize callus cultures also have a high frequency of variation, at least as high as through organogenesis. Another possibility is to manipulate the medium. The methylation effect seems to be modulated by auxin in the media [there's evidence from Italy (Lo Sciavo and coworkers there) indicating that 2,4-D levels in the medium affect degrees of methylation of DNA], so one could envision manipulating the hormone level. Genotypes will likely make a difference as well. In terms of cryptic changes, one line of thought is that, when plants are regenerated, they come from a mixture of cells. Therefore, the original plants are likely to be sectored. If you self-pollinate a regenerated maize plant, for example, the mutation may not segregate in the first generation; but, when you self-pollinate those R_1 plants, segregation of new mutation may occur in the next generation. The implication is that the original plant was heterozygous and chimeric.   [R. L. Phillips]

If we will be able to create new genes from model systems, are we wasting our efforts on storing huge germplasm collections in gene banks?

A person on my right is saying "Go ahead and say yes," but I have to say "no" for a large number of very good reasons. We've yet to discover how many genes can be manipulated by the molecular biologist to be adapted to different genetic backgrounds. Such adaptation really defines the value of a gene and the value of the subsequent germplasm. Of course, in the germplasm banks alleles already exist that have gone through some sort of adaptation. Furthermore, and most important, there are complexes of genes that have gone through some sort of selection and evaluation. I would hate to see somebody ignoring that and throwing the adapted gene complexes away when we have virtually no information on the extent to which we can put in a gene designed in a test tube and maximize the value of that gene in the genetic background. We have a long, long way to go to accumulate enough information to give an authoritative view on this point.   [R. B. Flavell]

The value of the gene bank should be enhanced tremendously with some of the newer technologies. Obviously, when you try to utilize gene bank material and you make the cross, the hybrid has 50% of its genes from the adapted material and 50% from the unadapted; and the laws of probability make it very difficult to extract the genes you really want in a clean fashion. With RFLPs you have the possibility of flanking markers to specifically transfer genes of interest. Transformation offers similar opportunities. Other techniques such as chromosome isolation and transfer may be valuable as well. I can give you an example with oats. If you make a cross, you're bringing in 21 chromosomes from each parent. The probability in one generation of getting a gamete produced in that F_1 that contains 20 chromosomes from the adapted source and one chromosome from the unadapted source is roughly one in a million. If you could isolate chromosomes *en masse*, sort them, and add one chromosome at a time, then you could create an addition line or trisomic. Depending on the behavior of the chromosomes, the probability of obtaining the desired gamete may be one in four. The point I am making is that being able to utilize genes in a gene bank may be greatly enhanced by some of these technologies.   [R. L. Phillips]

Is it accurate to state that RFLP marker loci are not influenced by the environment?

It is a pretty good assumption that over most conditions the DNA sequence is relatively constant and unaffected by the environment. Certainly we try to steer away from enzymes and loci that might be affected by the environment - for example, a transposable element sequence or loci that may be affected by methylation. [T. G. Helentjaris]

In experiments on wheat, looking at methylation changes and the inheritance thereof, it's interesting to see that the level of variation is very high compared with the levels of polymorphism that one sees looking at other target sites not capable of varying in methylated cytosine. The environment might be responsible for this high level of polymorphism. In these special circumstances, doing RFLP analysis to make use of variation in methylation (which does seem to be occurring at a high level in wheat) might be valuable where there isn't indigenous polymorphism in a cross. But, I add a note of caution - the rate of change seems sufficiently high that, if you are not dealing with the actual parents as controls, you can be misled. Environmentally induced changes then, if that's what is behind the methylation changes, could be useful as well as a problem. [R. B. Flavell]

Are there any techniques for the direct cloning (by transposons) of recessive alleles?

I'm not sure I understand the question. If the question is "can you knock out a recessive allele?", the answer is sometimes the recessive allele has a different phenotype from a true knockout or deletion, But, you wouldn't know that unless you already had the genetic variant. However, if you just wanted a particular allele, you could knock out any allele that is easy to score in a tagging experiment. Then, with your probe, you can easily clone virtually any other allele. For the purposes of obtaining an allele, you don't need to start with a particular one. You want to start with the one that will work best with the tagging experiment and then clone out your desirable alleles.

[S. P. Briggs]

Your comment concerning a strategy to reduce or eliminate culture-induced variation is to reduce the time a callus is in culture. If this is the case, what is your strategy for long-term storage of clonal lines?

I don't have a strategy for long-term storage. If the long-term storage involves callus, obviously you have initially to generate that callus and later induce growth to bring it out of storage; so certainly there is ample time for variation to be induced. One might have variations generated during storage itself; we don't understand such mechanisms well. For long-term storage, in my opinion, one wants to keep the callus aspect to a minimum.

[R. L. Phillips]

Is the technology available (has it actually been done) to transform a known gene from *Arabidopsis thaliana* back into *A. thaliana* to see if it goes back to its original or a random site, i.e., once you have a gene and you want to knock out its endogenous homolog, can you do it by directed transformation?

Arabidopsis genes have been cloned back into *Arabidopsis* and shown to complement a mutant gene, but the sites into which they go are usually not of homologous sequence. Rarely, however (1 in 5000), the insertion does appear to be via homologous DNA recombination (preliminary results). There is a need, in a very large number of scenarios, to remove genes as well as add them. The efficiency of doing this is exceedingly low and, therefore, remains a huge technical problem.   [R. B. Flavell]

Directed gene insertion is a priority in plant breeding groups, and there are many reasons for doing it. You have position effects when you insert a gene and, if you are testing a promoter, then how strong is the promoter? If it goes into site A, it might be real strong; but, if it goes into site B, it might not work at all and, depending on what you pick, you might conclude something about the promoter when actually you are looking at it from the position in which the gene went. So, you want to be able to always target the gene. For regulatory purposes, I think that it's going to be important to be able to well characterize where the gene goes and have it defined and consistent.   [S. P. Briggs]

We usually think about trying to add in a recessive allele (or usually we think of transformation as adding a function) but, in many cases, we want to subtract a function. Besides just homologous recombination, there are a couple of techniques that look like they will be successful (1) either in using an antisense technology to knock out an existing gene function or (2) perhaps the ribozyme technology - where you might also knock out the gene function. We might be able to do that additively so that we can also subtract and control the rate. So, there might be other ways besides homologous recombination to delete functions.   [T. G. Helentjaris]

The comment has been made several times that transformation will be useful from the standpoint of bringing in genes from other sources where we can't introduce them by sexual means. I'm not at all convinced that's going to be the most prominent use of transformation. I think what Tim just said about knocking out functions by antisense mutations is very important. If you want to convert a line to *waxy*, for example, you may want to introduce an antisense sequence and knock out the waxy function so you don't have to go through the backcross procedure which may bring along adverse linkage blocks. But, if we look ahead in time, from an efficiency point of view, even something as simple as transferring a single gene from within the species may be a very common use of that technology.   [R. L. Phillips]

What role with RFLP technology have in breeding programs as a routine screening tool where the number of segregating populations exceeds 500?

If we narrowly define RFLP in their utility as indirect selection criteria, then we're not at the point of being able to use them on a large scale. However, I think our initial experiments should to be to look at the quality of data that come out of this analysis. Too often I talk to people and they say, "Well, I can do this faster and easier with traditional means" and, in a sense, I wouldn't argue. But at the same time, if we look at the quality of the data and types of insights we obtain, then perhaps our initial use of this technology will be to understand how we got to where we are today. Clearly, I'm very interested in the idea of using these as indirect selection criteria, and I think it's possible that we may be able to develop technologies that will handle very large numbers and even be applicable in a field station. I know that companies are building a machine that can spit out DNA preps at one a minute and they are looking at doing 60,000 analyses per day by PCR. That's on the scale that gets to be fairly relevant.   [T. G. Helentjaris]

In the natural genetic system where corn inbred lines are maintained sexually for many generations and have been shown to exhibit genetic drift, would you speculate whether natural changes in methylation might, or might not, be involved?

An inbred line developed by self-pollination is not completely homozygous. We know, when inbred lines are maintained at separate locations and at some point crosses are made, heterosis can ensue. Also, for doubled haploids that we know are homozygous, it doesn't take long until there is variation in the material; and the people doing those studies argue that it is too high for conventional mutation. It seems to me that non-traditional kinds of mutational events may be occurring; one can envision DNA methylation changes, unequal crossing-over of repeated sequences that would give increased or decreased copy numbers in otherwise homozygous materials, etc. People have commonly done selections within inbred lines to improve them for their own purposes such as altering maturity within inbred lines. So, there is variation. There are various types of plasticity in the genome. We seem to forget that and even in the realm of plant molecular biology, there still hasn't been much done on genome plasticity. So, in terms of relevance to plant breeding *per se*, I think plasticity is a very important area to investigate. Breeders may be able to take advantage of such variation.

[R. L. Phillips]

Given the shift in resources within the seed industry, from plant breeding efforts toward biotechnology research, when do you expect seed companies to realize a net profit from their investment in biotechnology?

I don't know the answer to that. We have a biotech product on the market now, and it's not making a whole lot of money for us. However, I think that we're going to see a gradual introduction of incremental improvements and products, and I think we are going to see things the consumer might not notice but that will have significant impact on our production costs and, therefore, our profitability. I think most of our biotech improvements are going to be gradual and incremental - not spectacular like insect resistance. By the year 2000, anybody who is really seriously into biotechnology will be seeing a significant profit from their efforts. Just one success will pay for a biotech program for 20 years. So, it doesn't take many successes to pay off research costs.   [S. P. Briggs]

I would echo that and I'm not sure I could put a finite date on it. But, I think within the next 10 to 15 years we should see some products coming out, and I agree that it doesn't take many successes to be profitable. For instance, if we learn more about how to define heterotic groups in corn and, if you were to use these approaches to develop a new heterotic group, then a significant impact would be realized that would stretch for a long time.

[T. G. Helentjaris]

Chapter 19
New enthusiasm for microbial products?

Winston J. Brill
Winston J. Brill & Associates, 4134 Cherokee Drive, Madison, WI 53711

INTRODUCTION

It seems that public and private research activities, aimed to apply specific strains of bacteria and fungi as commercial inoculants for agriculture, has decreased over the past decade. For instance, many large chemical companies began substantial inoculant programs in the early 1980s, but very few of these programs remain active. Is this decrease due to realization that such products do not have much potential to help the farmer, dissatisfaction with research progress, realization that the required work is too expensive and long term, or is the loss of activity due to regulatory pressures? Certainly, the fact that many non-performing inoculant products have been sold (fortunately, their commercial lives are short) has decreased farmer's enthusiasm for future inoculants. In fact, most *Rhizobium* products do not seem to be useful to farmers (due to competition by indigenous *Rhizobium* strains). I believe, however, that much of the work on microbial inoculants may have been prematurely terminated due to lack of consideration for many of the research and development criteria that should be carefully thought through. In this chapter, I will point out various general problems and opportunities that may have been overlooked with prior research activities.

OVERVIEW

We continue to see increased pressure for agriculture to be more efficient and environmentally compatible; thus, there is great incentive to search for new ways to increase crop yield. Advantages of microorganisms as agricultural products include the fact that they may be able to proliferate (e.g., along a developing root) as the plant grows; thus, relatively few organisms in the original product may be needed. An inoculant's activity can be quite specific. From an ecological point of view, contamination problems with bodies of water or on food or feed will be far less than that customarily observed with current chemicals. Also, it is quite possible that the pests will not readily become resistant to antagonistic microorganisms, since experience has shown that it usually is much easier to gain resistance to an antagonistic chemical than to an antagonistic microorganism. *Bacillus thuringiensis* insecticide is an example. There are a number of strategies available to minimize a resistance problem; for instance, an inoculant containing microorganisms with different antagonistic activities would further decrease the incidence of resistance by the pest.

Many mechanisms are possible for microorganisms to increase yield. Nitrogen fixation, specific mineral uptake, mineral solubilization, insecticidal, fungicidal or nematicidal activity, and plant growth hormone production are examples. This chapter will not review information related to inoculants. Good reviews are available (i.e., Lambert and Joos, 1989; Nakas and Hagedorn, 1990). The chapter should be read as a challenge to take a broader view when embarking on a program to develop microbial inoculants.

A variety of inoculants already are available. The best known examples are *Rhizobium* and *B. thuringiensis*. During the past century, there have been hundreds of claims of other microbial genera that increase yield. An example is the decades-long widespread practice in the Soviet Union of inoculating soils with *Azotobacter* (this practice ceased in the early 1960s). Most claims from laboratory or greenhouse experiments with *Azotobacter* and many other microorganisms have not been confirmed with repeated field tests. When small field tests provide an indication of efficacy, larger trials frequently do not show a yield differential by treatment with the microorganism. Are there reasons why scaled-up experiments commonly fail? Most importantly, are we overlooking aspects that have potential to render some of these organisms as useful inoculants in the future? Lastly, are there opportunities to improve inoculants through genetic engineering and mutation? Modern biotechnological methods are extremely powerful and will certainly play an important role to develop future microbial products for agriculture. However, a project that does not also take into account the complexity of the relevant microenvironment in the field and the physiology of the recombinant microorganism will have a good chance of reaching a dead end.

DEVELOPING INOCULANT PRODUCTS

Use of a sensitive screening assay

We must first ask why *Rhizobium* and *B. thuringiensis* are so well developed as commercial products. The easiest system to demonstrate efficacy in the laboratory or field is the nitrogen-fixing ability of *Rhizobium*. At early stages in a legume's life cycle the differences between nodulated and non-nodulated plants in a nitrogen-poor environment are easy to determine. *Bacillus thuringiensis* insecticidal activity also is relatively easy to assay. In both systems, the effect of inoculation is dramatic. Will there be microbial inoculants that have potential value as future products, but do not cause such dramatic effects?

To be commercially applied in the future, the product should be useful if it's mechanism is more subtle than *Rhizobium* or *B. thuringiensis*. For instance, the microorganism may only debilitate a pest rather than kill it. While this effect is more difficult to test, it may still provide sufficient protection to satisfactorily influence yield. If an inoculant produces a plant growth hormone, the result may only be an average yield increase of 5% (for example), which may be difficult to detect in an initial laboratory/greenhouse screening program. **Thus, screening programs need to be sufficiently rigorous to detect such activities.** Otherwise, potentially valuable strains will be bypassed. It is also important to screen for activity with the microorganisms growing conditions mimicking, to some extent, the expected conditions that the organism will face in the field.

Depending on the target, there are many strategies to isolate strains with potential to become useful microbial inoculants. One method could involve isolating strains from plants that seem to be more resistant during field pest infestation. Another method could involve large-scale screening programs similar to those used in the pharmaceutical industry to find new antibiotics. However, unlike antibiotic production in fermenters where most growth parameters are readily controlled, agricultural inoculants have to express activity in the field. That means they have to survive for a sufficiently long period and produce products (e.g., antibiotic, plant growth hormone) in amounts to give the microorganism its beneficial effect. The concentration of the microorganism in the field will be far less than the concentration of an antibiotic producer in a pharmaceutical fermentation vessel.

Minimize spontaneous mutation

Because survival of the microorganism is as important to agricultural pro-ductivity as its beneficial activity, both have to be considered throughout development of a product. It is crucial, therefore, to minimize spontaneous mutation that decreases survival on the plant or the soil. The mere act of isolation selects for mutants, and it is easy to imagine that these mutants will lose some of their ability to survive well in the soil. This may be seen as a benefit for the producers of a product since the farmer will have to repur-chase inoculants annually, but it is also a pitfall in that mutants may not be as effective as their parent strain. Similarly, transfer and storage of strains cause other selective pressures. Thus, a potentially useful organism may produce the active product (e.g., plant growth hormone), but not be suffi-ciently viable or active on the plant root. **Laboratory manipulation of the strain in early stages of a study should be minimized.** Multiple stocks should be made and screening should occur as soon as possible after isola-tion. After a promising candidate strain is chosen for further studies, it may also be important to select for better growth on the plant. Perhaps cycling the organism on the plant in the greenhouse or field will be useful to regain survivability, or even to improve survivability over the parent strain. It will be important to retest such strains to be sure they retain their beneficial activity. In fact, testing for both viability at the plant target site and for activity should occur at each step in the development program.

Test strains under a variety of conditions

In order to study a strain that seems to have some beneficial activity, it must be grown in a culture medium. For well-studied systems such as *Rhizobium*, *B. thuringiensis*, or pharmaceutically valuable antibiotic producers, the beneficial properties are only observed when the organism is in a specific physiological state. For instance, *Bradyrhizobium japonicum* nodulates and fixes nitrogen for *Glycine max* (L.) Merr. and only nodulates when it is in a specific phase of growth. To be active, it must be harvested at a specific time during the growth cycle in aerated liquid media. Different strains seem to be optimum for nodulation at different stages. Likewise, antibiotic-producing bacteria make the antibiotic only during the sporulation stage. With these examples in mind, it is easy to see how an activity may not be detected. **Therefore, strains should be grown under a variety of different conditions to minimize the chance of missing a positive activity response.** Even if some activity is observed during an initial evaluation, the results may not be repeated during scale-up because the phase of the microorganism's activity in the growth cycle may be different. For instance, oxygen available to the organism may vary greatly between a shaking test tube to a shaking flask or small fermenter. The physiological status of the strain must continually be monitored. What commonly happens is that researchers aim for large numbers of viable cells instead of physiologically (with respect to the beneficial activity) active cells.

Initial studies with a new strain probably will use high numbers of cells in a pure culture grown on an agar medium or in shake flask with liquid medium. The inoculant then will be applied directly to the soil, seed or leaves. The literature is filled with examples of positive effects on plants from such inocula. However, the work rarely is advanced to determine reasons why such activity is most commonly never observed in extensive field trials. It is very possible that scale-up renders the organisms inactive for the desired activity. In the antibiotic industry, careful monitoring of activity is required during each of many scale-up steps. Such concern usually is not considered for agricultural inoculants, and one wonders how many times researchers have discarded a strain that could have become an economical product, but was not grown by the appropriate method to maintain activity.

Screen several cultivars

Complexity is increased by the fact that the beneficial activity of microorganisms may be dependent on the physiological state of the target plant.

That state may depend on age of the plant, plant vigor or soil properties - all of which change during the growing season. In addition, different cultivars of the same crop may be affected to different degrees by the microorganism's activity or different cultivars may affect the physiological status of the microorganism. **Therefore, several cultivars should be used for the initial screening.** Obviously, cultivars planned for widespread purchase at the time predicted for commercial production of the inoculant should be used.

Optimize growth conditions

Some microorganisms are difficult to grow in culture. **Thus, attention should be aimed towards developing conditions to allow for efficient growth (yet maintain activity).** This can be a difficult task. Again, I wonder how many strains that showed some interesting activity were not aggressively followed up because of difficulty in growing large amounts of inoculum for further tests. A major research effort will be required just to learn how to grow a particular organism in culture.

Use a practical formulation

Once a strain can repeatedly be shown to benefit the plant species of interest, it is important to consider formulations that will allow the microorganism to become a commercial product. Cost of production, farmer acceptance, compatibility with currently used farm practices and shelf life are examples of factors that must be considered. It is obvious that the fewer microbial cells required per hectare, the less costly the product will be. In some cases, the microorganisms will be natural colonizers. That is, they will multiply as the root grows. Therefore, if a certain population density is necessary for the desired activity, that population will develop from relatively few cells.

Another opportunity to develop high populations of microorganisms is through coating large numbers of cells on the seed as it is planted in the furrow, or through a planter box, or precoated on the commercially purchased seed. If the strains then do not colonize, they probably will only influence the plant during germination and early growth. In some cases (e.g., seedling disease suppression), activity during this period may be sufficient to exhibit a beneficial effect. It is critical, then, for the beneficial effect of the strains to be as active as possible. Otherwise, too many cells per plant will be needed to use the inoculant in a commercial venture. Still, there are other methods that may be useful to increase populations in the soil (discussed in GENETIC ALTERATIONS, below).

Field testing must be initiated early in the program

While the beneficial activity (e.g., enhancing plant vigor or debilitating a pest) needs to be routinely tested at each stage of a program, it is equally important continually to field test the organism with the target crop lines. Such tests should be performed in fields that mimic the soils, climate, and farm practices where the product is expected to be used. Experiments must be performed on many different sites, with each experiment having sufficient replicates and/or large hectarages to obtain statistically valid data. Information about the fields, climate, etc. should be collected and correlations made with the effectiveness of the inoculant. Thus, early trials should continue to provide information useful to the development strategy of the program.

Field trials that concentrate on only a single microorganism strain may yield inconsistent results. However, by analyzing field data from many different strains, one may find that one strain is more effective under one set of conditions (e.g., drier soils) and another exhibits optimal activity under another set of conditions (e.g., moist soils). Thus, a mixture of the two or more strains could lead to greater consistency - a requirement for a good product. If both soil types are common in the normal growing region of the crop, then both strains should be included in the inoculum product. In fact, future products may be composed of many strains. Currently used commercial *Rhizobium* products commonly contain two or more strains.

GENETIC ALTERATIONS

Mutation

In many cases, the exact cause of the yield increase may not be understood at the biochemical genetic level. However, through a screening program it should be possible to increase the valuable activity of the microorganism. For instance, it should be possible to isolate mutants producing more of an antibiotic, or mutants that produce more of a metal-sequestering compound. One could also select mutants that allow the microorganism to survive longer in the commercial inoculant package, or ones which survive or cluster more on the root surface.

Genetic engineering

The techniques of genetic engineering have tremendous potential to advance use of microbial inoculants. This potential probably is why so many large chemical companies became excited about inoculant programs in the early 1980s. It is relatively easy to genetically engineer microorganisms. Potentially useful single-gene products, such as the *B. thuringiensis* endotoxin, can be introduced and the resulting recombinant strain then can be used as an inoculant. Proteins that may help bind the inoculant to roots (e.g., lectins) also could be added to a microorganism through recombinant techniques in order to concentrate the activity on the roots. Genetic engineering also can be used to delete an activity from the microorganism. For instance, if genes required for disease by a pathogen are deleted, then the resulting strain may be able to compete for plant sites that the natural pathogen would occupy and, thus, protect the plant from the pathogen. Using "Ice-Minus" *Pseudomonas* for frost protection is an example.

As we learn more about the molecular biology of both pathogens and beneficial microorganisms, it should be easier to design strategies to enhance or add new properties to potentially useful strains to allow them to bind or colonize the root (or leaf). This is an especially exciting area of research. Another possible use (but more difficult) of genetic engineering is to transform the plant to produce a metabolite that inoculant strains prefer and, in this way, give desirable strains a competitive advantage. Similarly, a microorganism can be modified to efficiently utilize a specific chemical that is added to the soil or in the inoculant package. Again, this gives the strain an ecological advantage, thus increasing the population of inoculant organisms.

Many beneficial effects of inoculants will, most likely, be the result of multi-gene pathways that synthesize compounds such as antibiotics, plant growth hormones, or nutrient sequestering compounds. These pathways can be introduced into microorganisms much easier than into plants. A concerted effort would be required to achieve favorable results, and attention has to be placed on possibly modifying regulatory regions for each of the desired genes. The work, while difficult, is quite straightforward.

Genetic alterations must be integrated into the system

When one selects a mutant originating either spontaneously or through a mutagen, it is possible that a secondary consequence of the mutation will make the inoculant less useful. For instance, the organism's survivability may be decreased. Thus, it is critical that activity tests and field tests be performed on these mutants as early as possible in the development process to learn if a mutant, in fact, has lost its value as an inoculant.

The same concern as above exists for strains altered through recombinant DNA technology. However, another problem arises through the multiple reisolation of strains from a single cell through genetic engineering techniques. For instance, markers such as β-galactosidase may have to be introduced to select or screen for a plasmid containing the important gene. The strain has to be grown on special media for selection or screening the desired recombinant colony. This additional selection certainly introduces new spontaneous mutations to the cell - some of which may negate the value of the introduced gene. To minimize secondary problems associated with genetic manipulation, it may be useful to cycle strains through the plant - perhaps in the greenhouse - in order to attempt to regain (or even increase) properties essential for the valuable activity.

Regulatory disincentives

Unfortunately, a long-term genetic engineering program to produce inoculants will be expensive and has to be considered in light of public concern about released recombinant organisms - especially microorganisms. The recent public response to the use (even testing) of recombinant bovine growth hormone certainly provides disincentive to pursue such projects, even if there is no scientific basis to be concerned about health or the environment. Of the many thousands of microbial inoculant experimental trials, worldwide during the past century, there has not been a single recorded case of an environmental problem. Hundreds of different strains and mutants have been used covering dozens of genera. In the food industry, many mutant strains have been developed for yogurt, cheese, etc. No special regulations have been needed, or previously asked for, to proceed with experimental field tests of indigenous organisms for agriculture or the food industry. The 1989 National Research Council study states that genetically engineered organisms should be no more dangerous than those same organisms modified by traditional means. Will this negative public perception actually keep ideas, that may have real benefit to agriculture and the environment, from being developed into products? Many strains previously described as increasing crop vigor or yield would have to undergo rigorous regulatory scrutiny that would take months of preparation by the investigator and probably cost several hundred thousand dollars - just to do a 1 m^2 field test. No wonder interest in recombinant organisms has decreased!

Crop breeding to maximize inoculant effectiveness

Once desirable strains are developed, there may be other opportunities to increase the inoculant's value by specifically breeding crop plants to be more effective with the inoculant. For instance, there are examples of *G. max* cultivars that fix more nitrogen and increase yield better with certain *Bradyrhizobium* strains than with others.

CONCLUSION

The purpose of this chapter is to re-stimulate interest in the possibility of developing useful microorganisms for agriculture. Many university, government, and industrial inoculant programs may have been abandoned prematurely due to insufficient rigor in assay methods, problems with secondary spontaneous mutations, scale-up problems, insufficient field testing or formulation problems. An effective program will require persistence and expertise from a variety of disciplines. Will such activities provide a new generation of agricultural tools, or will they end up as non-practical dead-ends? My guess is the former. The timing and quality of these products will depend on the level of public and private research expenditures, on the scientist's enthusiasm for the potential of new types of inoculants and, finally, on the public's acceptance of genetically altered microorganisms.

REFERENCES

Lambert, B.; Joos, H. (1989) Fundamental aspects of rhizobacterial plant growth promotion research. *TIBTECH* 7:215-219.

Nakas, J. P.; Hagedorn, C. (eds.) (1990) *Biotechnology of plant-microbe interactions.* McGraw-Hill, New York, 348 pp.

National Research Council (1989) *Field testing genetically modified organisms: Framework for decisions.* National Academy Press, Washington, DC, 170 pp.

Chapter 20
Marker-assisted selection in relation to traditional methods of plant breeding

Russell Lande
Department of Biology, University of Oregon, Eugene, OR 97403

INTRODUCTION

Experienced plant breeders are well aware that biotechnology can never replace traditional methods of plant breeding, but must be integrated with them to achieve the maximum improvement in crop yield and quality. One promising approach for integrating biotechnology with traditional plant breeding is marker-assisted selection (MAS).

Efforts to construct high-density linkage maps of molecular genetic polymorphisms (marker loci) are currently underway for several crops (Helentjaris *et al.*, 1986; O'Brien, 1990). Soon it should be possible to routinely score large numbers of such polymorphisms (e.g., allozymes, RFLPs, DNA fingerprints, oligonucleotide probes) on many individuals in a population. Statistical associations between alleles at molecular marker loci and alleles at quantitative trait loci (QTLs) can be used to select indirectly, but with potentially very high accuracy, for DNA segments containing favorable QTL alleles, effectively increasing the heritability of economically important characters (Stuber *et al.*, 1982; Soller and Beckmann, 1983).

When selecting for characters with low individual heritability, such as yield, plant breeders typically use some form of family selection to effectively increase heritability. In the context of selecting among highly inbred lines with large family sizes, for example, there clearly is little opportunity for increasing selection efficiency using molecular genetic information. However, if we consider a wider range of breeding strategies, including all possible family sizes, it may be that marker-assisted selection with small families, or at the individual plant level, can produce a better response, or a comparable response with less effort, than traditional methods. Here I explore some of the possibilities and limitations in utilizing molecular data as an aid in selection for crop improvement.

A statistical methodology for MAS combines information on molecular genetic markers with standard phenotypic measurements to construct optimal selection indices. Under certain usual assumptions, all of the molecular information can be condensed into a single molecular score for each individual or family, which is then treated simply as another character correlated with the trait(s) of economic importance to be included in a selection index. Retaining phenotypic information in the selection index increases its efficiency by exerting selection on additive genetic variance that can not be accounted for using molecular markers.

Plant breeders have been reluctant to use selection indices, in part because of difficulties in assigning relative economic values to different plant characters (Hallauer *et al.*, 1988). This is not an obstacle for MAS, at least in the improvement of a single character, because the molecular markers have no intrinsic economic value. The weights in the selection index can be computed directly from basic quantitative genetic parameters concerning the phenotypic character(s) and the associated molecular markers.

The theory described here and in Lande and Thompson (1990) is an extension of earlier work in animal breeding by Neimann-Sorensen and Robertson (1961) and Smith (1967) who developed selection indices to incorporate loci with known, direct effects on a character of interest. Current approaches instead rely on non-random associations (linkage disequilibria) between alleles at molecular marker loci and QTLs.

The efficiency of MAS relative to traditional breeding methods is analyzed here for individual plant selection, pedigree selection, and recurrent selection programs. If a large fraction of the additive genetic variance in a character can be explained using the molecular markers, MAS can produce substantial increases in selection response, although this may entail working with small family sizes or individual plant selection. MAS may be especially useful in the early generations of pedigree selection, and in selection for characters that are difficult or expensive to measure. The cost of MAS is currently prohibitive for most commercial applications, although it is likely to become more economical in the near future.

LINKAGE DISEQUILIBRIA AND NUMBER OF MARKER LOCI

The sources of linkage disequilibria most important for MAS are random genetic drift and hybridization between genetically different lines. Random genetic drift is likely to create substantial linkage disequilibrium between a pair of loci only if their recombination rate, r, is less than about $1/(4N_e)$, where N_e is Wright's effective population size (Hill and Robertson, 1968). Following hybridization between two inbred lines, in a large population t generations after the F_2 or S_1 generation, the expected linkage disequilibrium between a pair of selectively neutral loci with r less than about 10 cM is proportional to approximately e^{-rt} under random mating, or to $e^{-2r(1 - 1/2^t)}$ under complete self-fertilization.

Except in populations of very small effective size, hybridization is, therefore, generally a more powerful mechanism for producing linkage disequilibria than is random genetic drift. However, recombination gradually erodes the initial linkage disequilibria produced by hybridization. In a genome with a total recombination map length of L Morgans, the minimum number of randomly placed molecular markers to make likely the detection of substantial linkage disequilibria with QTLs located at any position is roughly $2Lt +$ c under random mating, or $4L(1 - 1/2^t) +$ c under complete selfing. The factor c represents the haploid (gametic) number of chromosomes. These formulas, which neglect the influence of selection, are evaluated for typical crop plant genomes in Table 1.

Table 1 Minimum numbers of molecular marker loci needed for the likely detection of substantial linkage disequilibria with QTLs at different times after crossing two inbred lines in typical diploid, tetraploid and hexaploid crop plants.

Map length (Morgans)	No. of chromosomes	Breeding system	Generations since hybridization		
			1	*5*	*10*
10	10	Outcrossing	30	110	210
20	20	Outcrossing	60	220	420
30	30	Outcrossing	90	330	630
10	10	Selfing	30	49	50
20	20	Selfing	60	98	100
30	30	Selfing	90	146	150

Natural or artificial selection for epistatic gene combinations can also produce linkage disequilibria. Artificial selection on a combination of molecular and phenotypic information is expected to decrease the useful associations between marker loci and QTLs by fixing marker alleles and by altering linkage disequilibria. The action of recombination and selection may necessitate re-evaluation of the associations between alleles at marker loci and QTLs every few generations.

INDIVIDUAL SELECTION IN AN OUTCROSSED POPULATION

To illustrate the basic principles of MAS, consider individual plant selection to improve a single character in an open-pollinated population or the F_2 or S_1 generation of a cross between two inbred lines. Suppose that the individual phenotypes and the allelic states at a number of molecular marker loci are measured on a sample of individuals from the population.

From the classical assumption of small gene frequency changes each generation, the phenotypic selection response is based on the additive effects of QTLs. The extent to which these are associated with variation at molecular marker loci can be estimated by multiple regression of individual phenotypic value, z, on the number of copies (0, 1 or 2) of particular alleles at the polymorphic marker loci. For diallelic marker loci, one allele can be chosen arbitrarily to enter the regression; with multiple alleles, all but one can be entered separately (Falconer, 1981; Kempthorne, 1969). If a linkage map of the marker loci is available, a separate multiple regression can be performed for each chromosome, because linkage disequilibria among loci on different chromosomes is likely to be small.

The multiple regression methodology accounts for linkage disequilibria among the marker loci, as well as among the markers and the QTLs to which they are linked. This allows information from more than one marker locus per QTL to be utilized, which is important to maximize the proportion of additive genetic variance in the character explained by the markers, because with recombination any single marker is likely to have only an imperfect correlation with the linked QTLs. Performing a separate multiple regression for each chromosome will help to avoid statistical problems that occur as the number of independent marker alleles approaches the number of individuals in the sample.

Coefficients in the multiple regressions estimate the apparent additive effects on the character associated with alleles at the marker loci. Without loss of information, these coefficients can be combined into a net molecular score, m, for each individual, which is the sum of the additive effects associated with all of its marker alleles (Lande and Thompson, 1990). Inclusion of only those marker alleles having coefficients that are statistically significant at some level will reduce the noise from marker loci containing no real information, but will tend to overestimate the additive effects associated with the remaining marker alleles. Such biases can be avoided if marker alleles used for selection are chosen *a priori* based on information from previous generations, and their effects are reestimated from current data (Lande and Thompson, 1990).

The phenotype, z, and the net molecular score, m, of an individual can be combined into the linear selection index

$$I = b_z z + b_m m \tag{1}$$

where the relative weights, b_z and b_m, are chosen to maximize the rate of improvement in the mean phenotype. Because the net molecular score has no intrinsic economic value, the relative weights are

$$b_m/b_z = (1/h^2 - 1)/(1 - p) \tag{2}$$

where h^2 is the individual plant heritability of the character and p is the proportion of the additive genetic variance explained by the marker loci. Thus, the net molecular score in the selection index is weighted heavily for characters with low heritability and when the proportion of additive genetic variance explained by the markers is high (Neimann-Sorensen and Robertson, 1961; Smith, 1967; Lande and Thompson, 1990).

The efficiency of MAS on this index can be expressed relative to traditional phenotypic selection of the same intensity. Assuming that both the phenotype, z, and the selection index, I, are normally distributed, a given proportion of the population saved in selection corresponds to the same intensity of selection in both cases, and the relative efficiency is

$$[p/h^2 + (1 - p)^2/(1 - h^2p)]^{1/2} \tag{3}$$

(Smith, 1967; Lande and Thompson, 1990). The first part of this expression, $(p/h^2)^{1/2}$, is the efficiency of selection only on the markers, relative to that only on the phenotype. The remaining part represents the additional gains made from including the phenotype in the selection index along with the net molecular score. For characters with low heritability it can be seen that when **p** is at least comparable to h^2, substantial increases in efficiency can be achieved by MAS. The maximum relative efficiency, achieved when **p** = 1, is 1/h (see Figure 1). Evidently, for characters with low heritability, the efficiency of MAS can be quite large compared to traditional phenotypic

Figure 1 Efficiency of marker-assisted selection on individuals, using the optimal selection index combining molecular and phenotypic data, relative to traditional phenotypic selection.

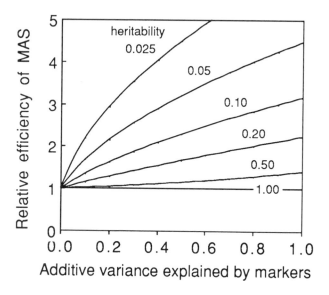

selection, if a large fraction of the additive genetic variance can be associated with the molecular marker loci.

ADDITIVE GENETIC VARIANCE EXPLAINED BY MARKERS

The ability to detect QTLs depends on the availability of marker loci in linkage disequilibria with them, as discussed above. The proportion of additive genetic variance in the character explained by the marker loci also depends on the magnitude of effects of the QTLs and the population sample size (number of individuals) used to estimate associations with the markers.

Suppose that each QTL contributes only a small fraction of the total phenotypic variance, and that an adequate number of closely linked marker loci are available to account for nearly all of the additive genetic variance contributed by a particular QTL. A particular QTL is likely to be detected at the α level of significance if the proportion it contributes to the additive genetic variance in the character is greater than $4x_\alpha^2/h^2N$ (Smith, 1967; Lande and Thompson, 1990). Here, N is the sample size, and x_α is the number of standard deviations above the mean of a normal distribution needed to achieve the α level of significance (e.g., $x_{0.05} = 1.65$ and $x_{0.01} = 2.33$). Thus, rather large samples may be needed to detect QTLs in characters with low heritability.

Genetic variation in quantitative characters generally is caused by a few loci with relatively large effects and many loci having progressively smaller effects (Thompson, 1975; Edwards *et al.*, 1987; Paterson *et al.*, 1988; Shrimpton and Robertson, 1988). Lande and Thompson (1990) developed a model of a geometric distribution of QTL effects to analyze the proportion of the total additive genetic variance that could be detected using a large number of marker loci in a sample of a given size. They assumed that the several QTLs of largest effect were either unlinked or near linkage equilibrium with each other. If each successive QTL contributes an additive genetic variance that is a fraction, a, of the previous QTL in the series, an effective number of quantitative genetic factors can be defined as $n_E = (1 + a)/(1 - a)$. This is analogous to Wright's (1968) effective factor number but is based on a formula for additive genetic variance within, rather than among, populations (Lande, 1981).

With realistic values of n_E in the range of about 2 to 10, and individual plant heritabilities as low as $h^2 = 0.1$, Lande and Thompson (1990) showed that sample sizes, N, on the order of a few hundred to a few thousand individuals would be required for the marker loci to explain a substantial fraction of the additive genetic variance for a character (the larger samples being required for characters with high n_E and low h^2). This analysis is based on the assumption of random sampling of individuals on which phenotypic and molecular data are obtained. An alternative procedure for reducing sample size, suggested by Lander and Botstein (1989), is to score the molecular markers only on individuals with extreme phenotypes, which reduces the effort in mapping QTLs. However, this also produces overestimates of QTL effects, which should be avoided when using a selection index for MAS (as above) because, unless the bias is corrected, it would place too much weight on the molecular information and decrease selection efficiency.

PEDIGREE SELECTION

Much of the commercial plant breeding is based on the method of pedigree selection, in which elite inbred lines are crossed and then self-fertilized for several generations to produce a large number of recombinant inbred lines that are tested to select a new set of elite inbreds. Evaluation in the early generations (S_1 and S_2) is usually weak or absent, often being exerted by culling of the worst families based on visual criteria. Pedigree selection methods are thought to give the best short-term responses to selection (Hallauer, 1981).

Accurate testing of inbred lines can be accomplished by averaging the performance of many plants per line, which increases the heritability among lines. It might, therefore, appear that pedigree selection allows little scope for improving selection efficiency using molecular genetic markers. However, this argument ignores the early generations following a cross. The application of MAS among individual plants would be well suited to increase the accuracy and strength of selection in the S_1 and S_2 generations to augment the non-existent or weak "visual" selection usually practiced. The increased early selection intensity could permit a reduction in the number of lines tested in later generations or to increase the intital number of lines created in the S_1 and S_2 generations.

MAS may be especially useful in selecting for characters that are difficult or expensive to measure, such as drought or pest resistance. Exposure of the S_1 (or F_2) generation to the appropriate environment would allow associations between the marker loci and QTLs to be evaluated. MAS could be practiced on individuals in the S_1 generation, and, perhaps more importantly, selection on the markers alone could be performed in a normal environment, that is in the absence of drought or pests, at the seedling stage in the S_2 generation or among mature plants in the S_2 and/or S_3 generations. The cost of scoring the molecular markers would be offset by the savings from the relative ease of selection on markers and/or from testing a reduced number of lines in later generations.

It is also worth noting from Table 1 that, in a cross of two inbred lines with continued self-fertilization, the number of marker loci needed to explain a substantial fraction of the genetic variance is not very large; hence, a sparse linkage map with a recombination distance of about 10 cM between adjacent markers should be sufficient. However, to fully utilize the linkage disequilibria generated in pedigree selection, associations between the marker loci and the QTLs will have be assessed separately for each independent cross between inbred lines.

Even in the later generations of pedigree selection, when selection is focused on distinguishing genetic values among a set of improved inbred lines, there is an opportunity for MAS to substantially increase selection efficiency, if breeders are willing to consider selection among a larger number of inbred lines using smaller numbers of plants per line. To illustrate the possibilities, let us analyze the following situation, which might reflect a typical breeding program for self-fertilizing crops such as wheat (*Triticum aestivum* L.), barley (*Hordeum vulgare* L.), oats (*Avena sativa* L.), or rice (*Oryza sativa* L.) in a single environment.

Suppose that facilities are available to test 50,000 plants and that the best 50 inbred lines will be selected. We can then consider situations where the number of plants per line ranges from 1 to 1000, and the corresponding number of inbred lines ranges from 50,000 to 50. With a given number of plants tested and a given number of lines selected, increasing the number of plants per line increases the heritability estimates among lines but decreases the selection intensity. There is, therefore, an optimum number of plants per line to maximize the response to selection. This is a classical problem in selection theory (Robertson, 1957, 1960; Wricke and Weber, 1986). The optimum number of plants per line depends on a combination of the total number of plants tested, the number of lines selected, and the heritability of the trait among individual plants in different inbred lines. Including information from molecular markers in a selection index for MAS could, of course, change the optimal number of plants per line. Formulas for the calculations appear in the Appendix.

To represent the situation where previous culling has already eliminated the worst inbred lines, we suppose that the heritability of the character among individual plants from different inbred lines is low, $h^2 = 0.2$. Figure 2 shows that under traditional phenotypic selection a nearly maximal selection response is achieved by testing 5 to 20 plants per line. At family sizes of 20 or larger, there is indeed little room for improvement using MAS. Nevertheless, if a large fraction of the total genetic variance among lines can be associated with the molecular marker loci, then MAS at the individual plant level (or with a small number of plants per line) can substantially increase the selection response.

RECURRENT SELECTION

The original form of recurrent selection for hybrid yield in crops such as maize (*Zea mays* L.) occurs in cycles of three generations (Hull, 1945). In the first generation, individuals in a population are (1) crossed to an inbred tester and (2) self-pollinated. In the second generation, parental plants are

Figure 2 Response per generation to selection among inbred lines, plotted as a function of the number of plants per line, **n**, for various values of **p**, the proportion of the total genetic variance explained by molecular marker loci. Parameters are individual plant $h^2 = 0.2$, and total plants tested = 50,000, with the best 50 lines selected. The selection response is scaled in units of individual plant phenotypic standard deviations for the population as a whole.

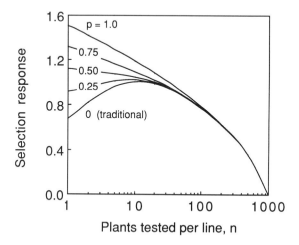

selected based on the performance of their hybrid progeny. In the third generation, selfed seeds from the selected parent are grown and intercrossed to produce the parents for the next cycle. Reciprocal recurrent selection (Comstock *et al.*, 1949) involves simultaneous improvement of hybrid performance between two populations, samples from each of which are used as testers for the other. Recurrent selection methods are thought to give the best long-term responses to selection (Hallauer, 1981).

Let us analyze a program of recurrent selection with an inbred tester in a single environment. The following example provides a simplified description of a long-term selection experiment for yield in maize, which has been conducted for many years at Iowa State University (Hallauer *et al.*, 1988). This actually is a reciprocal recurrent selection experiment; but the results of our simplified analysis, nevertheless, reveal the range of possibilities for conventional selection and MAS methods.

Assume that a total of 15,000 hybrid plants are tested per cycle, and that the best 15 parental plants are selected. The heritability of individual plant yield is about $h^2 = 0.2$, and the typical test-cross family size actually used in a single environment is 50, so that about 300 parental plants are tested per cycle (K. Lamkey, 1991, pers. comm.). The response per cycle as a function of test-cross family size, using a selection index for MAS, is illustrated in Figure 3 and details of the calculations appear in the Appendix.

Figure 3 Response per cycle of recurrent selection with an inbred tester, plotted as a function of the test-cross family size, **n**, for various values of **p**, the proportion of total genetic variance in the performance of individual hybrid progeny explained by molecular marker loci. Parameters are individual plant h^2 = 0.2, and total hybrid progeny tested per cycle = 15,000, with the best 15 parental plants selected per cycle. The selection response is scaled in units of individual plant phenotypic standard deviations in the outbred parental population.

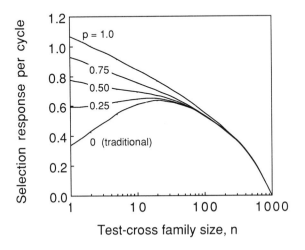

Under traditional phenotypic selection, a nearly maximal response to selection is achieved with test-cross family sizes in the range of 10 to 50 (Figure 3), which includes the numbers actually used in the Iowa experiments. If a large fraction of the genetic variance in hybrid performance among individual progeny can be explained by association with molecular markers inherited from the outbred parents, then MAS on small test-cross families (as small as a single individual) can produce a substantial increase in the selection response per cycle.

DISCUSSION

The traditional plant breeding methods of pedigree selection and recurrent selection involve phenotypic selection among inbred lines, or among outbred parents based on test-cross performance. For characters of low heritability (such as yield) in experiments with a given total number of plants measured, there is an optimum range of family sizes that maximize the response per generation or cycle of selection. With typical family sizes used in breeding programs for crop plants, from about 15 to 50 individuals per environment, the heritability of family mean phenotypes is usually intermediate or high, leaving little opportunity for increasing selection efficiency through the use of molecular genetic markers.

However, if sufficient linkage disequilibria between marker loci and QTLs are created by hybridization between inbred lines or by random genetic drift, then the markers can account for a large proportion of the genetic variance in a character, effectively increasing its heritability among individuals. In this situation, MAS among individual plants (or among small families), using a selection index combining molecular and phenotypic information, can produce substantial increases in selection efficiency in comparison to traditional breeding methods (see Figures 2 and 3).

MAS also creates opportunities for increasing selection intensity in the early generations of pedigree selection, when traditionally either there is no selection or weak "visual selection" is applied to eliminate the worst lines. MAS may be especially useful in pedigree selection for characters that are difficult or expensive to measure, such as drought or pest resistance. Associations between marker loci and QTLs can be established in the drought or pest environment during the S_1 generation. In subsequent generations, selection can be performed on seedlings or adults that are grown in a normal environment, thus reducing the number of derived lines per cross to be tested in stress environments during later generations.

To explain a large fraction of the genetic variance using molecular markers, it will usually be necessary initially to score 100 to a few hundred marker loci on many individuals. For annual crops grown in the U.S.A., the yearly cost of traditional breeding programs ranges from about \$0.05 to \$0.50 per plant for various species. In contrast, the cost of scoring molecular markers ranges from roughly \$0.10 to \$1.00 per marker locus per plant, depending on the type of marker. Thus, the current cost of scoring one hundred or more molecular marker loci is on the order of 100 to 1000 times as expensive as measuring standard phenotypes in most crops. The expense of performing molecular genetic assays on this scale is clearly prohibitive for most commercial applications. Although it can be anticipated that assays of molecular markers will become much cheaper within the next several years, economic considerations will continue to play a major role in practical applications of MAS.

The expense of scoring molecular markers dictates that the initial applications of MAS in commercial plant breeding will be in situations where the largest increases in selection efficiency are expected, for characters that are difficult or expensive to measure, or in species where individual plants have high value.

ACKNOWLEDGMENTS

I thank P. S. Baenziger, A. R. Hallauer, K. R. Lamkey, L. Mets, H. T. Stalker, E. A. Wernsman and A. M. Wood for helpful discussions. This work was partially supported by U.S. Public Health Service grant GM27120.

REFERENCES

Comstock, R. E.; Robinson, H. F.; Harvey, P. H. (1949) A breeding program designed to make maximum use of both general and specific combining ability. *Agronomy Journal* 41:360-367.

Edwards, M. D.; Stuber, C. W.; Wendel, J. F. (1987) Molecular-marker-facilitated investigations of quantitative-trait loci in maize. I. Numbers, genomic distribution and types of gene action. *Genetics* 116:113-125.

Falconer, D. S. (1981) *Introduction to quantitative genetics.* 2nd ed. Longman, London.

Hallauer, A. R. (1981) Selection and breeding methods. In: Frey, K. E. (ed.), *Plant breeding II.* Iowa State University Press, Ames, pp. 3-55.

Hallauer, A. R.; Russell, W. A.; Lamkey, K. R. (1988) Corn breeding. In: *Corn and corn improvement.* Agronomy Monograph no. 18, 3rd ed. American Society of Agronomy, Crop Science Society of America, and Soil Science Society of America, Madison, WI, pp. 463-564.

Helentjaris, R.; Slocum, M.; Wright, S.; Shaeffer, S.; Nienhuis, J. (1986) Construction of genetic linkage maps in maize and tomato using restriction fragment length polymorphisms. *Theoretical and Applied Genetics* 72:761-769.

Hill, W. G.; Robertson, R. (1968) Linkage disequilibrium in finite populations. *Theoretical and Applied Genetics* 38:226-231.

Hull, F. H. (1945) Recurrent selection for specific combining ability in corn. *Journal of the American Society of Agronomy* 37:134-145.

Kempthorne, O. (1969) *An introduction to genetic statistics.* Iowa State University Press, Ames.

Lande, R. (1981) The minimum number of genes contributing to quantitative variation between and within populations. *Genetics* 99:541-553.

Lande, R.; Thompson, R. (1990) Efficiency of marker-assisted selection in the improvement of quantitative traits. *Genetics* 124:743-756.

Lander, E. S.; Botstein, D. (1989) Mapping Mendelian factors underlying quantitative traits using RFLP linkage maps. *Genetics* 121:185-199.

Neimann-Sorensen, A.; Robertson, A. (1961) The association between blood groups and several production characters in three Danish cattle breeds. *Acta Agriculturae Scandinavica* 11:163-196.

O'Brien, S. (1990) *Genetic maps. Book 6. Plants.* Cold Spring Harbor Laboratory Press, Cold Spring Harbor, NY.

Paterson, A. H.; Lander, E. S.; Hewitt, J. D.; Peterson, S.; Lincoln, S. E.; Tanksley, S. D. (1988) Resolution of quantitative traits into Mendelian factors by using a complete linkage map of restriction fragment length polymorphism. *Nature (London)* 335:721-726.

Robertson, A. (1957) Optimum group size in progeny testing and family selection. *Biometrics* 13:442-450.

Robertson, A. (1960) On optimum family size in selection programmes. *Biometrics* 16:296-298.

Shrimpton, A. E.; Robertson, A. (1988) The isolation of polygenic factors controlling bristle score in *Drosophila melanogaster*. II. Distribution of the third chromosome bristle effects within chromosome sections. *Genetics* 118:445-459.

Smith, C. (1967) Improvement of metric traits through specific genetic loci. *Animal Production* 9: 349-358.

Soller, M.; Beckmann, J. S. (1983) Genetic polymorphism in varietal identification and genetic improvement. *Theoretical and Applied Genetics* 67:25-33.

Stuber, C. W.; Goodman, M. M.; Moll, R. H. (1982) Improvement of yield and ear number resulting from selection at allozyme loci in a maize population. *Crop Science* 22:737-740.

Thompson, J. N. (1975) Quantitative variation and gene number. *Nature (London)* 258:665-668.

Wricke, G.; Weber, W. E. (1986) *Quantitative genetics and selection in plant breeding*. Walter de Gruyter, New York.

Wright, S. (1968) *Evolution and the genetics of populations. Vol. 1. Genetic and biometric foundations*. University of Chicago Press, Chicago.

APPENDIX

Pedigree selection

Consider a population of inbred lines. Assume that all variation within lines is environmental or developmental noise, and that there are no consistent environmental differences among lines. Let the heritability of a trait among individual plants (in the case of one plant per line) be denoted as h^2. The heritability of line mean phenotypes, denoted as H^2 with n plants per line, can be written as

$$H^2 = nh^2/[(n - 1)h^2 + 1] \qquad (A1)$$

The ratio of the variance among lines (with n plants per line) to the individual variance among lines (with one plant per line), signified as v^2, is

$$v^2 = h^2 + (1 - h^2)/n \qquad (A2)$$

The optimal selection index for MAS among inbred lines has the same form as that for individual selection in text equations (1) and (2), but with H^2 in place of h^2 and with p denoting the fraction of the total genetic variance among lines explained by the marker loci. When scaled in units of the phenotypic standard deviations among individual plants (with one plant per line) the expected response to MAS among lines of size n is

$$ivH^2[p/H^2 + (1 - p)^2/(1 - H^2p)]^{1/2} \qquad (A3)$$

in which i is the standardized selection differential among families. For a character with a normal distribution among families, $f(x)$, scaled to unit variance, i equals the height of the unit normal curve at the truncation point, f(y), divided by the corresponding proportion saved in selection, **Q**. In a population of **N** inbred lines with **n** plants per line, if the total number of plants measured and the number lines selected are given, say **k** lines selected from a total of nN = **k** x **1000** plants, then **Q** = k/N = n/1000 and

$$i = 1000 \ f(y)/n \qquad\qquad (A4)$$

For any value of **n** the corresponding value of **Q**, the truncation point **y**, and the height **f(y)** can be found in tables of the standard normal distribution. The discussion of Figure 2 specifies that **k** = 15, but the results are essentially independent of **k**, provided **k** > > 1.

Recurrent selection

The trait under selection in an outbred parental population is the performance of their progeny in crosses with an inbred (homozygous) tester strain. The hybrid progeny phenotype may be due in part to dominance interactions between alleles from the outbred parents and the inbred tester. Nevertheless, neglecting epistasis, all of the genetic variance within the outbred parental population for the performance of their hybrid progeny is effectively additive, because at any locus the progeny contain only one allele from an outbred parent against a constant genetic background from the inbred tester. Assuming that the QTLs in the parental population are in Hardy-Weinberg equilibrium and linkage equilibrium among themselves, with a very large (hypothetically infinite) number of progeny per test-cross, half the total genetic variance in test-cross performance by individual progeny is among parents and half is within parents.

The heritability of the average performance of **n** progeny per test-cross, denoted as H^2, can be written as

$$H^2 = nh^2/[(n - 1)h^2 + 2] \qquad\qquad (A5)$$

in which h^2 is the total genetic variance divided by the total genetic plus environmental variance. For characters in which the genetic variance in parental performance itself is largely additive, such as individual plant yield in maize (Hallauer, 1981), h^2 can be roughly equated to the conventional heritability of the individual plant character within the parental population.

The optimal selection index for marker-assisted recurrent selection has the same form as in text equations (1) and (2) but with H^2 instead of h^2 and with **p** denoting the proportion of the total genetic variance in the performance of individual hybrid progeny explained by the markers. The expected response to MAS among parents based on the average performance of **n** progeny per test-cross, scaled in units of the individual phenotypic standard deviation (square root of the total genetic variance plus environmental variance) is

$$i(hH/\sqrt{2})[p/H^2 + (1 - p)^2/(1 - H^2p)]^{1/2} \qquad (A6)$$

If a total of $k \times 1000$ hybrid plants are tested and the best k are selected, then i is the same as in formula (A4). The discussion of Figure 3 specifies that $k = 50$, but the results are essentially unchanged for any value of $k \gg 1$.

Discussion

Arthur K. Weissinger, Moderator

Is it possible that some soil microorganisms which have beneficial effects when applied to crop "A" will have dramatic negative effects on crop "B" which is planted at the same location the next growing season?

I could never say it's impossible, but I don't recall any examples. Certainly this situation would be discovered during testing and would end the use of the microorganism as an experimental/development tool or as a product.

[W. J. Brill]

As a corollary to the above question, is it common to test on multiple plants (i.e., more than one species) if you are developing an inoculant, for example?

It depends on what you are really after. If you are after something that is antagonistic to a certain pathogen that hits a number of plants, then a good initial first experiment may be multiple testing. In my opinion, it would be better to pick a target plant and work with it. Later you can determine if the microbe works on other plant species. [W. J. Brill]

What is the status of work to enhance the *Rhizobium*/legume symbiosis via genetic engineering of plants?

I can't think of any programs where people are trying to enhance the system by manipulating the plant. There certainly has been a variety of efforts where the microorganism has been modified through *nif* genes, glutamine synthetase genes, etc. As far as I know, nothing has been shown to be very useful in the field. It's still difficult to engineer plants, and there isn't much known about genes that are involved. However, one can imagine manipulating leghemoglobin genes, perhaps, by increasing the amount or adding different species to the plant. These are things that are certainly going to be tried in the future. The genes are being discovered and isolated, but I don't know any that have been put into plants. [W. J. Brill]

Have you given any thought to the side effects of over-expression of chitinases in leguminous plants, as it might affect the *Rhizobium* symbiosis?

Because it is such a tight symbiosis and you must deal with it in nature, you have to be very careful about over-expression of an inocuous gene. Over-expression obviously takes energy away from the plant, or from the microorganism, and can debilitate its ability to survive or to be competitive.

[W. J. Brill]

As the number of individuals per family goes down, does that imply that the number of families goes up? That is, if efficiency is maximum at one per family, do you need 15,000 independent families?

That was the assumption for the numerical example where the constraint of dealing with 15,000 total individuals was used. However, it's not necessarily the case. It may be that one could avoid a lot of the expense by considering an even broader range of possibilities. If one goes to individual plant marker-assisted selection, it may be possible to obtain essentially the same amount of gain if one uses only 1000 individuals instead of some form of family selection. There are many possibilities that have yet to be explored.

[R. Lande]

There are *Rhizobium* strains with enhanced motility. Do these have promise as improved inoculants that will better infect the entire root system?

There were several papers a number of years ago that indicated that motility was important. But, I think the wealth of the data now says that even non-motile mutants are just as effective as motile ones. Organisms can move without motility in the soil (without flagella) - for example, water movements.

[W. J. Brill]

It seems that Dr. Lande has set up a "straw man" in that most of the comparisons are based on a complete substitution of marker-assisted selection for field analysis; and, yet, I don't think I or a lot of other people would argue for that case. Rather than a "pick-the-winner" approach based entirely on markers, I would see this technology as being a "get-rid-of-the-losers" scheme where you alternate molecular with field selection to pick the winners. What do you think of this idea in light of the calculations you made?

I don't see any advantage in separating things in that way. I consider molecular markers to be just like another character that's used to increase the efficiency of selection. People selecting for yield often consider component characters or correlated characters to improve yield, and this is no different. In that context, I don't think a change in strategy is necessary.

What I'm advocating is a type of index selection that uses both molecular and phenotypic information. This information must be obtained in the field under appropriate environments. I'm not in any way suggesting that markers would be a substitute for evaluation and field testing.   [R. Lande]

For most quantitative traits, the plant breeder's problem is rarely generating adequate levels of variability, but rather to identify specific desirable gene combinations. Marker-assisted selection assumes that you can associate molecular markers with the trait in question - for example, yield. What would be the practicality of doing this in the near future, especially considering costs? Secondly, because many of our crop species are polyploids, will QTLs ever be useful in this group of plants?

I think polyploids are amenable to marker-assisted selection, only it will take a considerably larger amount of work to do the mapping and to distinguish the duplicate loci from each other. The cost of scoring one molecular marker using current technology usually exceeds the cost of growing the plant and scoring its phenotype. Considering that you would usually need to score hundreds of markers per plant to apply these techniques, this is not going to be economically feasible for most companies or university programs. Now, if we were dealing with some nationalized industry for corn improvement, where it is all under one economic system, then I think even a small gain in yield would well offset the cost of even a few million dollars that it might take to apply this technology. But, that isn't the current situation, at least in the United States. In contrast, many animal breeding programs in other countries have this situation because animals are more expensive on a per-individual basis than plants. The animal breeders are somewhat behind because animals have larger genomes (with the exception of several polyploid plants) and less capital input from industry. But, there are indications that within several years the cost of scoring RFLPs or other types of polymorphisms may go down by a factor of a hundred or even a thousand. The cost of obtaining the molecular information may then be comparable to getting the phenotypic information on a per-plant basis. In that case, I think it will be feasible and economically valuable even for middle-sized companies to begin employing this type of selection. [R. Lande]

It almost sounded like you were talking yourself out of a job when you were answering the question. When you look at something like yield that could have 50 or 100 genes, should we really be targeting traits like yield with QTLs, or should you be targeting quantitative traits for insect or disease resistance, where there may be a major impact or where it may be easier to select phenotypes by using QTLs?

Yield is the ultimate character of interest, of course, but yield is also a low heritability trait. There are many difficulties with scoring traits like insect or pest resistance. The economic factors have to be figured in with the heritability of the character. I don't see any reason why yield can't be a target for this type of technology. If we're talking about resistance, then in the right environment it may have a higher heritability and be easier to detect associations with QTLs. Thus, it may be better, at least in the initial stages when we're just getting over the hump of having things being economically feasible, to work with low to middle heritability traits instead of the really low ones. Only when it becomes truly cheap and easy to score large numbers of these markers will we go for the really low heritability traits.   [R. Lande]

What is the potential of *Pseudomonas fluorescens* as an antagonistic agent against plant pathogenic organisms?

I think the potential is pretty high. There are many examples of *P. fluorescens* strains that have been isolated and a couple of them have been commercialized. Again, I think consistency is the problem with them and perhaps use of a mixture of strains will be necessary. *Pseudomonas fluorescens* is an organism that's commonly found on roots in relatively high numbers. A variety of techniques have been used to isolate various useful strains to plants. I think there is potential for it, but most people have given up on these strains.   [W. J. Brill]

I have talked with Steve Lindow about his "Ice-Minus" work. One of the greatest problems with the work is achieving persistence and repeatable efficacy. What kinds of experiments speak to that difficulty?

There are certainly many possibilities. To achieve persistence, one can select just on the plants in the field to continually get better and better strains. In fact, *Rhizobium* strains used to be improved just by cycling them through the soil to improve competitiveness. Other ways of making them potentially more persistent include growing strains in different kinds of media of sucrose, mannitol, and a variety of other compounds that have been shown to stabilize organisms. The literature has many examples of things added to a microorganism that will allow it to survive desiccation, freezing, etc. a little longer. There's a whole field that should be re-opened.
                                          [W. J. Brill]

In your simulations where you had many QTLs with small effects, did you determine the frequency of the total number of QTLs that you were able to detect?

We haven't done any simulations. These are all analytical calculations, and what we have assumed in much of the chapter was that one had as many QTLs available as could be desired, thousands, let's say. Then, we were asking questions like: "What is the maximum amount of additive genetic variance you could explain with a given sample size?" We were attempting only to outline the limits of possibilities, but a lot of people are starting to do simulations of this kind. This is a necessary addition to the sort of calculations that we were doing which may help to guide the types of things one needs to simulate. I can't answer questions about details of gene frequencies at this time. One point that should be made clear, however, is that we're assuming that you could use multiple markers associated with a single QTL in order to get the maximum explanatory power. Especially if one is dealing with situations where the linkage disequilibrium is produced by random genetic drift, even very tightly linked markers may be only partially associated with a given QTL. By using multiple regression or multivariate methodology you can explain a larger fraction of the variance. If you have enough very tightly linked markers that have associations by random drift, then you should, in theory, be able to explain all of the additive genetic variance associated with the QTL even if any one of the markers explains only a certain fraction. This is assuming a very large sample size of individuals in the population.   [R. Lande]

Why do you assume that QTL effects follow a geometric progression in additive effects?

We're not assuming purely additive effects. What I was dealing with in this theory is the response to a single generation or cycle of selection. Classical theory tells us that the response is generally based on additive effects unless we're changing the gene frequency at any one locus by a very large amount (which may be the case if we are just picking out one or two QTLs of a relatively large effect). I was assuming that the additive fraction of the variance associated with each QTL fell off in a geometric series and that is just a rough approximation that seemed to fit much of the data coming from QTL mapping, and earlier data or thoughts by people like Sewall Wright. Generally there's going to be a large number of loci that have larger effects and some that have smaller ones. Something like a geometric series roughly approximates a wide range of different situations in which you would either have a small number of additive factors - that is, the first factor explains some large fraction - and successive factors explain some much smaller fraction, possibly, or the effects of the factors may be much more similar. That encompasses a wide range of possibilities. One situation that is not encompassed is when one factor has a major effect and then 10 additional factors all have equal but much smaller effects (this seems to be fairly unrealistic). With just the simplest of realistic models, the data reasonably represent the range of results that one has to be considering in dealing with characters of different degrees of complexity.   [R. Lande]

What is the status of research on microorganisms for nitrogen fixation in non-legumes?

It's not very active. First, *nif* genes were found to be all clustered together. There was certainly a lot of activity in France, Israel, and the United States trying to put this cluster into plants and yeast, but nothing has ever been accomplished. As we learn more about these genes, we realize that just putting the gene into a plant isn't sufficient. You must also worry about the physiology and metabolism of the system. When all of these genes have been put into yeast, there wasn't a trace of nitrogen fixation. That's partially to be expected because of regulatory regions.

Years ago I tried to get *Azotobacter* (which is free-living and doesn't enter into any kind of symbiosis) to somehow associate with corn. Hundreds of corn lines were used to see if corn would supply carbon to the *Azotobacter* to give energy for nitrogen fixation. The association was heritable and we could put the gene into commercial lines (the level of N fixed was about 1% of the total plant's N if you could extrapolate onto a mature plant). When the project was finished, it showed breeding potentials for forming these kinds of associations. This program could have been pursued further by breeding to select *Azotobacter* for binding and growing along the corn roots.   [W. J. Brill]

About a year ago, Ted Cocking gave a seminar in which he showed a wheat plant with a single nodule hanging from it. Are you aware of any progress for getting nodulation in non-legumes?

An ex-postdoc of mine from China has also found nodules on rice by using mutated *Rhizobium*, but there's no evidence that they fix any nitrogen.
                          [W. J. Brill]

In a crop like maize where you find a high level of polymorphisms for isozymes and RFLPs, do you see the need to use the system like RAPD in your future analysis? Why do we need that?

There are certain concessions you make when you move to something like RAPD, but there is an advantage that it's a plus/minus-type analysis which might remove the need for electrophoresis, radioactivity, etc. On the other hand, you also move to a two-allele system which causes a problem with the plus/minus condition. I think in maize it's particularly a problem because we have so many alleles that we rely on that for much of the power of our analysis. For some other species which only reveal two polymorphisms at

most loci, it's not much of a problem. But, I think that's something we have
to be aware of. I'm really glad we're trying these things because we clearly
need to get to a replacement technology that is at least non-radioactive and,
hopefully, non-electrophoretic.   [T. G. Helentjaris]

**While the linear relationship between percent RFLP heterozygosity
and yield of corn hybrids was strong over a wide range of heterozygosity values, the two would be much less strongly related in the
range of 80% or greater. In that range, is there evidence that specific allelic combinations are important?**

Most of the published information on mapping to date isn't for yield *per se*,
but I think that it is about to come out. There are definitely chromosome
segments that are associated with higher yield. As T. Helentjaris was
saying, if you had some sort of a multivariate index, you could get a tighter
association or better explanation of the amount of heterosis or yield. I think
that a more detailed level of analysis is going to prove useful for practical
applications. The graph of yield vs. genetic similarity averaged across many
marker loci is essentially the same as classical graphs that have been produced for inbreeding depression. Yield is a function of inbreeding. In that
case, you're really just using markers as a kind of substitute for the probability of identity by descent which is the inbreeding coefficient. In that sense,
averaging is losing all of the really useful information and one has to consider these markers separately to make use of their maximum possibilities.

[R. Lande]

I agree completely with Dr. Lande. The idea that you could plot yield vs.
dissimilarity and get a straight line is diametrically opposed to the idea that
there might be major loci. The fact that it breaks down fits with the evidence that we do find major loci. I think the most revealing thing about that
curve is not necessarily that we would use it predictively, but to provide
insight about the makeup of hybrid yield.   [T. G. Helentjaris]

PART FIVE

STRATEGIES FOR UTILIZING UNADAPTED GERMPLASM

Chapter 21
Varied roles for the haploid sporophyte in plant improvement

Earl A. Wernsman

Department of Crop Science, North Carolina State University, Raleigh, NC 27695-7620

Inbreeding procedures are of primary importance to most plant breeding programs. These processes result in pronounced effects on the magnitudes of additive genetic variances in plant populations and permit breeders to partition additive genetic variation into family structures. Total additive genetic variance in a diploid population increases with inbreeding as a function of $1+F$, where F is the coefficient of inbreeding and the additive genetic variation can be partitioned into among families and among plants within families components whose coefficients are 2F and 1-F, respectively. Under completely inbred conditions, genotypes are precisely repeatable, heritabilities of quantitative traits on family bases are increased, and these characteristics lead to improved selection efficiencies. Breeding procedures permitting rapid and efficient development of inbred populations are of great value in plant breeding as evidenced by the popularity of the modified pedigree method or single-seed-descent technique (Brim, 1966) for breeding many self-pollinated plant species. The production of haploids, sporophytic plants with the gametophytic chromosome number for the species, and their subsequent chromosome doubling to produce doubled haploid lines permits complete inbreeding in a single step. Although haploid plants were first reported almost 70 years ago (Blakeslee *et al.*, 1922) and have been observed since in most plant species, only in the last two decades have techniques been developed sufficiently to permit the production of haploid plants in sufficient numbers to be more than genetic novelties. *In vitro* culture of anthers and pollen (Guha and Maheshwari, 1964; Maheshwari *et al.*, 1982), ovule culture (D'Halluin and Keimer, 1986), and interspecific hybridization techniques (Kasha and Kao, 1970; Burk *et al.*, 1979) permit the production of large haploid populations in a limited number of crop species.

Tobacco (*Nicotiana tabacum* L.) has been used as a model system for doubled haploid research. Haploid plants can be produced by *in vitro* anther culture (Bourgin and Nitsch, 1967), by pollination of tobacco females with pollen of *N. africana* Merxmüller and Buttler (Burk *et al.*, 1979), or fertilization of tobacco females with irradiation-impaired pollen (Kumashiro and Oinuma, 1985) of an alien species. Treatments of small haploid plantlets with aqueous colchicine solutions result in a low percentage of chromosome doubling (Burk *et al.*, 1972), or haploid plants may be chromosome-doubled reliably by *in vitro* tissue culture techniques (Murashige and Nakano, 1966; Kasperbauer and Collins, 1972). Doubled haploid genotypes produced from the same sources by these different methods have led to unexpected results. Hence, haploids of this species produced by differing methodologies may be utilized in varied roles for the improvement of this species. It is the purpose of this chapter to review the results and characteristics of haploid and doubled haploid genotypes produced from different nuclei, and to demonstrate that they have different roles in this model organism.

ANTHER CULTURE-DERIVED DOUBLED HAPLOIDS

Bourgin and Nitsch (1967) demonstrated that anthers collected in the mitotic metaphase stage of microspore development could be cultured *in vitro* and induced to form haploid plantlets. In such cultures the vegetative cell of pollen undergoes division and an embryo is formed from the mass of haploid cells (Dunwell and Sunderland, 1975; Sunderland, 1980). Tobacco pollen is dimorphic and consists of large, densely staining, normal or "N" grains and smaller, more opaque "S" grains. It is the latter grains that undergo embryogenesis and haploid plant formation (Horner and Street, 1978).

Initial investigations on the agronomic characterization of anther culture-derived doubled haploid lines (ADHs) were generally conducted on lines produced from plants of conventionally inbred, near-homozygous cultivars (Collins *et al.*, 1974; Burk and Matzinger, 1976). In ADH lines from cultivar Coker 139, Burk and Matzinger reported a mean leaf yield reduction of approximately 10%, and significant genetic variability among the lines. The inferior agronomic performance of ADH lines compared to selfed progenies of near-homozygous cultivars was confirmed in other investigations involving ADH lines and populations developed from both near-homozygous and heterozygous population sources (Arcia *et al.*, 1978; Schnell *et al.*, 1980; Deaton *et al.*, 1986). When first-cycle ADH lines (first cycle of anther culture and chromosome doubling) were again subjected to anther culture and chromosome doubling, additional genetic changes and differences between the second-cycle ADH lines and their first-cycle homozygous ADH parents were noted (Brown *et al.*, 1983; Kasperbauer *et al.*, 1983). The mean reduction in productivity among the ADH lines from the second cycle of anther culture was only slightly lower than that observed from the parental cultivar to the first-cycle ADH lines. Following two cycles of anther culture, ADH lines have been obtained which are only 50% as productive as the original cultivar (Figure 1). Although *N. sylvestris* Speg. and Comes, the maternal parent of allotetraploid tobacco, is not grown as a crop plant, De Paepe *et al.* (1981, 1983) noted similar results with ADH lines from this species and recorded the capacity of additional anther culture cycles to continue to induce genetic changes.

Figure 1 After two cycles of anther culture of an inbred cultivar, ADH lines can be identified whose leaf yields are approximately 50% of those of the cultivar. Partially harvested plots of two second cycle ADH lines from Coker 139 (left and center) and the parental cultivar (right).

Brown and Wernsman (1982) found the genetic changes to be of nuclear origin, and maternal and/or cytoplasmic effects were not of principal importance. A lack of cytoplasmic and maternal effects was also found by Kumashiro and Oinuma (1985). Nevertheless, tobacco anther culture can result in genetic changes in chloroplasts, as Matzinger and Burk (1984) found normal, temperature-insensitive ADH lines when a temperature-sensitive line possessing a genetic lesion in the chloroplast was subjected to anther culture and chromosome doubling.

Heritable nuclear DNA changes have been found in ADH lines of *N. sylvestris* and *N. tabacum* (De Paepe *et al.*, 1983; Dhillon *et al.*, 1983; Reed and Wernsman, 1989). Total nuclear DNA in many (if not all) ADH lines is increased compared to the amount in the parents and this occurs without changes in chromosome numbers. The amplified DNA appears to be in the highly repetitive fraction (De Paepe *et al.*, 1983) in *N. sylvestris*.

Cytological data obtained by Reed *et al.* (1991) suggest that DNA amplification is not totally restricted to specific tobacco chromosomes but may be dispersed throughout the genome. Nevertheless, sites in specific chromosomes which have undergone DNA amplification are evident in meiocytes of

F_1 hybrids between ADH lines and parental cultivars from which they were obtained. One or two atypical quadrivalents were observed when selected ADH lines derived from cultivars Coker 139 and NC 95 were hybridized with the parental cultivars. Figure 2 shows an example of an atypical quadrivalent formed in a microsporocyte of an F_1 hybrid of NC 95 x NC 95 SCDHL 12, a second-cycle ADH line from this cultivar. The chromosomes involved in the amplification process may be the same or different in the different cultivars (Reed *et al.*, 1992).

Figure 2 An atypical quadrivalent formed in a microsporocyte of an F_1 hybrid of NC 95 x NC 95 SCDHL 12, a second-cycle ADH line from this cultivar. Chromosome pairing in DNA-amplified regions in homeologous chromosomes is thought to be the basis for the unusual cytological figure (courtesy of J. A. Burns, N. C. State University).

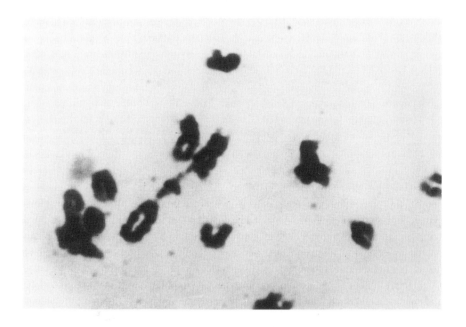

Reed and Wernsman (1989) selected an ADH line array with approximately equal descending increments of yield reduction from a single cultivar and isolated the nuclei from leaf lamina. These were stained with a DNA-specific fluorochrome and the relative quantities of nuclear DNA were estimated by flow cytometry for the parental cultivar and ADH genotypes. Total nuclear DNA in the low-yielding ADH lines was increased above that of the parental cultivar. The negative relationship between reduced yielding ability and nuclear DNA quantity was not high, however. The authors suggested that a poor association between DNA amplification and yield reductions might result from differential effects of DNA amplification in different chromosomes or regions within chromosomes.

ANTHER CULTURE AS A SOURCE OF GENETIC VARIATION

Qualitative traits

In vitro anther culture has been documented to induce large amounts of genetic variation, but the value of this variability for plant improvement has not been extensively investigated. Witherspoon *et al.* (1991) demonstrated that genetic variability for resistance or tolerance to potato virus Y (PVY) could be recovered from *in vitro* anther cultures of a virus-susceptible genotype. Gooding and Tolin (1973) identified a strain of PVY, designated Strain III or NN, that was highly necrotic to all American flue-cured germplasm. Small seedlings inoculated with this virus and grown at cool temperatures express a severe necrotic reaction and are usually killed within 3 weeks following inoculation. Witherspoon *et al.* (1991) subjected a single plant of flue-cured cultivar McNair 944 to *in vitro* anther culture and inoculated a resulting population of 545 haploid plants with Strain III of PVY. One plant, although infected with the virus, survived, was chromosome-doubled, and self-pollinated to give rise to a breeding line designated NC 602. Gametoclonal variant, NC 602, was shown to differ from selfed progenies of the McNair 944 parental plant by the presence of a single gene expressing additive effects which provided resistance to the necrotic reaction produced by PVY Strain III. This gametoclonal variant was resistant or tolerant to two other PVY strains from an international collection but was highly susceptible to others and to a second potyvirus, tobacco etch virus (TEV) (Yung *et al.*, 1991).

Yung *et al.* (1991) cultured anthers of NC 602 and produced an array of 192 ADH lines (second-cycle doubled haploids) from this gametoclonal variant. The ADH lines, as well as over 1000 haploid plants from NC 602, were inoculated with a strain of TEV that produces severe necrosis and plant stunting on NC 602. The authors were unsuccessful in obtaining ADH lines that were highly resistant to TEV, but three ADH lines were identified that gave significantly delayed disease symptom development. The three ADH lines showing delayed symptoms also expressed significantly less plant stunting than NC 602.

NC 602 was decidedly inferior to McNair 944 in yielding ability and produced approximately 19% less cured leaf than the parental cultivar (Yung *et al.*, 1991). Nevertheless, the reduction in yielding ability was not associated with the gene for PVY resistance. By backcrossing the gametoclonal variant to the parental cultivar and practicing selection among lines in advanced generations of inbreeding, it was possible to identify PVY-resistant lines of equal yielding ability to McNair 944. Two breeding lines possessing this gene for PVY resistance, NCTG 51 and NCTG 52, are currently in the final year of evaluation in a regional testing program prior to cultivar release.

In spite of the demonstration by Witherspoon *et al.* (1991) that the gene for PVY resistance in NC 602 arose in a haploid plant derived from anther culture, the possibility that the genetic change occurred spontaneously and was unrelated to anther culture *per se* cannot be eliminated.

Nichols (1991) identified ADH tobacco lines with high levels of resistance to *Phytophthora parasitica* (Dust.) var. *nicotianae* (B. de Haan) Tucker from highly susceptible F_1 hybrids. Cultivars highly susceptible to *Phytophthora* were hybridized, and arrays of ADH and single-seed-descent lines were developed from the F_1s. No single seed descent-derived, disease-resistant, inbred lines were obtained from the hybrids, while approximately 5% of the ADH lines from the same F_1s expressed a high level of resistance to this disease. Information is not available on the number of genes involved in the *Phytophthora* resistance reaction or their mode of inheritance. Collectively, these studies demonstrate that genetic variability conferring resistance to plant diseases can be recovered from haploid plants produced by *in vitro* anther cultures of disease-susceptible genotypes, while conventionally inbred genotypes from the same sources do not provide resistant types.

Quantitative traits

The value of anther culture-induced gametoclonal variation for the improvement of quantitative traits in tobacco was investigated by Schnell and Wernsman (1986) in ADH germplasm generated from cultivar NC 95. Sixty-four ADH lines were produced from a single plant of the pure line cultivar by anther culture and colchicine chromosome doubling. The ADH lines were subdivided into eight sets of eight lines each. Lines were crossed in a North Carolina factorial mating Design II (Comstock and Robinson, 1952), and full-sib (FS) families were evaluated for five agronomic and two chemical composition characters in two environments. Significant genetic variation was detected for all measured characters and the variance was primarily of the additive type; dominance variances were negligible for all characters.

The two FS families producing superior leaf yields in each set (16 FSs total) were intercrossed to provide population, C_1. An ADH composite population consisting of equal contributions from each of the 64 ADH lines, population C_0 (equal contributions from each FS), C_1, sublines of NC 95, and selfed progenies of the original plant anther cultured were evaluated in multiple environments and selection progress was measured. The ADH composite yielded 13.6% less leaf mass than the cultivar. Approximately 25% of this loss in yield was recovered when the ADH lines were intercrossed to produce FS families. Schnell and Wernsman (1986) suggested that this value might represent inbreeding depression resulting from heterozygosity in the original plant anther cultured, followed by restoration of Hardy-Weinberg equilibrium in population C_0. A single cycle of FS family selection resulted in a C_1 population which yielded 5.8% more leaf than C_0. Multiple cycles of FS family selection were estimated to be needed before the entire loss in yielding ability resulting from anther culture of NC 95 would be recovered.

Yung and Wernsman (1990) continued the FS family selection program for two additional cycles, evaluated all selected populations, and estimated genetic variances in population C_3 in a comparable study to the one described above. Following three cycles of FS family selection, the mean yielding ability of population C_3 was equal to that of NC 95, and significant genetic variability for leaf yields remained in the population. The magnitude of estimated genetic variance for leaf yields in C_3 was not greatly different from estimates in the original C_0 population. Genetic variability for other agronomic and leaf chemical constituents had decrease significantly.   Additive genetic variance for leaf yields among maternal half-sib families in population C_3 was very large, while that for paternal half-sibs did not differ from zero. The genetic variability in the population appeared to be transmitted differently through male and female parents. The authors speculated that this observation could be related to DNA amplification described by Reed and Wernsman (1989) and pollen possessing DNA amplified sequences might be at a competitive disadvantage in effecting fertilization.

These studies demonstrate that genetic variability for the improvement of quantitative genetic characters such as leaf yields can be generated by the *in vitro* anther culture-derived doubled haploid methods. In view of readily available genetic variation for this trait in traditional tobacco populations, and the immense effort required to "sort out" genes with unfavorable effects on plant productivity, the use of anther culture for the generation of genetic variability for leaf yields cannot be highly recommended.

ANTHER CULTURE AS AN AID TO INTERSPECIFIC GENE TRANSFER

Interspecific hybridization and alien gene transfer has been widely employed in tobacco breeding in the development of disease resistant genotypes. Examples of genes for disease resistance of interspecific origin that are widely employed in burley tobacco cultivar development include (1) resistance to tobacco mosaic virus (from *N. glutinosa* L.), (2) wildfire disease caused by *Pseudomonas syringae* pv. *tabaci* (Wolf and Foster) (from *N. longiflora* Cav.), and (3) black root rot caused by *Thielaviopsis basicola* (Berk. and Br.) Ferraris (from *N. debneyi* Domin). Numerous other examples of successful gene transfer have been realized, yet these genes are not as widely deployed (Wernsman and Rufty, 1987).

Production of chromosome addition lines

In many interspecific hybridization and gene transfer programs, the ploidy of recipient tobacco is first changed to a higher level from normal, allotetraploid ($2n = 4x = 48$) to the auto-allo-octoploid ($2n = 8x = 96$) state (Chaplin and Mann, 1961). Researchers unacquainted with the species frequently find tobacco literature confusing with regards to polyploidy. Much of the published literature in interspecific transfers refers to tobacco as a diploid because of the species amphidiploid origin. Hybridizations of the $8x$ tobacco with an alien diploid *Nicotiana* species possessing the gene to be transferred yields a pentaploid F_1 interspecific hybrid with the $4x$ genomic condition for tobacco and a single copy of the donor species genome. The pentaploid interspecific hybrid (also termed sesquidiploid) usually possesses sufficient female fertility to be backcrossed with pollen of normal tobacco, and resulting progenies segregate for germplasm from the donor species (Wernsman and Matzinger, 1966). Through continued backcrossing to tobacco, each backcross accompanied by selection among progenies for the trait to be transferred, the breeder can obtain an alien chromosome addition line possessing the gene of interest on the donor species chromosome. Obtaining recombination between the alien chromosome and a tobacco chromosome can be very difficult and is a major limitation to interspecific transfer within the species. Moreover, alien chromosome addition lines are frequently difficult to maintain; pollen possessing the alien chromosome may be inviable or non-competitive with normal pollen in effecting fertilization, and the alien chromosome may be transmitted at a low frequency through the egg.

In a classic example of interspecific gene transfer in tobacco, Holmes (1938) obtained a line of tobacco containing a pair of chromosomes from *N. glutinosa* possessing the gene (*Nc*) conferring the local lesion reaction to tobacco mosaic virus. Further studies demonstrated that Holmes Samsoun was a homozygous chromosome substitution line, in which the *N. glutinosa* chromosome carrying the *Nc* gene, was substituted for H chromosome of tobacco (Gerstel, 1946). This event was followed by genetic exchange between tobacco and *N. glutinosa* H chromosomes (Gerstel, 1948).

Nicotiana africana ($2n = 46$) possesses resistance to potato virus Y (Lucas *et al.*, 1990) and tobacco etch virus. Wernsman and Gooding (unpublished data) hybridized potyvirus susceptible tobacco ($2n = 48$) cultivar McNair 944 as the female with *N. africana* and doubled the chromosome number of a few, viable, interspecific hybrids. The chromosome doubled amphiploids possessed sufficient fertility that selfed progenies could be obtained; the amphiploid progenies did not express disease symptoms when inoculated with PVY strain NN, a highly necrotic strain of the virus which kills small seedlings of all American flue-cured cultivars as described in a previous section. These observations were interpreted to indicate that the genes for potyvirus resistance in *N. africana* were dominant to genes for susceptibility in *N. tabacum*.

The *N. tabacum* x *N. africana* amphiploid was backcrossed as the female parent three times to McNair 944, and in each generation the backcross progenies were inoculated with PVY strain NN to identify segregants that possessed resistance to the virus. In the BC_3 generation, a 52-chromosome plant with a high level of resistance to PVY strain NN was subjected to *in vitro* anther culture. Resulting haploid and aneuhaploid plants possessing the tobacco genome plus one or more chromosomes from *N. africana* were inoculated with PVY strain NN (Figure 3). From a population of 256 plants, four individuals survived disease development and, although virus-infected, appeared to possess good resistance. Leaf midvein explants from each plant were cultured *in vitro* by the method of Kasperbauer and Collins (1972) to achieve chromosome doubling. Chromosome-doubled lines were self-pollinated and checked for virus resistance. Three of the ADH lines (designated as NC 58, NC 147, and NC 152) have proven to be genetically stable through several generations of sexual reproduction. The fourth line (NC 4) is genetically unstable in advanced generations. NC 147 and NC 152 each possess 50 chromosomes, with 24 chromosome pairs from tobacco and one pair from *N. africana*; both lines possess resistance to the necrotic reaction caused by PVY strain NN. NC 58 possessed two chromosome pairs from *N. africana* and was virus resistant as well.

Chromosomal recombinant types in which a tobacco and an *N. africana* chromosome had recombined were never found. Witherspoon (1987) was also unsuccessful in obtaining chromosomal recombinant haploid plants from anther culture of tobacco plants possessing additional chromosomes from the same alien species. His results confirmed, however, that aneuhaploid microspores or pollen, readily undergo embryogenesis in *in vitro* anther cultures and form aneuhaploid plants. Chromosome doubling of the aneuhaploid plants results in disomic chromosome addition lines which may be genetically stable in advanced generations of inbreeding.

Procedures for obtaining non-homologous chromosome recombination and genome reorganization

The initial step in any interspecific gene transfer program is the production of hybrids between recipient and donor species. The immense reproductive potential of tobacco has permitted the production of numerous "chance" hybrids that would have not been possible in a species with less fecundity. Advances in protoplast fusion techniques and somatic hybridization research in the past two decades have increased the number of interspecific hybrids within the genus, yet even these advances have done little to obviate the

Figure 3 Haploid and aneuhaploid plants produced from anther culture of a 52-chromosome tobacco plant with four chromosomes from *N. africana*. The aneuhaploid plant on the left survived the inoculation with PVY because it possesses an alien chromosome carrying a major gene for potyvirus resistance; tobacco plants without this extra chromosome were killed by the virus.

N. AFRICANA DERIVED HAPLOIDS
PVY STRAIN NN

most limiting problem. As emphasized in an earlier section, obtaining recombination between tobacco and alien *Nicotiana* species chromosomes (which possess limited homology) has been a severely limiting factor in interspecific gene transfer and utilization of the germplasm reservoir of this genus.

Information on chromosomal recombination events which have led to successful transfers of loci from alien species into tobacco, except for the Nc gene from *N. glutinosa*, is very limited. Moav (1958) investigated the relationship of residual chromosome homology and interspecific gene transfer for a dominant gene for green plant color. The white seedling character in tobacco is conditioned by recessive alleles (*ws*) at duplicate loci (Clausen and Cameron, 1950), but many genotypes possess a recessive allele in homozygous condition at one locus. In crosses of white seedling heterozygotes with such genotypes, the white seedling character segregates as a single Mendelian gene. A dominant gene, *Ws(pbg)* from *N. plumbaginifolia* Viv., confers green color to a white seedling tobacco genotype; but the alien *N. plumbaginifolia* chromosome carrying this gene is mitotically unstable in tobacco. A white seedling tobacco carrying the alien chromosome with *Ws(pbg)* can be recognized by a green-white mottling of the leaves. Moav repeatedly transferred and incorporated the *Ws(pbg)* gene to *ws ws* tobacco and then determined the tobacco chromosome in which newly transferred *Ws(pbg)* resided. *Ws(pbg)* was incorporated in the same tobacco chromosome in eight of 14 independent transfers, which suggested that residual chromosome homology between the two species was probably very important in the recombinational event. In the remaining six transfers, the gene was equally distributed among three other tobacco chromosomes, and residual chromosome homology might have been less important in these cases.

In subsequent investigations, Moav (1962) suggested the use of an autotriploid ploidy level as a means of enhancing recombination between non-homologous chromosomes. Using Moav's terminology, a simplex green autotetraploid (auto-allo-octoploid) tobacco (*Ws ws ws ws*) was hybridized with a mottled diploid (allotetraploid) alien addition plant [*ws ws* + *Ws(pbg)*] where genes *Ws* and *Ws(pbg)* were employed to designate green alleles from tobacco and *N. plumbaginifolia*, respectively. Autotriploid alien addition plants exhibiting the "mottled" green color characterized by the instability of the alien chromosome were identified and crossed with white seedling tobacco (*ws ws*). In autotriploid meiosis only two chromosomes may pair at any one point, and Moav suggested that the total length of the unpaired tobacco chromosomes must sum to a complete unpaired genome. The transfer of the *Ws(pbg)* allele from the alien chromosome to tobacco was determined and compared to the transfer rate in self-pollinations of the diploid *ws ws* + *Ws(pbg)* alien addition line. Results of the study demonstrated that a significantly increased rate of interspecific chromosomal exchanges occurred at the autotriploid level over the diploid control and suggested that the presence of unpaired chromosomal sequences in the recipient genome could enhance alien gene transfer.

Recent investigations in our laboratory have been centered upon the effects of the anther culture process upon meiotic chromosome behavior in haploid plants. Meiosis in haploid sporophytes derived from the egg nucleus of a standard cultivar, such as NC 95, is characterized by the presence of 24 univalents; homeologous chromosomes from the two progenitor species genomes do not pair. As NC 95 was subjected to multiple cycles of anther culture, numbers of chromosome pairs increased steadily and were 10 times as frequent in haploid plants resulting from four cycles of anther culture (Wernsman *et al.*, unpublished data). Moreover, multivalents were observed in the anther culture derived haploids. Bivalent pairs could result from pairing of homeologous chromosomes in the two progenitor species genomes. Multivalent associations could only result from pairing of chromosomal segments in non-homoeologous chromosomes. Causes of this change in chromosome pairing are not known but might result from suppression of function of a homeologous chromosome pairing gene similar to that in *Triticum aestivum* L. (Riley and Chapman, 1958), pairing among homologous DNA amplified sequences in different chromosomes, or combinations thereof.

Haploid sporophytes developed from egg nuclei of tobacco cultivars via parthenogenesis exhibit extremely high pollen and ovule sterility. Conversely, pollination of anther culture-derived haploid plants by the same parental plant used to produce the haploid results in a high frequency of fruit development; and one or more viable seed is usually produced in each capsule. Progenies from haploid x doubled haploid or cultivar crosses vary in chromosome numbers but may provide opportunities for non-homologous chromosome recombination and genomic rearrangements. Hence, the use of anther culture to obtain alien chromosome addition plants as described above, the capacity of this type of haploid plant to undergo non-homologous chromosome pairing, and their limited female fertility provides a potentially new route to achieve interspecific gene transfer. These plants would also possess an entire, unpaired genome and possess the same advantages of autotriploidy as described by Moav (1962), yet they would be much easier to produce.

Characteristics of ovule-derived doubled haploids

Haploid plants of parthenogenetic origin have been known in tobacco for decades, but only in the past decade have they been available in sufficient numbers to be of plant breeding value. Burk *et al.* (1979) hybridized tobacco as the maternal parent with *N. africana* and found the cross to produce abundant seed, but the germinating seedlings were highly lethal in the cotyledonary stage of development. Approximately 0.1% of the seedlings were viable and consisted of mixtures of gynogenetic haploids and aneuploid interspecific hybrids in more or less equal numbers. Tobacco haploid and the viable aneuploid interspecific hybrids are easily distinguished phenotypically when the investigator is acquainted with the species. Haploids result from a parthenogenetic event and do not arise by a chromosome elimination phenomenon (Wernsman, unpublished data). Although the frequency of haploid plants is low, tobacco flowers are large, easy to pollinate, and one to three haploid plants are commonly produced from a single capsule. Hence, large haploid populations are possible.

Kumashiro and Oinuma (1985) demonstrated the capacity to produce significant numbers of egg-derived haploids by fertilization of tobacco ovules with irradiation-impaired pollen of *N. alata* Link and Otto. In a limited number of studies, haploid plants have been produced from vegetative nuclei of pollen via anther culture and from ovules of the same highly inbred genotypes. Kumashiro and Oinuma (1985) cultured the anthers of a single, doubled haploid plant, N 34, of "Bright Yellow 104", crossed the plant with X-irradiated pollen of *N. alata* and produced 17 S_1 families from the plant by conventional self-pollination. Seventeen S_1 lines, 38 ADH lines, and 12 PDH (parthenogenic-derived lines) were compared for six characteristics. Genetic variation was detected among ADH lines for all characters. When compared to N 34, the ADH lines flowered earlier, were shorter in plant height, had less leaves per plant, and were lower yielding. Variation among PDH lines was detected only for yield. The PDH lines differed from N 34 for leaf width and yielding ability; leaves of PDH lines were narrower than the check but were higher yielding. ADH and PDH lines differed for four of six characters measured.

Wernsman *et al.* (1989) developed ADH, MDH (maternal-doubled haploid), and selfed lines (controls) from single plants of each of three burley tobacco cultivars - KY 10, KY 15, and KY 17. ADH lines differed from the cultivars for eight of nine characters measured; MDH lines differed from respective controls for three of nine traits. Both groups of lines were significantly lower yielding than the cultivars, but the MDH lines were of superior yielding ability compared to ADHs and were more similar to the conventionally inbred genotypes. A sample of 16 second-cycle MDH lines was produced from a random first-cycle MDH of KY 10, and the second-cycle lines compared with the original cultivar and the first-cycle MDH line. As a group the mean of the second cycle lines was significantly taller than their first-cycle parent but did not differ from the first-cycle parent for any of six other traits measured, including leaf yields. Although variability for yielding ability was not detected among the second-cycle MDH lines, variation was detected for five other traits. Second-cycle MDH lines have also been developed from first-cycle MDH lines from flue-cured cultivars Coker 139 and NC 95 and evaluated for mean yielding ability and variation among lines (Wernsman, unpublished data). Genetic changes from the first cycle to the second cycle of MDH line production were found to be negligible.

Nielsen and Collins (1989) compared 20 ADH lines with 18 MDH lines developed from four plants of KY 17. Both ADH and MDH lines were of inferior yielding ability to S_1 lines of the parental plants, but ADH- and MDH-derived lines from three of the four KY 17 plants were generally not different from selfed lines for other traits. For each type of DH line, one line in each group was unusual and yielded approximately 18% less than the check. The cause of the poor performance of the "outlier" lines was unknown.

Bowman and Wernsman (1989) developed a random MDH line from each of 11 flue-cured cultivars and compared the environmental stability of the MDH lines with the parental cultivars for 11 traits over nine environments with three replications per location. Yielding ability comparisons of a MDH line with its respective cultivar were not considered because of inadequate sample of MDHs from each cultivar. Nevertheless, as a group of genotypes, the MDH lines did not differ from the cultivars for any trait except concentrations of reducing sugars in cured leaf. Although absolute homozygotes, the phenotypic stability of the MDH lines was equal to the conventionally inbred cultivars.

Generalizations on the performance of ovule-derived DH lines suggests that there does not appear to be a "genetic penalty" associated with the development of ovule-derived lines. When developed from highly inbred genotypes developed by self-pollination, these MDH lines are not always equal to selfed progenies of the plants from which they are derived; but any depressed performance and variability among these lines may be expressions of inbreeding depression.

EFFICIENCY OF SELECTION AMONG HAPLOID AND DOUBLED HAPLOID GENOTYPES

A serious limitation to most plant breeding procedures is genotype evaluation. Doubled haploid tobacco breeding methods encounter an additional limitation - chromosome doubling of haploid sporophytes. Although spontaneous chromosome doubling can occur, this event is highly influenced by genotype, and the frequency is very low (less than 1% of haploids double spontaneously). Numerous reports on chromosome doubling of haploid by colchicine treatments suggest that the methods would be suitable for large-scale tobacco breeding programs (Burk *et al.*, 1972; Oka *et al.*, 1977; Kumashiro and Oka, 1978). In our experience, none of the methods provide high percentages of chromosome-doubled plants. Furthermore, maternal haploids produced by the *N. africana* method cannot be distinguished precisely from viable interspecific hybrids until the plants develop a few true leaves. Colchicine treatment of larger plants is not as effective in achieving chromosome doubling as when plants are very small (Oka *et al.*, 1977). Deaton *et al.* (1986) found an array of colchicine-doubled ADH lines to give inferior performance compared to an array of *in vitro* chromosome-doubled genotypes from the same population; the authors suggested that the colchicine treatment could be been responsible for genetic changes.

In vitro culture of tobacco permits ploidy changes (Murashige and Nakano, 1966) in regenerated plants, and the leaf midvein culture procedure of Kasperbauer and Collins (1972) is highly reliable in producing chromosome-doubled plants. These procedures are slow and laborious, however.

Numerous studies have considered selection theory in doubled haploid populations and the relative efficiency of such methods compared to similar procedure under conventional (Nei, 1963; Griffing, 1975) methods of reproduction. Experimental data are very limited, however. In view of chromosome doubling limitations, research in our laboratory has been focused upon opportunities for selection in the haploid, as well as the doubled haploid state. This is feasible because (1) the species is grown for leaf, rather than seed; (2) three bud primordia are present in each leaf axil and the buds (suckers) can be rooted and used to propagate or rescue a genotype; and (3) the species is perennial. Using haploid selection techniques, resources for chromosome doubling can be restricted to genotypes with merit; and resulting DH populations have been subjected to a preliminary selection. Using the characteristics of the species and available technology of plant manipulation, a single cycle of tobacco breeding can be divided into haploid and DH selection phases as shown in Figure 4.

Figure 4 A single cycle of plant breeding using haploid and doubled haploid techniques may consist of application of selective forces to a haploid population, chromosome doubling of selected haploids, evaluation of MDH population from selected haploids, intermating of selected MDHs to produce an improved population and/or consideration of MDH lines for release as new cultivars.

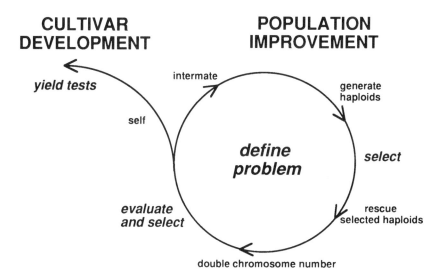

Selection for qualitative traits in haploid populations

Several disease resistance mechanisms used in commercial tobacco production are under the control of single genes. Examples include resistance to tobacco mosaic virus, resistance to the root-knot nematode, *Meloidogyne incognita* (Kofoid and White) Chitwood, tobacco wildfire caused by *P. syringae* pv. *tabaci*, resistance to black root rot caused by *T. basicola*, and resistance to potyviruses, potato virus Y, tobacco etch virus, and tobacco vein mottle virus (Wernsman and Rufty, 1987). Simple detached leaf tests for some of these diseases have been developed to permit the classification of haploid genotypes for resistance to multiple pathogens without introducing the organisms into the plants. Leaves from haploid plants are detached, inoculated with pathogens, the petioles immersed in water-filled test tubes or stored in plastic "water boxes" until disease development is sufficient to classify the genotypes (Rufty *et al.*, 1987).

Tobacco genotypes with the *Nc* gene from *N. glutinosa* exhibit a hypersensitive reaction when inoculated with tobacco mosaic virus (TMV) (Holmes, 1938). Detached leaves from haploid plants with the *Nc* genotype exhibit the local lesion or hypersensitive leaf spotting in 3 to 5 days following TMV inoculation, while leaves from susceptible (*nc*) genotypes do not express symptoms. Using the same technique, haploid populations can be classified for resistance or susceptibility to tobacco wildfire disease by inoculating a detached leaf from each haploid plant with the disease producing bacterium. Resistant vs. susceptible genotypes can be discerned by lesion size and presence or absence of the halo characteristic of this disease. Precision of the TMV resistance classification was found to be 100%, while wildfire classification was slightly less accurate; and Rufty *et al.* (1987) recommended replicating the test by inoculating more than one detached leaf.

Only one source of resistance to the root-knot nematode, *M. incognita*, is available in tobacco; and the single dominant gene conferring resistance produces a pleiotropic effect when inoculated with a specific strain of potato virus Y (Henderson and Troutman, 1963; Rufty *et al.*, 1983). A severe necrotic reaction is expressed when root-knot resistant genotypes are inoculated with PVY strain II (MN strain), while nematode-susceptible plants exhibit a mild mosaic reaction (Gooding and Tolin, 1973). The necrotic reaction and its severity and time of development in detached leaves can be used to classify haploid genotypes known to be segregating for gene conferring resistance to root knot (Rk) and the recessive allele for potyvirus resistance (*va*). Detached leaves from nematode-susceptible plants do not express symptoms when inoculated with PVY strain II (Figure 5). Leaves from plants resistant to nematodes and susceptible to potyviruses (*Rk Va*) express severe necrotic spotting in 5 days when inoculated with PVY strain II. Necrotic spotting on leaves from genotypes with genes for nematode and potyvirus resistance (*Rk va*) is delayed 2 to 3 days, and the intensity of the necrotic reaction is attenuated (Wernsman and Gooding, unpublished data). Hence, genotypes with genes for nematode and potyvirus resistance can be determined by the same test.

The *N. africana* method produces haploid plants from seed; consequently, large populations of haploid sporophytes can be produced in field plant beds and transplanted to soil-borne disease nurseries. Campbell and Wernsman (1990) transplanted a haploid population segregating for resistance to *P. parasitica* var. *nicotianae*, a trait not under the control of a single gene, to a soil-borne disease nursery and permitted the disease to kill the most

Figure 5 Detached leaves from haploid plants segregating for resistance to root-knot nematode (*Meloidogyne incognita*) in a potyvirus-susceptible background. Leaves developing necrotic spots from inoculation with PVY strain II are from root-knot resistant genotypes (*Rk Va*), while asymptomatic leaves are from nematode susceptible genotypes (*rk Va*).

susceptible plants. Surviving, disease-free plants were rescued, chromosome-doubled, and the disease reaction of the doubled haploid lines evaluated. The procedure appeared to be very effective in producing homozygous genotypes with high levels of resistance to black shank.

These studies demonstrate that, in practical tobacco breeding programs, much of the selection of disease resistance genotypes can be accomplished at the haploid level. Moreover, individual plants can be evaluated for resistance to multiple diseases; and resources devoted to chromosome doubling, seed production and handling etc. can be restricted to genotypes of interest. Although MDH lines developed in one cycle of selection based upon haploid plants may not result in any one line being suitable for release as a new cultivar, intermating the selected MDH lines would be expected to increase rapidly the frequency of genes for disease resistance. Desirable alleles for qualitative disease resistance traits can be fixed in a single cycle of selection.

Selection for quantitative traits in haploid populations

Four cycles of mass selection for increased fresh leaf weight in a flue-cured tobacco population resulted in a linear increase of 4.29% cured leaf yield per cycle of selection (Matzinger and Wernsman, 1968). An even more dramatic response was observed by Gupton (1981) in a burley tobacco population; a correlated increase in cured leaf weight of 7.2% per cycle of selection for fresh leaf weight over three cycles of mass selection was found. Phenotypic comparisons between haploid and doubled haploid populations from the same sources may not be relative because of the inherent "weakness" or smaller size of haploid plants. Nevertheless, plant-to-plant genetic variability in a haploid population should be of a type "fixable" among homozygous DH lines, and genetic variability in haploid populations might be considerably larger than respective genetic variability in a random-mating diploid population assuming an additive genetic model. In view of the impressive rates of gain from practicing mass selection in diploid tobacco populations, mass selection for quantitative traits in haploid populations is of considerable interest.

Witherspoon and Wernsman (1989) examined the correlation of performance over haploid and diploid ploidy levels of an array of tobacco genotypes known to vary widely in their performance for quantitative traits. Genotypic rank orders across the two ploidy levels were in good agreement; correlations of performance were highly significantly different from zero for green and cured leaf yield, number of leaves per plant, plant height, and total alkaloid concentrations in cured leaf. Correlations were not different from zero for days from transplanting to flowering and reducing sugar concentrations in cured leaf. The authors concluded that selection in haploid populations for quantitative traits should be feasible.

Wilhite (1990) conducted a cycle of mass selection for increased green leaf weight in a population of 200 haploid sporophytes generated from an F_1 hybrid of "Coker 176" x breeding line NC DH 62. Following the completion of total green leaf weight determinations on each plant in the haploid population, the 20 highest yielding plants were rescued from basal-stalk axillary buds, chromosome-doubled, and intercrossed to produce a C_1 haploid-selected population. Two cycles of mass selection for improved green leaf weight were also conducted in a random-mated population from the same biparental cross. The parental genotypes, the 20 MDH lines, and populations improved by mass selection in haploid and diploid populations were evaluated for cured leaf yields. Two cycles of mass selection in the diploid population were more effective in improving cured leaf yields than one cycle of mass selection in the haploid state. Mass selection in the haploid population was effective in identifying individual plants which, when chromosome-doubled and intercrossed, gave a population of superior yielding ability to that of the original C_0 population. Moreover, several DH lines resulting from chromosome doubling of the haploid plants, identified as producing the highest green leaf yields in the population, were superior to the highest yielding parental cultivar, Coker 176, used to produce the population and were of equal yielding ability to the mean of the C_2 diploid mass-selected population.

Selection for quantitative traits in doubled haploid populations

Corbin (1987) crossed 100 random plants from a large random-mated flue-cured tobacco population with pollen of *N. africana*, self-pollinated each plant to produce S_1 families, and each of the 100 plants was outcrossed to a random plant in the same population to generated 100 full-sib (FS) families. A random haploid plant from each of the 100 plants in the population was chromosome-doubled to produce 100 MDH lines. The MDH lines were intercrossed to provide an unselected, random-mated synthetic population that had been subjected to the *N. africana* haploid production procedure and the *in vitro* chromosome doubling method. The 100 MDH lines, the 100 S_1 lines, and the 100 FS families were evaluated for leaf yields, and the 10 families exhibiting the highest performance by each method were inter-crossed to provide populations selected on the basis of MDH line, S_1, and FS family performance. Because of flooding at the site of the initial evaluation of the MDH lines, a second, independent test of the 100 MDH lines was conducted, and the 10 superior lines in the second test were intercrossed as well. The original C_0 population, the MDH C_0 synthetic, a MDH composite of equal numbers of seed from all 100 lines, and C_1 populations selected by FS, S_1, and two MDH evaluations were evaluated in multiple environments.

Mean leaf yields of the original C_0 population, the MDH composite, and the C_0 MDH synthetic were not different. These data were interpreted by Corbin (1987) as supportive of a hypothesis that the *N. africana* method and the *in vitro* chromosome-doubling procedure were not producing genetic changes or selection - or if they were, the changes had no effects on leaf yields. The data suggested that gene frequencies were constant from C_0 to the C_0 MDH synthetic, and Hardy-Weinberg equilibrium had been maintained. The data were also interpreted as indicative that dominance variance and dominant types of epistatic interactions in this population were negligible and that inbreeding depression from the random mating population was not great. Predicted and realized gains in kg ha^{-1} and in percentages of the population mean from practicing selection for improved leaf yields via the three methods are shown in Table 1. The relative magnitude of additive genetic variance among the FS families, S_1 lines, and MDH lines was expected to be X, 2X, and 4X, respectively. Because predicted gains are highly related to the magnitude of additive genetic variance among the family units being selected, the MDH line recurrent selection procedure was

Table 1 Responses from one cycle of recurrent family selection for improved leaf yields by three methods.†

Selection method	Predicted gain		Realized gain	
	kg ha^{-1}	%	kg ha^{-1}	%
Full-sib	144	5.7	61	2.4
S$_1$ line	221	8.7	197	7.8
MDH 84‡	383	15.1	246*	9.7
MDH 85‡	487	19.2	341*	13.4

†Data from Corbin (1987) and Corbin and Wernsman (unpublished data).

‡MDH lines evaluated in separate years and selected lines intercrossed from each year's evaluation.

*Significantly different from zero.

expected to result in greater gains than the other two recurrent selection procedures. These results were realized.

Available data, although limited, demonstrate that haploid and doubled haploid phases of the selection scheme shown in Figure 4 could be combined into a comprehensive, highly flexible breeding program. In the haploid phase, selection might be practiced for multiple disease resistance traits in a segregating population, and chromosome doubling of selected haploids would result in a disease-resistant MDH population genetically variable for unselected traits. Confirmation of disease resistance could be accomplished in tests of the MDH lines at the same time the population is being evaluated for agronomic traits. Following evaluation of the MDH population, selected genotypes can be intermated to produce an improved population for quantitative traits, yet gene frequencies for qualitative traits may be fixed. Homozygosity of selected MDH lines would permit further evaluation and consideration for release as cultivars.

SUMMARY

Large populations of haploid tobacco sporophytes can be produced from the vegetative nucleus of pollen via *in vitro* anther culture or from egg nuclei via pollination of tobacco females with *N. africana*. Chromosome doubling of anther culture haploids results in ADH lines which generally have little cultivar potential because of their low productivity concomitant with genetic changes. Associated with the reduced productivity and genetic changes is an amplification of total nuclear DNA without chromosome number changes. In spite of these problems, anther culture haploids have value as sources of gametoclonal variation, in production of homozygous alien chromosome addition lines and their unusual chromosome pairing and fertility characteristics provide opportunities for non-homologous chromosome recombination and genomic reorganization. Chromosome-doubled gynogenetic haploids developed by the *N. africana* method are more competitive agronomically with conventionally inbred genotypes than ADH lines. Haploid production from seed permits synchronization of field plantings of large haploid populations in disease nurseries or for agronomic evaluations, followed by rescue of selected genotypes for chromosome doubling. Complete inbreeding of the MDH lines maximizes additive genetic and additive x additive epistatic variances in the population and partitions this variation completely among lines. A cycle of tobacco breeding can be partitioned into selection in haploid and doubled haploid phases. This has been shown to enhance gains from selection and to speed plant breeding progress in this model system.

REFERENCES

Arcia, M. A.; Wernsman, E. A.; Burk, L. G. (1978) Performance of anther-derived-dihaploids and their conventionally inbred parents as lines, in F_1 hybrids, and in F_2 generations. *Crop Science* 18:413-418.

Blakeslee, A. F.; Belling, J.; Farnham, M. E.; Bergner, A. D. (1922) A haploid mutant in the jimson weed, *Datura stramonium. Science (Washington, DC)* 55:646-647.

Bourgin, J. P.; Nitsch, J. P. (1967) Obtention de *Nicotiana* haploïdes a partir d'étamines cultivées *in vitro. Annales Physiologie Vegetale* 9:377-382.

Bowman, D. T; Wernsman, E. A. (1989) Stability of doubled haploid lines of flue-cured tobacco. *Tobacco Science* 33:74-76.

Brim, C. A. (1966) A modified pedigree method of selection in soybeans. *Crop Science (Washington, DC)* 6:220.

Brown, J. S.; Wernsman, E. A. (1982) Nature of reduced productivity of anther-derived dihaploid lines of flue-cured tobacco. *Crop Science* 22:1-5.

Brown, J. S.; Wernsman, E. A.; Schnell, R. J. II. (1983) Effect of a second cycle of anther culture on flue-cured lines of tobacco. *Crop Science* 23:729-733.

Burk, L. G.; Gerstel, D. U.; Wernsman, E. A. (1979) Maternal haploids of *Nicotiana tabacum* L. from seed. *Science (Washington, DC)* 206:585.

Burk, L. G.; Gwynn, G. R.; Chaplin, J. F. (1972) Diploidized haploids from aseptically cultured anthers of *Nicotiana tabacum*: A colchicine method applicable to plant breeding. *Journal of Heredity* 63:355-360.

Burk, L. G.; Matzinger, D. F. (1976) Variation among anther-derived doubled haploids from an inbred line of tobacco. *Journal of Heredity* 67:381-384.

Campbell, K. G.; Wernsman, E. A. (1990) Plant breeding in the haploid state: *Phytophthora* resistance in tobacco. *Agronomy Abstracts* 1990:83.

Chaplin, J. F.; Mann, T. J. (1961) Interspecific hybridization, gene transfer and chromosome substitution in *Nicotiana*. North Carolina Agricultural Experiment Station Technical Bulletin Number 145.

Clausen, R. E.; Cameron, D. R. (1950) Inheritance in *Nicotiana tabacum*. XXIII. Duplicate factors for chlorophyll production. *Genetics* 35:4-10.

Collins, G. B.; Legg, P. D.; Litton, C. C. (1974) The use of anther-derived haploids in *Nicotiana*. II. Comparison of doubled haploid lines with lines obtained by conventional breeding methods. *Tobacco Science* 18:40-42.

Comstock, R. E.; Robinson, H. F. (1952) Estimation of average dominance of genes. In: Gowen, J. (ed.), *Heterosis*. Iowa State College Press, Ames, IA, pp. 494-516.

Corbin, T. C. (1987) Evaluation of a doubled haploid breeding procedure for simultaneous recurrent selection and inbred line development. Unpublished Ph.D. Dissertation, North Carolina State University, Raleigh, NC. (Dissertation Abstracts 88-014788).

Deaton, W. R.; Collins, G. B.; Nielsen, M. T. (1986) Vigor and variation expressed by anther derived doubled haploids of burley tobacco (*Nicotiana tabacum* L.). I. Comparison of sexual and doubled-haploid populations. *Euphytica* 35:33-40.

De Paepe, R.; Bleton, D.; Gnangbe, F. (1981) Basis and extent of genetic variability among doubled haploid plants obtained by pollen culture in *Nicotiana sylvestris*. *Theoretical and Applied Genetics* 59:177-184.

De Paepe, R.; Prat, D.; Huguet, T. (1983) Heritable nuclear DNA changes in doubled haploid plants obtained by pollen culture of *Nicotiana sylvestris*. *Plant Science Letters* 28:11-28.

D'Halluin, K.; Keimer, B. (1986) Production of haploid sugarbeets (*Beta vulgaris* L.) by ovule culture. In: Horn, W.; Jensen, C. J.; Odenbach, W.; Schieder, O. (eds.), *Genetic manipulation in plant breeding. Proceedings International Symposium Organized by EUCARPIA*, September 8-13, 1985, Berlin, Germany. W. de Gruyter, Germany, pp. 307-309.

Dhillon, S. S.; Wernsman, E. A.; Miksche, J. P. (1983) Evaluation of nuclear DNA content and heterochromatin changes in anther-derived haploids of tobacco (*Nicotiana tabacum*) cv. Coker 139. *Canadian Journal of Genetics and Cytology* 25:169-173.

Dunwell, J. M.; Sunderland, N. (1975) Pollen ultrastructure in anther cultures of *Nicotiana tabacum*. III. The first sporophytic division. *Journal of Experimental Botany* 26:240-252.

Gerstel, D. U. (1946) Inheritance in *Nicotiana tabacum*. XXI. The mechanism of chromosome substitution. *Genetics* 31:421-427.

Gerstel, D. U. (1948) Transfer of the mosaic resistance factor between H chromosomes of *N. glutinosa* and *N. tabacum*. *Journal of Agricultural Research* 76:219-223.

Gooding, G. V., Jr.; Tolin, S. A. (1973) Strains of potato virus Y affecting flue-cured tobacco in the southeastern United States. *Plant Disease Reporter* 57:200-204.

Griffing, B. (1975) Efficiency changes due to use of doubled-haploids in recurrent selection methods. *Theoretical and Applied Genetics* 46:367-386.

Guha, S.; Maheshwari, S. C. (1964) *In vitro* production of embryos from anthers of *Datura*. *Nature (London)* 204:497.

Gupton, C. L. (1981) Phenotypic recurrent selection for increased leaf weight and decreased alkaloid content of burley tobacco. *Crop Science* 21:921-925.

Henderson, R. G.; Troutman, J. L. (1963) A severe virus disease of tobacco in Montgomery County, Virginia. *Plant Disease Reporter* 47:187-189.

Holmes, F. O. (1938) Inheritance of resistance to tobacco-mosaic in tobacco. *Phytopathology* 28:553-561.

Horner, M.; Street, H. E. (1978) Pollen dimorphism - Origin and significance in pollen formation by anther culture. *Journal of Experimental Botany* 29:217-226.

Kasha, K. J.; Kao, K. N. (1970) High frequency haploid production in barley (*Hordeum vulgare* L.). *Nature (London)* 225:874-876.

Kasperbauer, M. A.; Collins, G. B. (1972) Reconstitution of diploids from leaf tissue of anther-derived haploids in tobacco. *Crop Science* 12:98-101.

Kasperbauer, M. A.; Legg, P. D.; Sutton, T. G. (1983) Growth, development, and alkaloid content of doubled haploids vs. inbreds of burley tobacco. *Crop Science* 23:965-969.

Kumashiro, T.; Oinuma, T. (1985) Comparison of genetic variability among anther-derived and ovule-derived doubled haploid lines of tobacco. *Japanese Journal of Breeding* 35:301-310.

Kumashiro, T.; Oka, M. (1978) Studies of the haploid method of breeding by anther-culture in tobacco VII. Colchicine treatment of haploid plantlets. *Bulletin of the Iwata Tobacco Experiment Station* 10:31-39.

Lucas, G. B.; Gooding, G. V., Jr.; Sasser, J. N.; Gerstel, D. U. (1980) Reaction of *Nicotiana africana* to black shank, Granville wilt, root knot, tobacco mosaic virus, and potato virus Y. *Tobacco Science* 24:141-142.

Maheshwari, S. C.; Rashid, A.; Tyagi, A. K. (1982) Haploids from pollen grains - retrospect and prospect. *American Journal of Botany* 69:865-879.

Matzinger, D. F.; Burk, L. G. (1984) Cytoplasmic modification by anther culture in *Nicotiana tabacum* L. *Journal of Heredity* 75:167-170.

Matzinger, D. F.; Wernsman, E. A. (1968) Four cycles of mass selection in a synthetic variety of an autogamous species *Nicotiana tabacum* L. *Crop Science* 8:239-243.

Moav, R. (1958) Inheritance in *Nicotiana tabacum*. XXIX. Relationship of residual-chromosome homology to interspecific gene transfer. *American Naturalist* 92:267-268.

Moav, R. (1962) Inheritance in *Nicotiana tabacum*. XXX. Autotriploidy, a possible means of increasing the rate of interspecific gene transfer. *Heredity* 17:373-379.

Murashige, T.; Nakano, R. (1966) Tissue culture as a potential tool in obtaining polyploid plants. *Journal of Heredity* 75:115-118.

Nei, M. (1963) The efficiency of haploid method of plant breeding. *Heredity* 18:95-100.

Nichols, W. A. (1991) Anther culture as a probable source of resistance to *Phytophthora parasitica* var. *nicotianae* in tobacco. M.S. Thesis, Library, North Carolina State University, Raleigh, NC.

Nielsen, M. T.; Collins, G. B. (1989) Variation among androgenic and gynogenic doubled haploids of tobacco (*Nicotiana tabacum* L.). *Euphytica* 43:263-267.

Oka, M.; Nakamura, A; Yamada, T; (1977) An efficient colchicine method for chromosome doubling of haploid tobacco plantlets. *Sabrao Journal* 9:108-110.

Reed, S. M.; Burns, J. A.; Wernsman, E. A. (1992) A cytological comparison of amplified chromosome segments in four tobacco doubled haploids. *Crop Science* 32:(in press).

Reed, S. M.; Wernsman, E. A. (1989) DNA amplification among anther-derived doubled haploid lines of *Nicotiana tabacum* L. and its relationship to agronomic performance. *Crop Science* 29:1072-1076.

Reed, S. M.; Wernsman, E. A.; Burns, J. A. (1991) Aberrant cytological behavior in tobacco androgenic doubled haploid x parental cultivar hybrids. *Crop Science* 31:97-101.

Riley, R.; Chapman, V. (1958) Genetic control of the cytologically diploid behavior of hexaploid wheat. *Nature (London)* 182:713-715.

Rufty, R. C.; Wernsman, E. A.; Gooding, G. V., Jr. (1987) Use of detached leaves to evaluate tobacco haploids and doubled haploids for resistance to tobacco mosaic virus, *Meloidogyne incognita*, and *Pseudomonas syringae* pv. *tabaci. Phytopathology* 77:60-62.

Rufty, R. C.; Wernsman, E. A.; Powell, N. T. (1983) A genetic analysis of the association between resistance to *Meloidogyne incognita* and a necrotic response to infection by a strain of potato virus Y in tobacco. *Phytopathology* 73:1413-1418.

Schnell, R. J.,; Wernsman, E. A. (1986) Androgenic somaclonal variation in tobacco and estimation of its value as a source of novel genetic variability. *Crop Science* 20:84-88.

Schnell, R. J., III.; Wernsman, E. A.; Burk, L. G. (1980) Efficiency of single-seed-descent vs. anther-derived dihaploid breeding methods in tobacco. *Crop Science* 20:619-622.

Sunderland, N. (1980) Anther and pollen culture, 1974-1979. In: Davies, D. R.; Hopwood, D. A. (eds.), *The plant genome. Proceedings of the Fourth John Innes Symposium, The Plant Genome, and Second International Haploid Conference* held in Norwich, England, Sept. 1979. John Innes Charity, John Innes Institute, Norwich, England, pp. 171-183.

Wernsman, E. A.; Matzinger, D. F. (1966) A breeding procedure for the utilization of heterosis in tobacco-related species hybrids. *Crop Science* 6:298-300.

Wernsman, E. A.; Matzinger, D. F.; Rufty, R. C. (1989) Androgenetic vs. gynogenetic doubled haploids of tobacco. *Crop Science* 29:1151-1155.

Wernsman, E. A.; Rufty, R. C. (1987) Tobacco. In: Fehr, W. F. (ed.), *Principles of cultivar development.* Vol. 2. Macmillan Publishing Company, New York, pp. 669-698.

Wilhite, F. M. (1990) Selection and linkage investigations in maternally-derived doubled haploid populations of tobacco. Ph.D. Dissertation, North Carolina State University, Raleigh, NC.

Witherspoon, W. D., Jr. (1987) Utilization of the haploid sporophyte as the selection unit in tobacco breeding. Ph.D. Dissertation, North Carolina State University, Raleigh, NC. (Dissertation Abstracts No. DA 8712560).

Witherspoon, W. D., Jr.; Wernsman, E. A. (1989) Feasibility of selection for quantitative traits among haploid tobacco sporophytes. *Crop Science* 29:125-129.

Witherspoon, W. D., Jr.; Wernsman, E. A.; Gooding, G. V., Jr.; Rufty, R. C. (1991) Characterization of a gametoclonal variant controlling virus resistance in tobacco. *Theoretical and Applied Genetics* 81:1-5.

Yung, C. H.; Wernsman, E. A. (1990) The value of gametoclonal variation in breeding for quantitative traits in flue-cured tobacco (*Nicotiana tabacum* L.). *Theoretical and Applied Genetics* 80:381-384.

Yung, C. H.; Wernsman, E. A.; Gooding, G. V., Jr. (1991) Characterization of potato virus Y resistance from gametoclonal variation in flue-cured tobacco. *Phytopathology* 81:887-891.

Chapter 22
Techniques for introgressing unadapted germplasm to breeding populations

Stanley J. Peloquin and Rodomiro Ortiz

Departments of Genetics and Horticulture, University of Wisconsin-Madison, Madison, WI 53706

INTRODUCTION

The use of exotic germplasm to broaden the genetic base of the cultivated gene pool will be illustrated with potato (*Solanum tuberosum* L.) and its tuber-bearing relatives. However, recent results indicate that the ploidy manipulation techniques used with potatoes are being used in other crops such as alfalfa (*Medicago sativa* L.), cassava (*Manihot esculenta* Crantz), and sweet potatoes (*Ipomoeas batata* L.), and should be applicable in yams (*Dioscorea* spp.). It is our strong opinion that wherever non-additive genetic variance for yield is most important (polysomic polyploids) these techniques will be succesful.

Attributes of the potato

Several features related to the genetics and reproductive biology of the tetraploid (4x) cultivated potato, need to be emphasized. They provide background information and the rationale for germplasm transfer techniques in the potato, which is a tetrasomic polyploid with four sets of similar chromosomes.

Many related species

The potato is very fortunate in having many wild and cultivated tuber-bearing and non-tuber-bearing relatives. More than 150 species form a polyploid series of species from diploids (2x) to hexaploids (6x), with about 70% of the species being diploid. They are distributed from southern U.S. to southern Chile, from sea level to 4000 m, and by far the largest number of species and the greatest diversity occur in the Andean regions of Peru and Bolivia (Hawkes, 1945). A worldwide collection of *Solanum* germplasm is available in the Inter-Regional Potato Introduction Project at Sturgeon Bay, Wisconsin (Hanneman and Bamberg, 1986) and from other potato "gene banks". The extent and value of this germplasm have been documented by

Hanneman (1989), and the success in using exotic germplasm to improve cultivars by Ross (1986) and Plaisted and Hoopes (1989).

Two points about *Solanum* germplasm resources need particular emphasis.

1. Any desirable trait we need as breeders appears to be available in the related species; this includes resistance/tolerance to major biotic and abiotic stresses, important tuber quality characteristics, and allelic diversity significantly greater than that in cultivated forms.

2. The potato is unexcelled, among economic plants, in the abundance of related germplasm, and the ease of incorporating this germplasm into cultivated forms.

Haploids easy to obtain

Haploids ($2n = 2x = 24$) of cultivars and advanced selections ($2n = 4x = 48$) are relatively easy to obtain through interploid crosses ($4x \times 2x$) where haploids originate from unfertilized eggs (Peloquin *et al.*, 1990). The $4x$ seed parents are selections of *S. tuberosum* Group Tuberosum and Group Andigena, and the $2x$ pollen source is a Group Phureja clone with appropriate genetic markers such that potential haploids can be detected in the seed or early seedling stage.

The haploids perform as normal diploids with bivalent pairing and disomic genetics. Two major advantages of haploids are the simple genetic ratios compared to a tetrasomic polyploid and the opportunity to cross directly with the $2x$ species. Haploids can be hybridized with most 24-chromosome tuber-bearing species. The hybrids are vigorous, variable, and many possess good fertility, and improved tuberization as compared to the wild species, which normally do not tuberize in the field under the long day conditions of northern temperate regions.

The haploid species hybrids generally have normal chromosome pairing and crossing over. Cytological investigations indicate very little differentiation between the chromosomes of the cultivated potato and those of most wild tuber-bearing relatives.

Wide occurrence of $2n$ gametes

Inherited variations in the meiotic process that result in $2n$ gametes (gametophytes or gametes with the sporophytic chromosome number) are common in potato (Peloquin *et al.*, 1989b). In microsporogenesis, these variations result in $2n$ pollen, in megasporogenesis in $2n$ eggs. Thus, $4x$ progeny can be obtained from either $4x \times 2x$ ($2n$ pollen) or $2x$ ($2n$ eggs) $\times 4x$, and from $2x$ ($2n$ eggs) $\times 2x$ ($2n$ pollen). Screening of wild and cultivated species for $2n$ pollen and $2n$ eggs indicates they occur in some plants of almost all tuber-bearing species.

The ability to obtain sporophytes with the gametophytic chromosome number (haploids) and gametophytes with the sporophytic chromosome number ($2n$ gametes) is the basis for the ease of manipulating sets of chromosomes in potatoes. In fact, the potato is the best organism in which to manipulate whole sets of chromosomes.

Triploid block

Triploids are rarely recovered from $4x$ x $2x$, $2x$ x $4x$, and $2x$ x $2x$ crosses where the $4x$ parent is Tuberosum or Andigena and the $2x$ parent is a tuber-bearing species or a haploid species hybrid (Johnston *et al.*, 1980). Exceptions to this generalization occur in a few $4x$ wild species. The basis of the "triploid block" resides in the endosperm. Triploids are not recovered because the associated endosperms do not differentiate normally and the developing seeds usually abort.

An explanation of the basis of this "triploid block" is provided by the endosperm balance number (EBN) hypothesis. It assumes that normal endosperm development occurs when the EBN of the endosperm is a balance of 2 EBN from the female to 1 EBN from the male. Deviations from the 2:1 ratio result in faulty endosperms and seed failure.

Non-additive genetic variance for yield

The genetic variance for yield in potatoes is predominantly non-additive (Mendiburu and Peloquin, 1977a). This was explained by Wright (1956) and Crow and Kimura (1965) as the result of evolution in asexual populations, in which epistatic combinations were accumulated to a greater extent than in the evolution of sexual populations. Therefore, intralocus interactions (heterozygosity) and interlocus interactions (epistasis) are very important in relation to breeding methods for increased yield. Thus, one must strive to obtain maximum heterozygosity (more than two alleles per locus) to increase these possible interactions. In this regard, the genetic architecture of the potato is significantly different from that of diploids and disomic polyploids. This difference must be recognized in the choice of breeding methods in potato improvement.

Breeding strategy

The breeding strategy for germplasm utilization in potatoes includes three components: (1) the species are the source of valuable traits and allelic diversity, (2) haploids of cultivars and advanced selections provide a method of capturing the genetic diversity (putting the unadapted germplasm in usable form), and (3) $2n$ gametes are an effective and efficient method of transmitting genetic diversity from unadapted germplasm to cultivated potatoes.

Types of ploidy manipulations

Three main types of ploidy manipulations are being used in transferring germplasm from the 2x to the 4x ploidy level: (1) obtaining 4x progeny from 4x x 2x crosses, (2) obtaining 4x progeny from 2x x 4x crosses, and (3) obtaining 4x progeny from 2x x 2x crosses. Scheme 1 requires functioning of 2n pollen, scheme 2 requires functioning of 2n eggs, and scheme 3 requires functioning of both 2n pollen and 2n eggs. All breeding schemes are dependent on a highly effective triploid block. This chapter concentrates on schemes 1 and 3.

Meiotic mutants

Meiotic mutants that result in formation of 2n pollen are the tools for transferring germplasm from the 2x to the 4x level. The most useful is parallel spindles in the second meiotic division (Mok and Peloquin, 1975a) which is inherited as a Mendelian recessive gene, ps (Mok and Peloquin, 1975b). It is genetically equivalent to a first division restitution (FDR) mechanism. The genetic consequence of 2n pollen formation by FDR is that about 80% of the heterozygosity and a large fraction of the epistasis are transmitted from the 2x parent to the 4x offspring. Fortunately, the ps allele occurs in a high frequency in both wild and cultivated potatoes. Parallel spindles, similar to other meiotic mutants, is characterized by variable expressivity and incomplete penetrance. Both genetic and environmental factors can modify the frequency of 2n pollen.

The most common meiotic variation that results in 2n eggs is omission of the second meiotic division (Werner and Peloquin, 1987). Results of genetic studies indicate a single recessive gene, os, controls the production of 2n eggs by this mechanism (Werner and Peloquin, 1990). Omission of the second meiotic division is genetically equivalent to a second division restitution (SDR) mechanism. It is estimated that SDR transmits less than 40% of the heterozygosity of the 2x parent to the 4x offspring.

Endosperm balance number (EBN)

The role of the endosperm in the "triploid block" can be illustrated with 4x x 2x and 2x x 4x crosses. The EBN hypothesis states that normal endosperm development occurs with a balance 2 female:1 male EBN. In a 4x (4EBN) x 2x (2 EBN) cross with normal n pollen, the EBN ratio will be 4 female:1 male and viable seeds with 3x embryos will rarely be obtained. However, if 2n pollen from the 2x parent function the 2 female:1 male EBN ratio is restored and plump, viable seeds with 4x embryos occur. Similarly, in a 2x (2 EBN) x 4x (4 EBN) cross, the EBN ratio in the hybrid is 1 female:1 male and the developing seeds abort. If 2n eggs function in the 2x parent, the resulting EBN ratio in the endosperm is 2 female:1 male, and normal seeds with 4x embryos are recovered. In the preceding examples, the EBN number and number of chromosomes sets were identical. However, 4x species with 2 EBN and 2x species with 1 EBN have been identified. In these, the EBN is more important than the number of sets of chromosomes in predicting the success of a cross and the ploidy level of the offspring.

RESULTS AND DISCUSSION

Unilateral sexual polyploidization

The essential ingredients of a potato breeding method to transfer unadapted germplasm into the 4x cultivated germplasm pool through the recovery of 4x progeny from 4x x 2x crosses (unilateral sexual polyploidization) are illustrated in Figure 1. Desirable 2x haploid species hybrids can be obtained through use of disomic inheritance. The 2x hybrids, then, can be crossed with standard 4x cultivars to obtain 4x progeny.

Figure 1 A potato breeding method for germplasm utilization using species, haploids and 2n pollen to obtain 4x progeny from 4x x 2x crosses.

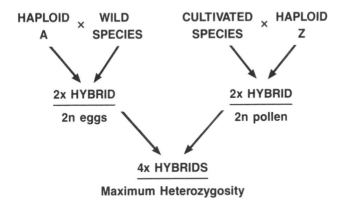

4x progeny from 4x x 2x crosses

Preliminary evaluation of this method was done by crossing 22 4x cultivars or advanced selections with 108 Phureja-haploid *Tuberosum* hybrids (Phureja is a cultivated 2x group from the Andean Region of South America). The number of seeds per fruit was less than two, indicating a severe triploid block and a low frequency of 2n pollen. The progeny were vigorous, mainly

4x, and the mean tuber yield of many 4x families was similar or better than the 4x parent (Hanneman and Peloquin, 1968).

Three of the Phureja-haploid hybrid 2x parents were unusual in the following respects: (1) high seed set of 30-50 seeds per fruit; (2) progeny were more than 99% 4x; (3) they were the parents of the highest yielding families; and (4) the progeny within a family were relatively uniform, which was unexpected because both 2x and 4x parents are highly heterozygous. Cytological observations of microsporogenesis in the three exceptional clones indicated that 2n pollen was formed by parallel spindles in the second meiotic division, a FDR mechanism (Mok and Peloquin, 1975a).

Yield of 4x progeny from 4x x 2x (FDR) crosses

The breeding value of 2x FDR clones and standard cultivars was compared by crossing nine cultivars with four FDR clones and eight cultivars (Mok and Peloquin, 1975c). The average yield in kilogram per hill of the means of 36 families from 4x cultivar x 2x FDR was more than 3.1 compared to 2.2 for the 76 families from intermating between 4x cultivars. Several results (Peloquin et al., 1990) indicated that high yielding 4x clones can be obtained from 4x x 2x (FDR) crosses. This breeding scheme also has wide adaptation because the best locally adapted cultivars from a particular area can be used in crosses with highly heterozygous adapted FDR 2x hybrids. Thus, what started out to be a method of transferring exotic germplasm from the 2x to 4x ploidy level ended up in being a method of increasing yield.

Cultivars released with Phureja germplasm

Several cultivars have been released from use of the 4x-2x breeding method where a Phureja-haploid hybrid is the 2x parent. Thirty-one parallel spindle-derived 4x clones were sent to the International Potato Center (CIP) from the University of Wisconsin Potato Breeding project. Several were early and high yielding in the lowland tropics. Three - "DTO-2", "DTO-28", and "DTO-33" - have been released as cultivars for use in countries with lowland tropics. Either the 4x x 2x (FDR) breeding scheme is very efficient or we were very lucky.

Another joint effort between the University of Wisconsin and CIP involves the transfer of bacterial wilt (*Pseudomonas solanacearum* E.F. Smith) resistance (BWR) found in unadapted 2x Phureja to the 4x level (Sequeira, 1979). The project has been very successful in the development of BWR cultivars through use of the ploidy manipulations approach. Several BWR cultivars have been released that make it possible to grow potatoes in areas of several countries were it was previously not possible due to the presence of the causal organism of bacterial wilt (International Potato Center, 1984). The Phureja source of BWR is very effective in tropical

highlands, but not in areas with higher temperatures. A new source of resistance in the wild species *S. sparsipilum* (Bitt.) Juz. & Buk. has been
identified and is being incorporated into 4*x* clones by CIP.

Collaboration between North Carolina State University and CIP has resulted in the identification of early blight (*Alternaria solanii* Sorauer) resistance in Phureja. Through the use of 2*n* gametes, the resistance has been
transferred to the cultivated 4*x* gene pool (Herriot *et al.*, 1990).

Phureja has thus provided allelic diversity for yield, adaptation to
the lowland tropics and resistance to bacterial wilt and early blight to the
cultivated 4*x* potato. However, there are two problems with Phureja. First,
the tubers of the Phureja-haploid hybrids tend to be rough due to deep eyes
and raised internodes. Further, hybrids with Phureja also produce too many
small tubers and highly variable sized tubers per hill. Second, when Phureja
is used as a male in crosses with *Tuberosum* haploids, almost all the F_1
hybrids are male sterile. The sterility results from the interaction of the
sensitive cytoplasm of *Tuberosum* with the dominant male sterility gene from
Phureja. The reciprocal cross usually produces male-fertile F_1 progeny, but
because about 98% of the haploids are male sterile, the male sterility of the
haploid x Phureja F_1 hybrids severely restricts the range of haploids that can
be used. The recent finding of a restorer gene for genetic-cytoplasmic male
sterility in many advanced selections and cultivars (Iwanaga *et al.*, 1991)
helps remedy this problem. Haploids, with the restorer gene can be extracted from these 4*x* clones, used in crosses with Phureja as the male parent and
male fertile haploid x Phureja hybrids recovered.

Haploid x wild species hybrids

The male sterility problem of haploid x Phureja F_1 hybrids was alleviated by
use of the wild, tuber-bearing species *S. chacoense* Bitt. (Leue and Peloquin,
1980). It can be crossed with male-sterile haploids to give male-fertile F_1
hybrids, allowing a wide range of male-sterile haploids to be used as parents. The results with *S. chacoense* lead to increased investigations with
haploid *Tuberosum* x wild species hybrids. Forty male-sterile haploids,
representing gametes of eight 4*x* clones were crossed with plants of eight 2*x*
wild species, *S. berthaultii* Hawkes, *S. boliviense* Dun., *S. canasense*
Hawkes, *S. infundibuliforme* Phil., *S. microdontum* Bitt., *S. raphanipholium*
Card. & Hawkes, *S. sanctae-rosae* Hawkes, and *S. tarijense* Hawkes.
Male-fertile F_1 hybrids were found in families with all species except *S.
infundibuliforme* (Hermundstad and Peloquin, 1985a). The results in regard
to tuberization in the haploid x wild species F_1 hybrid families were surprising but significant (Hermundstad and Peloquin, 1985b). Five major points
need particular emphasis: (1) there was large variation for tuberization
among full-sibs (some plants had no tubers and others had very good tuberization), (2) the F_1 hybrids that tuberized often outyielded their parents - the

best F_1 hybrids had three to four times the yield of the haploid parent (the wild species did not tuberize), (3) there was large variation among haploids in their ability to produce hybrids with good tuberization, (4) haploids obtained from the same $4x$ parent differed in their ability to produce hybrids that tuberized, and (5) tuberization in F_1 haploid x species hybrids was also influenced by the particular species and plants of a species used as parents. Similar results have been obtained with other haploids and with other species (Yerk and Peloquin, 1988).

Advantages of haploids

The value of haploids for germplasm enhancement is apparent from the previous results. In the seedling generation, haploid-wild species hybrids may be evaluated for percent tuberization to identify superior haploid and wild species parents. Further, tuber characteristics, fertility and $2n$ gamete production may also be studied at this stage. Selected hybrids may then be grown from tubers in replicated trials and evaluated for maturity, tuber set, specific gravity, reducing sugar level, tuber dormancy, and disease and pest resistance. Thus, one can evaluate the contribution of wild species to tuber characteristics through the haploid-species hybrids (Jansky et al., 1990).

Several other advantages also exist for using haploids for obtaining haploid-species hybrids. The good tuberization of many haploid-species hybrids allows for clonal maintenance of unadapted germplasm of the wild species which do not tuberize under the long days of the northern temperate regions. Tuberosum haploids provide the opportunity of adding only 25% unadapted germplasm to the cultivated $4x$ gene pool. This is important because crossing wild species directly with $4x$ cultivars, through use of $2n$ pollen from the species, results in $4x$ progeny with 50% exotic germplasm and poorly adapted material. Selection for adaptation in wild species before crossing to cultivars has so far proved very ineffective (Jacobsen and Jansky, 1989). Thus, one must cross the $4x$ clones with 50% exotic germplasm to cultivars and contend with all the complications of tetrasomic inheritance in order to obtain adapted $4x$ clones.

Preliminary evidence indicates that early maturing haploids are better than late maturing haploids for obtaining haploid-wild species hybrids with good tuberization and early maturity. We are currently evaluating this relationship through the use of early and late haploids from several cultivars and advanced selections. Through the use of haploid-wild species hybrids we are for the first time in a position to evaluate the contributions of various species, of different plant introductions (PIs) within a species and of plants within a PI, for tuber characteristics. This should provide guidelines for the breeders and geneticists to capture the wealth of genetic diversity present in wild species. The use of haploids to put wild species germplasm into a form that will tuberize and can, therefore, be planted with a planter and dug with a digger, should facilitate the use of this material in potato breeding programs.

Since potato cultivars are highly heterozygous, haploids extracted from them can be heterozygous at some loci and homozygous at other loci. In addition most 2*x* wild species, are self-incompatible, and results indicate that they are highly heterozygous. As a consequence, large variation in haploid-wild species hybrids occurred for male fertility, 2*n* pollen production, tuberization, tuber dormancy, specific gravity, reducing sugar content, and plant maturity (Yerk and Peloquin, 1988, 1989).

ps gene frequency

The exploitation of 4*x* x 2*x* crosses for germplasm transfer from the 2*x* to 4*x* level is dependent on the occurrence and frequency of the *ps* allele in both haploid and species parents. The frequency in haploids is related directly to the frequency of this allele in 4*x* clones from which they are derived. Analysis of 56 U.S. cultivars and advanced selections for their genotype at the *ps* locus indicated a *ps* gene frequency of 0.69 (Iwanaga and Peloquin, 1982). Further, about 65% of the 4*x* clones were simplex, *Pspspsps*. Simplex 4*x* clones, assuming chromosome segregation, produce haploids in a ratio of one *psps* to one *Psps*. With a duplex (*PsPspsps*) 4*x* clone, about 67% of its haploids would be *Psps* and 17% *psps*. An estimation of the *ps* gene frequency in Group Andigena (the cultivated 4*x* potato of the Andean Region) was 0.83 (Watanabe and Peloquin, 1989). Thus, haploids of Andigena either heterozygous or homozygous at the *ps* locus are easy to obtain. Plants with 2*n* pollen have been detected in almost all 2*x* wild species screened. Assuming Hardy-Weinberg equilibrium, the *ps* gene frequency varied from 0.15 to 0.40 in different wild species. It is important to select wild species plants with 2*n* pollen to cross with haploids. Fortunately, it is easy to screen for 2*n* pollen production by staining the pollen and identifying plants that produce large 2*n* pollen grains along with normal *n* pollen.

Yield of 4*x* families from 4*x* x 2*x* (haploid x wild species hybrids)

The tuber yield of 4*x* families from 4*x* x 2*x* crosses where the 2*x* parents were hybrids between haploids and wild species was evaluated. These hybrids were selected for improved tuber type and adaptation. The tuber yield of their 4*x* families was compared with that of 4*x* x 4*x* families and that of the 4*x* parents. Six 2*x* hybrids of *Tuberosum* haploid x *S. tarijense* or *S. berthaultii* that form 2*n* pollen by FDR were crossed to four 4*x* clones. Twenty-three 4*x* x 2*x* families were evaluated along with the 4*x* parents and seven 4*x* x 4*x* families for tuber yield, specific gravity, chip color, vine maturity, and general tuber appearance in three environments (Darmo and Peloquin, 1991). The mean tuber yield of 4*x* x 2*x* progeny was significantly higher than the mean of the 4x x 4x progeny. More importantly, it also exceeded the mean yield of the 4x parental clones (Figure 2). Compared with the 4x parents and 4x x 4x families, the 4x x 2x families also had higher specific gravity and better general tuber appearance, but darker chip color. The vine maturity of the 4x clones was significantly earlier than that of 4x x 2x and 4x x 4x families.

Figure 2 Tuber yield (kg hill^{-1}) of 4x families from 4x x 2x (left) and 4x x 4x (right) crosses and 4x parental clones (middle) grouped by 4x parent.

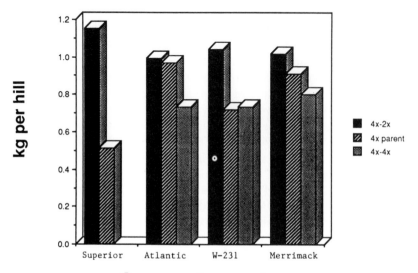

The performance of haploid x wild species 2x hybrids, involving five newly used species (*S. bukasovii* Juz., *S. gourlayi* Hawkes, *S. multidissectum* Juz., *S. vernei* Bitt. & Wittm., and *S. verrucosum* Schlechtd.), were evaluated in 4x x 2x families (Yerk and Peloquin, 1990). All 2x male parents formed 2n pollen by FDR. The mean yield of the 4x x 2x families significantly exceeded that of the 4x x 4x families used for comparison. General tuber apperance was similar in both groups, but vine maturity was significantly later in the 4x x 2x families. Two main points need emphasis: (1) it appears that high tuber yield can be obtained from 4x x 2x crosses irrespective of the species in the 2x parent as long as the haploid-species hybrids form 2n pollen by FDR and (2) one needs to select haploid-wild species 2x hybrids with early maturity in order to obtain 4x progeny with early maturity. This agrees with the results of Schroeder (1984) who found a very high correlation between the maturity of the 2x hybrid and of the 4x progeny where the 2x hybrid was a Phureja-haploid hybrid.

Yield, yield stability, and tuber quality of PSD 4x clones

Specific clones were selected from 4x families obtained from 4x-2x (FDR) crosses. These were identified as parallel spindle-derived (PSD) 4x clones. The results of evaluation of these selected clones from 4x-2x crosses were very promising (Darmo and Peloquin, 1990). Nine PSD 4x clones were compared with four commercial cultivars in six Wisconsin environments. The five PSD 4x clones with *S. tarijense* germplasm were selected from 53 seedlings and the four PSD 4x clones with Phureja germplasm were selected from several hundred seedlings with some prior selection, but little evaluation. The traits evaluated were marketable yield, specific gravity, chip color, and yield stability. Three PSD clones were more stable and not significantly different in yield from the best cultivar Atlantic. Four of the five PSD 4x clones with *S. tarijense* germplasm had specific gravity equal to or higher than Atlantic. Five of the nine PSD 4x clones had chip color ratings equal or better than the two standard chipping cultivars Atlantic and Norchip. Also of interest are the clones which combine high specific gravity with good chip color. Three of PSD 4x clones were similar to Atlantic for the combination of specific gravity and good chip color.

The specific gravity and chip color of 10 PSD 4x clones was further evaluated by comparing them with 12 cultivars and 14 advanced selections (from five state potato breeding programs) at three locations in Wisconsin. The ordered specific gravity of the best 14 clones is listed in Table 1. It is obvious that PSD 4x clones compared very favorably with the standard cultivars and advanced selections in this important quality trait. The ordered chip color score of the best 12 clones is presented in Table 2. Again the PSD 4x clones were outstanding in chip color compared to the two leading chipping cultivars Norchip and Atlantic and the best advanced selection W 855.

The ability to obtain clones which are equal or superior to cultivars from small initial populations indicates that the 4x x 2x (FDR) method may increase the efficiency of cultivar development.

Transfer of disease and pest resistance

The International Potato Center is one of the several institutions in the world applying the ploidy manipulation approach using haploids, species and 2n gametes in potato breeding for disease/pest resistance. The goal is to develop 2x clones from a wide genetic background, through the use of many wild species with specific resistances, that produce 2n pollen. Transfer of resistance to the 4x level from the 2x level has been successful mainly

Table 1 Ordered specific gravity of ten 4x clones from 4x-2x crosses, three cultivars (Atlantic, Norchip, and Russet Burbank), and the best advanced selection (W 855) evaluated at three Wisconsin locations in 1989.

Hancock		Antigo		Rhinelander	
S 482*	1.103	S 482	1.098	S 482	1.083
S 478	1.091	S 478	1.089	W 855	1.079
S 440	1.089	D 55	1.088	S 440	1.076
S 459	1.088	D 45	1.088	S 478	1.074
S 452	1.088	Atlantic	1.086	S 452	1.074
D 55	1.087	W 855	1.085	S 459	1.073
Atlantic	1.087	Norchip	1.085	S 465	1.072
D 45	1.086	S 440	1.084	D 45	1.072
R. Burbank	1.086	S 452	1.084	S 487	1.072
S 487	1.086	S 459	1.083	S 438	1.070
W 855	1.085	S 487	1.082	Atlantic	1.069
S 438	1.085	R. Burbank	1.080	D 55	1.068
S 465	1.082	S 465	1.079	Norchip	1.068
Norchip	1.076	S 438	1.074	R. Burbank	1.066
$LSD_{0.05}$	0.007		0.005		0.004
CV (%)	4.6		2.8		3.0

*S and D numbers are parallel spindle derived 4x clones.

through the use of FDR 2n pollen. The genetic background of the 2x clones includes haploids of Groups Andigena and *Tuberosum*, x cultivated Groups Phureja and Stenotomum, and the 2x wild species *S. chacoense*, *S. microdontum*, *S. sparsipilum*, and *S. vernei*.

A successful transfer of root-knot nematode, *Meloidogyne* spp. (RKN), resistance genes from wild species to the 4x cultivated gene pool has been achieved. Breeding for RKN is important for potato production in lowland tropics. The frequency and degree of resistance is very low in the cultivated potato; however, a high level of resistance exists in *S. sparsipilum*. Iwanaga *et al.* (1989) evaluated RKN resistance in the 2x breeding population, and transferred the resistance from the 2x to the 4x level by FDR 2n pollen. At the 2x level, 1170 seedlings from 23 families were screened, and 14% were

Table 2 Ordered chip color of nine 4x clones from 4x-2x crosses, two cultivars (Atlantic and Norchip), and the best advanced selection (W 855) evaluated at three Wisconsin locations in 1989.

Hancock		Antigo		Rhinelander	
S 440*	3.0	S 440	3.0	S 438	2.9
W 855	3.5	S 438	3.2	S 440	3.0
S 438	3.7	S 482	3.3	S 478	3.0
S 482	4.7	Norchip	3.4	S 482	3.1
D 55	4.7	D 55	3.5	W 85	3.1
D 478	5.2	W 855	3.5	S 459	3.2
D 45	5.2	S 465	3.5	S 465	3.2
S 465	5.2	D 45	3.7	Atlantic	3.2
S 459	5.2	S 459	3.8	D 55	3.2
S 452	5.7	S 478	3.9	S 452	3.3
Atlantic	5.7	Atlantic	4.0	D 45	3.3
Norchip	6.0	S 452	4.8	Norchip	3.3
$LSD_{0.05}$	0.9		1.3		0.9
CV (%)	7.4		12.6		11.7

*S and D numbers are parallel spindle derived 4x clones.

RKN resistant. Estimates of narrow-sense heritability (0.62 and 0.48 for average root gall index and percentage of RKN-resistant progeny, respectively) indicates that phenotypic recurrent selection is appropriate to continue increasing the levels of resistance at the 2x level. At the 4x level, 2183 seedlings were screened, and 14.6% were RKN resistant. There was RKN resistance only in the progeny of crosses between RKN-susceptible 4x female x RKN-resistant FDR 2n pollen producing 2x. There was no RKN resistance in the progeny of 4x x 4x crosses.

Another important storage pest in developing countries is the potato tuber moth (PTM), *Phthorimoea opercullela* (Zeller). Resistance has been found mainly in *S. sparsipilum*. Ortiz *et al.* (1990) generated 62 2x families and evaluated them for PTM resistance under natural infestation in storage and in laboratory tests. They were able to transfer the high level of resistance from the original accessions into an advanced 2x population using phenotypic recurrent selection in the haploid x species hybrid population. Their results suggest that 4x x 2x crosses can be used to transfer the resistance into the cultivated 4x gene pool.

More than 100 clones which produce FDR 2n pollen and have known pest and/or disease resistance - including resistance to cyst nematode (CN, *Globodera pallida* Stone), bacterial wilt, early blight, late blight [LB, *Phytophthora infestans* (Mont). de Bary], potato virus Y (PVY), potato virus X (PVX) and potato leaf roll virus (PLRV) - have been developed at CIP and are available for use by breeders of national programs.

The Institute for Potato Research (Poland) has also used the ploidy manipulation approach to transfer the resistance genes from the wild species to the 4x cultivated gene pool (Zimmoch-Guzowska, 1986). Hybrids were produced between *Tuberosum* haploids and the following species: *S. chacoense, S. gourlayi, S. phureja* Juz. & Buk., *S. yungaense* Hawkes, *S. vernei*, and *S. microdontum*. Hybrid clones with extreme resistance to PVX, field resistance to PVY and resistant to potato wart (*Synchytrium endobioticum* Schilb.) were identified. Other hybrids were resistant to nematodes, potato viruses M, LB, PLRV, and soft and dry rot (*Erwinia* spp.). The transfer of the desirable gene(s) from the 2x level via 2n gametes is underway.

Bilateral sexual polyploidization

A breeding method for germplasm utilization using species, haploids, 2n eggs and 2n pollen to obtain 4x progeny from 2x x 2x crosses is illustrated in Figure 3. The 2x parents are unrelated haploid-species hybrids, one of which produces 2n eggs and the other 2n pollen.

Yield of 4x from 2x x 2x crosses

Mendiburu and Peloquin (1977b) obtained both 4x and 2x progeny from crosses between 2x haploid x Phureja hybrids that produce 2n pollen and 2n eggs. The 4x progeny were more vigorous and had significantly higher tuber yields than the 2x "full sibs" if the 2n pollen was formed by FDR.

Sexual and asexual polyploidization

The 4x progeny from 2x x 2x crosses also significantly outyielded 4x clones obtained from colchicine doubling of 2x parents (Peloquin *et al.*, 1989c). Production of 4x clones through colchicine doubling (asexual polyploidization) did not increase yield over the undoubled 2x clones. But production of 4x clones from 2n gametes (sexual polyploidization) significantly increased yield in the 4x progeny compared to the 2x parental clones.

Figure 3 A potato breeding method for germplasm utilization using species, haploids, 2n eggs and 2n pollen to obtain 4x progeny from 2x x 2x crosses.

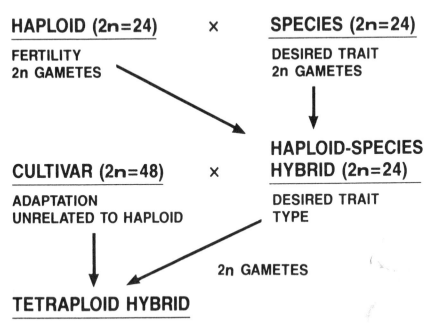

HAPLOID (2n=24)   x   SPECIES (2n=24)

FERTILITY   **DESIRED TRAIT**
2n GAMETES   **2n GAMETES**

HAPLOID-SPECIES
CULTIVAR (2n=48)   x   HYBRID (2n=24)

ADAPTATION   **DESIRED TRAIT**
UNRELATED TO HAPLOID   **TYPE**

                    2n GAMETES

TETRAPLOID HYBRID

ADAPTATION AND HORTICULTURAL TYPE FROM CULTIVAR
DESIRED TRAIT AND ALLELIC VARIATION FROM HAPLOID-SPECIES HYBRID

These results can be explained in terms of number of alleles per locus and number of first-order interallelic interactions (Watanabe *et al.*, 1991). Assuming the parents are unrelated (A^1A^2 and A^3A^4) following colchicine doubling, the 4x clones will be diallelic, $A^1A^1A^2A^2$ and $A^3A^3A^4A^4$, with one first-order interaction. In contrast, sexual polyploidization between the 2x parents can result in four alleles ($A^1A^2A^3A^4$) per locus and six first-order interactions in the 4x progeny. Also, 4x progeny from sexual poly-ploidization of the 2x clones (A^1A^2 x A^1A^3) have three alleles per locus and three first-order interactions.

The results for yield in 4x progeny from bilateral sexual polyploidization were confirmed and extended by Chujoy (1987) with different haploid x species hybrids. He obtained 4x and 2x progeny from many crosses between *Tuberosum* haploid x *S. chacoense* hybrids that produce 2n eggs by SDR and Phureja x *Tuberosum* haploid hybrids that produce 2n pollen by FDR. The means and ranges in tuber yields of 2x and 4x families from 2x x 2x crosses are presented in Table 3. The 4x families significantly outyielded the 2x families at both locations. It is interesting that the inbreeding coefficient

Table 3 Means and ranges of tuber yield (kg hill^{-1}) of 2x and 4x families from 2x x 2x crosses in two locations.

Group	No. families	Hancock (1984)	Rhinelander (1984)
4x	44	1.38 [0.53-2.48]*	0.64 [0.00-1.33]
2x	42	0.88 [0.11-1.68]	0.27 [0.00-0.76]
Comparisons			
4x vs. 2x full-sibs		**	**

*Range in brackets.

**Indicates that group means compared are significantly different at the 0.01 probability level.

was always higher in the 4x families than that of the 2x families, but sexual polyploidization still resulted in higher yielding 4x families. Surprisingly, he found that after crosses between *Tuberosum* haploid-*S. chacoense* full-sib hybrids, the 4x progeny had significantly higher tuber yield than the 2x parents (heterosis), but the 2x progeny had similar yield to the 2x parents. We think this finding has important implications in regard to the role of sexual polyploidization in the evolution of polyploids.

Recently, Werner and Peloquin (1991) compared the tuber yields and other tuber characteristics of nineteen 4x families obtained from 2x (SDR) x 2x (FDR) crosses with seven 4x check cultivars at two Wisconsin locations (Table 4). The mean tuber yield of the 4x families significantly exceeded that of the 4x cultivars at both locations. However, the tuber set per hill was two to three times larger in the 4x families than in 4x cultivars. This, and the relatively rough tubers of the 4x from 2x x 2x crosses, prevents immediate exploitation of this method. We have, however, identified haploid x species hybrids with desirable tuber type and tuber number per hill, plus either 2n eggs or 2n pollen. New 4x clones from these 2x hybrids will be generated and tested to further evaluate the breeding method of obtaining 4x clones from 2x x 2x crosses.

Table 4 Means and ranges of nineteen 4x families from 2x x 2x crosses and of seven cultivars for tuber yield (kg hill^{-1}) and tuber set at two locations.

Group	Tuber yield		Tuber set*	
	Hancock	Rhinelander	Hancock	Rhinelander
4x families	1.80	1.04	3.00	2.89
	[1.16-2.54]†	[0.69-1.77]	[3.00]	[2.00-3.00]
Cultivars	1.41	0.80	1.32	1.20
	[1.19-1.59]	[0.57-1.03]	[1.00-1.75]	[1.00-1.50]
Families vs. cultivars	**	**	**	**

*1 to 3 scale (1 = below 10; 3 = above 30 tubers hill^{-1}).

†Range in brackets.

**Indicates that group means compared are significantly different at the 0.01 probability level.

Complementation of SDR and FDR

The previous results demonstrate that 2x (SDR) x 2x (FDR) crosses can result in vigorous, high yielding 4x progeny. This was somewhat of a surprise, but is certainly worthy of attempted explanations. It is possible to envisage how SDR and FDR can complement each other from the standpoint of transmission of heterozygosity. With FDR, 100% of the parental heterozygosity from the centromere to the first crossover and 50% from the first to second crossover is transmitted from the 2x parent to the 4x offspring. With SDR, 0% of the heterozygosity from the centromere to the first crossover and 100% from the first to second crossover is transmitted from the 2x parent to the 4x progeny. Thus, FDR x FDR is superior to SDR x FDR from the centromere to the first crossover, but SDR x FDR is superior to FDR x FDR from the first to second crossover.

It is also worthwhile to consider these two mating systems in regard to the concept of maximum heterozygosity (more than two alleles per locus is advantageous for yield). As an example, crosses between unrelated 2x hybrid clones (A^1A^2 x A^3A^4) would generate different genotypic frequencies in the 4x progeny depending on proximity to the centromere. For loci near the centromere (no single exchange tetrads), 100% would be tetrallelic following FDR x FDR and 100% triallelic after SDR x FDR. In contrast, when the locus is very distal to the centromere (100% single exchange tetrads), FDR x FDR results in 25% tetrallelic, 50% triallelic, and 25% diallelic and SDR x FDR results in 50% tetrallelic, and 50% triallelic. If we assume that tetra- and triallelic loci are superior to mono- and diallelic loci for yield, then SDR x FDR could result in slightly better yielding progeny than FDR x FDR.

Bilateral sexual polyploidization and transfer of pest resistance

Population improvement at the $2x$ level using haploid species hybrids to improve agronomic traits and to combine multiple resistances to diseases and pests as well as tolerance to abiotic stress seems to be feasible in potato breeding. After combining the desirable attributes with a high frequency of $2n$ gamete production in selected individuals, they will be used for the production of $4x$ hybrids from $2x$ x $2x$ crosses. A hypothetical example illustrates the proposal. Haploids will be extracted from cultivars resistant to viruses; e.g., the cultivar Pirola is immune to PVY, and Atlantic is immune to PVX. After screening for virus resistance, resistant haploids will be crossed with species carrying a specific attribute; e.g., PI 218215 (*S. tarijense* from Argentina) with BW resistance and the clone CIP 760147.7 (*S. sparsipilum* from Peru) with resistance to RKN, CN and PTM. The haploid-species hybrids will then be screened for the specific combined resistances, and the production of SDR $2n$ eggs and FDR $2n$ pollen. Thus, individuals which combine two resistances and production of $2n$ gametes in each population will be crossed to produce highly heterozygous and resistant $4x$ hybrids. At minimum, all the individuals can be triallelic for the loci located close to the centromere, and 50% of them could be tetrallelic for the loci located after the first crossover. Therefore, high yield will be expected due to maximum heterozygosity in the resultant $4x$ progeny.

Production of potatoes from true potato seed (TPS)

This alternative potato production method is particularly suited to areas that cannot produce virus-free tubers and cannot afford to import clean tubers. The main advantages of TPS are (1) economic considerations, importing clean tubers can represent 40-80% of total production costs in contrast to 2-5% for TPS, (2) viruses are not often seed transmitted, (3) efficient transportion (100 g TPS vs. 2000 kg of tubers needed to plant a hectare), (4) refrigerated storage is not necessary for TPS, and (5) all potato tubers can be consumed, because none are needed for next year planting.

The best method of obtaining high yielding and uniform TPS progeny appears to be the use of $4x$ families from $4x$ x $2x$ crosses where both parents are adapted to the area where TPS will be grown. These families had better tuber yield, seedling vigor and tuber uniformity than hybrid families obtained from 4x x 4x crosses or progeny obtained from open-pollinated seed (Peloquin *et al.*, 1989c).

The production of 4x x 2x hybrid seed requires emasculation, if the 4x parent is male fertile, collection of pollen and hand pollination. We have developed a scheme that could reduce seed production costs. The new haploid-wild species hybrids with desirable tuber type and high 2n gamete production provide the opportunity for the production of inexpensive 4x hybrid TPS from 2x x 2x crosses. The system utilizes plants producing either 2n eggs or 2n pollen. One haploid-species hybrid (female parent) would be selected for male fertility (to attract bumblebees), self-incompatibility (to avoid selfing), high frequency of 2n eggs (to obtain high seed set) and no 2n pollen (to avoid self-compatibility); an unrelated haploid-species hybrid (male parent) would be selected for male fertility and high 2n pollen production. In addition, both parents would be selected for profuse flowering and other desirable characteristics (e.g., proper maturity). The male and female parents will be planted in alternate hills with bumblebees collecting the pollen and doing the pollinations. The elimination of emasculation, pollen collection, and hand pollination could reduce 4x hybrid TPS costs by more than 50%.

Ploidy manipulations for germplasm transfer in other crops

The transfer of exotic germplasm into the cultivated gene pool using ploidy manipulations has been used in alfalfa, cassava, sweet potato, banana (*Musa* spp.) (Rowe, 1984), blackberry (*Rubus* spp.) (Hall, 1990), peanuts (*Arachis hypogaea* L.) (Singh *et al.*, 1990), sugar cane (*Saccharum officinarum* L.) (Bremer, 1961), strawberry (*Fragaria* spp.) (Bringhurst and Voth, 1984), and appears promising in other fruit species (Sanford, 1983) and yams.

Cultivated alfalfa, is a tetrasomic polyploid (4x). Germplasm from 2x *Medicago falcata* L. was used in the development of eight U.S. cultivars including Ladak and Vernal (Bolton, 1962). However, only a fraction of the available *M. falcata* germplasm has been evaluated for cultivar improvement. Bingham (1968) indicated that the transfer of germplasm from 2x *M. falcata* to cultivated alfalfa via 4x x 2x crosses was efficient. All 4x hybrids were vigorous and fertile. Further, hybrids between haploids of *M. sativa* and *M. falcata* can be used as 2x parents in 4x x 2x crosses (Bingham and McCoy, 1979). Recently, Bingham (1990) proposed backcrossing of the 4x cultivated alfalfa into diploid *M. falcata* using 2n eggs.

The sweet potato [*Ipomoea batatas* (L.) Poir.] is a hexasomic polyploid (6x). The section *Batatas* of the genus *Ipomoea* consists of the cultivated sweet potato and 11 closely related wild species. The wild species *I. trifida* (HBK) G. Don has been utilized successfully in sweet potato breeding programs in Asia (Sakamoto, 1970) and by CIP (Iwanaga *et al.*, 1991). *Ipomoea trifida* has 2x, 4x, and 6x cytotypes. One 6x cytotype has been used extensively in Japan as a source of resistance to nematodes, leading to the release of successful cultivars (Sakamoto, 1976). The Japanese cultivar Minamiyutaka has one-eighth of its genes from $6x$ *I. trifida*, and it is resistant to root-lesion nematode, and has high starch content and high yield. The latter is possibly the result of heterosis caused by the introgression of exotic germplasm from *I. trifida* to the cultivated gene pool. However, $6x$ cytotypes of *I. trifida* are rare in contrast to the numerous $2x$ and $4x$ accessions of *I. trifida* available. The latter cannot be used directly due to the difference in ploidy levels. However, the presence of $2n$ gametes in $2x$ and $4x$ *I. trifida* (Orjeda *et al.*, 1990) provides a way for the efficient utilization of this germplasm.

Freyre *et al.* (1991) found $2n$ pollen production in $3x$ individuals obtained from $4x$ x $2x$ crosses in *I. trifida*. Then, $3x$ with $2n$ pollen production were successfully crossed with *I. batatas*. In this way, the transfer of desirable genes from $2x$ and $4x$ cytotypes of *I. trifida* to the cultivated gene pool of sweet potato was achieved. Moreover, Freyre *et al.* (1991) indicated that synthetic $6x$ (obtained by colchicine doubling of $3x$ from $4x$ x $2x$) had reduced vigor due to inbreeding depression. This also limits the heterozygosity of the *I. batatas* x *I. trifida* progeny. Therefore, the use of FDR $2n$ pollen from $3x$ *I. trifida* in crosses with $6x$ sweet potato seems more attractive. High yielding progeny are expected from such crosses due to maximum heterozygosity as a consequence of an increase in allelic diversity. It is also interesting to note that $2n$ eggs function in $3x$ *I. trifida*, since $6x$ plants can be obtained from $3x$ genotypes.

Another obstacle for using wild *Ipomoea* species for sweet potato improvement is that none of the wild species of section *Batatas* produce storage roots. Thus, it is impossible to evaluate desirable storage root traits (earliness, eating quality, starch content, nutritive value, storage capability and weevil resistance) in the wild *Ipomoea* species. Orjeda (1989) developed an alternative path for the screening of wild $2x$ and $4x$ *Ipomoea* for desirable storage root traits. She produced a $4x$ interspecific population between *I. batatas* ($6x$) and *I. trifida* ($2x$) with storage-root production capability. This population is being used in crosses with other $2x$ and $4x$ wild accessions to obtain progeny that can be screened for root traits.

An interesting example of bilateral sexual polyploidization has recently been reported in cassava by Hahn *et al.* (1990). A total of 27 *Manihot* species examined cytologically all have the chromosome number, $2n = 2x = 36$, the same as cassava. Interspecific crosses of five cultivated cassava varieties were made with two related *Manihot* species, *M. epruinosa* Pax & K. Hoffmann and *M. glaziovii* Muell. From these diploid interspecific crosses, four tetraploids ($2n = 4x = 72$) were identified, suggesting the functioning of both $2n$ eggs and $2n$ pollen. Cytological examination of pollen from eight families of $2x$ interspecific hybrids indicated that from one-third to one-half of the $2x$ plants in each family produced $2n$ pollen. The most interesting result was the vigor and yields of the 4*x* clones. One of them performed as well as the best variety in uniform yield trials conducted in Nigeria. Obtaining tetraploids from 2*x* x 2*x* crosses is also an effective method of incorporating disease and pest resistance and other desirable traits from wild species to cultivated cassava.

The results from cassava and sweet potato support the concept developed in potato of the importance of allelic diversity in relation to yield in polysomic polyploids.

ACKNOWLEDGMENTS

Paper No. 3199 from the Laboratory of Genetics. Research supported by the College of Agricultural and Life Sciences; International Potato Center, USDA-CRGO-88-37234 3619, and Frito Lay, Inc.

REFERENCES

Bingham, E. T. (1968) Transfer of diploid *Medicago* spp. germplasm into tetraploid *M. sativa* L. in 4*x*-2*x* crosses. *Crop Science* 8:760-762.

Bingham, E. T. (1990) Backcrossing tetraploidy into diploid *Medicago falcata* L. using 2*n* eggs. *Crop Science* 30:1353-1354.

Bingham, E. T.; McCoy, T. J. (1979) Cultivated alfalfa at the diploid level: Origin, reproductive stability and yield of seed and forage. *Crop Science* 19:97-100.

Bolton, J. L. (1962) *Alfalfa*. Interscience Publishers, New York.

Bremer, G. (1961) Problems in breeding and cytology of sugarcane. *Euphytica* 10:59-78.

Bringhurst, R. S.; Voth, V. (1984) Breeding octoploid strawberries. *Iowa State University Journal of Research* 58:371-382.

Chujoy, J. E. (1987) Tuber yields of 2*x* and 4*x* progeny from 2*x* x 2*x* crosses in potatoes; barriers to interspecific hybridization between *Solanum chacoense* Bitt. and *S. commersonii* Dun. Ph.D. Dissertation, University of Wisconsin-Madison. Dissertation Abstracts 47(1):2B Order No. DA85284.

Crow, J. F.; Kimura, M. (1965) Evolution in sexual and asexual populations. *American Naturalist* 99:439-450.

Darmo, E.; Peloquin, S. J. (1990) Performance and stability of nine 4*x* potato clones from 4*x*-2*x* crosses compared with that of four commercial cultivars. *Potato Research* 33:357-364.

Darmo, E.; Peloquin, S. J. (1991) The use of 2*x* *Tuberosum* haploid-wild species hybrids to improve yield and quality in 4*x* cultivated potatoes. *Euphytica* 53:1-9.

Freyre, R.; Iwanaga, M.; Orjeda, G. (1991) Use of *Ipomoea trifida* (HBK) G. Don germplasm for sweet potato improvement. 2. Fertility of synthetic hexaploids and triploids with 2*n* gametes of *I. trifida*, and their interspecific crossability with sweet potato. *Genome* 34:209-214.

Hall, H. K. (1990) Blackberry breeding. *Plant Breeding Reviews* 8:244-312.

Hahn, S. K.; Bai, K. V.; Asiedu, R. (1990) Tetraploids, triploids and 2*n* pollen from diploid interspecific crosses with cassava. *Theoretical and Applied Genetics* 79:433-439.

Hanneman, R. E., Jr. (1989) Potato germplasm resources. *American Potato Journal* 66:655-668.

Hanneman, R. E., Jr.; Bamberg, J. B. (1986) Inventory of tuber-bearing *Solanum* species. University of Wisconsin-Madison Research Bulletin 533.

Hanneman, R. E., Jr.; Peloquin, S. J. (1968) Use of Phureja and haploids to enhance the yield of cultivated tetraploid potatoes. *American Potato Journal* 46:436.

Hawkes, J. G. (1945) The indigenous American potatoes and their value in plant breeding. Part I. Resistance to disease. Part II. Physiological properties, chemical composition and breeding capabilities. *Empire Journal of Experimental Agriculture* 13:11-40.

Hermundstad, S. A.; Peloquin, S. J. (1985a) Male fertility and 2n pollen production in haploid x wild species hybrids. *American Potato Journal* 62:479-482.

Hermundstad, S. A.; Peloquin, S. J. (1985b) Germplasm enhancement with potato haploids. *Journal of Heredity* 76:463-467.

Herriot, A. B.; Haynes, F. L.; Shoemaker, P. B. (1990) Inheritance of resistance to early blight disease in tetraploid x diploid crosses of potato. *HortScience* 25:224-226.

International Potato Center (1984) *Potatoes for the developing world*. Lima, Peru.

Iwanaga M.; Peloquin, S. J. (1982) Origin and evolution of cultivated 4x potatoes via 2n gametes. *Theoretical and Applied Genetics* 78:329-336.

Iwanaga, M.; Freyre, R.; Orjeda, G. (1991) Use of *Ipomoea trifida* (HBK) G. Don germplasm for sweet potato improvement. 1. Development of synthetic hexaploids of *I. trifida* by ploidy-level manipulations. *Genome* 24:201-208.

Iwanaga, M.; Jatala, P.; Ortiz, R.; Guevara, E. (1989) Use of FDR 2n pollen to transfer resistance to root-knot nematodes into cultivated 4x potatoes. *Journal of the American Society of Horticultural Science* 114:1008-1013.

Iwanaga, M.; Ortiz, R.; Cipar, M. S.; Peloquin, S. J. (1991) A restorer gene for genetic-cytoplasmic male sterility in cultivated potatoes. *American Potato Journal* 68:19-28.

Jacobsen, T. L.; Jansky, S. H. (1989) Effects of pre-breeding wild species on tuberization of *Solanum tuberosum* haploid wild species hybrids. *American Potato Journal* 66:803-811.

Jansky, S. H.; Peloquin, S. J.; Yerk, G. L. (1990) The use of potato haploids to put 2x wild species germplasm in a usable form. *Plant Breeding* 104:290-294.

Johnston, S. A.; den Nijs, T. P. M.; Peloquin, S. J.; Hanneman, R. E., Jr. (1980) The significance of genic balance to endosperm development in interspecific crosses. *Theoretical and Applied Genetics* 57:5-9.

Leue, E. F.; Peloquin, S. J. (1980) Selection for 2n gametes and tuberization in *S. chacoense*. *American Potato Journal* 57:189-195.

Mendiburu, A. O.; Peloquin, S. J. (1977a) The significance of 2n gametes in potato breeding. *Theoretical and Applied Genetics* 49:53-61.

Mendiburu, A. O.; Peloquin, S. J. (1977b) Bilateral sexual polyploidization in potatoes. *Euphytica* 26:573-583.

Mok, D. W. S.; Peloquin, S. J. (1975a) Three mechanisms of 2n pollen formation in diploid potatoes. *Canadian Journal of Genetics aand Cytology* 17:217-225.

Mok, D. W. S.; Peloquin, S. J. (1975b) The inheritance of diplandroid (2n pollen) formation in diploid potatoes. *Heredity* 35:295-302.

Mok, D. W. S.; Peloquin, S. J. (1975c) Breeding value of 2n pollen (diplandroids) in 4x x 2x crosses in potatoes. *Theoretical and Applied Genetics* 46:307-314.

Orjeda. G. (1989) Development of 4x interspecific sweet potato hybrids as storage root initiators of diploid and tetraploid wild species. *CIP Circular* 17:12-13.

Orjeda, G.; Freyre, R.; Iwanaga, M. (1990) Production of 2n pollen in diploid *I. trifida*, a putative wild ancestor of sweet potato. *Journal of Heredity* 81:462-467.

Ortiz, R.; Iwanaga, M.; Raman, K. V.; Palacios, M. (1990) Breeding for resistance to potato tuber moth, *Phthorimoea opercullela* (Zeller), in diploid potatoes. *Euphytica* 50:119-126.

Peloquin, S. J.; Jansky, S. H.; Yerk, G. L. (1989a) Potato cytogenetics and germplasm utilization. *American Potato Journal* 66:629-638.

Peloquin, S. J.; Werner, J. E.; Yerk, G. L. (1990) The use of potato haploids in genetics and breeding. In: Gupta, P. K.; Tsuchiya, T. (eds.), *Chromosome engineering in plants. Part B.* Elsevier, Essex, England, pp. 79-92.

Peloquin, S. J.; Yerk, G. L.; Werner, J. E. (1989b) Ploidy manipulations in potato. In: Adolph, K. W. (ed.), *Chromosomes: Eukaryotic, prokaryotic and viral. Vol. II.* CRC Press, Boca Raton, FL, pp. 167-178.

Peloquin, S. J.; Yerk, G. L.; Werner, J. E.; Darmo, E. (1989c) Potato breeding with haploids and 2n gametes. *Genome* 31:1000-1004.

Plaisted, R. L.; Hoopes, R. W. (1989) The past record and future prospects for the use of exotic germplasm. *American Potato Journal* 66:603-628.

Ross, H. (1986) *Potato breeding - Problems and perspectives.* Verlag Paul Parey, Berlin, Germany.

Rowe, P. (1984) Breeding banana and plantains. *Plant Breeding Reviews* 2:135-155.

Sakamoto, S. (1970) Utilization of related species on breeding of sweet potato in Japan. *Japanese Journal of Agricultural Research Quarterly* 5:1-4.

Sakamoto, S. (1976) Breeding a new sweet potato variety, Minamiyutaka, by the use of wild relatives. *Japanese Agricultural Research Quarterly* 10:183-186.

Sanford, J. C. (1983) Ploidy manipulations. In: Moore, J. N.; Janick, J. (eds.), *Methods in fruit breeding.* Purdue University Press, West Lafayette, IN, pp. 100-123.

Schroeder, S. H. (1984) Parental value of 2x, 2n pollen clones and 4x cultivars in 4x x 2x crosses in potatoes. Ph.D. Dissertation, University of Wisconsin-Madison. Dissertation Abstracts 44(5):1300B Order No. DA 8417048.

Sequeira, L. (1979) Development of resistance to bacterial wilt derived from *Solanum phureja.* In: *International Potato Center report of a planning conference on development in control of potato bacterial diseases.* CIP, Lima, Peru, pp. 55-62.

Singh, A. K.; Moss, J. P.; Smartt, J. (1990) Ploidy manipulations for interspecific gene transfer. *Advances in Agronomy* 43:199-240.

Watanabe, K.; Peloquin, S. J. (1989) Occurrence of 2n pollen and ps gene frequencies in cultivated groups and their related wild species in tuber-bearing Solanums. *Theoretical and Applied Genetics* 78:329-336.

Watanabe, K.; Peloquin, S. J.; Endo, M. (1991) Genetic significance of mode of polyploidization: Somatic doubling or 2n gametes? *Genome* 34:28-34.

Werner, J. E.; Peloquin, S. J. (1987) Frequency and mechanisms of 2n egg formation in haploid *Tuberosum* wild species F$_1$ hybrids. *American Potato Journal* 64:641-654.

Werner , J. E.; Peloquin, S. J. (1990) Inheritance of two mechanisms of 2n egg formation in 2x potatoes. *Journal of Heredity* 81:371-374.

Werner, J. E.; Peloquin, S. J. (1991) Yield and tuber characterization of 4x progeny from 2x x 2x crosses. *Potato Research* (in press).

Wright, S. (1956) Modes of selection. *American Naturalist* 90:5-24.

Yerk, G. L.; Peloquin, S. J. (1988) 2n pollen in eleven 2x, 2 EBN wild species and their haploid x wild species hybrids. *Potato Research* 31:581-589.

Yerk, G. L.; Peloquin, S. J. (1989) Evaluation of tuber traits of 10, 2x (2 EBN) wild species through haploid x wild species hybrids. *American Potato Journal* 66:731-739.

Yerk, G. L.; Peloquin, S. J. (1990) Performance of haploid x wild species (involving five newly evaluated species) in 4x x 2x families. *American Potato Journal* 67:405-417.

Zimmoch-Guzowska, E. (1986) Breeding of diploid potatoes and associated research in the Institute for Potato Research in Poland. In: Beekman, G. B.; Louwes, K. M.; Dellaert, L. M. W.; Neele, A. E. F. (eds.), *Potato research of tomorrow.* Pudoc, Wageningen, The Netherlands, pp. 115-119.

Discussion

J. Paul Murphy, Moderator

What are the speakers' conclusions concerning the biological mechanisms behind the differences between anther- and egg-derived haploid plants?

The anther-derived embryo arises from the nucleus of the vegetative cell of pollen that gives rise to the tube nucleus; it is a "deadend"-differentiated nucleus. Conversely, the maternally derived embryo arises from an egg nucleus that normally participates in sexual reproduction. I speculate that the vegetative nucleus of pollen is a key to the problem, but I have no definitive data to prove it because I have no way of obtaining plants from the vegetative nucleus without anther culture. So, the effects of anther culture and any differentiated state that might occur in this particular nucleus are always confounded.

Pollen is shed in the binucleate condition in most dicots whereas, in monocots, pollen is shed in the trinucleate condition. In dicots the generative nucleus does not divide until the pollen germinates and grows down the style. Measurements of DNA in these nuclei show that DNA synthesis in the generative nucleus is complete prior to anthesis, although the chromosomes have not yet divided. Surprisingly, one can also show a DNA amplification in the vegetative nucleus which does not undergo further division; this may be related to the amplification phenomenon.   [E. A. Wernsman]

I think it is biologically naive to think that plants originating from microspores and eggs would be identical. Besides Earl Wernsman's work, there are similar examples in rice, barley, diploid alfalfa and diploid tobacco where egg haploids are superior to anther-derived haploids, so one needs to be aware of this phenomenon and be cautious. We could speculate that imprinting is the basis of differences in the two types of haploids.

If we could obtain potato haploids from anther culture in large numbers, it would offer an excellent opportunity to study these comparisons because we could asexually propagate the gamete and conduct yield trials. It is fortunate in a way that we were unable to use anther culture in potato because the egg-derived haploids are extremely valuable.   [S. J. Peloquin]

The data are unequivocal that the means of anther-derived materials are lower, but there are two ways for that mean to be lower - (1) a distribution with a skewed deleterious tail, as we have seen in wheat, but with higher yielding types adequately represented at a low frequency, or (2) a totally skewed distribution with no overlap between

508

the anther-derived and maternally derived doubled haploids. Do you find the deleterious changes in tobacco uniform, or is there a range in the derived material?

There is a range in varietal materials; it is worse in some than in others but, in general, it is consistent across genotypes. If skewing results from the loss of residual heterozygosity in the parental plant from which the haploids were derived, when the doubled haploids derived from the same parent are intermated, the heterozygosity reduced though the inbreeding process should be restored. In our studies where we compared such hybrid progenies with selfed progenies from the same original parent, we observed a 20-25 % recovery of the loss in mean, but the other 75 % was not recovered. I conclude that this is due to the genetic changes wrought by the tissue culture process. [E. A. Wernsman]

What do you believe the organelles are doing relative to anther and egg culture? (Conifers might be interesting to study because, unlike most angiosperms, most conifers have patroclinal chromosome inheritance and matroclinal mitochondrial inheritance. That might be a nice contrast?)

In *Nicotiana*, the case of organellar or maternal effects has been investigated and generally found to be negative except where D. F. Matzinger and L. G. Burke of N. C. State University found changes in chloroplasts in materials subjected to anther culture. [E. A. Wernsman]

In rice and wheat anther culture, there is a high frequency of albino progeny. Is that a problem in tobacco anther culture?

No. [E. A. Wernsman]

Would the tetraploid potato have evolved in the absence of the *ps* gene? How important is this gene in the evolution of other polysomic tetraploid crops?

I do not believe the tetraploid potato would have originated without 2*n* pollen and egg formation. In the case of potato, we are convinced that the mechanism of 2*n* pollen formation is parallel spindles; we are not as sure about the 2*n* egg mechanism in nature. The asexual polyploids resulting from somatic doubling would have no advantage over the diploids, but the polyploids resulting from sexual polyploidization would have an advantage over the diploids. We made full-sib crosses between haploid species hybrids and the resulting tetraploids had great vigor. This has important evolutionary consequences in that diploid full-sibs could form tetraploids from 2*n* gametes that were superior to the existing diploids. In addition, the *ps* frequency in the progenitors of *S. tuberosum* is one-half the *ps* frequency in *S. tuberosum andigena* and, if somatic doubling had occurred, the frequencies would have been the same at both ploidy levels. There has been continual introgression from diploids to tetraploids involving 2*n* pollen, and this also increases the allelic diversity. A recent computer simulation study by Netanabe *et al.* (*Genome* 34:28-34) made a convincing case for the advantages of sexual over non-sexual polyploidization, particularly when you need allelic diversity at the locus. I believe the *ps* gene was very important in the evolution of other polysomic tetraploid crops.   [S. J. Peloquin]

Why do doubled haploids of barley and oilseed rape perform better than doubled haploids from tobacco?

Doubled haploids from barley can be of two origins - (1) from anther culture and (2) from the "*bulbosum*" method which offers the opportunity to obtain doubled haploid lines from the egg nucleus that normally participates in fertilization. One can cite data to support or negate differences among doubled haploid lines developed by both methods. I do not know what the prevailing opinion is at this time. With respect to canola, there does not seem to be any major problem with anther culture, but I do not know from which nucleus the haploid plant originates.   [E. A. Wernsman]

Does anther culture affect DNA methylation patterns?

This is an excellent and very pertinent question, but I do not have any answers.   [E. A. Wernsman]

When wild relatives have been utilized in potato breeding programs, has there been carry-over of deleterious traits such as high glyco-alkaloid levels in the tubers?

No. Glycoaldehydes are highly qualitative in their inheritance, and they are no higher in haploid species hybrids than in *S. tuberosum* varieties. The tetraploid USDA variety "Lenape" was taken off the market because of its high glycoaldehyde level, but a full-sib of Lenape plus clones derived from the cultivar have been released as varieties, all with low glycoaldehydes. So, they are no problem - just measure the levels. Several years ago Hans Ross at the Max Planck Institute compared a series of varieties containing wild germplasm with a series containing no wild germplasm, and he found they had the same glycoaldehyde distribution in both series. We have found the same results in our program at Wisconsin.   [S. J. Peloquin]

Keeping in mind that much of the literature for other crops is in agreement with Dr. Peloquin's observations that any trait needed for potato improvement is available in potatoes or related species, and given the alarming statistics on population growth we have heard during this symposium, isn't it logical to invest most of our limited resources in the coming decade in germplasm enhancement by sexual means, which has a proven track record, rather than in grandiose genome mapping projects and experiments in asexual gene transfer?

We need both. I am a cytogeneticist and we are almost a dead race. I would like to know where to get money to do cytogenetic, germplasm, or breeding methods research. We need to share the monies. Both types of research need to be done and they have different futures. The molecular work is more long range, and this basic information should be accumulated; further, some will be worthwhile. But I am reminded of somaclonal variation and the time when the National Science Foundation (NSF) said that somaclonal variation was the whole future of plant breeding. It has come to nothing, except in ornamentals. To quote Jeff Schell speaking to breeders of Clanse Seed Company: "You better continue to develop elite cultivars while we are developing methods to transform those elite cultivars later." In potato we test advanced selections for 8 years before release, and I know of no molecular substitute for field testing at this time.

It is very disturbing to me that we spend a lot of money on collecting, taxonomic evaluation, and storage of germplasm; but I don't know where to go for money to make use of that germplasm. I think we have really been lacking in leadership at the national level in this regard; but new leadership in this area, Henry Shands in particular, shares my feelings. But, whether he can put money into this area is uncertain. One cannot go to NSF or Competitive Grants for help; the only way I know is through a congressman who will write a line item in the budget for you. I am against this approach in principle but am tempted to go that route in order to support research in this area.   [S. J. Peloquin]

The *Nicotiana tabacum* species has been the recipient of numerous genes from wild relatives primarily for disease resistance. The positive/negative/neutral effects of these interspecific gene transfers must be considered on a gene-by-gene basis. In terms of utilization, they have not been highly successful in bright leaf tobacco because of deleterious effects on quality. In burley tobacco, it is exactly the opposite, as there is not a cultivar planted that does not have one or more genes of interspecific origin.

In bright leaf tobacco, the "glutinosa" resistance factor reduces yield by 4%; in burley, this is not the case. My question is this: With technology for transformation available, are the long, laborious transfer systems in interspecific research relative anymore? Is it the genes themselves or the linkage blocks that accompany them that cannot be broken that cause the effects?   [E. A. Wernsman]

I think the three species potato, tomato and oat probably have made more use of germplasm collections, if you look at all traits involved, than any of the other species. I believe, however, that we cannot blame the USDA entirely for not taking on germplasm enhancement, as this is something the breeders have to tackle themselves. In oats, we have recently acquired enhancement money for the first time; yet, for 25 years, we have been using genes from the collection because we had to have them. The breeders bear partial blame here for not using the germplasm we've collected, stored and, in some cases, evaluated. So, we should look at ourselves in the mirror as we attempt to find money for some large programs for systematic introgression.   [K. J. Frey]

SYMPOSIUM OVERVIEW

Chapter 23
Plant breeding: The next ten years

Peter Day

Center for Agricultural Molecular Biology, Rutgers University, New
Brunswick, NJ 08903

I thank the organizers for inviting me to take part in this meeting, and the
other speakers who sent in advance copies of their manuscripts. My plan is
to review briefly my expectations for this meeting and then discuss the contributions that were to me the highlights of the meeting. I have set down my
views on some of the issues we face as plant breeders in the 1990s and have
tried to summarize what I believe are some important "take-home" messages
from this symposium.

EXPECTATIONS

As a one-time plant breeder coming to North Carolina State University at
Raleigh, my expectations were high. The program seemed well balanced and
likely to cover most, if not all, of the important technical issues faced by plant
breeding in the 1990s. The invited speakers were well known for their research contributions and accomplishments. I looked forward eagerly to
hearing them. There were some gaps which I mention below.

I wondered if the graduate students and postdoctoral researchers here
would come away as enthusiastic and interested as we might hope. I believe
these meetings should be designed with their needs in mind. The older
members of the audience have many opportunities to talk to each other.
However, I was mildly disappointed. I heard our hosts telling students to
corner the speakers, but none got hold of me. Perhaps a "student/speaker"
mixer on Wednesday afternoon or Tuesday evening might have helped to
show that speakers don't bite. I also know that the Raleigh students worked
hard in helping to run the sessions and this left them less time to talk with
us.

Would I leave Raleigh feeling that the profession of plant breeding is
alive and kicking and in a vigorous, healthy state? What changes have taken
place over the last few years in our expectations of what biotechnology has
to offer? Are they realistic? I certainly have some observations on this.

What do we have to tell the developing world? Is there a place for biotechnology or is it too sophisticated still? Are we addressing the problem of
sustainability or do we still think of it merely as the flavor of the month? I
will attempt to answer these and a number of other questions.

THE PAPERS

Dean Bateman's welcome struck several chords that reverberated through the symposium. One was that modern plant breeding brings together so many disciplines that they cannot all be included in one graduate program. As Dick Flavell and other speakers pointed out, plant breeding is increasingly a collaborative activity. Six years ago, when plant molecular biology was newer than it is now, I suggested training for a third profession (Day, 1986). This would produce people who understood the new vocabulary of molecular biology but who were also familiar with plant breeding. They could work with both groups adapting the techniques and methods of the first discipline to serve the needs of the second. I no longer think it is that clear-cut. The challenge to us now is to educate students who can develop in either direction. Whether their subsequent training is as plant breeders or molecular biologists, they can communicate with each other. In the early 1980s this was very difficult but it is now getting easier all the time.

We also tend to take advances in other technologies for granted in plant breeding. Improvements in chemical analysis include new, rapid methods such as infra-red reflectance spectroscopy for measuring protein and oil content in milled grain samples. Engineering and electronics have provided micromalting tests for barley, strain gauges for on-combine weighing, barcode labels and readers, computer-printed field notebooks, dataloggers, and trials analyzed on the day of harvest, not to mention the array of seeders and planters now available. All these have made the breeder more productive by increasing the sizes of the populations that he or she can manage each season.

Ken Frey in his excellent and comprehensive introduction said that the tools of biotechnology were powerful for basic studies but would not be a part of plant breeding until beyond the 1990s. However, transgenic plants with commercial potential are already with us. Ken referred to the judgments that are made of the breeder's products by groups outside the profession. One area of some concern is the public's acceptance of genetically engineered food crops.

In November 1990, Calgene filed an application with the Food and Drug Administration for approval of tomato, oilseed rape, and cotton, expressing the bacterial gene for neomycin phosphotransferase II which confers resistance to the antibiotic kanamycin. When normal plant tissue is sensitive to kanamycin, this gene is useful as a selectable marker to recover transformants carrying introduced foreign DNA. Because this resistance gene is so widely used in plant transformation, Calgene, Oakland, CA (1990, unpublished data) made the case that its presence in an engineered food presents no hazard to human health.

They showed first that it is extremely unlikely that it would be transferred from a raw transgenic tomato fruit to the resident microflora in the human gut. Even if this were to happen, it would be of little significance because kanamycin resistance is common among the soil bacteria commonly ingested with uncooked foods. In any case, kanamycin is little used in medicine.

The question of possible pleiotropic effects resulting from the introduced DNA itself are a little more difficult to deal with. Following some suggested guidelines developed by the International Food Biotechnology Council (IFBC, 1990), it can be shown by biochemical analysis that known toxicants in the genus or species are not present in greater than normal amounts. For tomato, the analysis is largely concerned with solanine alkaloids such as tomatine. The inserted foreign DNA can also be shown to be stably integrated and that it has not moved to other sites. Since each transposition event is independent, this may have to be shown for each construct. Of much greater concern to me is the agronomic or field performance of transgenic crop plants. The dangers were described by Bill Libby with his example of a micropropagated *Eucalyptus* clone which, after several years in the field, began to grow horizontally instead of vertically. The importance of thorough trials at a number of different sites over several years cannot be overstated.

On the one hand, the potential precision of changes brought about by transformation is reassuring, but it may raise the specter of closer regulation of the products of conventional breeding which are frequently far less precise.

An equally important but entirely different question relates to the opportunities for women in plant breeding. There were many women at this meeting, but there were none among the invited speakers. This must be a source of concern. What can be done? Perhaps most important of all is encouragement to help young women overcome the prejudice and obstacles that still persist and to work among our male colleagues and reduce that prejudice and those obstacles. Role models are important here and N. C. State is fortunate in having several outstanding ones. One hopeful sign is the increasing numbers of women now working in biotechnology and its application to plant breeding. Some universities and companies (not nearly enough) provide day-care facilities for young children. More universities could be flexible in their tenure policies so that mothers are not penalized.

PART 1. THE GENE BASE FOR PLANT BREEDING

The case for large germplasm collections made by T. T. Chang was dramatically illustrated by the International Rice Research Institute's (IRRI's) success in controlling rice pests and diseases with resistant germplasm. I, like others, was interested to learn from Gurdev Khush's talk that these genes are deployed, one at a time, against a pest or disease; although he did mention a successful rotation, in time, of different resistance genes for green leafhopper in Southeast Asia. In fact, we spent no time discussing the importance of strategies for deploying these genes that can maximize their useful life in agriculture. The gains we make in breeding resistant cultivars are so hard won it surprises me that at this meeting we paid so little attention to mixtures, multilines, pyramiding resistant genes directed at the same pest or disease, or reducing the selection pressure for new biotypes by downregulating the expression of resistance.

The question was asked whether the brave new world of biotechnology will soon make gene banks unnecessary. Dick Flavell and Tim Helentjaris emphatically denied this, saying that biotechnology will make gene banks even more useful. This is a point not lost on our own U.S. National Seed Storage Facility in Fort Collins which is increasing its investment in biotechnology.

T. T. Chang wisely stressed the importance of plant genetic resources workers becoming full partners in plant breeding teams. This will be essential if gene banks are to be more than museums or seed morgues. Many of you will know of the study of Global Genetic Resources being carried out by the Board on Agriculture of the National Research Council. A report on the U.S. National Plant Germplasm System came out last year and four more volumes on forestry, fish, animals, and plant germplasm will appear this year. This initiative was begun by the late Dr. Bill Brown and the project owes much to his enthusiasm and support along with the help and guidance of several people at this meeting.

Arnel Hallauer's advice to aspiring plant breeders was to learn how and then do it! Steve Baenziger and several other speakers said it doesn't much matter what selection and improvement system you adopt - if you are any good as a breeder, you will get the job done. However, Bruce Maunder reminded us of the importance of starting with the best material. At the Plant Breeding Institute (PBI) in Cambridge, this was summed up in the expression "Cross the best with the best and hope for the best". Dean Bateman referred to "the art of plant breeding"; an art we must never lose.

In forest breeding, Bill Libby introduced us to a new kind of genotype x environment interaction, one in which a climate shift occurs and a different environment visits an already planted site. If the global climate changes that are predicted come to pass, then many other breeders may have to re-evaluate the environments in which they are breeding plants. To the breeder of annual crops, trees are a daunting prospect. Rather, few forest species can be easily propagated clonally once they have reached maturity. I wonder why it has proved to be so difficult to set back the developmental clock in tissue explants from mature trees in order to recover juvenile tissue from which they could be more easily cloned. Several years ago there were attempts to correlate developmental maturity with DNA methylation. Clonal propagation is evidently more complex.

Bob Allard told us that rare alleles are very unlikely to be as useful as common alleles. This suggested to some of us that very large gene banks might be unnecessary - a view supported by Allard's observation that disease resistance alleles often have deleterious pleiotropic effects (a point debated by other speakers) and that useful alleles arise in cultivation. However, insect resistance is apparently not deleterious to humans (or only very rarely) according to de Ponti. In fact, food scientists are now interested in "designer foods" in which any known or potential carcinogen (or other toxic compound) might be removed by breeding and selection. In this context, Ames *et al.* (1990) have suggested that plant defense chemicals may pose more significant concerns for human health than pesticide residues. Whether such plants, if they can be produced, would ever survive in farmers' fields remains to be seen.

PART 2. STRESS TOLERANCE

In Tuesday's discussion of stress resistance, less talk about corn and more discussion of legumes, such as peanuts, and various minor crops would have been welcome. I began wondering why we spend so little time on breeding minor crops, or even new ones altogether. What will North Carolina tobacco growers plant when the Surgeon General finally has his or her way? In Britain, we spent many hours debating the virtues of such exotic possibilities as meadowfoam, evening primrose, and lupine. At the PBI we ran some small trials, and there was even a symposium on "New crops for food and industry" (Wickens *et al.*, 1989). Of course, new crops also require new markets if they are to be grown for profit. The kiwi fruit provides a good example of creating a market for a new product. New crop plant breeders, therefore, need to work with agricultural economists.

John Boyer reminded us how efficient the pineapple is in its use of water. Deep rooting is also an adaptation to drought stress, but how effective is it likely to be in a field crop where all the competing plants are equally deep rooted? John's work with his student, Mark Boyle, showed that it is possible to maintain reproductive development by feeding stems of dehydrated plants without significantly rehydrating them. This suggests that approaches aimed at maintaining supplies of critical constituents during reproductive development could lead to new ways of overcoming short periods of drought.

Jan Dvorák's paper on the genetic basis of tolerance to soil toxicity illustrated the importance of this technology in expanding the areas that can be planted to cereals, especially in soils that have severe aluminum toxicity. His paper called to mind our own interest at Rutgers in New Jersey for exploiting plants that hyperaccumulate metal ions to free soils of heavy metal contamination. I had expected that Jan would mention work in southern California on *Salicornia*, a halophyte, that can be irrigated with sea water. This was reported in a recent *New York Times* article as a potential new oilseed crop that has a protein-rich meal left after crushing to remove the oil.

In the discussion after J. Palta's paper on freezing resistance, we learned that ice injury is often due to supercooling and that there would be less damage if ice crystal formation occurred in plant tissue immediately on freezing. If this is so, why not engineer a plant to express the bacterial ice nucleation factor constitutively to test the hypothesis?

PART 3. DISEASE AND INSECT STRESS TOLERANCE

Jan Parlevliet's account of partial resistance was the first of four papers on biotic stresses caused by diseases and insects. He referred to high levels of partial resistance to powdery mildew in European barleys. However, in the presence of effective major gene resistance, it is impossible to select for partial resistance. It would be interesting to know why European barleys have high levels of partial resistance when they commonly carry major genes for resistance. Is it because the major genes are for the most part ineffective?

Gurdev Khush mentioned 16 genes for resistance to bacterial leaf blight, 14 for blast, 8 for the leafhopper vector of tungro virus, 9 for grassy stunt, 5 for whitebacked planthopper, 3 for zigzag leafhopper, and 3 for gall midge - a total of 58 resistance genes to choose from or contend with in rice. Although pest and pathogen biotype changes dictate the kind of responses a breeder must make, it strikes me that the management and deployment of so many resistance factors has become a formidable task that needs new tools and concepts.

Chris Lamb gave an elegant account of why it is so difficult to clone a gene for race-specific resistance. He has elected to concentrate on isolating the genes that control the transduction pathway producing the resistance response. Others are attempting instead to identify and clone genes that encode receptors for avirulent ligands. These are likely to be the resistance genes with which a breeder works. The two approaches are complementary in that they come at the problem from different ends.

Steve Briggs described progress in cloning a different kind of resistance gene controlling the response of maize to a pathotoxin produced by race 1 of *Helminthosporium carbonum*. As Steve later explained, diseases like *H. carbonum*, whose effects are produced by pathotoxins, show specificity for susceptibility rather than for avirulence. The latter kind of specificity is a feature of those systems governed by a gene-for-gene interaction. I learned from friends at the symposium that there is a new race of *H. carbonum* that causes severe damage to the corn inbred B73 and many of its derivatives. It will be interesting to know how its pathotoxin compares with that of race 1.

PART 4. BIOTECHNOLOGY

Not surprisingly, six of the speakers dealt exclusively with the impact of biotechnology on plant breeding. As Russell Lande put it, there is now a healthy, open-minded skepticism on this subject. I suppose that most of the skeptics with closed minds have retired, if they have not been converted at least to open mindedness.

Tim Helentjaris was refreshingly frank in admitting that there are at least five years of developmental work required to solve cost and technological problems before we have something analogous to a "pregnancy test kit" for marker-assisted selection in the field. It seems that marker-assisted selection, either using RFLPs or RAPD markers, is getting less, rather than more, user-friendly. It will likely remain a research tool until, as Russell Lande predicts, costs go down by 100- to 1000-fold. Even so, accomplishments so far are impressive. Tim's example of yield versus dissimilarity at 34 RFLP loci was especially interesting to those who have successfully "fixed" heterosis - for example, in selecting high yielding and vigorous inbreds.

Perhaps most remarkable of all biotechnology is the progress that has been made with quantitative trait loci (QTLs) and the prospect of manipulating the chromosome hot spots where many seem to be located. Having worked many years ago at the John Innes Institute, under the same roof where Kenneth Mather and Brian Harrison were busy counting bristles on *Drosophila* to develop the polygene theory of quantitative inheritance, I was especially interested to learn from Steve Briggs of the work showing that bristle number maps to the achaete/scute complex.

We now take transformation in dicots somewhat for granted. It is still far from routine in wheat and rice and other crops, but is perhaps getting that way in corn. Isolating genes of interest is one area where great progress has been made, and I especially enjoyed Steve Brigg's account of the methods being used and their relative merits.

Judging by the questions, many were surprised to learn that recombinational events, such as transposition, occur within rather than between genes. To use an outdated analogy, it seems the beads are more fragile than the string on which they are threaded!

I have always been somewhat skeptical of the claims made for somaclonal variation. Ron Phillips presented some compelling evidence for DNA methylation as an explanation for some of the stable epigenetic effects found in plants regenerated from tissue culture. I meant to ask him why the agronomically useful potato variants that several groups have selected are unstable. None that I know have survived as new cultivars.

Jerry Mikshe directs the USDA Plant Genome Project which Dr. Dean Plowman referred to in his after-dinner speech. Some two years ago, a group of us got together in Washington to pool ideas on what types of research this project should support. Many of us were very strongly in favor of USDA joining with the National Science Foundation (NSF) and National Institutes of Health (NIH) to support work on *Arabidopsis* rather than attempting to sequence completely the much larger genomes of one or more economically important crop plants. It seems our advice (although not well received at the time) took root, and that the USDA is assisting the NSF and will entertain proposals on *Arabidopsis*. As Dick Flavell said, work on *Arabidopsis* is not the distraction that some may think it is. It is becoming an enormously powerful tool to enhance our understanding of plants, including economic crop plants.

For 30 years I have complained about the lack of a model system for plant pathogenic fungi. I believed that we needed an analog of tobacco mosaic virus to advance their study. Most plant pathologists are hired to work on the problems of specific crops. As a result, most work on different plant systems. Pooling resources and building upon earlier contributions is much more difficult. However, molecular biology is changing this dramatically.

I enjoyed Winston Brill's review of what one may call the "snake oil" side of biotechnology. His account of unethical misrepresentation of tests was illuminating, but his important message was how easily one can be fooled by laboratory tests. Field tests are essential.

CONCLUSION

It is clear that plant breeding has much to contribute to feeding, sheltering, and clothing humankind in the 1990s. Fulfilling the expectations engendered by breeders' past successes will not be easy. In addition to supporting the ever-growing human population, there are newly emphasized objectives to be addressed - sustainable agriculture, greater tolerance of abiotic stress, environmental conservation, and genetic conservation.

The next 10 years are but a short time for most plant breeders. The original crosses that created most of today's newly released cultivars were made more than 10 years ago. To date, new technologies have improved the breeder's productivity through more rapid assays and chemical analyses, in-field data logging, and computerized data management that enables breeders to handle larger plant populations with greater efficiency. However, genetic transformation - while increasingly routine for some crops - has yet to speed up the hybridization process significantly. It is very clear that biotechnology cannot replace plant breeding. It is rather a complementary technology which provides new tools, genotypes and concepts to help solve the many pressing problems breeders face.

We cannot train the next generation of plant breeders with all they need to know in a single disciplinary graduate program. Rather, we must educate them in multidisciplinary programs that give them both the theoretical basis of the technology and some field experience so that they may become plant breeders who can exploit the new technologies, or molecular biologists who understand the needs of plant breeders in their further training after graduate school. I find the prospect exciting and the rewards for hard and imaginative work exhilarating. Plant breeding is alive and kicking vigorously.

ACKNOWLEDGMENTS

I thank my colleagues Laura Meagher and Judy Snow for their comments and suggestions.

REFERENCES

Ames, B. N.; Profet, M.; Gold, L. S. (1990) Dietary pesticides (99.99% all natural). *Proceedings of the National Academy of Science, USA* 87:7777-7781.

Day, P. R. (1986) The impact of biotechnology on agricultural research. *Diversity* 9:33-37.

IFBC (1990) Biotechnologies and food: Assuring the safety of foods produced by genetic modification. *Regulatory Toxicology and Pharmacology* 12(3, pt 2):196 pp.

Wickens, G. E.; Haq, N.; Day, P. (1989) *New crops for food and industry*. Chapman and Hall, London, U.K., 444 pp.

Discussion

H. Thomas Stalker, Moderator

The success of crop breeding depends to a great extent on decentralization and the genetic diversity to which it leads. Shouldn't we worry that diversion of public and private resources from breeding and germplasm work into "Big Science", along with putting germplasm exchange on a cash basis, could concentrate control of our genetic resources in the hands of a few powerful entities, with disastrous effects on both genetic progress and genetic diversity?

Cornering the market has always been a concern for many people. Fortunately, there are many bright scientists working for smaller companies and universities who find ways to circumvent this. Diverting resources from breeding to Big Science is, of course, to be deplored. However, funding agencies seem to tire of funding the same things for long periods. The International Agricultural Research Centers (IARCs) are experiencing this now and are moving their work more upstream in the expectation that national programs will take on some of the breeding programs the IARCs will drop. Clearly the products of commercial, and many public, breeding programs will be exchanged for cash. However, we must work hard to ensure that by recognizing the ownership of germplasm we do not restrict the ready availability of germplasm to all who need it. [P. Day]

Would the increased constraint of intellectual property rights pose a serious threat to the trend for increasing international and institutional collaboration of germplasm research, breeding, and technology?

Intellectual property rights are an important issue. The third world countries will want to share in the benefits of patented materials and at the same time safeguard their own germplasm. The IARCs have traditionally released germplasm to all national agricultural research organizations with no strings attached. Will this policy change as they make use of protected germplasm resulting from biotechnology? I hope that it will not. [P. Day]

You said that plant breeding is alive and kicking, but I believe this is true only for major crops (i.e., maize, wheat, and soybean); for minor crops there is a true funding crisis which leaves us just kicking (i.e., the loss of breeding positions at land-grant universities leaves little to no funding for cultivar or germplasm development). Do you have an opinion and who should support this work?

If you are kicking, then you are still alive. It is a competitive world. As other speakers have said, we have to sell what we do to receive support. Breeding minor crops surely must be continued by public institutions.

[P. Day]

In the light of the role of the state agricultural experiment stations, how would you allocate a budget to fund a balance between basic and applied plant research?

Regular plant breeding isn't as "sexy" as molecular biology. The opportunities for promotion in a public institution by developing varieties are not as good in a publish or perish environment. We need to convince agricultural experiment station administrators that they must evaluate people on more than just publishing, and that funding other than competitive grants must be implemented (for example, by check-off funds). [P. Day]

On a global scale, should the governments of the developing countries up their own investments in agricultural research, or even become small donors of the international centers, to counterbalance the changes in major donor's perspectives and in protecting their own sovereignty over germplasm, patents, etc.?

I think that is an excellent idea. An analogous situation occurred a few years ago in Kenya where that government wanted to build a molecular biology institute similar to ILRAD, an international livestock research center. The advice they received was only if you make an investment yourselves will you receive help. If work is to be done in a developing country, then the local government will have to support the efforts. The international community cannot do it just by donating money. What you pay for, you value. Even if developing countries donate a marginal amount of funds to international centers, they will have a vested interest in the work. [P. Day]

Given the history of the use of major gene resistance and accompanying pest evolution, is there any possibility that resistance conferred by genes coming out of biotechnology projects will be any more sustainable than the resistance in today's cultivars?

The answer to the question must be no. The myriad of pathogens are more complex than can be handled by biotechnology programs. Perhaps there will be a few magic bullets waiting to be found, but I don't think so. Dr. de Ponti says that the *Bt* gene effective against Coleoptera has already succumbed to new biotypes of Colorado beetle. However, companies are isolating thousands of *B. thuringiensis* strains from soil samples and are reporting new specificities. It is possible that site-directed mutagenesis may produce still more forms although in my view it is unlikely that they would be any more stable.   [P. Day]

I agree that it is dangerous to speculate in areas that we know nothing about, but I would also like to raise another point for consideration. Most molecular biologists have concentrated on the gene-for-gene types of resistance. These types of genes are characterized from the pathotoxin (or "simpler" gene systems) by their specificity for interaction with a specific host. Both pathotoxin genes and gene-for-gene types of resistance are race-specific, but in the gene-for-gene system the specificity of the interaction is for incompatibility. On the other hand, the specificity of the pathotoxin genes is for establishing compatibility. In general, genes that block the establishment of compatibility have been very stable, as opposed to genes that are specific for incompatibility.   [P. S. Baenziger]

You raised an excellent point. The question it raises, however, is how many host-parasite relations have a pathotoxin mechanism for inducing susceptibility that are amenable to such an approach?   [P. Day]

The question may not be how many interactions have a pathotoxin effect because assaying requires the target molecules to be toxic. We are deliberately selecting a subset of what may be part of a very large group of chemicals needed to establish compatibility. I believe that there must be a chemical basis to compatibility to allow the pathogen to infect the plant. However, most of these chemicals may not be toxic and, unless other assays are used, they will not be detected. Many of the toxins can only be detected under very special conditions, but they do reveal the fact that there are chemical inducers to susceptibility.   [P. S. Baenziger]

Will 5 years of protection justify the very large investment that isolation and transfer of each gene may require?

Five years of protection is not nearly long enough to recover costs but protection for 20 years, typically granted for plant variety rights, probably is.
                                                                         [P. Day]

Should plant breeders be concerned with the "sustainability" of the cultivars they release, or should the development of alternated cropping systems (i.e., polycultures) or deployment strategies be left to the agronomists or ecologists?

I think plant breeders should be very interested in sustainability. The contributions of agronomists and ecologists are important but those of breeders are equally if not more so. At the present time we emphasize the production of genetically uniform varieties for large-scale, intensive monoculture. Most of these varieties are unsuitable for less intensive, sustainable systems.

[P. Day]

Dr. Frey commented on possible future dangers of having fewer university programs in a position to train breeders. In addition, we may expect that fewer and fewer researchers will have any agricultural backgrounds. In other words, very few will have grown up on farms. Would you care to comment?

Training plant breeders who do not have a farm background is more difficult since a knowledge of farming helps in knowing what to seek in breeding new cultivars. However, these skills can be learned. Understanding the contribution of plant breeding to society should also be a part of every breeder's training.

[P. Day]

You suggested that the third person connecting molecular biology and applied plant breeding will not be needed in the future. You may be correct with respect to understanding theories; if, however, most molecular biology will be carried out on model plants as you suggest (which I think is a good idea), we need scientists to apply this knowledge to breeding programs. After all, *Arabidopsis* doesn't set too many tubers or grow 300 feet tall like *Sequoia*. Do you have any comments?

A number of plant molecular biologists are interested in finding practical applications for their ideas and are making the connection between molecular biology and plant breeding. If the plant breeder also knows something about molecular biology, which is increasingly the case, the connection can be more easily made. Five or six years ago there was little or no dialogue between breeders and molecular biologists except through third parties. This is no longer true. *Arabidopsis* indeed forms no tubers but, as an example, the exercise of introducing genes that control tuber formation into *Arabidopsis* could show us how to manipulate them in potato. [P. Day]

Bruce Ames' observation that plants carry substantial amounts of secondary compounds that might be toxic is correct, but we should be very careful in accepting his suggestion that breeding for resistance to insects leads by definition to higher levels of these compounds. There is very little experimental evidence.

This is probably true. On the other hand, we need to remember the example of Lenape potato which left a sour taste in a lot of mouths. All potato cultivars are now tested for their glycoalkaloid content because of that one incident.   [P. Day]

Where will plant breeders be trained in the future?

The best practical training in plant breeding is to work with an expert and to learn from him or her those things that are impossible to convey in a class-room or a laboratory. Graduate programs provide the theoretical basis for the technology, many of the technical skills needed, and an appreciation of the kinds of new tools provided by biotechnology. For most crops, the period of apprenticeship in a graduate program is not long enough for stu-dents to produce successful varieties themselves.   [P. Day]

Index

/